AN INTRODUCTION TO INFORMATION THEORY

Fazlollah M. Reza

DOVER PUBLICATIONS, INC.
New York

Published in Canada by General Publishing Company, Ltd., 30 Lesmill Road, Don Mills, Toronto, Ontario.

Published in the United Kingdom by Constable and Company, Ltd., 3 The Lanchesters, 162–164 Fulham Palace Road, London W6 9ER.

Bibliographical Note

This Dover edition, first published in 1994, is an unabridged and unaltered republication of the work first published by the McGraw-Hill Book Company, Inc., New York, in 1961 in the *McGraw-Hill Electrical and Electronic Engineering Series.*

Library of Congress Cataloging-in-Publication Data

Reza, Fazlollah M.
An introduction to information theory / Fazlollah M. Reza.
p. cm.
Originally published: New York : McGraw-Hill, 1961.
Includes bibliographical references and index.
ISBN 0-486-68210-2
1. Information theory. 2. Probabilities. I. Title.
Q360.R43 1994
003′.54—dc20 94-27222
CIP

Manufactured in the United States of America
Dover Publications, Inc., 31 East 2nd Street, Mineola, N.Y. 11501

PREFACE

Statistical theory of communication is a broad new field comprised of methods for the study of the statistical problems encountered in all types of communications. The field embodies many topics such as radar detection, sources of physical noise in linear and nonlinear systems, filtering and prediction, information theory, coding, and decision theory. The theory of probability provides the principal tools for the study of problems in this field.

Information theory as outlined in the present work is a part of this broader body of knowledge. This theory, originated by C. E. Shannon, introduced several important new concepts and, although a part of applied communications sciences, has acquired the unique distinction of opening a new path of research in pure mathematics.

The communication of information is generally of a statistical nature, and a current theme of information theory is the study of simple ideal statistical communication models. The first objective of information theory is to define different types of sources and channels and to devise statistical parameters describing their individual and ensemble operations.

The concept of Shannon's communication entropy of a source and the transinformation of a channel provide most useful means for studying simple communication models. In this respect it appears that the concept of communication entropy is a type of describing function that is most appropriate for the statistical models of communications. This is similar in principle to the way that an impedance function describes a linear network, or a moment indicates certain properties of a random variable. The introduction of the concepts of communication entropy, transinformation, and channel capacity is a basic contribution of information theory, and these concepts are of such fundamental significance that they may parallel in importance the concepts of power, impedance, and moment.

Perhaps the most important theoretical result of information theory is Shannon's fundamental theorem, which implies that it is possible to communicate information at an ideal rate with utmost reliability in the presence of "noise." This succinct but deep statement and its consequences unfold the limitation and complexity of present and future methods of communications. Mastery of the proof offers many fringe benefits to those interested in the analysis and synthesis of communication networks. For this reason we have included several methods of proof of this theorem (see Chapters 4 and 12). However, the impatient reader is forewarned that the proof entails much preparation which may prove to be burdensome.

This book originated about five years ago from the author's lecture notes on information theory. In presenting the subject to engineers, the need for preliminary lectures on probability theory was observed. A course in probability, even now, is not included in the curriculum of a majority of engineering schools. This fact motivated the inclusion of an introductory treatment of probability for those who wish to pursue the general study of statistical theory of communications.

The present book, directed toward an engineering audience, has a threefold purpose:

1. To present elements of modern probability theory (discrete, continuous, and stochastic)
2. To present elements of information theory with emphasis on its basic roots in probability theory
3. To present elements of coding theory

Thus this book is offered as an introduction to probability, information, and coding theory. It also provides an adequate treatment of probability theory for those who wish to pursue topics other than information theory in the field of statistical theory of communications.

One feature of the book is that it requires no formal prerequisites except the usual undergraduate mathematics included in an engineering or science program. Naturally, a willingness to consult other references or authorities, as necessary, is presumed. The subject is presented in the light of applied mathematics. The immediate involvement in technological specialities that may solve specific problems at the expense of a less thorough basic understanding of the theory is thereby avoided.

A most important, though indirect, application of information theory has been the development of codes for transmission and detection of information. Coding literature has grown very rapidly since it, presumably, applies to the growing field of data processing. Chapters 4 and 13 present an introduction to coding theory without recourse to the use of codes.

The book has been divided into four parts: (1) memoryless discrete schemes, (2) memoryless continuum, (3) schemes with memory, and (4) an outline of some of the recent developments. The appendix contains some notes which may help to familiarize the reader with some of the literature in the field. The inclusion of many reference tables and a bibliography with some 200 entries may also prove to be useful.

The emphasis throughout the book is on such basic concepts as sets, the probability measure associated with sets, sample space, random variables, information measure, and capacity. These concepts proceed from set theory to probability theory and then to information and coding theories. The application of the theory to such subjects as radar detection, optics, and linguistics was not undertaken. We make no pretension for "usefulness" and immediate application of information theory. From an educational standpoint, it appears, however, that the topics discussed should provide a suitable training ground for communication scientists.

The most rewarding aspect of this undertaking has been the pleasure of learning about a new and fascinating frontier in communications. By working on this book, I came to appreciate fully many subtle points and ingenious procedures set forth in the papers of the original contributors to the literature. I trust this attempt to integrate these many contributions will prove of value. Despite pains taken by the author, inaccuracies, original or inherited, may be found. Nevertheless, I hope the reader will find this work an existence proof of Shannon's fundamental theorem; that "information" can be transmitted with a high degree of reliability at a rate close to the channel capacity despite all forms of "noise."

At any rate, there is an eternal separation between what one strives for and what one actually achieves. As Leon von Montenaeken wrote,

La vie est brève,
Un peu d'espoir,
Un peu de rêve,
Et puis—bonsoir.

Fazlollah M. Reza

To the memory of
SHARIAT-MADAR RAFI
An Inspired Teacher

ACKNOWLEDGMENTS

The author wishes to acknowledge indebtedness to all those who have directly or indirectly contributed to this book. Special tribute is due to Dr. C. E. Shannon who primarily initiated information theory.

Dr. P. Elias of the Massachusetts Institute of Technology, has been generous in undertaking a comprehensive reading and reviewing of the manuscript. His comments, helpful criticism, and stimulating discussions have been of invaluable assistance.

Dr. L. A. Cote of Purdue University has been very kind to read and criticize the manuscript with special emphasis upon the material on probability theory. His knowledge of technical Russian literature and his unlimited patience in reading the manuscript in its various stages of development have provided a depth that otherwise would not have been attained.

Thanks are due to Dr. E. N. Gilbert of Bell Telephone Laboratories and Dr. J. P. Costas of General Electric Company for helpful comments on the material on coding theory; to Prof. W. W. Harman of Stanford University for reviewing an early draft of the manuscript; to Mr. L. Zafiriu of Syracuse University who accepted the arduous task of proofreading and who provided many suggestions.

In addition numerous scientists have generously provided reprints, reports, or drafts of unpublished manuscripts. The more recent material on information theory and coding has been adapted from these current sources but integrated in our terminology and frame of reference. An apology is tendered for any omission or failure to reflect fully the thoughts of the original contributors.

During the past four years, I had the opportunity to teach and lecture in this field at Syracuse University, International Business Machines Corp., General Electric Co., and the Rome Air Development Center. The keen interest, stimulating discussions, and friendships of the scientists of these centers have been most rewarding.

Special acknowledgment is due the United States Air Force Rome Air Development Center and the Cambridge Research Center for supporting several related research projects.

I am indebted to my colleagues in the Department of Electrical Engineering at Syracuse University for many helpful discussions and to Mrs. H. F. Laidlaw and Miss M. J. Phillips for their patient typing of the manuscript.

I am particularly grateful to my wife and family for the patience which they have shown.

CONTENTS

PREFACE iii

CHAPTER 1 Introduction

1-1. Communication Processes 1
1-2. A Model for a Communication System 3
1-3. A Quantitative Measure of Information 5
1-4. A Binary Unit of Information 7
1-5. Sketch of the Plan 9
1-6. Main Contributors to Information Theory 11
1-7. An Outline of Information Theory 14

Part 1: Discrete Schemes without Memory

CHAPTER 2 Basic Concepts of Probability

2-1. Intuitive Background 19
2-2. Sets 21
2-3. Operations on Sets 23
2-4. Algebra of Sets 24
2-5. Functions 30
2-6. Sample Space 34
2-7. Probability Measure 36
2-8. Frequency of Events 38
2-9. Theorem of Addition 40
2-10. Conditional Probability 42
2-11. Theorem of Multiplication 44
2-12. Bayes's Theorem 46
2-13. Combinatorial Problems in Probability 49
2-14. Trees and State Diagrams 52
2-15. Random Variables 58
2-16. Discrete Probability Functions and Distribution . . 59
2-17. Bivariate Discrete Distributions 61
2-18. Binomial Distribution 63
2-19. Poisson's Distribution 65
2-20. Expected Value of a Random Variable 67

CHAPTER 3 Basic Concepts of Information Theory: Memoryless Finite Schemes

3-1. A Measure of Uncertainty 76
3-2. An Intuitive Justification 78
3-3. Formal Requirements for the Average Uncertainty . . 80
3-4. H Function as a Measure of Uncertainty 82
3-5. An Alternative Proof That the Entropy Function Possesses a Maximum 86
3-6. Sources and Binary Sources 89
3-7. Measure of Information for Two-dimensional Discrete Finite Probability Schemes 91
3-8. Conditional Entropies 94
3-9. A Sketch of a Communication Network 96
3-10. Derivation of the Noise Characteristics of a Channel . 99
3-11. Some Basic Relationships among Different Entropies . 101
3-12. A Measure of Mutual Information 104
3-13. Set-theory Interpretation of Shannon's Fundamental Inequalities 106
3-14. Redundancy, Efficiency, and Channel Capacity . . 108
3-15. Capacity of Channels with Symmetric Noise Structures 111
3-16. BSC and BEC 114
3-17. Capacity of Binary Channels 115
3-18. Binary Pulse Width Communication Channel . . . 122
3-19. Uniqueness of the Entropy Function 124

CHAPTER 4 Elements of Encoding

4-1. The Purpose of Encoding 131
4-2. Separable Binary Codes 137
4-3. Shannon-Fano Encoding 138
4-4. Necessary and Sufficient Conditions for Noiseless Coding 142
4-5. A Theorem on Decodability 147
4-6. Average Length of Encoded Messages 148
4-7. Shannon's Binary Encoding 151
4-8. Fundamental Theorem of Discrete Noiseless Coding . 154
4-9. Huffman's Minimum-redundancy Code 155
4-10. Gilbert-Moore Encoding 158
4-11. Fundamental Theorem of Discrete Encoding in Presence of Noise 160
4-12. Error-detecting and Error-correcting Codes 166
4-13. Geometry of the Binary Code Space 168
4-14. Hamming's Single-error Correcting Code 171

4-15. Elias's Iteration Technique 176
4-16. A Mathematical Proof of the Fundamental Theorem of Information Theory for Discrete BSC. 180
4-17. Encoding the English Alphabet 183

Part 2: Continuum without Memory

CHAPTER 5 Continuous Probability Distribution and Density
5-1. Continuous Sample Space 191
5-2. Probability Distribution Functions 192
5-3. Probability Density Function 194
5-4. Normal Distribution 196
5-5. Cauchy's Distribution 198
5-6. Exponential Distribution 199
5-7. Multidimensional Random Variables 200
5-8. Joint Distribution of Two Variables: Marginal Distribution 202
5-9. Conditional Probability Distribution and Density . . 204
5-10. Bivariate Normal Distribution 206
5-11. Functions of Random Variables 208
5-12. Transformation from Cartesian to Polar Coordinate System. 214

CHAPTER 6 Statistical Averages
6-1. Expected Values; Discrete Case 220
6-2. Expectation of Sums and Products of a Finite Number of Independent Discrete Random Variables . . . 222
6-3. Moments of a Univariate Random Variable. . . . 224
6-4. Two Inequalities 227
6-5. Moments of Bivariate Random Variables 229
6-6. Correlation Coefficient 230
6-7. Linear Combination of Random Variables 232
6-8. Moments of Some Common Distribution Functions . 234
6-9. Characteristic Function of a Random Variable . . . 238
6-10. Characteristic Function and Moment-generating Function of Random Variables. 239
6-11. Density Functions of the Sum of Two Random Variables 242

CHAPTER 7 Normal Distributions and Limit Theorems
7-1. Bivariate Normal Considered as an Extension of One-dimensional Normal Distribution 248
7-2. Multinormal Distribution 250
7-3. Linear Combination of Normally Distributed Independent Random Variables 252

7-4. Central-limit Theorem 254
7-5. A Simple Random-walk Problem 258
7-6. Approximation of the Binomial Distribution by the Normal Distribution 259
7-7. Approximation of Poisson Distribution by a Normal Distribution 262
7-8. The Laws of Large Numbers 263

CHAPTER 8 Continuous Channel without Memory

8-1. Definition of Different Entropies 267
8-2. The Nature of Mathematical Difficulties Involved . . 269
8-3. Infiniteness of Continuous Entropy 270
8-4. The Variability of the Entropy in the Continuous Case with Coordinate Systems 273
8-5. A Measure of Information in the Continuous Case . . 275
8-6. Maximization of the Entropy of a Continuous Random Variable 278
8-7. Entropy Maximization Problems 279
8-8. Gaussian Noisy Channels 282
8-9. Transmission of Information in Presence of Additive Noise 283
8-10. Channel Capacity in Presence of Gaussian Additive Noise and Specified Transmitter and Noise Average Power 285
8-11. Relation between the Entropies of Two Related Random Variables 287
8-12. Note on the Definition of Mutual Information . . . 289

CHAPTER 9 Transmission of Band-limited Signals

9-1. Introduction 292
9-2. Entropies of Continuous Multivariate Distributions . 293
9-3. Mutual Information of Two Gaussian Random Vectors 295
9-4. A Channel-capacity Theorem for Additive Gaussian Noise 297
9-5. Digression 299
9-6. Sampling Theorem 300
9-7. A Physical Interpretation of the Sampling Theorem . 305
9-8. The Concept of a Vector Space 308
9-9. Fourier-series Signal Space 313
9-10. Band-limited Signal Space 315
9-11. Band-limited Ensembles 317
9-12. Entropies of Band-limited Ensemble in Signal Space . 320

9-13. A Mathematical Model for Communication of Continuous Signals 322
9-14. Optimal Decoding 323
9-15. A Lower Bound for the Probability of Error 325
9-16. An Upper Bound for the Probability of Error 327
9-17. Fundamental Theorem of Continuous Memoryless Channels in Presence of Additive Noise 329
9-18. Thomasian's Estimate 330

Part 3: Schemes with Memory

CHAPTER 10 Stochastic Processes

10-1. Stochastic Theory 338
10-2. Examples of a Stochastic Process 341
10-3. Moments and Expectations 343
10-4. Stationary Processes 344
10-5. Ergodic Processes 347
10-6. Correlation Coefficients and Correlation Functions. . 349
10-7. Example of a Normal Stochastic Process 352
10-8. Examples of Computation of Correlation Functions . 353
10-9. Some Elementary Properties of Correlation Functions of Stationary Processes 356
10-10. Power Spectra and Correlation Functions 357
10-11. Response of Linear Lumped Systems to Ergodic Excitation 359
10-12. Stochastic Limits and Convergence 363
10-13. Stochastic Differentiation and Integration 365
10-14. Gaussian-process Example of a Stationary Process . . 367
10-15. The Over-all Mathematical Structure of the Stochastic Processes 368
10-16. A Relation between Positive Definite Functions and Theory of Probability 370

CHAPTER 11 Communication under Stochastic Regimes

11-1. Stochastic Nature of Communication 374
11-2. Finite Markov Chains 376
11-3. A Basic Theorem on Regular Markov Chains . . . 377
11-4. Entropy of a Simple Markov Chain 380
11-5. Entropy of a Discrete Stationary Source 384
11-6. Discrete Channels with Finite Memory 388
11-7. Connection of the Source and the Discrete Channel with Memory 389
11-8. Connection of a Stationary Source to a Stationary Channel 391

Part 4: Some Recent Developments

CHAPTER 12 The Fundamental Theorem of Information Theory

PRELIMINARIES

12-1. A Decision Scheme 398
12-2. The Probability of Error in a Decision Scheme . . . 398
12-3. A Relation between Error Probability and Equivocation 400
12-4. The Extension of Discrete Memoryless Noisy Channels 402

FEINSTEIN'S PROOF

12-5. On Certain Random Variables Associated with a Communication System 403
12-6. Feinstein's Lemma 405
12-7. Completion of the Proof 406

SHANNON'S PROOF

12-8. Ensemble Codes. 409
12-9. A Relation between Transinformation and Error Probability 412
12-10. An Exponential Bound for Error Probability . . . 414

WOLFOWITZ'S PROOF

12-11. The Code Book 416
12-12. A Lemma and Its Application. 417
12-13. Estimation of Bounds 419
12-14. Completion of Wolfowitz's Proof 421

CHAPTER 13 Group Codes

13-1. Introduction 424
13-2. The Concept of a Group 425
13-3. Fields and Rings 428
13-4. Algebra for Binary n-Digit Words 429
13-5. Hamming's Codes 431
13-6. Group Codes 435
13-7. A Detection Scheme for Group Codes 437
13-8. Slepian's Technique for Single-error Correcting Group Codes 438
13-9. Further Notes on Group Codes 442
13-10. Some Bounds on the Number of Words in a Systematic Code 446

APPENDIX Additional Notes and Tables

N-1. The Gambler with a Private Wire 450
N-2. Some Remarks on Sampling Theorem 452
N-3. Analytic Signals and the Uncertainty Relation . . . 454

N-4. Elias's Proof of the Fundamental Theorem for BSC . 457
N-5. Further Remarks on Coding Theory 460
N-6. Partial Ordering of Channels 462
N-7. Information Theory and Radar Problems 464
T-1. Normal Probability Integral 465
T-2. Normal Distributions 466
T-3. A Summary of Some Common Probability Functions . 467
T-4. Probability of No Error for Best Group Code . . . 468
T-5. Parity-check Rules for Best Group Alphabets . . . 469
T-6. Logarithms to the Base 2 471
T-7. Entropy of a Discrete Binary Source 476

BIBLIOGRAPHY 481

NAME INDEX 491

SUBJECT INDEX 493

CHAPTER 1

INTRODUCTION

Information theory is a new branch of probability theory with extensive potential applications to communication systems. Like several other branches of mathematics, information theory has a physical origin. It was initiated by communication scientists who were studying the statistical structure of electrical communication equipment.

Our subject is about a decade old. It was principally originated by Claude Shannon through two outstanding contributions to the mathematical theory of communications in 1948 and 1949. These were followed by a flood of research papers speculating upon the possible applications of the newly born theory to a broad spectrum of research areas, such as pure mathematics, radio, television, radar, psychology, semantics, economics, and biology. The immediate application of this new discipline to the fringe areas was rather premature. In fact, research in the past 5 or 6 years has indicated the necessity for deeper investigations into the foundations of the discipline itself.

Despite this hasty generalization which produced several hundred research papers (with frequently unwarranted conclusions), one thing became evident. The new scientific discovery has stimulated the interest of thousands of scientists and engineers around the world.

Our first task is to present a bird's-eye view of the subject and to specify its place in the engineering curriculum. In this chapter a heuristic exposition of the topic is given. No effort is made to define the technical vocabulary. Such an undertaking requires a detailed logical presentation and is out of place in this informal introduction. However, the reader will find such material presented in a pedagogically prepared sequence beginning with Chap. 2. This introductory chapter discusses generalities, leaving a more detailed and precise treatment to subsequent chapters.* The specialist interested in more concrete statements may wish to forgo this introduction and begin with the body of the book.*

1-1. Communication Processes. Communication processes are concerned with the flow of some sort of information-carrying commodity in

* With the exception of Sec. 1-7, which gives a synopsis of information theory for the specialist.

some *network*. The commodity need not be tangible; for example, the process by which one mind affects another mind is a communication procedure. This may be the sending of a message by telegraph, visual communication from artist to viewer, or any other means by which *information* is conveyed from a transmitter to a receiver. The subject matter deals with the gross aspects of communication models rather than with their minute structure. That is, we concentrate on the over-all performance of such systems without being restrained to any particular equipment or organ. Common to all communication processes is the flow of some commodity in some network. While the nature of the commodity can be as varied as electricity, words, pictures, music, and art, one could suggest at least three essential parts of a communication system (Fig. 1-1):

1. Transmitter or source
2. Receiver or sink
3. Channel or transmission network which conveys the communiqué from the transmitter to the receiver

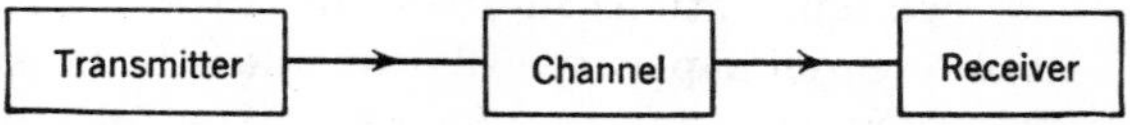

FIG. 1-1. The model of a communication system.

This is the simplest communication system that one can visualize. Practical cases generally consist of a number of sources and receivers and a complex network. A familiar analogous example is an electric power system using several interconnected power plants to supply several towns.

In such problems one is concerned with a study of the distribution of the commodity in the network, defining some sort of efficiency of transmission and hence devising schemes leading to the most efficient transmission.

When the communiqué is tangible or readily measurable, the problems encountered in the study of the communication system are of the types somewhat familiar to engineers and operational analysts (for instance, the study of an electric circuit or the production schedule of a manufacturing plant). When the communiqué is "intelligence" or "information," this general familiarity cannot be assumed. How does one define a measure for the amount of *information?* And having defined a suitable measure, how does one apply it to the betterment of the communication of information?

To mention an analog, consider the case of an electric power network transmitting electric energy from a source to a receiver (Fig. 1-2). At the source the electric energy is produced with voltage V_s. The receiver requires the electric energy at some prescribed voltage V_r. One of the problems involved is the heat loss in the channel (transmission line). In other words, the impedance of the wires acts as a parasitic receiver. One of the many tasks of the designer is to minimize the loss in the transmission lines. This can be accomplished partly by improving the quality of the transmission lines. A parallel method of transmission improvement is to increase the voltage at the input terminals of the line. As is well known, this improves the efficiency of transmission by reducing energy losses in the line. A step-up voltage transformer installed at the input terminals of the line is appropriate. At the output terminals another transformer (step-down) can provide the specified voltage to the receiver.

Without being concerned about mathematical discipline in this introductory chapter, let us ask if similar procedures could be applied to the transmission of *information*. If the channel of transmission of information is a lossy one, can one still improve the *efficiency* of the transmission

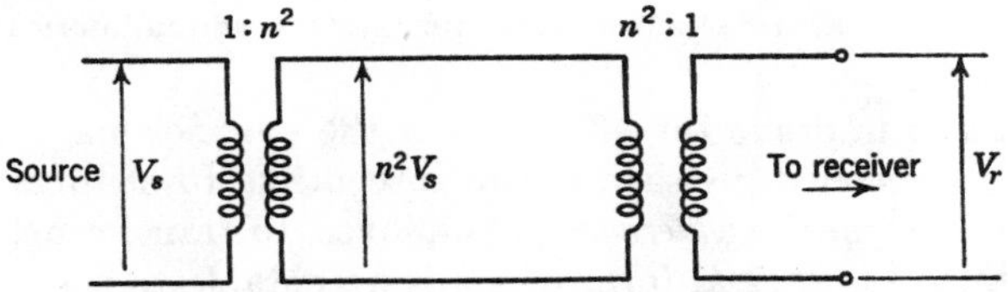

FIG. 1-2. An example of a communication system.

by procedures similar to those in the above case? This of course depends, in the first place, on whether a measure for the efficiency of transmission of information can be defined.

1-2. A Model for a Communication System. The communication systems considered here are of a statistical nature. That is, the performance of the system can never be described in a deterministic sense; rather, it is always given in statistical terms. A source is a device that selects and transmits sequences of symbols from a given alphabet. Each selection is made at random, although this selection may be based on some statistical rule. The channel transmits the incoming symbols to the receiver. The performance of the channel is also based on laws of chance. If the source transmits a symbol, say A, with a probability of $P\{A\}$ and the channel lets through the letter A with a probability denoted by $P\{A|A\}$, then the probability of transmitting A and receiving A is

$$P\{A\} \cdot P\{A|A\}$$

The communication channel is generally lossy; i.e., a part of the transmitted commodity does not reach its destination or it reaches the destination in a distorted form. There are often unwanted sources in a communication channel, such as *noise* in radio and television or passage of a vehicle in the opposite direction in a one-way street. These sources of disturbance are generally referred to as *noise sources* or simply *noise*. An important task of the designer is the minimization of the loss and the optimum recovery of the original commodity when it is corrupted by the effect of noise.

In the deterministic electrical model of Fig. 1-2, it was pointed out that one device which may be used to improve the efficiency of the system is called a *transformer*. In the vocabulary of information theory a device

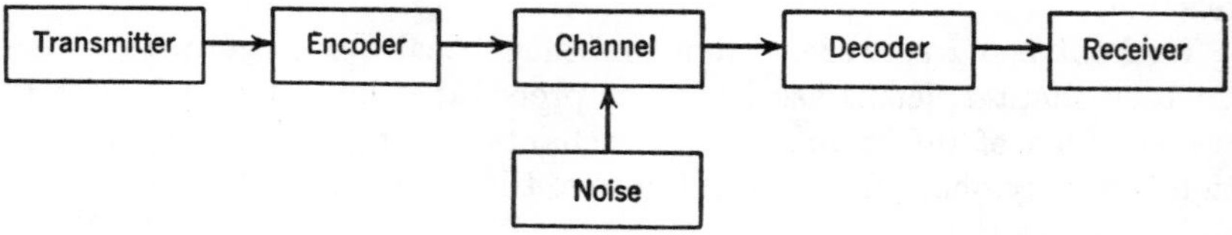

FIG. 1-3. General structure of a communication system used in information theory.

that is used to improve the efficiency of the channel may be called an *encoder*. An encoded message is less susceptible to channel noise. At the receiver's terminal a *decoder* is employed to transform the encoded messages into the original form which is acceptable to the receiver. It could be said that, in a certain sense, for more "efficient" communication, the encoder performs a one-to-one mathematical mapping or an operation F on the input commodity I, $F(I)$, while the decoder performs the inverse of that operation, F^{-1}.

$$\begin{array}{llll} \text{Encoder:} & F & I & F(I) \\ \text{Decoder:} & F^{-1} & F(I) & I \end{array} \tag{1-1}$$

This perfect procedure is, of course, hypothetical; one has to face the ultimate effect of noise which in physical systems will prevent perfect communication. This is clearly seen in the case of the transmission of electrical energy where the transformer decreases the heat loss but an efficiency of 100 per cent cannot be expected. The step-up transformer acts as a sort of encoder and the step-down transformer as a decoding apparatus.

Thus, in any practical situation, we have to add at least three more basic parts to our mathematical model: source of noise, encoder, and decoder (Fig. 1-3). The model of Fig. 1-3 is of a general nature; it may be applied to a variety of circumstances.

A novel application of such a model was made by Wiener and Shannon in their discussions of the statistical nature of the communication of messages. It was pointed out that a radio, television, teletype, or speech transmitter selects sequences of messages from a known transmitter vocabulary *at random* but with specified probabilities. Therefore, in such communication models, the source, channel, encoder, decoder, noise source, and receiver must be *statistically* defined. This point of view in itself constitutes a significant contribution to the communication sciences. In light of this view, one comes to realize that a basic study of communication systems requires some knowledge of probability theory. Communication theories cannot be adequately studied without having a good background of probability. Conversely, readers acquainted with the fundamentals of probability theory can proceed most efficiently with research in the field of communication.

In the macroscopic study of communication systems, some of the basic questions facing us are these:

1. How does one measure information and define a suitable unit for such measurements?

2. Having defined such a unit, how does one define an information *source*, or how does one measure the rate at which an information source supplies information?

3. What is the concept of channel? How does one define the rate at which a *channel* transmits information?

4. Given a source and a channel, how does one study the joint rate of transmission of information and how does one go about improving that rate? How far can the rate be improved?

5. To what extent does the presence of noise limit the rate of transmission of information without limiting the communication reliability?

To present systematic answers to these questions is our principal task. This is undertaken in the following chapters. However, for the benefit of those who wish to acquire a heuristic introduction to the subject, we include a brief discussion of it here.

1-3. A Quantitative Measure of Information. In our study we deal with ideal mathematical models of communication. We confine ourselves to models that are statistically defined. That is, the most significant feature of our model is its unpredictability. The source, for instance, transmits at random any one of a set of prespecified messages. We have no specific knowledge as to which message will be transmitted next. But we know the probability of transmitting each message directly, or something to that effect. If the behavior of the model were predictable (deterministic), then recourse to measuring an amount of information would hardly be necessary.

When the model is statistically defined, while we have no concrete assurance of its detailed performance, we are able to describe, in a sense, its "over-all" or "average" performance in the light of its statistical description. In short, our search for an amount of information is virtually a search for a statistical parameter associated with a probability scheme. The parameter should indicate a relative measure of uncertainty relevant to the occurrence of each particular message in the message ensemble.

We shall illustrate how one goes about defining the amount of information by a well-known rudimentary example. Suppose that you are faced with the selection of equipment from a catalog which indicates n distinct models:

$$[x_1,x_2, \ . \ . \ . \ ,x_n]$$

The desired amount of information $I(x_k)$ associated with the selection of a particular model x_k must be a function of the probability of choosing x_k:

$$I(x_k) = f(P\{x_k\}) \tag{1-2}$$

If, for simplicity, we assume that each one of these models is selected with an equal probability, then the desired amount of information is only a function of n.

$$I_1(x_k) = f\left(\frac{1}{n}\right) \tag{1-2a}$$

Next assume that each piece of equipment listed in the catalog can be ordered in one of m distinct colors. If for simplicity we assume that the selection of colors is also equiprobable, then the amount of information associated with the selection of a color c_j among all equiprobable colors $[c_1,c_2, \ . \ . \ . \ ,c_m]$ is

$$I_2(c_j) = f(P\{c_j\}) = f\left(\frac{1}{m}\right) \tag{1-2b}$$

where the function $f(x)$ must be the same unknown function used in Eq. (1-2a).

Finally, assume that the selection is done in two ways:

1. Select the equipment and then select the color, the two selections being independent of each other.

2. Select the equipment and its color at the same time as one selection from mn possible equiprobable choices.

The search for the function $f(x)$ is based on the intuitive choice which requires the equality of the amount of information associated with the selection of the model x_k with color c_j in both schemes (1-2c) and (1-2d).

$$I(x_k \text{ and } c_j) = I_1(x_k) + I_2(c_j) = f\left(\frac{1}{n}\right) + f\left(\frac{1}{m}\right) \tag{1-2c}$$

$$I(x_k \text{ and } c_j) = f\left(\frac{1}{mn}\right) \tag{1-2d}$$

Thus
$$f\left(\frac{1}{n}\right) + f\left(\frac{1}{m}\right) = f\left(\frac{1}{mn}\right) \tag{1-3}$$

This functional equation has several solutions, the most important of which, for our purpose, is
$$f(x) = -\log x \tag{1-4)*}$$

To give a numerical example, let $n = 18$ and $m = 8$.
$$I_1(x_k) = \log 18$$
$$I_2(c_j) = \log 8$$
$$I(x_k \text{ and } c_j) = I_1(x_k) + I_2(c_j)$$
$$I(x_k \text{ and } c_j) = \log 18 + \log 8 = \log 144$$

Thus, when a statistical experiment has n equiprobable outcomes, the average amount of information associated with an outcome is $\log n$. The logarithmic information measure has the desirable property of additivity for independent statistical experiments. These ideas will be elaborated upon in Chap. 3.

1-4. A Binary Unit of Information. The simplest case to consider is a selection between two equiprobable events E_1 and E_2. E_1 and E_2 may be, say, head or tail in a throwing of an "honest" coin. Following Eq. (1-4), the amount of information associated with the selection of one out of two equiprobable events is
$$-\log \tfrac{1}{2} = \log 2$$

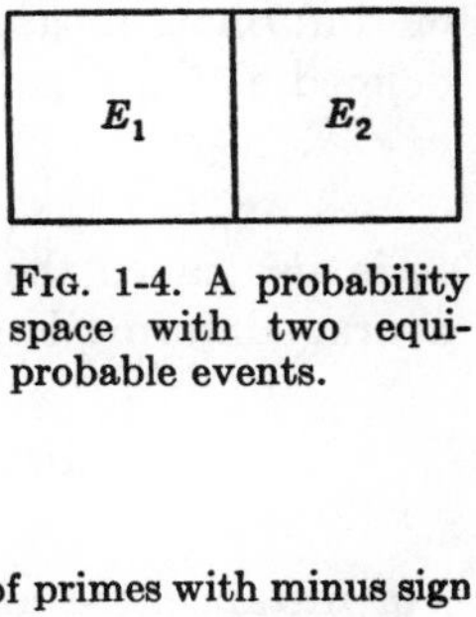

FIG. 1-4. A probability space with two equiprobable events.

An arbitrary but convenient choice of the base of the logarithm is 2. In that case, $-\log_2 \frac{1}{2} = 1$ provides a unit of information. This unit is commonly known as a *bit*.†

* Another solution is

$f(x)$ = number of factors in decomposition of $\frac{1}{x}$ in product of primes with minus sign

For example, let $n = 18$, $m = 8$; then
$$n = 2 \cdot 3 \cdot 3 \qquad f\left(\frac{1}{n}\right) = -3$$
$$m = 2 \cdot 2 \cdot 2 \qquad f\left(\frac{1}{m}\right) = -3$$
$$mn = 2 \cdot 2 \cdot 2 \cdot 2 \cdot 3 \cdot 3 \qquad f\left(\frac{1}{mn}\right) = -6$$

In information theory we require that $f(x)$ be a decreasing function of the probability of choices. This narrows down the solution of Eq. (1-4) to $k \log x$, where k is a constant multiplier. [For an axiomatic derivation, see Sec. 3-19 or A. Feinstein (I).]

† When the logarithm is taken to the base 10, the unit of information corresponds to the selection of one out of ten equiprobable cases. This unit is sometimes referred to as a *Hartley* since it was suggested by Hartley in 1928. When the natural base is used, the unit of information is abbreviated as nat.

Next consider the selection of one out of $2^2, 2^3, 2^4, \ldots, 2^N$ equally likely choices. By successively partitioning a selection into two equally likely selections, we come to the conclusion that the amounts of information associated with the previous selection schemes are, respectively, $2, 3, 4, \ldots, N$ bits.

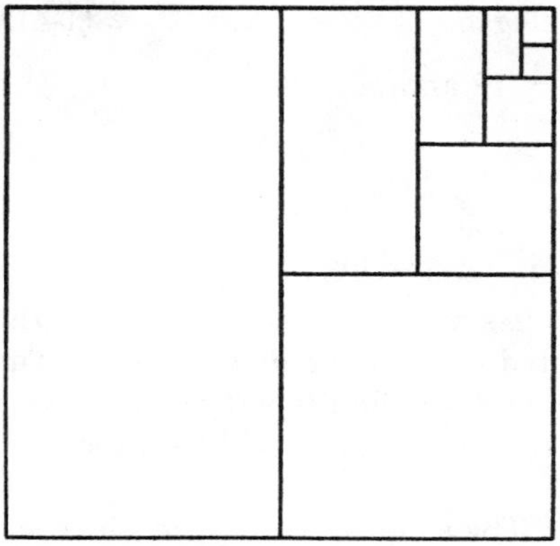

FIG. 1-5. Successive partitioning of the probability space.

In a slightly more general case, consider a source with a finite number of messages and their corresponding transmission probabilities.

$$[x_1,x_2, \ldots ,x_n]$$
$$[P\{x_1\},P\{x_2\}, \ldots ,P\{x_n\}]$$

The source selects at random each one of these messages. Successive selections are assumed to be *statistically independent*. The probability associated with the selection of message x_k is $P\{x_k\}$. The amount of information associated with the transmission of message x_k is defined as

$$I_k = -\log P\{x_k\}$$

I_k is also called the amount of self-information of the message x_k. The average information per message for the source is

$$I = \text{statistical average of } I_k = -\sum_{k=1}^{n} P\{x_k\} \log P\{x_k\} \qquad (1\text{-}5)$$

For instance, the amount of information associated with a source of the above type, transmitting two symbols 0 and 1 with equal probability, is

$$I = -(\tfrac{1}{2} \log \tfrac{1}{2} + \tfrac{1}{2} \log \tfrac{1}{2}) = 1 \text{ bit}$$

If the two symbols were transmitted with probabilities α and $1 - \alpha$, then the average amount of information per symbol becomes

$$I = -\alpha \log \alpha - (1 - \alpha) \log (1 - \alpha) \qquad (1\text{-}6)$$

The average information per message I is also referred to as the *entropy* (or the communication entropy) of the source and is usually denoted by the letter H. For instance, the entropy of a simple source of the above type is

$$H(p_1,p_2, \ldots ,p_n) = -(p_1 \log p_1 + p_2 \log p_2 + \cdots + p_n \log p_n)$$

where $(p_1,p_2, \ldots ,p_n)$ refers to a *discrete complete* probability scheme. Figure 1-6 shows the entropy of a simple binary source for different message probabilities.

Next, consider a second similar source having m symbols, and designate the amount of information per symbol of the two sources by $H(n)$ and $H(m)$, respectively. If the two sources transmit their symbols independently, their joint output might be considered as a source having mn distinct pairs of symbols. It can be shown that for two such independent sources the average information per joint symbol is

$$H(mn) = H(m) + H(n)$$

The formal derivation of this relation is given in Chap. 3.

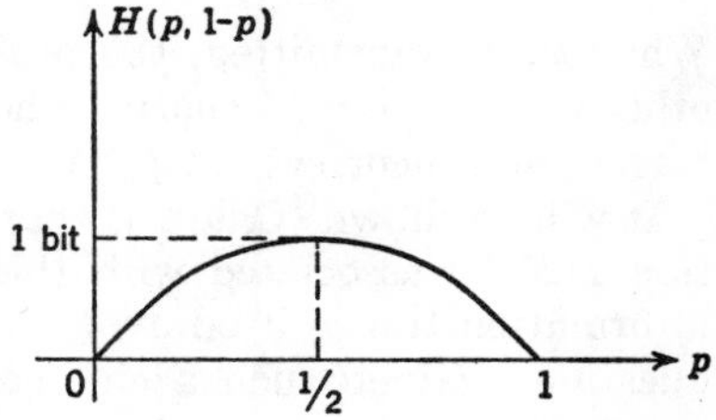

FIG. 1-6. The entropy of an independent discrete memoryless binary source.

1-5. Sketch of the Plan. From a mathematical point of view, the heuristic exposition of the previous two sections is somewhat incomplete. We still need to formalize our understanding of the basic concepts involved and to develop techniques for studying more complex physical models. It was suggested that, given an *independent source* S which transmits messages x_k from a finite set

$$[x_1,x_2, \ldots ,x_n]$$
$$[P\{x_1\},P\{x_2\}, \ldots ,P\{x_n\}]$$

there is an average amount of information $I(x)$ associated with the independent source S.

$$I(x) = \text{expected value or average of } I(x_k) \text{ for all messages}$$

Our next step is to generalize this to the case of random variables with two or more not necessarily statistically independent dimensions, for instance, to define the amount of information per symbol of a scheme having pairs of statistically related symbols (x_k,y_k). This investigation in turn will lead to the study of a channel driven by the source supplying information to that channel. It will be shown that the average information for such a system is

$$\text{Expected value of } I(x_k,y_k) = I(X;Y)^* \tag{1-7}$$

* This consideration will be further generalized to the study of the transinformation in discrete and continuous channels. We shall study the upper bound for transinformation under a variety of plausible circumstances. Previously we mentioned our use of many undefined technical terms such as event, probability of an event, discrete random variable, expected-value source, statistical independence, channel, code, encoding and decoding, channel capacity, etc. These terms as well as many other technical terms will be clearly defined as they are introduced later in the book.

From a physical point of view, the above model may be viewed in a simpler fashion. Consider a source transmitting any one of the two messages x_1 and x_2 with respective probabilities of α and $1 - \alpha$. The output of this source is communicated to a receiver via a noisy binary channel. The channel is described by a stochastic matrix:

$$\begin{bmatrix} a & 1 - a \\ 1 - b & b \end{bmatrix}$$

When x_1 is transmitted, the probability of a correct reception is a and otherwise $1 - a$. Similarly, when x_2 is transmitted, the probability of correct and incorrect receptions are b and $1 - b$, respectively.

It will be shown (Chap. 3) that there is an average amount of information $I(X;Y)$ associated with this model which exhibits the rate of the information transmitted over the channel. This, in turn, raises a basic question. Given such a channel, what is the highest possible rate of transmission of information over this channel for a specified class of sources? In this manner, one arrives in a natural way at the concept of channel capacity and efficiency of a statistical communication model.

In the above example, the capacity of the channel may be computed by maximizing the information measure $I(X;Y)$ over all permissible values of α.

In short, with each probability scheme we associate an entropy which represents, in a way, the average amount of information for the outcomes of the scheme. When a source and a receiver are connected via a channel, several probability schemes such as conditional and joint probabilities have special significance. An important task is to investigate the physical significance and the interrelationships between different entropies in a communication system. The formal treatment of these relations and the concept of channel capacity is presented in several chapters of the text.

The reader acquainted with probability theory may regard information theory as a new branch of that discipline. He can grasp it at a fair speed. The reader without such a background has to move much more slowly. However he will find the introductory material of Chap. 2 of substantial assistance in the study of Chaps. 3 and 4. An introductory treatment of a random variable assuming a continuum of values is given in Chap. 5. Chapter 6 presents a general study of averaging and moments. The reader with such a background will readily recognize the entropy functions that form the nucleus of information theory as moments of an associated logarithmic random variable: $-\log P\{X\}$. Thus the entropy appears to be a new and useful form of moment associated with a probability scheme. This idea will serve as an important link in the integration of information and probability theories. Chapter 7 gives a concise introduction to multinormal distributions, laws of large numbers, and central-limit theorems. These are essential tools for the proof of the main theorems of information theory.

Chapters 8 and 9 extend the information-theory concept to random variables assuming a continuum of values (also continuous signals). The probability background of Chaps. 2, 5, 6, 7, and 10 is in most part indispensable for the study of information theory. However, a few additional topics are included for the sake of completeness, although they may not be directed toward an immediate applicaton.

Chapter 10 presents a bird's-eye view of stochastic theory, followed by Chap. 11, which studies the information theory of stochastic models. A slightly more advanced consideration (but perhaps the heart of the subject) appears in Chap. 12.

A main application of the theory thus far seems to be in the devising of an efficient matching of the information source and the channel, the so-called *coding theory*. The elements of this theory appear in Chaps. 4 and 13. The Appendix is designed to introduce the reader to a few of the many topics available for further reading in this field.

1-6. Main Contributors to Information Theory. The historical background of information theory cannot be covered in a few pages. Fortunately there are several sources where the reader can find a historical review of this subject, e.g., The Communication of Information, by E. C. Cherry (*Am. Scientist*, October, 1952), and "On Human Communication," by the same author (John Wiley & Sons, Inc., 1957). (In Chap. 2 of the latter book, Cherry gives a very interesting historical account of developments leading to the discovery of information theory, particularly the impact of the invention of *telecommunication*.)

As far as the communication engineering profession is concerned, it seems that the first attempt to define a measure for "the amount of information" was made by R. V. L. Hartley* in a paper called Transmission of Information (*Bell System Tech. J.*, vol. 7, pp. 535–564, 1928).

Hartley suggested that "information" arises from the successive selection of symbols or words from a given vocabulary. From an alphabet of D distinct symbols we can select D^N different words, each word containing N symbols. If these words were all equiprobable and we had to select one of them at random, there would be a quantity of information I associated with such a selection. Furthermore, Hartley suggested the

* The work of Hartley was greatly influenced by a law which was simultaneously discovered in 1924 by Nyquist in the United States and Kupfmuller in Germany. The Nyquist-Kupfmuller law states that for transmitting telegraph signals at a given rate a definite-frequency *bandwidth* is required. Among those who contributed to the refinement of this law, which has been closely interwoven with the concept of a measure of information, are D. Gabor (1946) and D. M. Mackay (1948).

logarithm to the base 10 of the number of possible different words D^N as the quantity of information $I = N \log D$.

The main contributions, which really gave birth to the so-called information theory, came shortly after the Second World War from the mathematicians C. E. Shannon and N. Wiener. Wiener's mathematical contributions to the field of Fourier series and later to time series, plus his genuine interest in the field of communication, led to the foundation of communication theories in general. His two books, "Cybernetics" and "Extrapolation, Interpolation, and Smoothing of Stationary Time Series" (1948 and 1949), paved the way for the arrival of new statistical theories of communication. In a paper entitled The Mathematical Theory of Communication (*Bell System Tech. J.*, vol. 27, 1948), Shannon made the first integrated mathematical attempt to deal with the new concept of the amount of information and its main consequences. Shannon's first paper, along with a second paper, laid the foundation for the new science to be named *information theory*. Shannon's earlier contribution may be summarized as follows:

1. Definition of the amount of information from a semiaxiomatic point of view.
2. Study of the flow of information for discrete messages in channels with and without noise (models of Figs. 1-1 and 1-3).
3. Defining the capacity of a channel, that is, the highest rate of transmission of information for a channel with or without noise.
4. In the light of 1, 2, and 3, Shannon gave some fundamental encoding theorems. These theorems state roughly that for a given source and a given channel one can always devise an encoding procedure leading to the highest possible rate of transmission of information.
5. Study of the flow of information for continuous signals in the presence of noise, as a logical extension of the discrete case.

Subsequent to his earlier work, Shannon has made several additional contributions. These have considerably strengthened the position of the original theory.

Following Wiener's and Shannon's works an unusually large number of scientific papers appeared in the literature in a relatively short time. A bibliography of information theory and allied topics might now, 13 years after the publication of Shannon's and Wiener's works, contain close to 1,000 papers. This indicates the great interest and enthusiasm (perhaps overenthusiasm) of scientists toward this fascinating new discipline. Here it would be impossible to give a detailed account of the contributions in this field. The reader may refer to A Bibliography of Information Theory, by F. L. Stumpers, and also to *IRE Transactions on Information Theory* (vol. IT-1, no. 3, pp. 31–47, September, 1955).

Even though a historical account has not been attempted here, the names of some of the contributors should be mentioned in passing. Bell Telephone Laboratories appears to be the birthplace of information and coding theory. Among the contributors from Bell Labs are E. N. Gilbert, R. W. Hamming, J. L. Kelley, Jr., B. McMillan, S. O. Rice, C. E. Shannon, and D. Slepian. P. Elias, R. M. Fano, A. Feinstein, D. Huffman, C. E. Shannon, N. Wiener, and J. A. Wozencraft of the Massachusetts Institute of Technology have greatly contributed to the advancement of information and coding theory. Information theory has received significant stimuli from the works of several Russian mathematicians. A. I. Khinchin, by employing the results of McMillan and Feinstein, produced one of the first mathematically exact presentations of the theory. Academician A. N. Kolmogorov, a leading man in the field of probability, and his colleagues have made several important contributions. A few of the other Russian contributors are R. L. Dobrushin, D. A. Fadiev, M. A. Gavrilov, I. M. Gel'fand, A. A. Kharkevich, V. A. Kotelnikov,* M. Rozenblat-Rot, V. I. Siforov, and I. M. and A. M. Iaglom.

The afore-mentioned names are only a few of a long list of mathematicians and communication scientists who have contributed to information theory. Some other familiar names are D. A. Bell, A. Blanc-Lapierre, L. Brillouin, N. Abramson, D. Gabor, S. Goldman, I. J. Good, N. K. Ignatyev, J. Loeb, B. Mandelbrot, K. A. Meshkovski, W. Meyer-Eppler, F. L. Stumpers, M. P. Schutzenberger, A. Perez, W. Peterson, A. Thomasian, R. R. Varsamov, J. A. Ville, P. M. Woodward.

A list of those actively engaged in the field would be too long to be included here. Reference to some of the current work will be found in the text and in the bibliography at the end of the book.

For a comprehensive list, the reader is referred to existing bibliographies such as those by Stumpers, Green, and Cherry. Recent contributions to information theory have been aimed at providing more exact proofs for the basic theorems stated by earlier contributors. A state of steady improvement has been prevailing in the literature.

McMillan, Feinstein, and Khinchin have greatly enhanced the elegance of the theory by putting it on a more elaborate mathematical basis and providing proofs for the central theorems as earlier stated by C. E. Shannon. These contributors have confirmed that under very general circumstances, it is possible to transmit information with a high degree of reliability over a noisy channel at a rate as close to the channel capacity as desired.

J. Wolfowitz derived a strong converse of the fundamental theorem of information theory. Among other important theorems, he proved that reliable transmission at a rate higher than the channel capacity is not pos-

* Kotelnikov is known particularly for the development of the theory of potential noise immunity in the presence of white gaussian noise.

sible. In the past 2 or 3 years a large number of scientists have become interested in integrating some of the work on encoding theory within the framework of classical mathematics. Reference will be made to their work in Chaps. 13 and 14.

S. Kullback has described the growth of information theory from its statistical roots and emphasized the interrelation between information theory and statistics (Kullback).

The study of time-varying channels has also received considerable attention. Among those who have contributed are C. E. Shannon, R. A. Silverman and S. H. Chang, and V. I. Siforov and his colleagues.

To sum up, the present trend in information theory seems to be as follows: From an engineering point of view, a search for applications of the theory (radar detection, speech, telephone and radio communication, game and decision theory, and particularly implementation of codes) is evident, while the mathematician is still seeking for more rigor in the foundation of the theory and elegance par excellence.

1-7. An Outline of Information Theory. If we were to make a two-page résumé of information theory for those scientists with a broad background of probability theory, the following could be suggested.

1. The average amount of information conveyed by a discrete random variable Y about another discrete random variable X is suggested by C. E. Shannon.

$$I(X;Y) = \sum_{i=1}^{n} \sum_{k=1}^{m} P\{X = x_i, Y = y_k\} \log \frac{P\{X = x_i, Y = y_k\}}{P\{X = x_i\}P\{Y = y_k\}} \tag{1-8}$$

This definition can be generalized to cover not only the case of two or more random variables assuming a continuum of values but also the more general case of random vectors, generalized functions, and stochastic processes (Gel'fand and Iaglom*).

2. The channel is specified by $P\{Y = y_k|X = x_i\}$ for all encountered integers i and k. The largest value of the transinformation $I(X;Y)$ obtained over all possible source distribution $P\{X = x_i\}$ is called the *capacity* of the channel [Shannon (I)].

The definition of the channel capacity can be subjected to generalizations similar to those suggested in 1.

3. Let X and Y be two finite sets of alphabets with $x \in X$ and $y \in Y$. The simplest channel is specified by $P\{\quad |x \in X\}$. Now consider words of n symbols selected from the X alphabet. These words will be denoted

* I. M. Gel'fand and A. M. Iaglom, Calculation of the Amount of Information about a Random Function Contained in Another Such Function, *Uspekhi Mat. Nauk S.S.S.R.* (N.S.), vol. 12, no. 1(73), pp. 3–52, 1957; or *Am. Math. Soc. Transl.*, ser. 2, vol. 12, 1959.

by $u \in U$ and their corresponding received pairs by $v \in V$. This is an nth-order extension of the channel.

4. Given a source $P\{X = x_i\}$, a channel $P\{\quad |X = x_i\}$, and their respective nth-order extensions, then to a specified message ensemble U, we may associate a partitioning of the V space such that

$$\begin{array}{ll} u_k \rightarrow B_k & k = 1, 2, \ldots, N \\ B_k \cap B_j = \emptyset & \text{for } k \neq j \quad k = 1, 2, \ldots, N \\ P\{B_k|u_k\} \geq 1 - \lambda & k = 1, 2, \ldots, N \\ & \lambda \text{ a specified positive} \\ & \quad \text{number usually very small} \end{array}$$

This is a decision scheme which in turn specifies a code (N,n,λ) [A. Feinstein (I)].

5. The central theme of information theory is the following so-called fundamental coding theorem. Given a source, a channel with capacity C, and two constants

$$0 \leq H \leq C \qquad 0 < \lambda < 1$$

it can be shown that there are an integer $n = g(\lambda,H)$ and a code (N,n,λ) with (N = function of λ and n) $\geq 2^{nH}$. This is the coding theorem stating the possibility of transmitting information at a rate $H \leq C$ over a noisy channel under specified circumstances.

6. Further elaborate mathematical treatment of the concepts of information theory was presented by B. McMillan, who extended the definition of source and channels from a Markov chain to stationary processes. A proof of the fundamental theorem of 5 as well as a clear understanding of the concepts involved in 5 is due to Feinstein. Khinchin considerably improved the status of the art in general and gave a proof of the fundamental theorem of 5 for the case of stationary processes. A converse of the fundamental theorem of 5 is due to J. Wolfowitz, who also gave sharper estimates than those given in 5. Remaining questions include the search for more general encoding theorems along the lines suggested in 1. A recent step in this direction was taken by C. E. Shannon.* The search for engineering applications, particularly low-error probability codes, is ever increasing.

PROBLEMS

1-1. An alphabet consists of four letters A, B, C, D with respective probabilities of transmission $\frac{1}{3}$, $\frac{1}{4}$, $\frac{1}{4}$, $\frac{1}{6}$. Find the average amount of information associated with the transmission of a letter.

* C. E. Shannon, Probability of Error for Optimal Codes in a Gaussian Channel, *Bell System Tech. J.*, vol. 38, no. 3, pp. 611–656, 1959.

1-2. An independent, discrete source transmits letters selected from an alphabet consisting of three letters A, B, and C, with respective probabilities

$$p_A = 0.7 \qquad p_B = 0.2 \qquad p_C = 0.1$$

(*a*) Find the entropy per letter.

(*b*) If consecutive letters are statistically independent and two-symbol words are transmitted, find all the pertinent probabilities for all two-letter words and the entropy of the system of such words.

1-3. Plot the curve $y = -x \log_2 x$ for

$$0 \leq x \leq 1$$

1-4. A pair of dice are thrown. We are told that the sum of the faces is 7. What is the average amount of information contained in this message (that is, the entropy associated with the probability scheme of having the sum of the faces equal to 7)?

1-5. An alphabet consists of six symbols A, B, C, D, E, and F which are transmitted with the probabilities indicated below:

A	0	$\frac{1}{2}$
B	01	$\frac{1}{4}$
C	011	$\frac{1}{8}$
D	0111	$\frac{1}{16}$
E	01111	$\frac{1}{32}$
F	011111	$\frac{1}{32}$

(*a*) Find the average information content per letter.

(*b*) If the letters are encoded in a binary system as shown above, find $P\{1\}$ and $P\{0\}$ and the entropy of the binary source.

1-6. A bag contains 100 white balls, 50 black balls, and 50 blue balls. Another bag contains 80 white balls, 80 black balls, and 40 blue balls. Determine the average amount of information associated with the experiment of drawing a ball from each bag and predicting its color. The result of which experiment is, on the average, harder to predict?

1-7. There are 12 coins, all of equal weight except one, which may be lighter or heavier. Using information-theory concepts, show that it is possible to determine which coin is the odd one and indicate whether it is lighter or heavier in not more than three weighings with an ordinary balance.

1-8. Solve Prob. 1-7 when the number of coins is N. What is the minimum number of weighings?

1-9. There are seven coins, five of equal weight and the remaining two also of equal weight but lighter than the first five coins. Find the minimum number of weighings necessary to locate these two coins.*

*For a general discussion of coin-weighing problems, see A. M. Iaglom and I. M. Iaglom, "Probabilité et information" (translated from Russian), Dunod, Paris, 1959.

PART 1

DISCRETE SCHEMES WITHOUT MEMORY

. . . choose a set of symbols, endow them with certain properties and postulate certain relationships between them. Next, . . . deduce further relationships between them. . . . We can apply this theory *if* we know the "exact physical significance" of the symbols. That is, if we can find objects in nature which possess exactly those properties and inter-relations with which we endowed the symbols. . . . The "pure" mathematician is interested only in the inter-relations between the symbols. . . . The "applied" mathematician always has the problem of deciding what is the exact physical significance of the symbols. *If* this is known, then at any stage in the theory we know the physical significance of our theorems. But the strength of the chain depends on the strength of the weakest link, and on occasion the link of "physical significance" is exceedingly fragile.

J. E. Kerrich, "An Experimental Introduction to the Theory of Probability"
Belgisk Import Co., Copenhagen

CHAPTER 2

BASIC CONCEPTS OF DISCRETE PROBABILITY

2-1. Intuitive Background. Most of us have some elementary intuitive notions about the laws of probability, and we may set up a game or an experiment to test the validity of these notions. This procedure is much like the so-called classical approach to the theory of probability, which was commonly used by mathematicians up to the 1930s. However, this approach has been subjected to considerable criticism; indeed, the literature on the subject contains many contradictions and controversies in the writings of the major authors. These arise from the intuitive background used and the lack of well-defined formalism and rigor. Thus, the experiment or game is usually defined by assuming certain symmetries and by accepting certain results a priori, such as the idea that certain possible outcomes are equally likely to occur. For example, consider the following problem: Two persons, A and B, play a game of tossing a coin. The coin is thrown twice. If a head appears in at least one of the two throws, A wins. Otherwise, B wins. Intuitively, it seems that the four following possible outcomes are equally probable:

$$(HH),\ (HT),\ (TH),\ (TT)$$

where H denotes head and T denotes tail. A may assume that his chances of winning the game are $\frac{3}{4}$, since a head occurs in three out of four cases (to his advantage). On the other hand, the following reasoning may also seem logical. If the outcome of the first throw is H, A wins; there is no need to continue the game. Accordingly, only three possibilities need be considered, namely:

$$(H),\ (TH),\ \text{and}\ (TT)$$

where the first two cases are favorable to A and the last one to B. In other words, the probability that A wins is really $\frac{2}{3}$ instead of $\frac{3}{4}$. The intuitive approach in this problem thus seems to lead to two different estimates of probability.*

The twentieth century has witnessed enormous advances in the rigorous axiomatic treatment of many branches of mathematics. It is true

* The reasoning which assumes the equiprobable outcomes is incorrect. See also Prob. 2-38.

that the axiomatic approach is essentially present in the familiar euclidean geometry and is, in a way, a very old principle. But it was not until the early twentieth century, when the formal and logical structure of mathematics was given serious, systematic study, that its fundamental and profound implications were recognized. Actually, however, the groundwork for the axiomatic treatment was laid by mathematicians such as Peano, Cantor, and Boole during the middle of the nineteenth century. The later efforts of Hilbert, Russell, Whitehead, and others led to a complete reorientation of the basic formulations, bringing mathematics to its present level.

Although consideration of the axiomatic treatment is not our subject here, it may be interesting to point out its general nature. First, a necessary set of symbols is introduced. Then certain inference or operation rules are given for the desired formal manipulation of the symbols, and a proper set of axioms is determined. The formal system thus created must be consistent; that is, the axioms must be independent and noncontradictory. Strictly speaking, the derivation of the theorems is manipulation of symbols without content, using axioms as a starting point and applying the rules of operation. The fundamental nature of a formal system is by no means obvious, and the limitations are even today under very careful study.

A rather new branch of mathematics exists which deals in an axiomatic manner with properties of various abstract spaces and functions defined over these spaces. This is the so-called "measure theory." In the late 1930s and early 1940s attempts were made to put the probability calculus on an axiomatic basis. The work of Kolmogorov, Doob, and many others has contributed greatly toward this aim. Today formal probability theory is an important branch of measure theory (in a strictly formal sense), although the epistemological meaning of probability itself is subject to philosophical discussion. This latter aspect has been studied by several profound thinkers (von Neumann, Carnap, Russell, Fisher, Neyman, and many others).

Today engineers and research scientists recognize that they must have a working knowledge of the powerful tools of twentieth-century mathematics. Although completely axiomatic and rigorous treatment of this subject is far beyond the scope of this discussion, a classical presentation would be out of date, as it would completely forgo the important modern contributions to the theory. Under these circumstances, it seems that a survey of the modern theory of probability at a nonprofessional level will be a reasonable compromise. Most engineering students are not very familiar with concepts of probability, and it is important that they gain some appreciation of them.

In what follows, some elementary concepts of the theory of sets or so-called "set algebra" must first be introduced. Then these concepts are used to introduce the fundamental definitions of the theory of probability. Such a presentation allows a much wider application of the probability theory than does the older approach, which is inadequate for attacking a large class of modern problems.

2-2. Sets. The word *set*, in mathematics, is used to denote any collection of objects specified according to a well-defined rule. Each object in a set is called an *element*, a *member*, or a *point* of the set. If x is an element of the set X, this relationship is expressed by

$$x \in X \qquad x \text{ belongs to } X \tag{2-1}$$

When x is not a member of the set X, this fact is shown by

$$x \notin X \qquad x \text{ does not belong to } X \tag{2-2}$$

For example, if X is the set of all positive integers, then

$$\begin{aligned} 5 &\in X \\ \sqrt{2} &\notin X \\ -3 &\notin X \end{aligned}$$

A set can be specified by either giving all its elements or stating the requirements for the elements belonging to the set. If a, b, c, and d are the only members of a set X, then we may write either

$$X = \{a,b,c,d\} \tag{2-3}$$

or

$$X = \{x\} \tag{2-4}$$

In the latter case x designates a general element of X with the understanding that the rule for identifying the members of X is known. For example, if the set X consists of the number of dots on the faces of a die, then we may write

$$X = \{1,2,3,4,5,6\}$$

If the set X consists of all rectangles with an area of 1 square foot we may write $X = \{x\}$, denoting by x any general rectangle having the specified area.

When every element of a set A is a member of a set B, we say that A is a *subset* of B. This relationship is expressed by either of the forms

$$A \subset B \qquad A \text{ is contained in } B \tag{2-5}$$

or

$$B \supset A \qquad A \text{ is a subset of } B \tag{2-6}$$

For example, if A is the set of positive integers and B the set of all rational numbers, then A is a subset of B.

The sets A and B are said to be *equal* if they have exactly the same elements, that is, if

$$A \subset B$$

and

$$A \supset B \tag{2-7}$$

then

$$A = B$$

For instance, if the set A consists of the roots of the equation

$$x(x + 1)(x^2 - 4)(x - 3) = 0$$

and $B = \{-2,-1,0,2,3\}$

$C = \{x\}$ $\quad x$ being any integer such that $|x| < 4$

then

$$C \supset A$$

$$C \supset B$$

$$\left.\begin{matrix} A \supset B \\ A \subset B \end{matrix}\right\} \quad A = B$$

In many instances, when dealing with specific problems, it is most convenient to confine the discussion to objects that belong to a fixed class of elements. This is referred to as a *universal set.* For example, suppose that, in a certain problem dealing with the study of numbers, it may be required to define the set of all integers I, or the set of positive numbers P, or the set of perfect square integers S. All these sets can be looked upon as subsets of the larger set of all real numbers. This latter set may be considered as the universal set U, a definition which is useful in dealing with the specific problem under discussion.

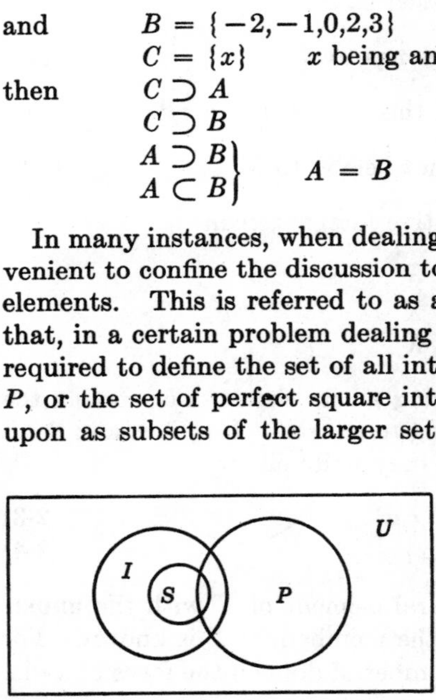

FIG. 2-1. Example of a Venn diagram.

In problems concerned with the interrelationship of sets, an illustrative diagram called a *Venn* diagram* is of considerable visual assistance. The elements of the universal set in a Venn diagram are generally shown by points in a rectangle. The elements of any set under consideration are commonly shown by a circle or by any other simple closed contour inside the universal set. The *universe* associated with the aforesaid example is illustrated in Fig. 2-1.

A set may contain a finite or an infinite number of elements. When a set has no element, it is said to be an *empty* or a *null* set. For example, the set of the real roots of the equation

$$2z^2 + 1 = 0$$

is a null set.

* Named after the English logician John Venn (1834–1923).

2-3. Operations on Sets. Consider a universal set U of any arbitrary elements. U contains all possible elements under consideration. The universal set may contain a number of subsets $A, B, C, D, \ldots$ which individually are well-defined sets. The operation of union, intersection, and complement is defined as follows:

The *union* or *sum* of two sets A and B is the set of all those elements that belong to A or B or *both.*

The *intersection* or *product* of two sets A and B is the set of all those elements that belong to both A and B.

The *difference* $B - A$ of any set A relative to the set B is a set consisting of all elements of B that are *not* elements of A.

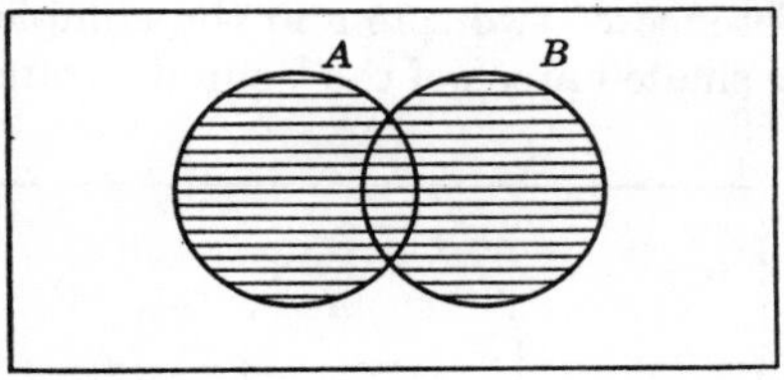

FIG. 2-2. Sum or union $A + B$.

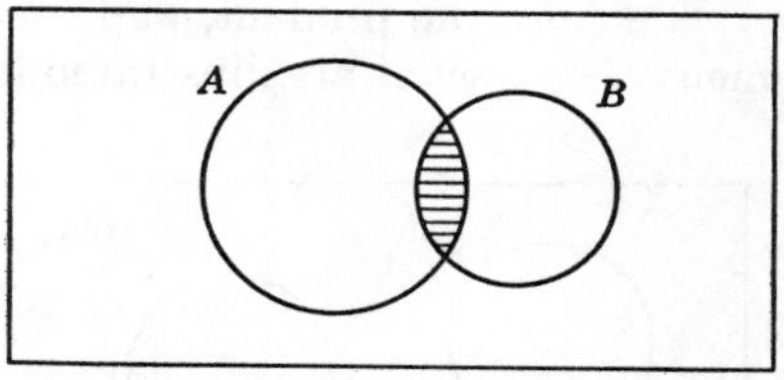

FIG. 2-3. Intersection or product $A \cdot B$.

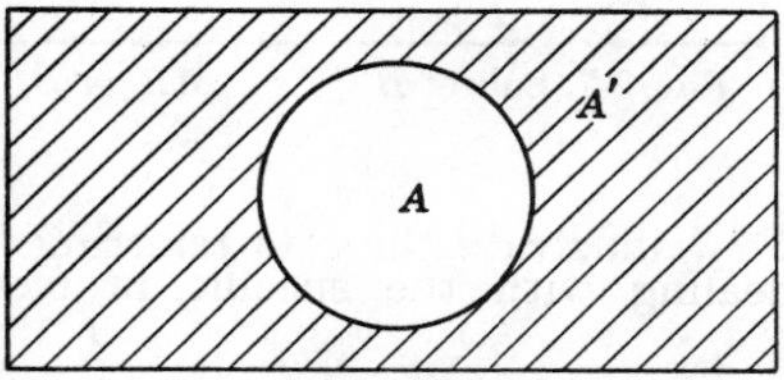

FIG. 2-4. Complement.

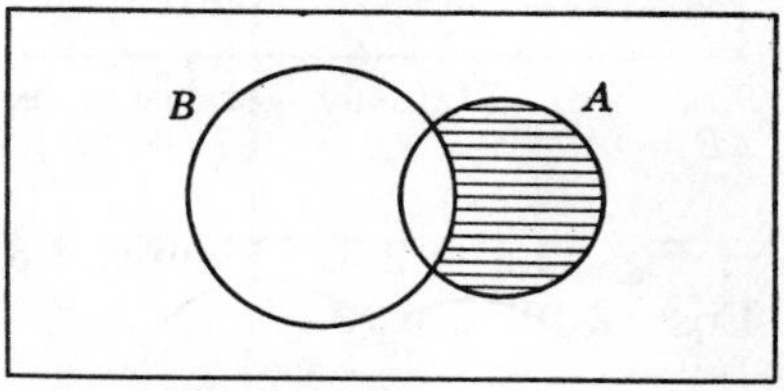

FIG. 2-5. Difference $A - B$.

The *complement* or *negation* of any set A is the set A' containing all elements of the universe that are *not* elements of A.

In the mathematical literature the following notations are commonly used in conjunction with the above definitions.

$A \cup B$	A union B, or A cup B	(2-8)
$A \cap B$	A intersection B, or A cap B	(2-9)
$A - B$	relative complement of B in A	(2-10)
$B \subset A$	B is contained in A	
$\sim A$	complement of A	(2-11)

In the engineering literature the notations given below are primarily used.

$A + B$	sum or union	(2-12)
$A \cdot B$ or AB	intersection or product	(2-13)
$A - B$	difference	(2-14)
A'	complement	(2-15)

For the convenience of the engineer we shall generally adhere to the latter notations. However, where any confusion in notation may occur we shall resort to mathematical notation.

The universe and the empty set will be denoted by U and $\emptyset$, respectively. When the product of two sets A and B is an empty set, that is,

$$A \cap B = \emptyset \tag{2-16}$$

the two sets are said to be *mutually exclusive*. When the product of the two sets A and B is equal to B, then B is a *subset* of A.

$$A \cap B = B \quad \text{implies} \quad B \subset A \tag{2-17}$$

The sum, the product, and the difference of two sets and the complement of any set A are illustrated in the shaded areas of the Venn diagrams

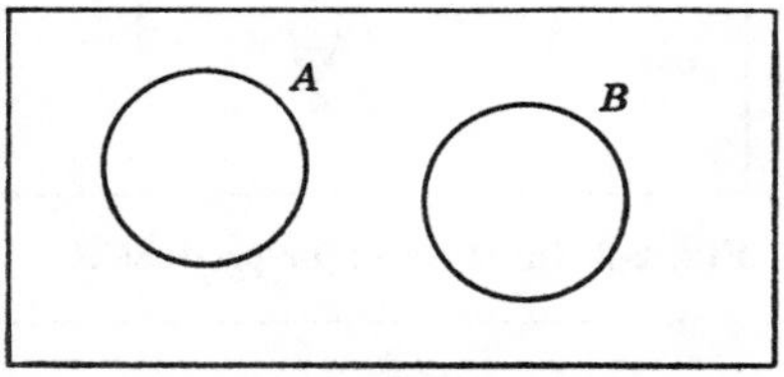

FIG. 2-6. Mutually exclusive sets. $AB = 0$.

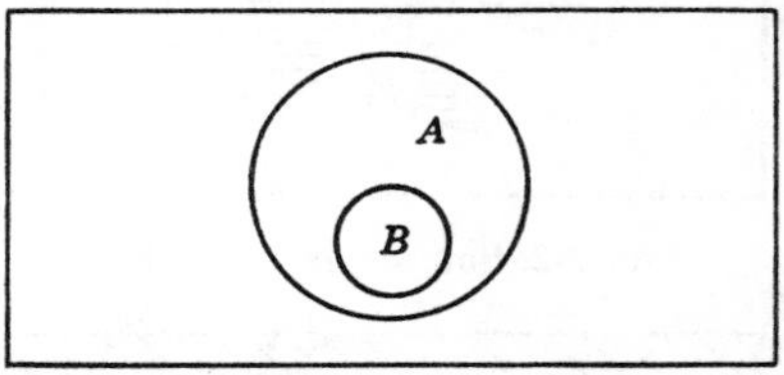

FIG. 2-7. Subset $B \subset A$. $AB = B$.

of Figs. 2-2 to 2-5. Figures 2-6 and 2-7 illustrate the sets referring to Eqs. (2-16) and (2-17).

Example 2-1. Let the universe consist of the set of all positive integers, and let

$$\begin{aligned} A &= \{1,2,3,6,7,10\} \\ B &= \{3,4,8,10\} \\ C &= \{x\} \end{aligned}$$

where x is any positive integer larger than 5.

Find $A + B$, $A \cdot B$, $A - B$, $A \cdot C$, $B \cdot C$, C', and $A + B + C$.

Solution

$$\begin{aligned} A + B &= \{1,2,3,4,6,7,8,10\} \\ A \cdot B &= \{3,10\} \\ A - B &= \{1,2,6,7\} \\ A \cdot C &= \{6,7,10\} \\ B \cdot C &= \{8,10\} \\ C' &= \{1,2,3,4,5\} \\ (A + B) + C &= U - \{5\} \end{aligned}$$

2-4. Algebra of Sets. We now state certain important properties concerning operations with sets. Let A, B, and C be subsets of a universal set U; then the following laws hold.

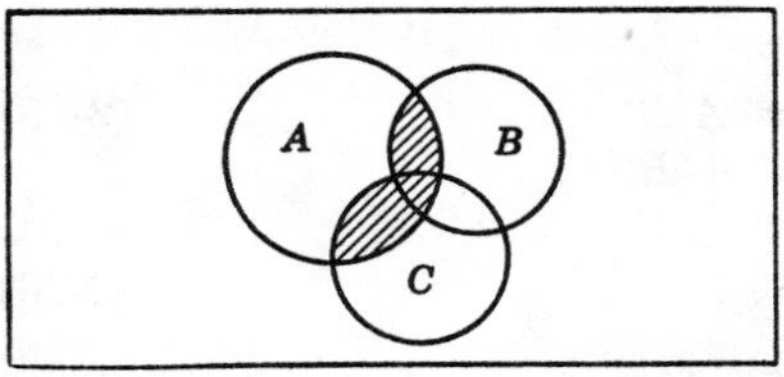

FIG. 2-8. Distributive law. $A(B + C) = AB + AC$.

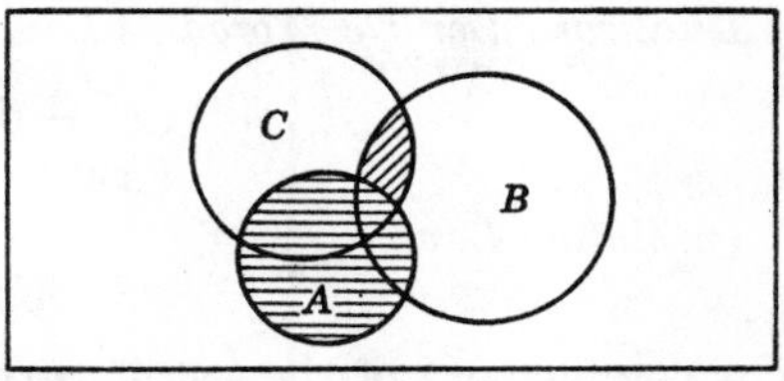

FIG. 2-9. Distributive law. $A + BC = (A + B)(A + C)$.

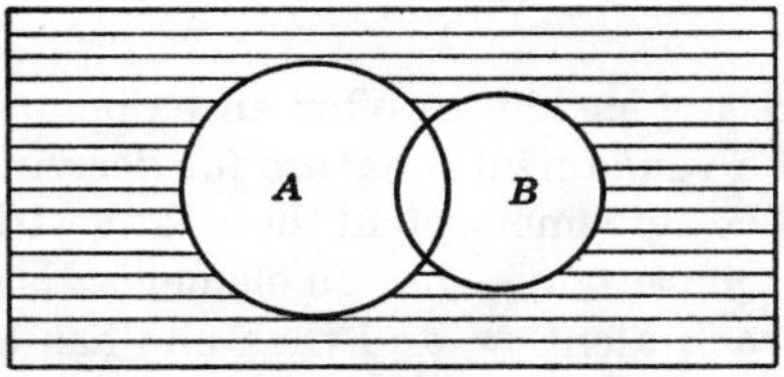

FIG. 2-10. Dualization. $(A + B)' = A'B'$.

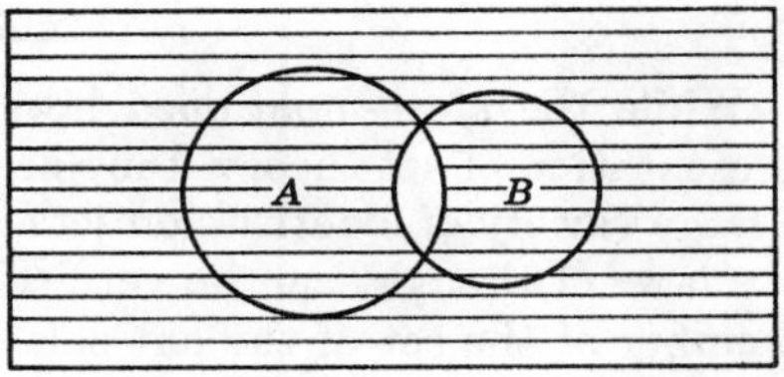

FIG. 2-11. Dualization. $(AB)' = A' + B'$.

Commutative Laws:

$$\begin{aligned} A + B &= B + A \\ AB &= BA \end{aligned} \tag{2-18}$$

Associative Laws:

$$\begin{aligned} (A + B) + C &= A + (B + C) \\ (AB)C &= A(BC) \end{aligned} \tag{2-19}$$

Distributive Laws:

$$\begin{aligned} A(B + C) &= AB + AC \\ A + BC &= (A + B)(A + C) \end{aligned} \tag{2-20}$$

Complementarity:

$$\begin{aligned} A + A' &= U \\ AA' &= \emptyset \end{aligned} \tag{2-21}$$

$$\begin{aligned} A + U &= U \\ AU &= A \end{aligned} \tag{2-22}$$

$$\begin{aligned} A + \emptyset &= A \\ A\emptyset &= \emptyset \end{aligned} \tag{2-23}$$

Difference Law:

$$\begin{aligned} (AB) + (A - B) &= A \\ (AB)(A - B) &= \emptyset \\ A - B &= AB' \end{aligned} \tag{2-24}$$

Dualization or De Morgan's Law:

$$(A + B)' = A'B' \qquad (2\text{-}25)$$
$$(AB)' = A' + B'$$

Involution Law:

$$(A')' = A \qquad (2\text{-}26)$$

The complement of the set A' is the set A.

Idempotent Law: For all sets A,

$$A + A = A \qquad (2\text{-}27)$$
$$AA = A$$

While the afore-mentioned laws are not meant to offer an axiomatic presentation of set theory, they are of a fundamental nature for deriving a large variety of identities on sets. The agreement of all these laws with the laws of thought can be verified. One assumes that an element x is a member of the set of the left side of each identity, and then one has to prove that x will necessarily be a member of the set of the right side of the same equation. For instance, in order to prove the distributive law [Eq. (2-20)], let

$$x \in A(B + C)$$

Then

$$x \in A$$
$$x \in (B + C)$$

Then at least one of the following three cases must be true:

(*a*) $x \in A$, $x \in B$ (*b*) $x \in A$, $x \in C$ (*c*) $x \in A$, $x \in B$, $x \in C$

These are in turn equivalent to

(*a*) $x \in AB$ (*b*) $x \in AC$ (*c*) $x \in ABC$

but

$$ABC \subset AB$$

Therefore it is sufficient to require

$$x \in AB + AC$$

Similarly, one can show that $x \in AB + AC$ implies $x \in A(B + C)$.

The Venn diagram is often a very useful visual aid. Its use is of valuable assistance in solving problems, as long as the formal proofs are not overlooked.

Example 2-2. Verify the following relation:

$$(A + B) - AB = AB' + A'B$$

Solution. By virtue of the third relation of Eqs. (2-24),

$$(A + B) - AB = (A + B)(AB)'$$

Application of De Morgan's law yields

$$(A + B)(AB)' = (A + B)(A' + B')$$
$$(A + B)(A' + B') = A'A + A'B + B'A + B'B = AB' + A'B$$

For an alternative proof, let

$$x \in [(A + B) - AB]$$

Then only one of the following two cases is possible:

(a) $x \in A$, $x \notin B$ (b) $x \in B$, $x \notin A$

These cases are equivalent to

(a) $\left.\begin{matrix} x \in A \\ x \in B' \end{matrix}\right\} x \in AB'$ (b) $\left.\begin{matrix} x \in B \\ x \in A' \end{matrix}\right\} x \in A'B$

Note that AB' and $A'B$ are mutually exclusive sets. Similarly, one can show that all the elements belonging to the set at the right side of the above equation also belong to the set of the left side. Thus the two sides present equivalent sets.

Example 2-3. Express the set composed of the hatched region of Fig. E2-3 in terms of specified sets.

Solution. The desired set A is

$$A = A_1A_2 + A_2A_3 + A_4A_5A_6$$

See Fig. E2-3.

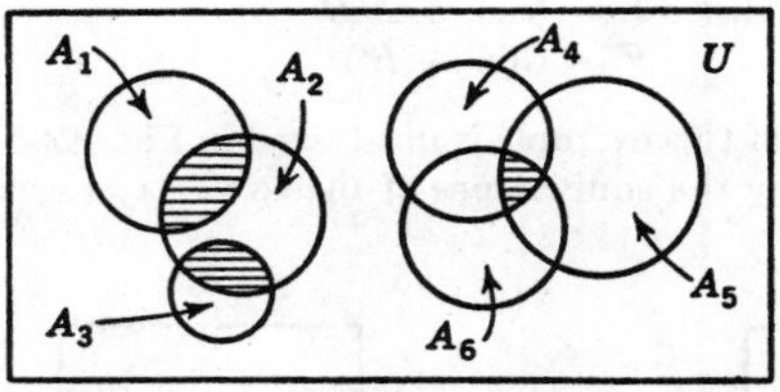

FIG. E2-3

Example 2-4. Verify the relation

$$(A + B)'C = C - C(A + B)$$

Solution. We may wish to verify the validity of this relation by using the Venn diagram of Fig. E2-4. The left side of this equation represents the part of the set C that is not in A or B. The right side represents $C - CA - CB$, that is, the part of C that is not included either in A or in B.

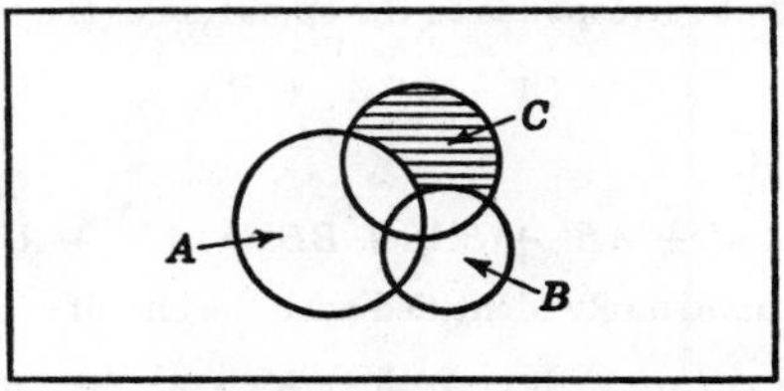

FIG. E2-4

Example 2-5. Consider the relay circuit of Fig. E2-5. The setup contains coils which must be activated for closing or opening the corresponding relay. A, B, and C are normally open relays and A', B', and C' are normally closed relays which are respectively activated by the same controlling source. For instance, when relay A is open because of the effect of its activating coil, A' is closed. In order to have a current flow between the terminals M and N, we must have the set of relay operations indicated by $ABC + AB'C + A'B'C$. With this in mind, the question is to replace the given network by a less complex *equivalent* circuit.

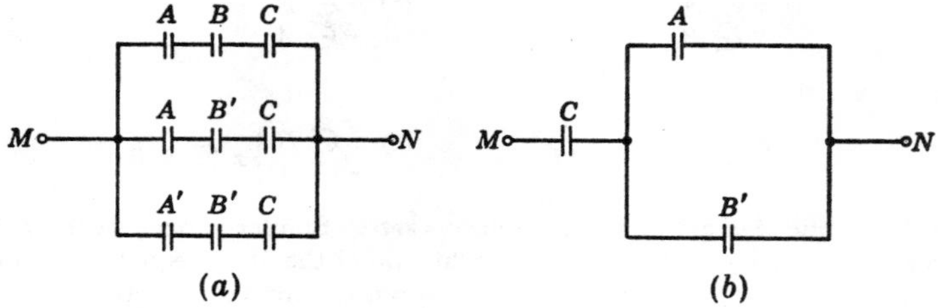

Fig. E2-5

Solution. A way of simplifying the above expression is the following:

$$F = C(AB + AB' + A'B')$$
$$F = C[A(B + B') + A'B']$$
$$F = C(AU + A'B')$$
$$F = C(A + A'B')$$
$$F = C(A + B')$$

A circuit presentation of this example is illustrated in Fig. E2-5*b*.

Example 2-6. Verify the equivalence of the two relay circuits of Fig. E2-6.

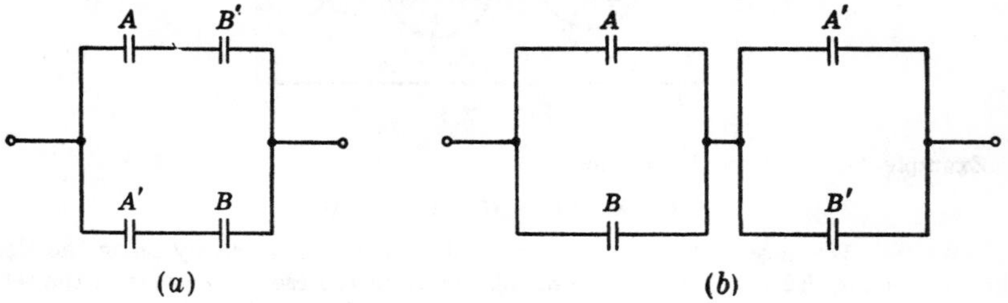

Fig. E2-6

Solution. The set that corresponds to the operation of the circuit in Fig. E2-6*b* is

$$(A + B)(A' + B')$$

Direct multiplication gives

$$AA' + AB' + BA' + BB' = AB' + A'B$$

The latter set can be immediately identified with the circuit of Fig. E2-6*b*.

Sheffer-stroke Operation. Examples 2-5 and 2-6 have illustrated some use of Boolean algebra in relay circuits. As another example of the use of

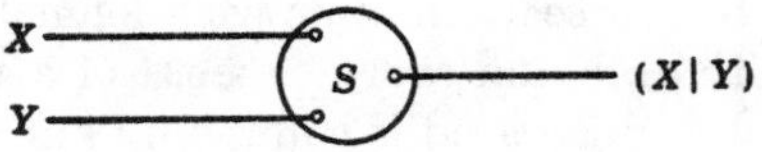

FIG. 2-12. Sheffer stroke.

Boolean algebra in engineering problems, we discuss briefly what is referred to as the *Sheffer-stroke operation.* This operation for two sets X and Y is denoted by $(X|Y)$ and is defined by the equation

$$(X|Y) = X' \cup Y' \qquad \text{not } X, \text{ or not } Y, \text{ or not } X \text{ and } Y$$

The Sheffer stroke commonly illustrated by the three-port diagram of Fig. 2-12 has the distinct property that it can replace all three basic

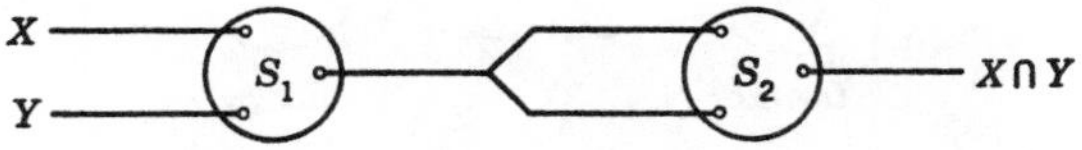

FIG. 2-13. Product operation by two Sheffer strokes.

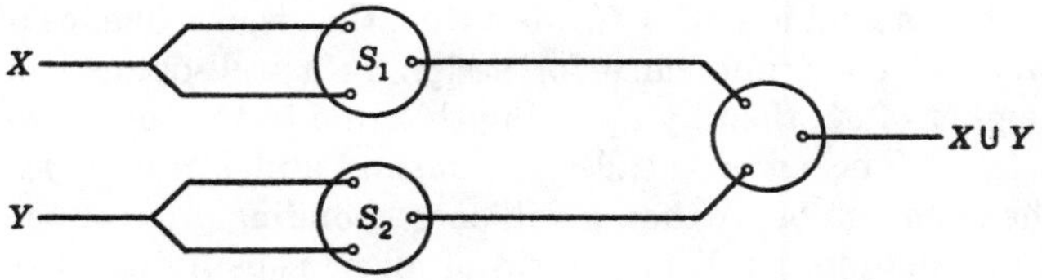

FIG. 2-14. Summing operation by three Sheffer strokes.

operations of Boolean algebra (sum, product, and negation). The validity of this statement can be exhibited in a direct manner.

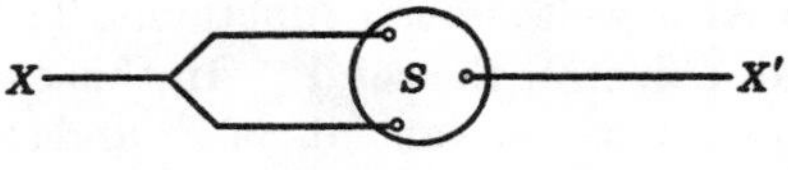

FIG. 2-15. Operation of negation with a Sheffer stroke.

PRODUCT OPERATION. Reference to the diagram of Fig. 2-13 suggests that

$$\begin{aligned}((X|Y)|(X|Y)) &= (X' \cup Y')' \\ &= ((X \cap Y)')' = X \cap Y\end{aligned}$$

SUMMING OPERATION. The diagram of Fig. 2-14 suggests

$$\begin{aligned}((X|X)|(Y|Y)) &= (X|X)' \cup (Y|Y)' \\ &= X \cup Y\end{aligned}$$

NEGATION. Reference is made to the diagram of Fig. 2-15.

$$(X|X) = X' \cup X' = X'$$

2-5. Functions. In this section, some well-defined objects or numbers will be associated with each and every element of a given set. The rule on which this relationship is based is commonly known as *function.*

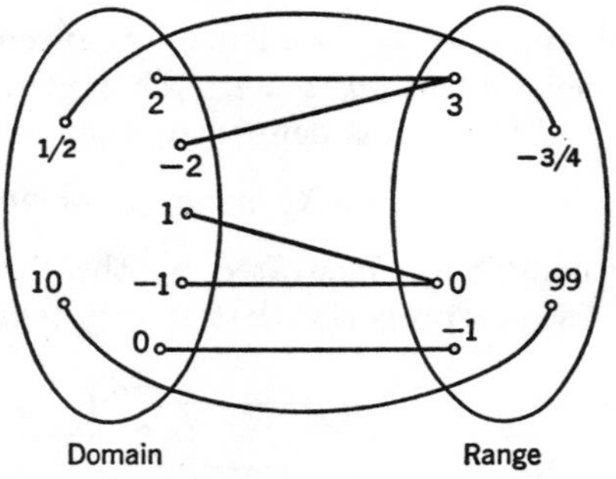

FIG. 2-16. Domain X and range Y.

If $X = \{x\}$ is a set and $y = f(x)$ is a rule, that is, a sequence of specified operations and correspondence for assigning a well-defined object y to every member of X, then by applying this rule to the set X, we obtain a set $Y = \{y\}$. The set X is called the *domain* and Y the *range*. When x covers the elements of X, then y will correspondingly cover the elements of Y. For example, let X be the set of all persons living in the state of California on January 1, 1959, and let the function be defined as follows: anyone who is the father of a person described by X and is in the state of Colorado on January 1, 1959. Assuming that all the words appearing in the rule, such as father, California, Colorado, are well-defined words, this may be considered as a well-defined function. To each member of X there corresponds an object in the set Y. In this example, element zero in Y corresponds to some of the elements of X, and several members of X might have a unique correspondent in Y.

As another simple example, consider the set

$$X = \{1,2,0,-2,-1,\tfrac{1}{2},10\}$$

and the function

$$f(x) = x^2 - 1$$

which lead to the set

$$Y = \{0,3,-1,3,0,-\tfrac{3}{4},99\}$$

The domain of x and the range of y are shown in Fig. 2-16, the correspondence being one-to-one from X to the Y set.

Example 2-7. A set of ordered pairs $s = \{(x,y)\}$, that is, a set of points in the rectangular coordinate system, is given in Fig. E2-7*a*.

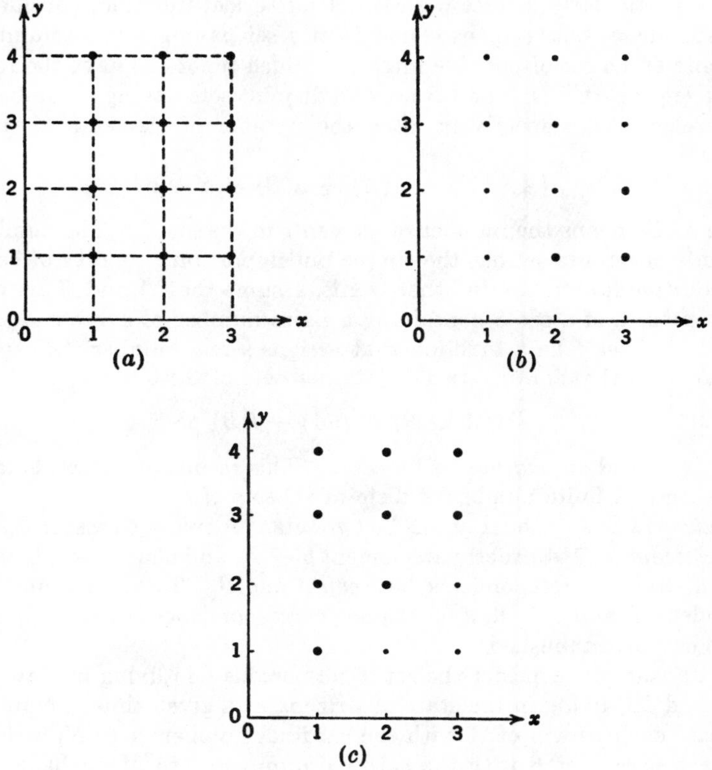

FIG. E2-7

(*a*) Describe the elements of the subset $a = \{(x,y)|y < x\}$.
(*b*) Describe the elements of the subset $b = s - a$.
Solution
(*a*) See Fig. E2-7*b*.
(*b*) See Fig. E2-7*c*.

Numerical Functions. Functions that have numerical values are the most common type. We can define the basic algebraic operations for a family of numerical functions defined over a specific domain $X = \{x\}$. For instance, if $f_1(x), f_2(x)$, and $f_3(x) = \text{const} = k$ have a common domain,

$$\begin{gathered} f_1(x) + f_2(x) \\ kf_1(x),\ kf_2(x) \\ f_1(x) \cdot f_2(x) \end{gathered} \tag{2-28}$$

are also defined over the same domain.

As a particularly interesting case of numerical function, consider the correspondence between the elements of a set having a finite number of elements and a set of positive integers. Such functions have the following basic property: If A and B are two disjoint sets having a number of a and b elements, respectively, then the number of elements of the set $A + B$ is

$$n(A \cup B) = n(A) + n(B) = a + b \tag{2-29}$$

where $n(X)$ means the number of elements in the set X. The number of elements of a finite set has the simple but important property of being a real additive function. In other words, assume that A and B are themselves subsets of a set S containing a finite number of subsets $A, B, C, D, \ldots$. Let f be a function that assigns a real number $f(X)$ to each $X \subset S$, such that for any two disjoint subsets of S we have

$$f(A \cup B) = f(A) + f(B)$$

Then f is called an *additive set function*. This result, of course, holds for the union of a finite number of disjoint subsets of S.

Equivalent Sets. Let A and B be two sets. A rule that associates with each element $a \in A$ exactly an element $b \in B$, and conversely, is said to be a one-to-one correspondence between A and B. Two sets A and B are equivalent if, and only if, a one-to-one correspondence between their elements can be established.

As an example, consider the set of all persons (A) living in New York State and (B) living in the state of Arizona at a given time. Now if we associate each person of A with the cardinal numbers 1 to N, inclusive, and each person of B with the cardinal numbers 1 to M, inclusive, it is clear that there is a one-to-one correspondence between the elements of $A + B$ and the set of cardinal numbers 1 to $M + N$, inclusive.

The number of elements in a set may or may not be finite. In the latter case, if the elements of the set can be placed in a one-to-one correspondence with the set of natural numbers

$$\{1,2,3, \ldots\} \tag{2-30}$$

we say that the set has a *denumerable* or countable number of elements. For example, the number of elements in the set

$$\{1,4,9,16, \ldots, n^2, \ldots\}$$

is denumerably infinite.

A common example of nondenumerable sets can be given by considering points on a straight line. Let x denote the abscissa of a point of the line segment between points A and B with respective abscissa a and b. The inequality

$$a < x < b$$

indicates a set of points on the line AB that does not contain the end points A and B. Such a set is termed an *open interval* and is denoted by

$$]a,b[\quad \text{open interval} \quad a < x < b \tag{2-31}$$

Similarly, a *closed interval* is defined and denoted as follows:

$$[a,b] \quad \text{closed interval} \quad a \leq x \leq b \tag{2-32}$$

It can be shown that the number of points in [0,1] are nondenumerable.* If the set A is equivalent to the set of points in [0,1], it is said that A has the power of continuum.†

The additive property of the function under consideration, i.e., the number of elements in finite sets, makes the following relations self-evident.

$$n(A \cup B) = n(A) + n(B) - n(AB) \tag{2-33}$$
$$n(A - B) = n(A) - n(AB)$$
$$n(A) + n(A') = n(U) \tag{2-34}$$

For a set containing three subsets A, B, and C one can derive

$$\begin{aligned} n(A \cup B \cup C) &= n[(A) \cup (B \cup C)] \\ n(A \cup B \cup C) &= n(A) + n(B \cup C) - n(AB \cup AC) \\ n(A \cup B \cup C) &= n(A) + n(B) + n(C) - n(BC) - n(AB) \\ &\quad - n(AC) + n(ABC) \end{aligned} \tag{2-35}$$

The following example is designed to employ the additive property of the afore-mentioned set functions.

Example 2-8. There are three radio stations A, B, and C which can be received in a town of 3,000 families. The following data are given:

(*a*) 1,800 families listen to station A.
(*b*) 1,700 families listen to station B.
(*c*) 1,200 families listen to station C.
(*d*) 1,250 families listen to stations A and B.
(*e*) 700 families listen to stations A and C.
(*f*) 600 families listen to stations B and C.
(*g*) 200 families listen to stations A, B, and C.

* See I. P. Natanson, "Theory of Functions of a Real Variable" (translated from Russian), p. 21, Frederick Ungar Publishing Co., New York, 1955.

† For a more complete mathematical treatment of probability, one has to examine in detail the operations on numerical functions associated with operations on denumerable and nondenumerable sets. Such a detailed undertaking is avoided here for the sake of brevity lest the average reader find the text too elaborate. Nonetheless, for the sake of logical completeness, we shall try now and then to remind the reader of missing links. Here are some of the theorems that had to be omitted in this introductory presentation.

1. The sum of a finite number of disjoint denumerable sets is denumerable.
2. The set of all rational numbers is denumerable.
3. The sum of a denumerable number of disjoint sets each with the power of continuum has itself the power of continuum. See T. M. Apostol, "Mathematical Analysis," pp. 31–33, Addison-Wesley Publishing Company, Reading, Mass., 1957.

Of course any family may listen to other stations besides the ones specified in each case. The problem is to obtain the number of families who are not listening to any station.

Solution. We draw the pertinent Venn diagram of Fig. E2-8 and, starting from the bottom of the above list, indicate the corresponding number of elements of each subset on the diagram. The number of families in set g is 200. Thus, the number of families listening to B and C but not to A is

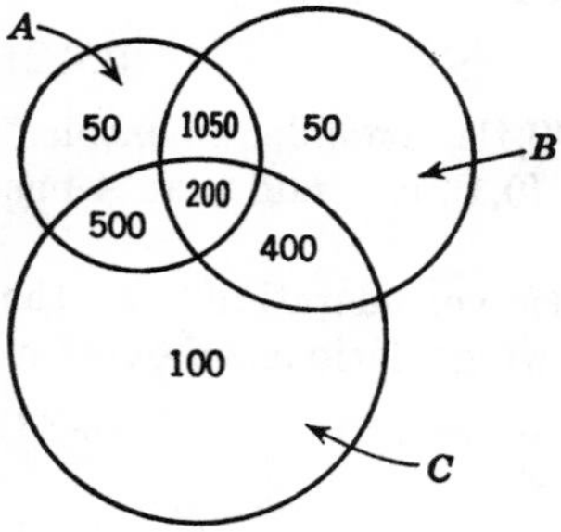

FIG. E2-8

$$n(BCA') = n(BC) - n(BCA) = 600 - 200 = 400$$

Following this procedure one can obtain all the numbers associated with each disjoint set in the Venn diagram. The total number of families listening to one or more stations is 2,350. This indicates that there are 650 families not listening to any of the above radio stations.

Similar questions can easily be answered by referring to the Venn diagram of Fig. E2-8. For example, the number of families who are not listening to A but are listening either to B or to C or to both is

$$n\{A'(B \cup C)\} = n(A'B) + n(A'C) - n(A'BC)$$
$$n\{A'(B \cup C)\} = 450 + 500 - 400 = 550$$

2-6. Sample Space. In this section we shall make preparations for applying the concept of set theory to probability. When talking about probability we usually have in mind what can be termed an *experiment* with certain *outcomes*. An outcome is any one of the possibilities that may be expected from the experiment. The totality of all these outcomes forms a universal set which is called the *sample space*. Each outcome is a *point* of the sample space.

For example, the throw of an ordinary die may be considered as an experiment having six possible outcomes. With this experiment we associate a universal set containing six points, each corresponding to one of the outcomes of the experiment:

$$\{1,2,3,4,5,6\}$$

If the die is thrown twice, the sample space associated with the experiment contains 36 points corresponding to the following outcomes:

11	12	13	14	15	16
21	22	23	24	25	26
31	32	33	34	35	36
41	42	43	44	45	46
51	52	53	54	55	56
61	62	63	64	65	66

A sample space may be finite or infinite, if it contains a finite or an infinite number of points, respectively. The sample space corresponding to a single throw of a die is finite. On the other hand, the sample space corresponding to an experiment of throwing the die until a 6 appears is an infinite space. It is possible to conceive a situation where one may have to throw the die infinitely many times without obtaining a 6. A sample space containing at most a denumerable number of elements is termed *discrete.* Sample spaces containing a nondenumerable number of elements include the so-called "continuous sample space." In this case the range of the elements covers a continuum of values in contrast with the discrete set of values in the discrete sample space.

A subset of a sample space is called an *event.* Thus, an event is a subset of a sample space containing any number of points or outcomes. (See Fig. 2-17.)

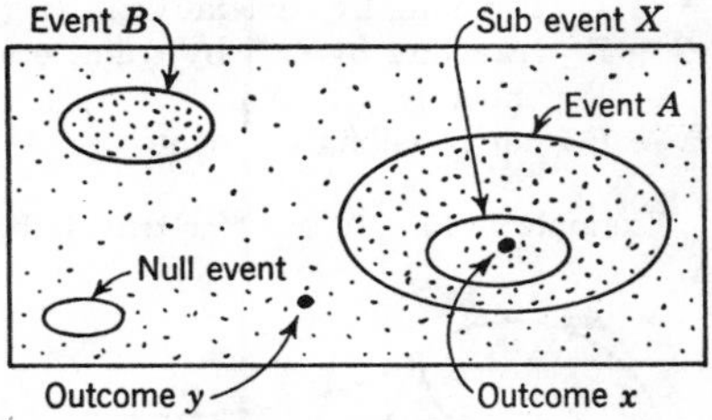

FIG. 2-17. A probability space.

An event containing no outcomes is a null set or an empty set and represents an event that is impossible. An event containing all sample points is an event that is certain to occur. This may be denoted by the universal set U, which means that the event under consideration is bound to occur. The outcome of an event implies the occurrence of any one of its possible outcomes. The following glossary of terms may be of assistance in the transition from the language of set theory to that of probability theory:

U	All possibilities.
$A \subset U$	A particular event.
$A = U$	The event A must occur (certain).
$A = \emptyset$	The event A is impossible.
A'	The event A does not occur.
$x \in X$ or $X \subset A$	x is any particular outcome of X. The occurrence of x implies the occurrence of A and X.
$y \notin A$	y is not an outcome of the event A.
$ABC \cdots D = S$	S is the event of the simultaneous occurrence of events $A,B,C, \ldots D$.
$A + B + C + \cdots + D = S$	S is the event of the occurrence of A or B or C or $\cdots$ or D, or any combination of these.
$ABC \cdots D = \emptyset$	The events $A,B,C, \ldots, D$ are incompatible.
$A + B + C + \cdots + D = U$	At least one of the events A, B, $C, \ldots, D$ must occur.

Example 2-9. A traveler has the choice of traveling by car, train, plane, or any combination of the three for a particular trip. Define the sample space and express some of the events of interest.

Solution. Let C, T, and P correspond to the fact of traveling by car, train, or plane, respectively. The following events are self-explanatory.

CTP	traveling by car, train, and plane
CTP'	traveling by car and train but not by plane
CT	traveling by car and train (with or without plane)
$C + T$	traveling by car, by train, or by car and train (may or may not take the plane)
$U - P$	not traveling by plane

Example 2-10. A traveler travels between cities M and N. The possible roads are shown in Fig. E2-10. Define the sample space and the events that the traveler goes through towns A, B, or both.

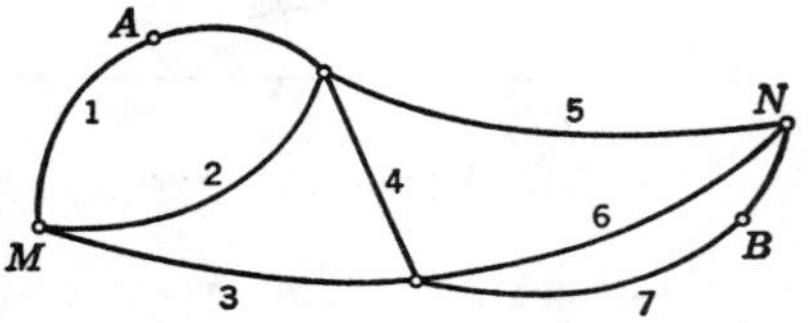

FIG. E2-10

Solution. Assuming that the traveler does not change direction while traveling, the following selection of roads is possible:

15, 25, 146, 147, 246, 247, 36, 37, 345

The sample space has nine points; i.e., our defined experiment may have nine distinct outcomes. The event of passing through the town A (event E_1) consists of any of the three points 15, 146, and 147. The event of passing through the town B (event E_2) consists of any of the three points 147, 247, and 37. Finally the event of passing through A and B (event E_1E_2) consists of a single point 147. Similarly, the following events can easily be identified:

E_1E_2'	15, 146
$E_1'E_2$	37, 247
$E_1 + E_2$	15, 146, 147, 247, 37
$E_1'E_2'$	25, 246, 36, 345

2-7. Probability Measure. In Sec. 2-5 on Functions we have associated arbitrary set functions with the elements of sets. In particular, we have outlined some numerical functions and observed certain rules such as the additivity relation of Eq. (2-33), when the set function was the number of elements in each set. The study of the mathematics of set functions has its place in a branch of mathematics known as *measure theory*. The *probability measure* is a specific type of function which can be associated with sets. When dealing with abstract mathematics, one may specify any arbitrary properties for the measure. However, the end result in this study is probability as applied to the physical world. It is mainly for this reason that we require our probability measure to fulfill the requirements that will be described below. These requirements are matters of convenience for our subsequent dealing with physical problems rather than a mathematical necessity.

An experiment is defined so that to each possible outcome of this experiment there corresponds a point in the sample space. The number of outcomes of this experiment is assumed to be at most denumerable. The outcomes are labeled by symbols a_k, and a single-valued real function $m\{a_k\}$ called the *probability measure* is defined. An event of interest A is considered as the set of the outcomes a_k giving rise to that event. The probability measure of an event is defined as the sum of the probability measures associated with all the outcomes a_k of that event. Two events A and B are termed *disjoint* if they contain no outcome in common. That is, two disjoint events cannot happen simultaneously. The probability measure has the following assumed properties:

$$0 \leq m\{A_k\} \tag{2-36}$$

*

$$m\{A \cup B\} = m\{A\} + m\{B\} \qquad \text{if } A \text{ and } B \text{ are disjoint} \tag{2-37}$$

$$m\{X\} = 0 \qquad \text{if } X = \emptyset \tag{2-38}$$

$$m\{X\} = 1 \qquad \text{if } X = U \tag{2-39}$$

For a more general case involving a continuous sample space one employs the concept of integration. This is not considered here as it requires rigorous mathematical treatment beyond the present scope of interest. The interested reader is referred to "Probability Theory" by M. Loève (Chap. 1).

The above measure-theory approach is certainly valid. Any measure satisfying the specified requirements, when applied to a problem involving sets, will lead to a consistent mathematical setup. For example, if A, B, and C are subsets of a universal set U with an additive probability measure, that is, the measure associated with the union of two disjoint sets is equal to the sum of their individual measures, the following relations are valid:

$$\left.\begin{aligned} m\{A\} &\leq m\{B\} \qquad (2\text{-}40) \\ m\{A\} &= m\{B\} - m\{B - A\} \qquad (2\text{-}40a) \end{aligned}\right\} \quad \text{if } A \subset B$$

$$m\{A'\} = m\{U - A\} = m\{U\} - m\{A\} = 1 - m\{A\} \tag{2-41}$$

$$m\{A \cup B\} = m\{(A - AB) \cup B\} = m\{A\} - m\{AB\} + m\{B\} \tag{2-42}$$

$$m\{A\} + m\{B\} \geq m\{AB\} \tag{2-43}$$

For three disjoint sets,

$$m\{A \cup B \cup C\} = m\{A\} + m\{B\} + m\{C\} \tag{2-44}$$

* It is certainly plausible to assign measures which have numerical values greater than one. While such measures can be consistently applied, they have no practical significance as far as probability is concerned in dealing with physical problems. Also note that the property stated in Eq. (2-38) can be derived from the other properties.

For three sets in general,

$$m\{A \cup B \cup C\} = m\{A\} + m\{B\} + m\{C\} - m\{AB\} - m\{BC\} - m\{CA\} + m\{ABC\} \quad (2\text{-}45)$$

Example 2-11. Consider a set of all intervals I contained in the closed interval [0,1]. With each and every interval I we associate a measure function $L(I)$ equal to the ordinary length of the same interval. See if such a measure satisfies the requirements of a probability measure.

Solution. The requirement of (2-36) is satisfied, as the length associated with each member of the set is a nonnegative number between 0 and 1. The condition (2-37) is fulfilled by nonoverlapping intervals (mutually exclusive sets). The requirements (2-38) and (2-39) are also met. For a more thorough discussion the reader is referred to Cramèr (Chap. 4, The Lebesgue Measure of a Linear Point Set).

2-8. Frequency of Events.* In Sec. 2-7 on Probability Measure an introductory axiomatic account of probability as a measure of a set was given. The object of this section is to supplement the set-theory point of view with some perhaps less formal discussion of the probability of occurrence of certain events of a defined experiment. In other words, we wish to make a transition from the suggested abstract mathematical measure to some empirical numerical function fulfilling the specified measure requirements.

The first step toward this objective is to define an experiment such as the tossing of a coin or the drawing of a card from a given deck of cards. Next, all the outcomes of this experiment must be specified. Now consider a specific event X_k among all the possible events of the experiment under consideration. If the basic experiment is repeated N times among which the event X_k has appeared $n(X_k)$ times, the ratio

$$\frac{n(X_k)}{N}$$

is defined as the *relative frequency* of the occurrence of the event X_k. In case N is increased indefinitely, intuitively speaking, the "limit" of

$$\frac{n(X_k)}{N} \quad (2\text{-}46)$$

as $N \to \infty$ is the probability $P\{X_k\}$ of the event X_k. This "definition" of probability is more elaborate than the classical definition of Laplace which defines the probability as the ratio of the number of favorable events to the total number of possible events. In the latter definition all events are considered to be equally likely, that is, throwing of a true

* This section has been inserted to accommodate those who feel more familiar with the old frequency concept. The intuitive concept of frequency requires a considerable amount of clarification before it becomes mathematically acceptable. This can be done in the light of *the laws of large numbers*. The section may be omitted by those who prefer mathematical accuracy to physical justification.

die by an honest person under prescribed circumstances. It is to be noted that

$$0 \leq n(X_k) \leq N \tag{2-47}$$

$$0 \leq \frac{n(X_k)}{N} \leq 1 \tag{2-48}$$

$$0 \leq \lim_{N \to \infty} \frac{n(X_k)}{N} \leq 1 \tag{2-49}$$

Equation (2-49) states that the probability of any event X_k is a real number in the real interval [0,1].

$$0 \leq P\{X_k\} \leq 1 \tag{2-50}$$

Considering an event that occurs in every observation yields the limiting case $P\{\check{X}\} = 1$, which is a *certain event*. Also, an event that never occurs will lead to the other limiting case, $P\{X\} = 0$, which is an *impossible event*.

We have thus far shown that this empirical definition of probability satisfies the requirements (2-36), (2-38), and (2-39). It remains to be seen whether the requirement (2-37) holds or not. In order to verify this, consider two particular events A and B among the events that result from the experiment. Let the experiment be repeated n times. Each observation can belong to only one of the four following categories (Fig. 2-18):

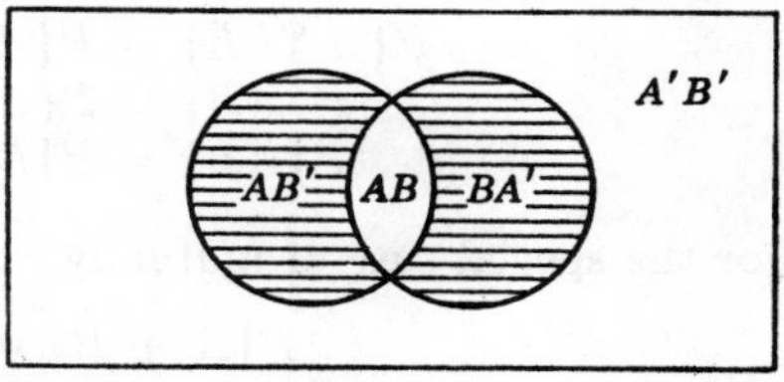

FIG. 2-18. Probability space of two events.

1. A has occurred but not B, the event AB'.
2. B has occurred but not A, the event BA'.
3. Both A and B have occurred, the event AB.
4. Neither A nor B has occurred, the event $A'B'$.

Note that

$$A = AB' \cup AB$$
$$B = BA' \cup AB$$
$$A \cup B = AB' \cup AB \cup BA'$$

If the number of events of each category is denoted by n_1, n_2, n_3, and n_4, respectively, the following equations are self-explanatory:

$$n_1 + n_2 + n_3 + n_4 = n \tag{2-51}$$

$$f\{A\}, \text{ relative frequency of } A \text{ independent of } B = \frac{n_1 + n_3}{n} \tag{2-52}$$

$$f\{B\}, \text{ relative frequency of } B \text{ independent of } A = \frac{n_2 + n_3}{n} \tag{2-53}$$

$f\{A + B\}$, relative frequency of either A, B, or both $= \dfrac{n_1 + n_2 + n_3}{n}$ (2-54)

$f\{AB\}$, relative frequency of A and B occurring together $= \dfrac{n_3}{n}$ (2-55)

$f\{A|B\}$, relative frequency of A under condition that B has occurred

$$= \frac{n_3}{n_2 + n_3} \tag{2-56}$$

$f\{B|A\}$, relative frequency of B under condition that A has occurred

$$= \frac{n_3}{n_1 + n_3} \tag{2-57}$$

When the number of experiments tends to infinity, these simple relations with proper interpretation lead to the addition law and multiplication law:

$$P\{A \cup B\} = P\{A\} + P\{B\} - P\{AB\} \tag{2-58}$$
$$P\{AB\} = P\{A\}P\{B|A\} \tag{2-59}$$
$$P\{AB\} = P\{B\}P\{A|B\} \tag{2-60}$$

For the special case of mutually exclusive events, $P(AB) = 0$,

$$P\{A + B\} = P\{A\} + P\{B\} \tag{2-61}$$

Equation (2-61) shows the validity of the requirement (2-37) for the chosen set function termed the relative frequency of the event.

More specifically, we have proved that the probability measure defined by Eq. (2-46) satisfies the following basic properties for all sets defined in sample space:

$$0 \leq P\{A\} \leq 1 \tag{2-62}$$
$$P\{A \cup B\} = P\{A\} + P\{B\} \qquad \text{for mutually exclusive } A \text{ and } B \tag{2-63}$$
$$P\{X\} = 0 \qquad \text{if, and only if, } X = \emptyset \tag{2-64}$$
$$P\{X\} = 1 \qquad \text{if, and only if, } X = U \tag{2-65}$$

Therefore the suggested definition of the frequency can serve as a probability measure. The implication of Eqs. (2-58) to (2-60) will be investigated in subsequent sections.

The frequency approach is a rather common approach for defining the probability when dealing with physical problems. Its mathematical concept relies on the tacit assumption of an equiprobable measure, that is, the equal likelihood of the outcome of the repeated experiments. We assume that the measure associated with an event, in the case of the repeated experiment, is proportional to the number of the outcomes in the event under consideration. In essence, this assumption makes the frequency definition somewhat too restrictive.

2-9. Theorem of Addition. It seems appropriate now to continue with our formalism without restriction to an immediately practical but slow procedure. For two events A and B of the sample space one has

$$A \cup (B - AB) = A \cup B \tag{2-66}$$

The additive property of the probability measure in Sec. 2-7 suggests that

$$\begin{aligned} m\{A + B\} &= m\{A\} + m\{B\} - m\{AB\} \\ P\{A + B\} &= P\{A\} + P\{B\} - P\{AB\} \leq P\{A\} + P\{B\} \end{aligned} \tag{2-67}$$

If two events A and B are mutually exclusive, then

$$P\{AB\} = P\{\emptyset\} = 0 \tag{2-68}$$

$$P\{A \cup B\} = P\{A\} + P\{B\} \tag{2-69}$$

For two opposite events A and A', one has $A + A' = U$, and since $AA' = 0$, then

$$\begin{aligned} P\{A \cup A'\} &= P\{A\} + P\{A'\} = P\{U\} = 1 \\ P\{A'\} &= 1 - P\{A\} \end{aligned} \tag{2-70}$$

For the three events A, B, and C, we may write

$$P\{A \cup B \cup C\} = P\{A \cup B\} + P\{C\} - P\{(A \cup B)C\} \tag{2-71}$$

$$P\{A \cup B \cup C\} = P\{A\} + P\{B\} + P\{C\} - P\{AB\} - P\{BC\} - P\{CA\} + P\{ABC\} \tag{2-72}$$

This is indeed made clear by employing a pertinent Venn's diagram, $P\{ABC\}$ being the probability of the simultaneous occurrence of the three events. If the events are mutually exclusive, then

$$P\{A \cup B \cup C\} = P\{A\} + P\{B\} + P\{C\} \tag{2-73}$$

More generally, for a number of events $A_1, A_2, \ldots, A_n$ one may write

$$\begin{aligned} P\{A_1 \cup A_2 \cup \cdots \cup A_n\} &= P\{A_1\} + P\{A_2\} + \cdots + P\{A_n\} \\ &\quad - P\{A_1A_2\} - P\{A_1A_3\} - \cdots - P\{A_{n-1}A_n\} + P\{A_1A_2A_3\} \\ &\quad + P\{A_1A_2A_4\} + \cdots + P\{A_{n-2}A_{n-1}A_n\} + \cdots \\ &\quad + (-1)^{n-1}P\{A_1A_2 \cdots A_n\} \end{aligned} \tag{2-74}$$

By extension of the relation in Eq. (2-66), it can be shown without difficulty that

$$P\{A_1 \cup A_2 \cdots A_n\} \leq P\{A_1\} + P\{A_2\} + \cdots + P\{A_n\} \tag{2-75}$$

The equality sign holds when the events A_k and A_j are mutually exclusive for all $k \neq j$.

Example 2-12. An urn contains 11 balls numbered from 1 to 11. If a ball is selected at random, what is the probability of having a ball with a number which is a multiple of either 2 or 3?

Solution. Let A and B be the events that the ball number is a multiple of 2 and 3, respectively. The event of interest is $A + B$.

$$P\{A\} = 5/11$$
$$P\{B\} = 3/11$$
$$P\{AB\} = 1/11$$
$$P\{A + B\} = 5/11 + 3/11 - 1/11 = 7/11$$

Example 2-13. One card is drawn from a regular deck of 52 cards. What is the probability of the card being either red or a king?

Solution. Let A be the event that the card is red, and B the event that the card is a king. The event of interest is $A + B$. Where A and B are not exclusive events, apply Eq. (2-67):

$$P\{A\} = 1/2$$
$$P\{B\} = 1/13$$
$$P\{AB\} = (1/13)(1/2) = 1/26$$
$$P\{A + B\} = 1/2 + 1/13 - 1/26 = 7/13$$

Example 2-14. An honest coin is tossed 10 times. What is the probability of having at least (*a*) one tail and (*b*) two tails?

Solution. The main assumption in this and in similar problems is the concept of independence of successive trials and the equally probable outcomes.

Let A and B be the events of getting no tail and exactly one tail, respectively. Then

$$P\{A\} = \left(\frac{1}{2}\right)^{10} = \frac{1}{1{,}024}$$
$$P\{B\} = 10\left(\frac{1}{2}\right)^{10} = \frac{10}{1{,}024}$$

The events of interest are

(*a*)
$$U - A = A'$$
$$P\{A'\} = 1 - \frac{1}{1{,}024} = \frac{1{,}023}{1{,}024}$$

(*b*)
$$U - (A + B) = (U - A) - B = A' - B$$
$$P\{A' - B\} = \frac{1{,}023}{1{,}024} - \frac{10}{1{,}024} = \frac{1{,}013}{1{,}024}$$

2-10. Conditional Probability. Consider two events A and B. The conditional probability of event A based on the hypothesis that event B has occurred is defined by the following relation:

$$P\{A|B\} = \frac{P\{AB\}}{P\{B\}} \qquad P\{B\} \neq 0 \tag{2-76}$$

The use of this definition can be justified by returning to the previously treated case of Sec. 2-8. The frequency of the occurrence of event A under the assumption that B has occurred is

$$f\{A|B\} = \frac{n_3}{n_2 + n_3} = \frac{f\{AB\}}{f\{B\}} \tag{2-77}$$

By the same token, the frequency of the occurrence of B, knowing that A has already occurred, is

$$f\{B|A\} = \frac{n_3}{n_1 + n_3} = \frac{f\{AB\}}{f\{A\}} \tag{2-78}$$

Increasing the number of trials indefinitely gives

$$P\{A|B\} = \frac{P\{AB\}}{P\{B\}} \qquad P\{B\} \neq 0 \tag{2-79}$$

$$P\{B|A\} = \frac{P\{AB\}}{P\{A\}} \qquad P\{A\} \neq 0 \tag{2-80}$$

The two events A and B are said to be *mutually independent* if

$$\begin{aligned} P\{A|B\} &= P\{A\} \\ P\{B|A\} &= P\{B\} \end{aligned} \tag{2-81}$$

Note that for mutually independent events

$$P\{AB\} = P\{A\} \cdot P\{B\} \tag{2-82}$$

Equations (2-81) and (2-82) are alternatively used as the defining relations for two mutually independent events.*

Example 2-15. Three boxes of identical appearance contain two coins each. In one box both are gold, in one box both silver, and in the third box one is a silver coin and the other is a gold coin. Suppose that a box is selected at random and, further, that a coin in that box is selected at random. If this coin proves to be gold, what is the probability that the other coin in the box is also gold?

Solution. Let

A_{gg} be the event that the other coin in the selected box is also gold (that is, the selected box is gg box)

B_g be the event that the first coin in the selected box is a gold coin

The desired probability is

$$\begin{aligned} P\{A_{gg}|B_g\} &= \frac{P\{B_g|A_{gg}\} \cdot P\{A_{gg}\}}{P\{B_g\}} \\ P\{B_g\} &= \tfrac{2}{3} \cdot \tfrac{3}{4} = \tfrac{1}{2} \\ P\{A_{gg}\} &= \tfrac{1}{3} \\ P\{B_g|A_{gg}\} &= 1 \\ P\{A_{gg}|B_g\} &= \frac{1 \cdot \frac{1}{3}}{\frac{1}{2}} = \tfrac{2}{3} \end{aligned}$$

* For two independent events, the defining equation (2-81) holds, but when $P\{A\} = 0$ or $P\{B\} = 0$, then $P\{B|A\}$ or $P\{A|B\}$ is not defined. For this reason some authors prefer to define mutual independence in such a way that Eq. (2-81) remains valid for all circumstances, including

$$\begin{aligned} P\{A\} &= 0 \qquad & P\{A\} &= 1 \\ P\{B\} &= 0 \qquad & P\{B\} &= 1 \end{aligned}$$

For this purpose, the following defining equations are suggested (Fortet, p. 85):

$$\begin{aligned} P\{AB\} &= P\{A\}P\{B\} \qquad & P\{A'B\} &= P\{A'\}P\{B\} \\ P\{AB'\} &= P\{A\}P\{B'\} \qquad & P\{A'B'\} &= P\{A'\}P\{B'\} \end{aligned}$$

Independent events are more specifically called statistically independent or stochastically independent.

Example 2-16. In a certain group of engineers, 60 per cent have insufficient background of information theory, 50 per cent have inadequate knowledge of probability, and 80 per cent are in either one or both of the two categories What is the percentage of people who know probability among those who have a sufficient background of information theory?

Solution. Let

A be those having insufficient background of information theory

B be those having inadequate knowledge of probability

Then

$$P\{A\} = 0.60 \qquad P\{A'\} = 0.40$$
$$P\{B\} = 0.50 \qquad P\{B'\} = 0.50$$
$$P\{A+B\} = 0.80 \qquad P\{A+B\}' = P\{A'B'\} = 0.20$$

It is required to find

$$P\{B'|A'\} = \frac{P\{A'B'\}}{P\{A'\}} = \frac{0.20}{0.40} = 50 \text{ per cent}$$

2-11. Theorem of Multiplication. The multiplication rule for the case of two events A and B can be obtained through the definition of the conditional probability.

$$\begin{aligned} P\{AB\} &= P\{A\}P\{B|A\} \\ P\{AB\} &= P\{B\}P\{A|B\} \end{aligned} \tag{2-83}$$

This rule can be extended to the case of more than two events. For instance, for three events A, B, and C, one writes

$$\begin{aligned} P\{ABC\} &= P\{AB\}P\{C|AB\} \\ &= P\{A\}P\{B|A\}P\{C|AB\} \end{aligned} \tag{2-84}$$

More generally,

$$P\{A_1,A_2, \ldots ,A_n\} = P\{A_1\}P\{A_2|A_1\}P\{A_3|A_1A_2\} \cdots P\{A_n|A_1,A_2, \ldots ,A_{n-1}\} \tag{2-85}$$

When a finite number or a countably infinite number of events A_1, A_2, . . . , A_n are mutually independent,* we have

$$P\{A_1,A_2, \ldots ,A_n\} = P\{A_1\}P\{A_2\} \cdots P\{A_n\} \tag{2-86}$$

* The events A_1, A_2, . . . , A_n are said to be statistically independent of each other when the probability of any of these events and the probability associated with the intersection of any number of these events do not depend on any other event except those occurring in the intersection. That is,

$$\begin{aligned} P\{A_iA_j\} &= P\{A_i\}P\{A_j\} \\ P\{A_iA_jA_k\} &= P\{A_i\}P\{A_j\}P\{A_k\} \\ &\cdots\cdots\cdots\cdots \\ P\{A_1,A_2, \ldots ,A_n\} &= P\{A_1\}P\{A_2\} \cdots P\{A_n\} \end{aligned}$$

for all combinations of i, j, and k satisfying

$$1 \le i < j < k < \cdots \le n$$

Example 2-17. In a small library there are 1,000 books, among which 500 are scientific. Among the scientific books are 100 which are devoted to engineering subjects. Three books are chosen at random, the chosen book being replaced each time. What is the probability of getting

(*a*) All three scientific books

(*b*) Three scientific books among which only one is an engineering book

(*c*) At least one of the three an engineering book

Solution. Let S and E stand for the event of selecting a scientific and an engineering book, respectively. The events of interest discussed in the problem are

(*a*) $$S_1S_2S_3$$

(*b*) $$(S_1E)(S_2E')(S_3E') + (S_1E')(S_2E)(S_3E') + (S_1E')(S_2E')(S_3E)$$

(*c*) $$U - E_1'E_2'E_3'$$

(*a*) $$P\{S_1S_2S_3\} = P\{S_1\}P\{S_2\}P\{S_3\} = (1/2)^3 = 1/8$$

(*b*)
$$P\{SE\} = P\{E|S\}P\{S\}$$
$$P\{SE\} = 1/5 \cdot 1/2 = 1/10$$
$$P\{SE'\} = P\{E'|S\} \cdot P\{S\}$$
$$P\{SE'\} = 4/5 \cdot 1/2 = 4/10$$
$$3P\{(S_1E)(S_2E')(S_3E')\} = 3 \cdot 1/10 \cdot 4/10 \cdot 4/10 = 0.048$$

(*c*)
$$P\{U - E_1'E_2'E_3'\} = 1 - P\{E_1'E_2'E_3'\}$$
$$P\{U - E_1'E_2'E_3'\} = 1 - (9/10)^3 = 0.271$$

Example 2-18. Four persons write their names on individual slips of paper and deposit the slips in a common box. Each of the four draws at random a slip from the box. Determine the probability of each person retrieving his own name slip.

Solution. Let E_k be the event that the kth person retrieves his own name slip. The event of interest is $E_1E_2E_3E_4$. Equation (2-85) yields

$$P\{E_1E_2E_3E_4\} = P\{E_1|E_2E_3E_4\} \cdot P\{E_2|E_3E_4\} \cdot P\{E_3|E_4\} \cdot P\{E_4\}$$
$$P\{E_1E_2E_3E_4\} = 1 \cdot 1/2 \cdot 1/3 \cdot 1/4 = 1/24$$

Example 2-19. The probability of the closing of each relay of the circuit of Fig. E2-19 is a given α. Assuming that all relays act independently, what is the probability of a current existing between terminals A and B?

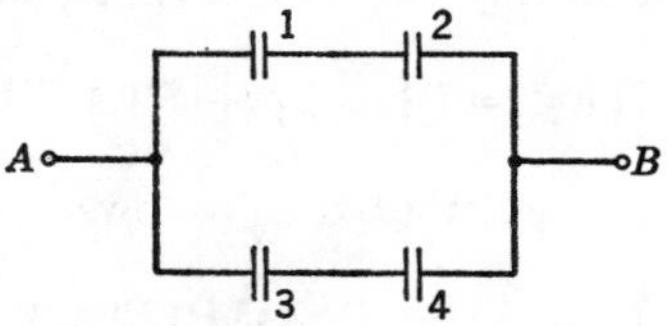

FIG. E2-19

Solution. Let the event of closing each relay 1, 2, 3, and 4 be E_1, E_2, E_3, and E_4, respectively. The four events are independent but not necessarily mutually exclusive. The event of interest is

$$E = E_1E_2 + E_3E_4$$
$$P\{E\} = P\{E_1E_2 + E_3E_4\} = P\{E_1E_2\} + P\{E_3E_4\} - P\{E_1E_2E_3E_4\}$$
$$P\{E\} = P\{E_1\}P\{E_2\} + P\{E_3\}P\{E_4\} - P\{E_1\}P\{E_2\}P\{E_3\}P\{E_4\}$$
$$P\{E\} = 2\alpha^2 - \alpha^4$$

Note that

$$P\{0\} = 0$$
$$P\{1\} = 1$$
$$0 \leq P\{E\} \leq 1 \qquad \text{for } 0 \leq \alpha \leq 1$$

2-12. Bayes's Theorem. In many problems we wish to concentrate on two mutually exclusive and exhaustive events of the sample space, that is, two events A_1 and A_2 such that

$$\begin{aligned} A_1A_2 &= \emptyset \\ A_1 + A_2 &= U \end{aligned} \tag{2-87}$$

The assumption is that each of these events has a subevent of special interest to us. If the subevents are indicated by EA_1 and EA_2, then the event of interest $E = EA_1 + EA_2$ can occur only when A_1 or A_2 occurs. The conditional probabilities $P\{E|A_1\}$ and $P\{E|A_2\}$ are assumed to be known; we are also given the information that E has occurred. The problem is to determine how likely it is that E has occurred because of the occurrence of either of the two events A_1 and A_2. In mathematical notation, given

$$\begin{aligned} P\{A_1\} &= \omega_1 & P\{A_2\} &= \omega_2 \\ A_1 + A_2 &= U & A_1A_2 &= \emptyset \\ P\{E|A_1\} &= p_1 & P\{E|A_2\} &= p_2 \end{aligned} \tag{2-88}$$

find $P\{A_1|E\}$ and $P\{A_2|E\}$.

The computation can be done in a direct way by applying the rule of addition and multiplication. Note that

$$E = A_1E \cup A_2E \tag{2-89}$$

As A_1E and A_2E are mutually exclusive events, we may write

$$P\{E\} = P\{A_1E\} + P\{A_2E\}$$

These probabilities can be calculated as follows:

$$\begin{aligned} P\{A_1E\} &= P\{A_1\}P\{E|A_1\} \\ P\{A_2E\} &= P\{A_2\}P\{E|A_2\} \end{aligned} \tag{2-90}$$

Therefore,

$$\begin{aligned} P\{E\} &= P\{A_1\}P\{E|A_1\} + P\{A_2\}P\{E|A_2\} \\ P\{A_1|E\} &= \frac{P\{A_1E\}}{P\{E\}} = \frac{P\{A_1\}P\{E|A_1\}}{P\{A_1\}P\{E|A_1\} + P\{A_2\}P\{E|A_2\}} \\ P\{A_2|E\} &= \frac{P\{A_2E\}}{P\{E\}} = \frac{P\{A_2\}P\{E|A_2\}}{P\{A_1\}P\{E|A_1\} + P\{A_2\}P\{E|A_2\}} \end{aligned} \tag{2-91}$$

Finally one finds

$$P\{A_1|E\} = \frac{\omega_1 p_1}{\omega_1 p_1 + \omega_2 p_2}$$
$$P\{A_2|E\} = \frac{\omega_2 p_2}{\omega_1 p_1 + \omega_2 p_2} \qquad (2\text{-}92)$$

The probabilities expressed in Eqs. (2-92) are called the a posteriori probabilities of A_1 and A_2, given E. The probabilities $\omega_1 p_1$ and $\omega_2 p_2$ are termed the a priori probabilities of E, given A_1 and A_2. Equations (2-92) provide a means for calculating the a posteriori probabilities from the a priori probabilities. Equations (2-92) are known as Bayes's rule. It is of interest to note that Bayes's rule applies to a partitioned sample space, as shown in Fig. 2-19. The events A_1 and A_2 may each consist of

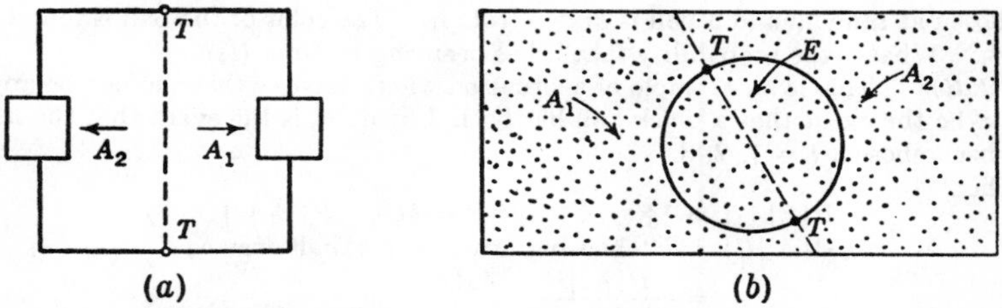

FIG. 2-19. (a) Thévenin's partitioning. A_1, a part of the network; A_2, the remainder network. (b) Bayes's partitioning.

sets containing a number of subevents. Electrical engineers may note that Bayes's rule is somewhat similar to Thévenin's theorem in network theory. Thévenin's theorem permits a partitioning of the network into two parts and a study of the system with respect to one pair of terminals of the partitioned boundary.

Bayes's theorem, like Thévenin's theorem, can be extended to a partitioning of the sample space into mutually exclusive and exhaustive parts. Suppose that an event E can occur as a result of the occurrence of several mutually exclusive and exhaustive events $A_1, A_2, \ldots, A_n$. Let the corresponding conditional probabilities be given as

$$P\{E|A_k\} = p_k \qquad k = 1, \ldots, n \qquad (2\text{-}93)$$

and let

$$P\{A_k\} = \omega_k$$

Then, by the law of addition, we have

$$E = A_1E + A_2E + \cdots + A_nE \qquad (2\text{-}94)$$
$$P\{E\} = P\{A_1E + A_2E + \cdots + A_nE\} \qquad (2\text{-}95)$$
$$P\{E\} = \omega_1 p_1 + \omega_2 p_2 + \cdots + \omega_n p_n \qquad (2\text{-}96)$$

The question is to find the a posteriori probability of the occurrence of event A_k, given the occurrence of E.

$$P\{A_k|E\} = \frac{P\{A_k\}P\{E|A_k\}}{P\{E\}} = \frac{P\{A_k\}P\{E|A_k\}}{\sum_{j=1}^{n} P\{E|A_j\}P\{A_j\}} \tag{2-97}$$

or, equivalently,

$$P\{A_k|E\} = \frac{\omega_k p_k}{\omega_1 p_1 + \omega_2 p_2 + \cdots + \omega_n p_n} \tag{2-98}$$

This equation comprises what is known as Bayes's theorem.

Example 2-20. Let U_1, U_2, U_3 be three urns with two red and one black, three red and two black, and one red and one black balls, respectively. One of the three urns is chosen at random and a ball is drawn from it. The color of the ball is found to be black. What is the probability that it has been chosen from U_3?

Solution. This is an example of a situation where Bayes's theorem can be applied. Let E be the event that a black ball has been drawn; A_i is the event that the ith urn has been chosen, $i = 1, 2, 3$.

Then

$$P\{E|A_1\} = \tfrac{1}{3} \qquad P\{E|A_2\} = \tfrac{2}{5} \qquad P\{E|A_3\} = \tfrac{1}{2}$$

Also,

$$\begin{aligned} P\{A_3|E\} &= P\{\text{choosing urn } U_3|\text{black ball drawn}\} \\ &= \frac{P\{A_3\}P\{E|A_3\}}{\sum_{i=1}^{3} P\{E|A_i\}P\{A_i\}} \\ &= \frac{\tfrac{1}{3}\cdot\tfrac{1}{2}}{\tfrac{1}{3}(\tfrac{1}{3}+\tfrac{2}{5}+\tfrac{1}{2})} = \frac{15}{37} \end{aligned}$$

Example 2-21. Three urns are given:

Urn 1 contains two white, three black, and four red balls.

Urn 2 contains three white, two black, and two red balls.

Urn 3 contains four white, one black, and one red ball.

One urn is chosen at random, and two balls are drawn from that urn. If the two balls happen to be white and red, what is the probability that they were drawn from urn 3?

Solution. Let A_i = event of choosing urn i, $i = 1, 2, 3$

RW = event of choosing a red and a white ball

We want $P\{A_3|RW\}$.

Using Bayes's rule,

$$P\{A_3|RW\} = \frac{P\{A_3\}P\{RW|A_3\}}{P\{A_1\}P\{RW|A_1\} + P\{A_2\}P\{RW|A_2\} + P\{A_3\}P\{RW|A_3\}}$$

But

$$P\{A_1\} = P\{A_2\} = P\{A_3\} = \tfrac{1}{3}$$

$$P\{RW|A_1\} = \frac{8}{\binom{9}{2}} = \frac{8}{36}$$

$$P\{RW|A_2\} = \frac{6}{\binom{7}{2}} = \frac{6}{21}$$

$$P\{RW|A_3\} = \frac{4}{\binom{6}{2}} = \frac{4}{15}$$

Therefore,
$$P\{A_3|RW\} = \frac{\frac{1}{3}\cdot\frac{4}{15}}{\frac{1}{3}(\frac{2}{9}+\frac{6}{21}+\frac{4}{15})} = \frac{21}{61}$$

Bayes's* theorem comprises one of the most used, and occasionally misused, concepts of probability theory. In many problems an event may occur as an "effect" of several "causes." From a number of observations on the occurrence of the effect, one can make an estimate on the occurrence of the cause leading to that effect. This rule is frequently applied to communication problems, particularly in the detection of signals from an additive mixture of signals and noise. When the detecting instrument indicates a signal, we have to make a decision whether the received signal is a true one or a false alarm due to undesired signals (noise) in the system. Such decisions are generally made possible by an application of Bayes's rule which is also called the rule of inverse probability. The decision criterion may be made more effective by introducing some kind of weighting coefficients called *loss matrix* and minimizing the over-all "loss."

2-13. Combinatorial Problems in Probability. In many problems involving choice and probability, the number of possible ways of arranging a given number of objects on a line is of interest. For example, if three persons A, B, and C are standing in a line, the probability that A remains next to B can be calculated as follows: There are six different arrangements possible:

$$ABC \quad ACB \quad BAC \quad BCA \quad CAB \quad CBA$$

Of these arrangements, there are four desirable ones. Thus, if the concept of equiprobable measures is assumed, the probability in question is $\frac{2}{3}$.

Combinatorial problems have a limited use in our subsequent studies. For this reason, we shall give only a review of the most pertinent definitions in this section. The reader interested in combinatorial problems will find a considerable amount of information in Feller (Chaps. 2 to 4).

Permutation: A permutation of the elements of a finite set is a one-to-one correspondence between elements of that set (such a correspondence is also called a mapping of the set onto itself). For example, if a set

* Reverend Thomas Bayes's article An Essay towards Solving a Problem in the Doctrine of Chances was published in *Philosophical Transactions of the Royal Society of London* (vol. 1, no. 3, p. 370, 1763). However, Bayes's work remained rather unknown until 1774, when Laplace discussed it in one of his memoirs.

contains only four objects A, B, C, and D, we may write two equivalent sets

$$A, B, C, D \qquad \text{and} \qquad B, C, A, D$$

The ordered sets $\begin{bmatrix} A & B & C & D \\ 1 & 2 & 3 & 4 \end{bmatrix}$ and $\begin{bmatrix} B & C & A & D \\ 1 & 2 & 3 & 4 \end{bmatrix}$ are two permutations of the elements of the original set, since

$$\begin{aligned} A &\rightarrow B \\ B &\rightarrow C \\ C &\rightarrow A \\ D &\rightarrow D \end{aligned} \tag{2-99}$$

The following definition is of considerable assistance in dealing with combinatorial problems.

Factorial: The factorial function for a positive integer n is defined as

$$n! = n(n-1)(n-2) \cdots 4 \cdot 3 \cdot 2 \cdot 1 \tag{2-100}$$

with the additional convention

$$0! = 1 \tag{2-101}$$

The number of different permutations of a set with n distinct elements is

$$P_n = n(n-1)(n-2) \cdots 4 \cdot 3 \cdot 2 \cdot 1 = n! \tag{2-102}$$

Combination: The number of different permutations of r objects selected from n objects is

$$P_r^n = n(n-1)(n-2) \cdots (n-r+1) = \frac{n!}{(n-r)!} \tag{2-103}$$

Every permutation of elements of a set contains the same elements but in different order. When two sets of objects are in one-to-one correspondence so that some of the elements of one do not appear in the other they are called different *combinations*. For example, if we combine the members of the set $\{A,B,C,D\}$ two by two, AB, AC, DB are different combinations but AB and BA are not.

The number of different combinations of n objects taken r at a time is

$$C_r^n = \frac{P_r^n}{r!} = \frac{n!}{r!(n-r)!} \tag{2-104}$$

When confusion will not result, one may use the notation $\binom{n}{r}$ for C_r^n. Note that

$$\binom{n}{n-r} = \binom{n}{r} \tag{2-105}$$

$$\binom{n}{1} = n \tag{2-106}$$

$$\binom{n+1}{r} = \binom{n}{r-1} + \binom{n}{r} \tag{2-107}$$

$$\binom{n}{0} + \binom{n}{1} + \cdots + \binom{n}{n} = \sum_{r=0}^{n} \binom{n}{r} = 2^n \tag{2-108}$$

The following theorem is often used in combinatorial problems. Let a set contain k mutually exclusive subsets of objects:

$$\{A_1, A_2, \ldots, A_k\}$$

with $\quad A_i = \{a_{i1}, a_{i2}, \ldots, a_{in_i}\} \qquad i = 1, 2, \ldots, k$

n_i being the number of elements in the set A_i. The number of permutations of the total number of elements n is

$$\frac{n!}{n_1! n_2! \cdots n_k!} \tag{2-109}$$

In fact, one has to divide the number of permutations of n objects by $n_i!$ (for $i = 1, 2, \ldots, k$) since the permutations of the identical objects of the A_i set cannot be distinguished from each other. For example, the number of color permutations of three black and two white balls is

$$\frac{5!}{3!2!} = \frac{5 \times 4}{2!} = 10$$

Binomial Expansion: Let n be a positive integer; then

$$(a + b)^n = a^n + \binom{n}{1} a^{n-1}b + \binom{n}{2} a^{n-2}b^2 + \cdots + \binom{n}{r} a^{n-r}b^r + \cdots + b^n \tag{2-110}$$

or

$$(a + b)^n = a^n + na^{n-1}b + \frac{n(n-1)}{2!} a^{n-2}b^2 + \frac{n(n-1)(n-2)}{3!} a^{n-3}b^3 + \cdots + b^n \tag{2-111}$$

A useful display of a binomial coefficient is given in a table which is called *Pascal's triangle:*

$$
\begin{array}{ccccccccc}
 & & & & \binom{0}{0} & & & & \\
 & & & \binom{1}{0} & & \binom{1}{1} & & & \\
 & & \binom{2}{0} & & \binom{2}{1} & & \binom{2}{2} & & \\
 & \binom{3}{0} & & \binom{3}{1} & & \binom{3}{2} & & \binom{3}{3} & \\
\binom{4}{0} & & \binom{4}{1} & & \binom{4}{2} & & \binom{4}{3} & & \binom{4}{4}
\end{array}
\qquad
\begin{array}{c}
1 \\ 1\ 1 \\ 1\ 2\ 1 \\ 1\ 3\ 3\ 1 \\ 1\ 4\ 6\ 4\ 1
\end{array}
\tag{2-112}
$$

In the following a number of simple examples dealing with permutations and combinations are presented. In these examples, the primary assumption is that the probability is given by the frequency of the event under consideration; that is, the concept of equiprobable measure prevails. Hence, such problems are reduced to a study of the ratio of the favorable cases to all possible cases. In this respect the formula of combinatorial analysis will be used.

Example 2-22. What is the probability of a person having four aces in a bridge hand?

Solution. The number of all possible different hands equals the combination of 13 from 52 cards. For the number of favorable cases one may think of first removing the four aces from the deck and then dealing all possible combinations of hands 9 by 9. The addition of the four aces to each one of these latter hands gives a favorable case.

$$\binom{48}{9} : \binom{52}{13} = \frac{10 \cdot 11 \cdot 12 \cdot 13}{49 \cdot 50 \cdot 51 \cdot 52} = \frac{11}{4{,}165}$$

Example 2-23. Two cards are drawn from a regular deck of cards. What is the probability that neither is a heart?

Solution. Let A and B be the events that the first and the second card are hearts, respectively; then we wish to know $P\{A'B'\}$.

$$P\{A'\} = 1 - P\{A\} = 1 - 1/4 = 3/4$$

$$P\{B'|A'\} = \frac{P\{A'B'\}}{P\{A'\}}$$

$$P\{B'|A'\} = 38/51$$

Therefore $$P\{A'B'\} = 38/51 \cdot 3/4 = 19/34$$

If we wish to apply combinatorial principles, we may say that the number of all possible cases of selecting two cards is $\binom{52}{2}$. The number of favorable cases is $\binom{39}{2}$. Therefore the probability in question is

$$\binom{39}{2} : \binom{52}{2} = \frac{39!}{2!37!}\,\frac{2!50!}{52!} = \frac{39 \cdot 38}{51 \cdot 52} = \frac{19}{34}$$

2-14. Trees and State Diagrams. The material of this section is intended to offer a graphical interpretation for certain simple problems of probability which arise in dealing with repeated trials of an experiment.

For example, suppose that a biased coin is tossed once; the outcome may be denoted by H and T and shown by the diagram of Fig. 2-20. Similarly, if the coin is tossed twice, the second set of outcomes may be shown in the same treelike diagram. If the probability of getting a head is denoted by p, then the probability of getting, say, HT can be directly computed from the weighted length of the associated tree path, that is,

$$p(1 - p)$$

If it is desired to obtain the probability of getting a head and a tail

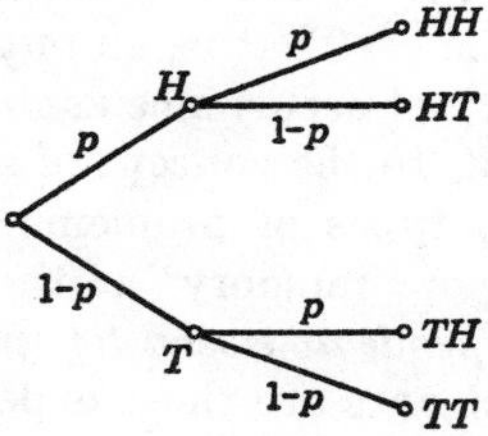

FIG. 2-20. A simple tree diagram.

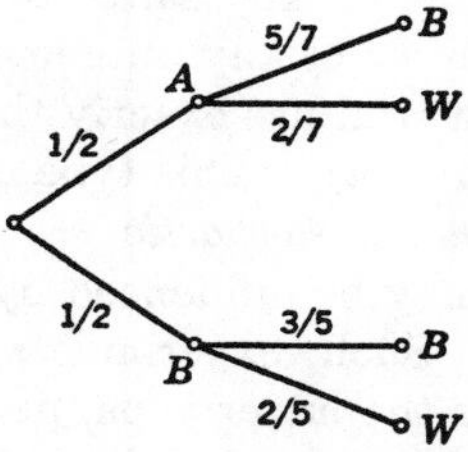

FIG. 2-21. An example of a probability tree.

irrespective of their order, then the answer to the problem is given by summing up the two weighted tree paths.

$$p(1 - p) + (1 - p)p = 2p(1 - p)$$

This simple graphical procedure can be used profitably in certain types of problems. The following are examples of such problems.

Example 2-24. The urn A contains five black and two white balls. The urn B contains three black and two white balls. If one urn is selected at random, what is the probability of drawing a white ball from that urn?

Solution. From the tree diagram of Fig. 2-21 one can see that the probability of the event of interest is the sum of the following measures:

$$\tfrac{1}{2} \cdot \tfrac{2}{7} + \tfrac{1}{2} \cdot \tfrac{2}{5} = \tfrac{12}{35}$$

Example 2-25. Find the probability that at least three heads are obtained in a sequence of four throws of an honest coin.

Solution. From a tree diagram or from the binomial expansion one obtains

$$\binom{4}{4} \cdot (\tfrac{1}{2})^4 + \binom{4}{3} \cdot (\tfrac{1}{2})^3(\tfrac{1}{2})^1 = \tfrac{1}{16} + \tfrac{4}{16} = \tfrac{5}{16}$$

If a coin is tossed n times, we note that the probability of getting, say, exactly r heads ($r < n$) is the sum of the tree measures of all tree paths leading to r heads and $n - r$ tails. Since there are $\binom{n}{r}$ such states, it is found that the desired probability is

$$\binom{n}{r} p^r \cdot (1 - p)^{n-r} \tag{2-113}$$

The tree diagram can easily be drawn for experiments with a finite number of outcomes. In the problems discussed thus far in this section, it is tacitly assumed that the outcomes of each experiment remain independent of the previous experiments. In engineering terminology such experiments are said to lack memory. For these experiments the probability of any outcome is always the same. That is, an outcome of the nth trial has exactly the same probability of occurrence as in the kth trial ($k \neq n$). This type of experiment leads to the concept of so-called *independent stochastic processes.* In certain types of problems an outcome may be influenced by the past history or "memory" of the experiment. Such experiments are termed *dependent stochastic processes.* Among the latter type, perhaps the simplest ones are those experiments in which the probability of an outcome of a trial depends on the outcome of the immediately preceding trial. These are called *Markov processes.*

Let an experiment have a finite number of n possible outcomes, a_1, a_2, . . . , and a_n, called *states.* We assume the process to be of the finite Markov type and initially in the state k. For a Markov process, we specify a table of probabilities associated with transitions from any state to any other state. This is called a *probability transition matrix.*

$$\begin{array}{c} \\ a_1 \\ a_2 \\ a_3 \\ \cdots \\ a_n \end{array} \begin{array}{c} \begin{array}{ccccc} a_1 & a_2 & a_3 & \cdots & a_n \end{array} \\ \begin{bmatrix} p_{11} & p_{12} & p_{13} & \cdots & p_{1n} \\ p_{21} & p_{22} & p_{23} & \cdots & p_{2n} \\ p_{31} & p_{32} & p_{33} & \cdots & p_{3n} \\ \cdots & \cdots & \cdots & \cdots & \cdots \\ p_{n1} & p_{n2} & p_{n3} & \cdots & p_{nn} \end{bmatrix} \end{array} \tag{2-114}$$

$p_{jk} = p\{a_k|a_j\}$ denotes the probability that the next outcome of the experiment will be the state k, given that the immediately preceding experiment led to the state j. Note that in a transition probability matrix the sum of all elements of each row must equal unity.

One of the most common problems associated with the Markov process is, given that it started with state j, to find the probability of reaching the state k after a specified number of steps r, that is, $p\{a_k|a_j\}^{(r)} = p_{jk}^{(r)}$. This question has a rather simple answer, namely, (1) draw the tree diagram, (2) select all tree paths connecting the node representing the state j to that of the state k in r steps, and (3) add the corresponding tree measures. This procedure is exemplified in the tree diagram of Fig.

2-22 for $r = 1$ and $r = 2$. When $r = 1$, the answer is obvious:

$$P\{a_k|a_j\}^{(1)} = P\{a_j\}P\{a_k|a_j\} \tag{2-115}$$

For $r = 2$ one has to add the probabilities of reaching state a_k from the state a_j in all possible ways, that is, the sum of the measures associated

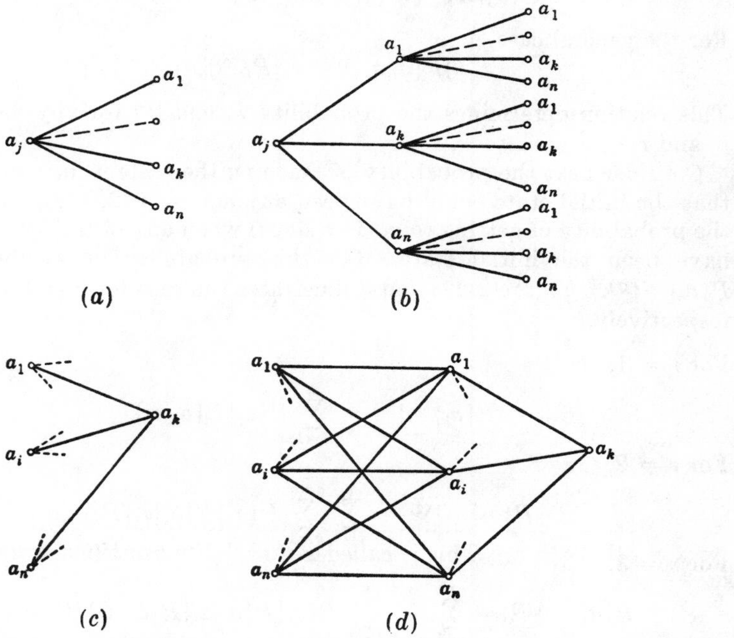

FIG. 2-22. Trees for a finite chain. (a) $r = 1$. (b) $r = 2$. (c) $r = 1$. (d) $r = 2$.

with all three paths connecting a_j to a_k in two steps.

$$P\{a_k|a_j\}^{(2)} = P\{a_j\}[P\{a_1|a_j\}P\{a_k|a_1\} + P\{a_2|a_j\}P\{a_k|a_2\} + \cdots + P\{a_n|a_j\}P\{a_k|a_n\}]$$

$$= P\{a_j\} \sum_{i=1}^{i=n} P\{a_i|a_j\}P\{a_k|a_i\} = P\{a_j\} \sum_{i=1}^{n} P_{ji} \cdot P_{ik}$$

For $r = 3$,

$$P\{a_k|a_j\}^{(3)} = P\{a_j\} \sum_{g=1}^{g=n} \sum_{i=1}^{i=n} P\{a_i|a_j\}P\{a_g|a_i\}P\{a_k|a_g\}$$

By defining the initial probability of different states as a diagonal matrix,

$$[P_D{}^{(0)}] = \begin{bmatrix} P\{a_1\} & 0 & \cdots & 0 \\ 0 & P\{a_2\} & \cdots & 0 \\ \cdots & \cdots & \cdots & \cdots \\ 0 & 0 & \cdots & P\{a_n\} \end{bmatrix}$$

we can sum up the above development in concise matrix notation. That is, for any states j and k we have

$$[P\{a_j|a_k\}^{(1)}] = [P_D^{(0)}][P]$$

Similarly,

$$[P\{a_j|a_k\}^{(2)}] = [P_D^{(0)}][P][P] = [P_D^{(0)}][P]^2$$

For the general case,

$$[P\{a_j|a_k\}^{(r)}] = [P_D^{(0)}][P]^r \tag{2-116}$$

This relation determines the probability $P\{a_j|a_k\}^{(r)}$ for any values of j, k, and r.

Consider next the probability of reaching the state a_k in r steps, given that the initial state could have been any a_i, $i = 1, 2, \ldots, n$, that is, the probability of getting to a_k (in r steps) when any of the n states could have been the initial state. Let this probability be symbolized by $P\{a_k| \quad\}^{(r)}$. Figure 2-22c and d illustrates the case for $r = 1$ and $r = 2$, respectively.

For $r = 1$,

$$P\{a_k| \quad\}^{(1)} = \sum_{i=1}^{n} P\{a_i\}P\{a_k|a_i\}$$

For $r = 2$,

$$P\{a_k| \quad\}^{(2)} = \sum_{i=1}^{n} \sum_{g=1}^{g=n} P\{a_i\}P\{a_g|a_i\}P\{a_k|a_g\}$$

For $r = 3$,

$$P\{a_k| \quad\}^{(3)} = \sum_{i=1}^{n} \sum_{g=1}^{n} \sum_{h=1}^{n} P\{a_i\}P\{a_g|a_i\}P\{a_h|a_g\}P\{a_k|a_h\}$$

The matrix formulation follows immediately. Let $[P^0]$ be a row matrix describing the initial probabilities $[P\{a_1\}, P\{a_2\}, \ldots, P\{a_n\}]$; then

$$[P\{a_k| \quad\}^{(1)}] = [P^{(0)}][P]$$
$$[P\{a_k| \quad\}^{(2)}] = [P^{(0)}][P]^2$$

For the general case,

$$[P\{a_k| \quad\}^{(r)}] = [P^{(0)}][P]^r \tag{2-117}$$

This relation determines the probability $P\{a_k| \quad\}^{(r)}$ for any values of positive integers k and r. Note that $P\{a_k| \quad\}^{(r)}$ will always be a row matrix since $[P^{(0)}]$ is a row matrix.

Example 2-26. A relay alternates between the open state denoted by 1 and the closed state designated by 0. The transition probability matrix is given as

$$\begin{matrix} & \begin{matrix} 1 & 0 \end{matrix} \\ \begin{matrix} 1 \\ 0 \end{matrix} & \begin{bmatrix} 2/3 & 1/3 \\ 1/3 & 2/3 \end{bmatrix} \end{matrix}$$

Assuming that the initial probability of the relay being in either state is ½, determine

(*a*) The probability of reaching state 1 via state 0 in one step, that is, $p_{01}^{(1)}$.

(*b*) $p_{00}^{(1)}$.

(*c*) $p_{01}^{(2)}$.

(*d*) $p_{11}^{(2)}$.

(*e*) $p\{1|\quad\}^{(2)}$, the probability of reaching state 1 in two steps.

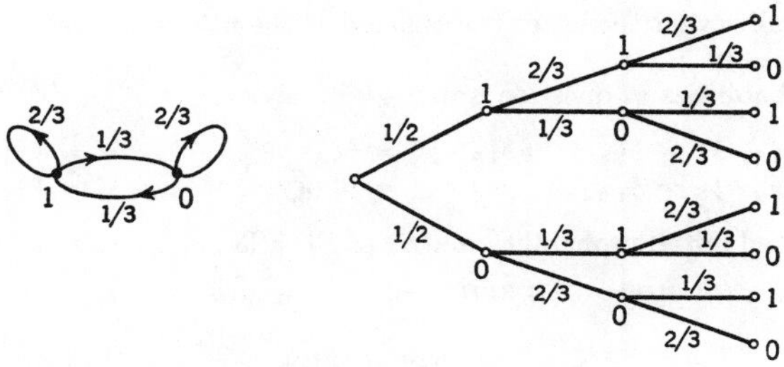

FIG. E2-26

Solution. The state diagram and the tree diagram are drawn in Fig. E2-26. According to the tree diagram,

$$
\begin{array}{ll}
(a) & p_{01}^{(1)} = \tfrac{1}{2}\cdot\tfrac{1}{3} = \tfrac{1}{6}\\
(b) & p_{00}^{(1)} = \tfrac{1}{2}\cdot\tfrac{2}{3} = \tfrac{1}{3}\\
(c) & p_{01}^{(2)} = \tfrac{1}{2}(\tfrac{1}{3}\cdot\tfrac{2}{3} + \tfrac{2}{3}\cdot\tfrac{1}{3}) = \tfrac{2}{9}\\
(d) & p_{11}^{(2)} = \tfrac{1}{2}(\tfrac{2}{3}\cdot\tfrac{2}{3} + \tfrac{1}{3}\cdot\tfrac{1}{3}) = \tfrac{5}{18}\\
(e) & p\{1|\quad\}^{(2)} = p_{11}^{(2)} + p_{01}^{(2)} = \tfrac{5}{18} + \tfrac{2}{9} = \tfrac{1}{2}
\end{array}
$$

An alternative solution for part (*e*) is given by the matrix relation of Eq. (2-117).

$$
[\tfrac{1}{2}\quad \tfrac{1}{2}]\begin{bmatrix}\tfrac{2}{3} & \tfrac{1}{3}\\ \tfrac{1}{3} & \tfrac{2}{3}\end{bmatrix}^2 = [\tfrac{1}{2}\quad \tfrac{1}{2}]\begin{bmatrix}\tfrac{5}{9} & \tfrac{4}{9}\\ \tfrac{4}{9} & \tfrac{5}{9}\end{bmatrix} = [\tfrac{1}{2}\quad \tfrac{1}{2}]
$$

Therefore, $$p\{1|\quad\}^{(2)} = p\{0|\quad\}^{(2)} = \tfrac{1}{2}$$

Finally we may answer the same question by using the materials of Secs. 2-9 (Theorem of Addition) and 2-10 (Conditional Probability).

$$
\begin{array}{lrl}
(a) & p\{01\} & = p\{0\}p\{1|0\} = \tfrac{1}{2}\cdot\tfrac{1}{3} = \tfrac{1}{6}\\
(b) & p\{00\} & = p\{0\}p\{0|0\} = \tfrac{1}{2}\cdot\tfrac{2}{3} = \tfrac{1}{3}\\
(c) & p\{001\} + p\{011\} & = p\{00\}p\{1|0\} + p\{01\}p\{1|1\}\\
 & & = \tfrac{1}{3}\cdot\tfrac{1}{3} + \tfrac{1}{6}\cdot\tfrac{2}{3} = \tfrac{2}{9}\\
(d) & p\{111\} + p\{101\} & = p\{11\}p\{1|1\} + p\{10\}p\{1|0\}\\
 & & = \tfrac{1}{2}\cdot\tfrac{2}{3}\cdot\tfrac{2}{3} + \tfrac{1}{2}\cdot\tfrac{1}{3}\cdot\tfrac{1}{3} = \tfrac{5}{18}\\
(e) & p\{1|\quad\}^{(2)} & = \tfrac{2}{9} + \tfrac{5}{18} = \tfrac{1}{2}
\end{array}
$$

Example 2-27. A communication source having a three-letter alphabet transmits sequences of messages. The transition probability matrix is given below:

$$
\begin{array}{c}
\begin{array}{ccc} A & B & C \end{array}\\
\begin{array}{c} A\\ B\\ C \end{array}\begin{bmatrix} 0 & \tfrac{1}{3} & \tfrac{2}{3}\\ \tfrac{1}{3} & \tfrac{1}{3} & \tfrac{1}{3}\\ \tfrac{5}{9} & 0 & \tfrac{4}{9} \end{bmatrix}
\end{array}
$$

For the beginning of each message, letters A, B, C occur with probabilities $3/18$, $8/18$, and $7/18$, respectively.

(a) Determine the probability of getting a message commencing with

$$AB, \quad BB, \quad CA,$$
$$ABA, \quad BBC, \quad CAC$$

(b) Find a set of initial probabilities which will produce a so-called "steady state," i.e., the probability that the letter transmitted at the nth state does not depend on n.

Solution

(a) The probabilities in question are, respectively,

$$\tfrac{3}{18} \cdot \tfrac{1}{3} = \tfrac{3}{54} \qquad \tfrac{8}{18} \cdot \tfrac{1}{3} = \tfrac{8}{54} \qquad \tfrac{7}{18} \cdot \tfrac{5}{9} = \tfrac{35}{162}$$
$$\tfrac{3}{54} \cdot \tfrac{1}{3} = \tfrac{3}{162} \qquad \tfrac{8}{54} \cdot \tfrac{1}{3} = \tfrac{8}{162} \qquad \tfrac{35}{162} \cdot \tfrac{2}{3} = \tfrac{35}{243}$$

(b) The desired initial probability matrix $[P^{(0)}] = [\alpha,\beta,\gamma]$ must satisfy the condition

$$[P^{(0)}][P]^{(n)} = [P^{(0)}][P]^{(n-1)} \qquad n \text{ a positive integer}$$

In particular,

$$[P^{(0)}][P] = [P^{(0)}]$$

Therefore

$$\alpha = \beta\tfrac{1}{3} + \gamma\tfrac{5}{9}$$
$$\beta = \alpha\tfrac{1}{3} + \beta\tfrac{1}{3}$$
$$\gamma = \alpha\tfrac{2}{3} + \beta\tfrac{1}{3} + \gamma\tfrac{4}{9}$$

These equations lead to

$$\alpha = \tfrac{1}{3} \qquad \beta = \tfrac{1}{6} \qquad \gamma = \tfrac{1}{2}$$

It can be shown that, if one considers very long messages, the frequency of the occurrence of the letter A will approach $\tfrac{1}{3}$, etc. For further comments on the Markov chain, see Chap. 11.

2-15. Random Variables. In the preceding sections the concept of an event and of sample space of an experiment played an important role. The discussion of the present section is aimed at an intuitive introduction of *random variables.*

Most experiments of practical interest have numerical outcomes; that is, the result of the experiment is a number, or a pair of numbers, etc. In other words, the results can be described by using a coordinate space, the coordinate space being in a correspondence with the sample space of the event.

A random variable is a real-valued function defined over the sample space of a random experiment. Restricting the random variable to assume only real values is quite natural, as one is interested in the numerical outcomes of an experiment (even though in various practical applications complex values of random variables are also considered). The word "random" stresses the fundamental fact that we are dealing with experiments governed by laws of chance rather than any deterministic law. The throws of a symmetrical die or coin under hypothetically symmetrical conditions represent random experiments. The salient feature of these experiments is that, even though they exhibit a certain kind of regularity when repeated over a long range of time, it is impossible to predict, with complete certainty, the outcome of any particular trial.

Let Ω be the sample space of a random experiment. Each point of Ω describes a possible outcome of the experiment. This outcome may not be a numerical result in itself but some numerical data can be assigned to it. For instance, if the experiment were the picking at "random" of a card out of a deck of 52, the number of possible outcomes at any particular trial would be 52, depending upon which one of the cards had been picked. Here, although the outcome does not furnish us with a numerical result, we can represent the possible outcomes by, say, the first 52 integers or by 52 points on a line.

The correspondence between a point of Ω and a point in the coordinate space is designated by a mathematical function. This function is termed a *random variable*. Generally, we shall denote random variables by capital letters such as X and Y, and their specific values by the same letters in lower case. A random variable X assumes different values $x_1, x_2, \ldots, x_n, \ldots$ which are points of the coordinate space. The coordinate space may be a one-dimensional or a multidimensional space. The random variable may take a finite number of n-tuple values or infinitely many. The sample space may be a space with finite or countably infinite points or even a continuous space, that is, with an uncountable number of points. The following practical examples illustrate some possibilities.

Example 2-28. The experiment is throwing an ordinary honest die. The sample space has six events of interest. The associated random variable takes only six possible numerical values, 1, 2, 3, 4, 5, and 6. Each of these real numbers corresponds to a specific event.

Example 2-29. The experiment is throwing three honest dice. The associated random variable takes on 6^3 different numbers of triads as values. The random variable may be conveniently represented by a point in the three-dimensional euclidean space.

2-16. Discrete Probability Functions and Distributions. Consider Ω the sample space of a random experiment. If the outcomes of this experiment can be put into one-to-one correspondence with the positive integers, the sample space will contain a countable number of points. Such a sample space is said to be a *discrete sample space*. In a discrete sample space, when the random variable X assumes values

$$[x_1, x_2, \ldots, x_k, \ldots]$$

the probability function $f(x)$ is defined as

$$[p_1, p_2, \ldots, p_k, \ldots]$$

where
$$f(x_k) = P\{X = x_k\} = p_k \tag{2-118}$$

The *probability distribution function* $F(x)$, known also as the *cumulative distribution function* (CDF), is defined as

$$F(x) = \sum_{x_j \leq x} f(x_j) \tag{2-119}$$

For example, the throw of an honest coin until a head appears is a random experiment. The sample space of this experiment is a discrete space. If X corresponds to the event of the appearance of the first head on the kth throw, then X assumes the following values:

$$[X] = [1,2,3, \ldots ,k, \ldots] \tag{2-120}$$

The probability function $f(x)$ and the CDF are

$$\begin{aligned} f(x) &= [2^{-1},2^{-2},2^{-3}, \ldots ,2^{-k}, \ldots] \\ F(x) &= 2^{-1} + 2^{-2} + \cdots + 2^{-x} \end{aligned} \tag{2-121}$$

These functions are plotted in Fig. 2-23*a* and *b*, respectively.

The definition of the probability function and CDF can be directly extended to the case of a multivariate random variable. For instance, in Example 2-29 the sample space is a three-dimensional euclidean space with 216 points. The random variable X assumes 216 triad values $X = (X_1,X_2,X_3)$ for any experiment. The corresponding probability function is

$$f(x_1,x_2,x_3) = P\{X_1 = x_1, X_2 = x_2, X_3 = x_3\}$$
$$f(x_1,x_2,x_3) = \tfrac{1}{6} \cdot \tfrac{1}{6} \cdot \tfrac{1}{6} = \tfrac{1}{216}$$

Here all permissible outcomes have equal probabilities.

The CDF gives the total probability of the set of points having each coordinate less than or equal to some specified value (x_1,x_2,x_3), that is,

$$F(x_1,x_2,x_3) = \Sigma f(x_i,x_j,x_k)$$

for

$$x_i \leq [x_1] \qquad x_j \leq [x_2] \qquad x_k \leq [x_3]$$

where [ ] denotes the greatest integer contained in the letter inside the brackets.

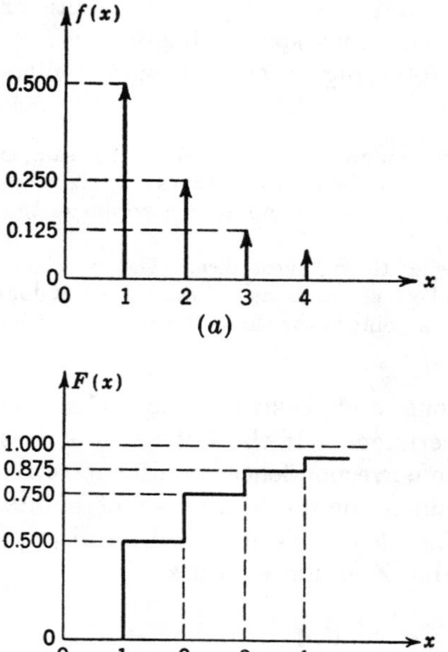

FIG. 2-23. (*a*) Probability function associated with Eq. (2-121). (*b*) CDF associated with Eq. (2-121).

2-17. Bivariate Discrete Distribution. The case of a random variable assuming pairs of values (x_j,y_k) is of particular interest. In fact, in most engineering problems the interrelation between two random quantities leads to a bivariate discrete distribution. The joint probability function and the CDF are defined as before:

$$\begin{aligned} f(x,y) &= P\{X = x,\ Y = y\} \\ F(x,y) &= P\{X \leq x,\ Y \leq y\} \end{aligned} \tag{2-122}$$

If the joint probability function $f(x,y)$ is known, say in the form of a matrix, then there are four additional quantities of interest which can be readily computed. These are marginal probability functions and marginal CDF's as defined below:

$$\begin{aligned} f_1(x_i) &= P\{X = x_i, \text{ all permissible } Y\text{'s}\} = \sum_y f(x_i,y) \\ f_2(y_j) &= P\{Y = y_j, \text{ all permissible } X\text{'s}\} = \sum_x f(x,y_j) \\ F_1(x_i) &= \sum_{x_k \leq x_i} f_1(x_k) \\ F_2(y_j) &= \sum_{y_k \leq y_j} f_2(y_k) \end{aligned} \tag{2-123}$$

The indices 1 and 2 in the marginal distributions are simply to indicate that $f_1(x)$ refers to the variable x, that is, the first variable, and $f_2(y)$ to the second variable. Now assume that all pairs of values (x_i,y_j) are written in a matrix form:

$$[X,Y] = \begin{bmatrix} (x_1,y_1) & (x_1,y_2) & \cdots & (x_1,y_n) \\ (x_2,y_1) & (x_2,y_2) & & (x_2,y_n) \\ \cdots & \cdots & \cdots & \cdots \\ (x_m,y_1) & (x_m y_2) & & (x_m,y_n) \end{bmatrix} \tag{2-124}$$

The corresponding probabilities can be written in a similar form:

$$[f(x_i,y_j)] = \begin{bmatrix} p_{11} & p_{12} & \cdots & p_{1n} \\ p_{21} & p_{22} & \cdots & p_{2n} \\ \cdots & \cdots & \cdots & \cdots \\ p_{m1} & p_{m2} & & p_{mn} \end{bmatrix} \tag{2-125}$$

The marginal probability $f_1(x_2)$ is the probability of the occurrence of events for which $X = x_2$ without regard to the value of Y. This is readily obtained by adding the terms appearing in the second row of the probability matrix.

$$f_1(x_2) = p_{21} + p_{22} + \cdots + p_{2n} \tag{2-126}$$

Similarly, the marginal distribution $f_2(y_k)$ can be obtained by adding the terms of the kth column of the joint probability matrix. For example,

$$f_2(y_2) = p_{12} + p_{22} + \cdots + p_{m2} \tag{2-127}$$

If the random variables X and Y are such that for all values of (x_i,y_j) we have

$$f(x_i,y_j) = f_1(x_i)f_2(y_j) \tag{2-128}$$

then the variables are said to be statistically independent of each other. For example, the simultaneous throw of two honest coins has the following outcomes:

$$[X,Y] = \begin{bmatrix} H_1H_2 & H_1T_2 \\ T_1H_2 & T_1T_2 \end{bmatrix} \qquad \text{and} \qquad [f(x,y)] = \begin{bmatrix} p_{11} & p_{12} \\ p_{21} & p_{22} \end{bmatrix}$$

Evidently, these two variables are independent of each other, since for any entry of the probability matrix we have

$$P\{X = H_1\} = p_{11} + p_{12} = \tfrac{1}{2}$$
$$P\{Y = T_2\} = p_{12} + p_{22} = \tfrac{1}{2}$$
$$P\{X = H_1,\ Y = T_2\} = (p_{11} + p_{12})(p_{12} + p_{22}) = \tfrac{1}{2} \cdot \tfrac{1}{2} = \tfrac{1}{4}$$

Conversely, a check for independence is to determine if Eq. (2-128) holds for all possible outcomes.

The conditional distributions can also be defined and obtained in a straightforward manner. The conditional probability $P\{X = x_i | Y = y_j\}$ is designated as $f(x_i|y_j)$. That is, if the computation of $f(x_i|y_j)$ is desired, then we concentrate on the jth column of the (x,y) matrix.

$$[X,Y = y_j] = \begin{bmatrix} x_1y_j \\ x_2y_j \\ \cdots \\ x_iy_j \\ \cdots \\ x_my_j \end{bmatrix} \tag{2-129}$$

Next the term x_iy_j is selected and its associated probability is obtained.

$$P\{X = x_i|Y = y_j\} = f(x_i|y_j) = \frac{f(x_i,y_j)}{f_2(y_j)} \tag{2-130}$$
$$f_2(y_j) \neq 0$$

It is to be noted that $f(x_i|y_j)$ is a permissible conditional probability function as all its terms are nonnegative and

$$\sum_{i=1}^{m} f(x_i|y_j) = \frac{\sum_{i=1}^{m} f(x_i,y_j)}{f_2(y_j)} = \frac{f_2(y_j)}{f_2(y_j)} = 1 \qquad (2\text{-}131)$$

$$f_2(y_j) \neq 0$$

Similarly, the conditional probability of Y, given $X = x_i$, is found to be

$$f(y_j|x_i) = \frac{f(x_i,y_j)}{f_1(x_i)} \qquad (2\text{-}132)$$

$$f_1(x_i) \neq 0$$

Example 2-30. Consider the simultaneous throw of two honest dice X and Y. Find $P\{3 \leq X \leq 5, 2 \leq Y \leq 3\}$ and the marginal probabilities.

Solution. The two-dimensional random variable assumes 36 pairs of values, each with an equal probability of $\frac{1}{36}$.

$$P_{jk} = \tfrac{1}{36} \quad \text{for each point of the sample space}$$
$$F(x,y) = P(X \leq x, Y \leq y)$$

The marginal CDF's are

$$F_1(x) = \sum_{j=1}^{[x]} \sum_{k=1}^{6} p(j,k) = \frac{[x]}{6}$$

$$F_2(y) = \sum_{k=1}^{[y]} \sum_{j=1}^{6} p(j,k) = \frac{[y]}{6}$$

Y \ X	1	2	3	4	5	6	$f_1(x)$
1	1/36	1/36	1/36	1/36	1/36	1/36	1/6
2	1/36	1/36	1/36	1/36	1/36	1/36	1/6
3	1/36	1/36	1/36	1/36	1/36	1/36	1/6
4	1/36	1/36	1/36	1/36	1/36	1/36	1/6
5	1/36	1/36	1/36	1/36	1/36	1/36	1/6
6	1/36	1/36	1/36	1/36	1/36	1/36	1/6
$f_2(y)$	1/6	1/6	1/6	1/6	1/6	1/6	

The probability of having $3 \leq X \leq 5$ and $2 \leq Y \leq 3$ is $\frac{6}{36} = \frac{1}{6}$. The marginal probabilities are $P\{3 \leq X \leq 5\} = \frac{1}{2}$ and $P\{2 \leq Y \leq 3\} = \frac{1}{3}$. Note that the two variables are independent, since, for all entries of the probability matrix, $\frac{1}{36} = \frac{1}{6} \cdot \frac{1}{6}$.

2-18. Binomial Distribution. Consider a random experiment with only two possible outcomes, E_1 and E_2. Let the probability of the occurrence of E_1 and E_2 be p and $q = 1 - p$, respectively. If the experiment is repeated n times and the successive trials are independent of each other, the probability of obtaining E_1 and E_2 r and $n - r$ times, respectively, is

$$\binom{n}{r} p^r q^{n-r} \tag{2-133}$$

This can be proved as follows: The probability of any sequence having r events E_1 and $n - r$ events E_2 is $p^r q^{n-r}$, as the successive trials are assumed to be independent of each other. Moreover, the number of such sequences is equal to the number of combinations of n objects r at a time. Hence the formula of Eq. (2-133) follows by the addition rule of probabilities.

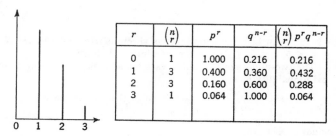

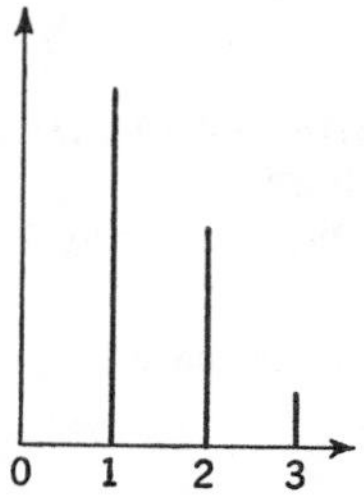

r	$\binom{n}{r}$	p^r	q^{n-r}	$\binom{n}{r} p^r q^{n-r}$
0	1	1.000	0.216	0.216
1	3	0.400	0.360	0.432
2	3	0.160	0.600	0.288
3	1	0.064	1.000	0.064

FIG. 2-24. Example of a binomial probability function.

Let us now define a random variable X which takes the values r if in a sequence of n trials there are exactly r E_1. Then by Eq. (2-133)

$$f(r) = P\{X = r\} = \binom{n}{r} p^r q^{n-r} \tag{2-134}$$

$$F(x) = P\{X \leq x\} = \sum_{r=0}^{[x]} \binom{n}{r} p^r q^{n-r}$$

The distribution function of the random variable X is a step function of the type shown in Fig. 2-23*b*. The corresponding probability density function is shown in Fig. 2-24.

Example 2-31. What is the probability of getting exactly three 1's in five throws of a die? What is the probability of obtaining at most two 1's?

Solution. According to Eq. (2-134), for $p = 1/6$, $q = 5/6$, $n = 5$, and $r = 3$, one writes

$$P\{X = r = 3\} = \binom{5}{3} \left(\frac{1}{6}\right)^3 \left(\frac{5}{6}\right)^2 = \frac{250}{7{,}776}$$

For the second part of the problem,

$$F(x) = P\{X \leq 2\} = \binom{5}{0} \left(\frac{1}{6}\right)^0 \left(\frac{5}{6}\right)^5 + \binom{5}{1} \left(\frac{1}{6}\right)^1 \left(\frac{5}{6}\right)^4 + \binom{5}{2} \left(\frac{1}{6}\right)^2 \left(\frac{5}{6}\right)^3 = 0.96$$

Example 2-32. In a game of n throws of a die, for what value of n is the probability of getting at least two 6's larger than ½?

$$P\{2,3, \ldots, n \text{ 6's}\} > \tfrac{1}{2}$$

Solution

$$\binom{n}{0}\left(\frac{1}{6}\right)^0\left(\frac{5}{6}\right)^n + \binom{n}{1}\left(\frac{1}{6}\right)\left(\frac{5}{6}\right)^{n-1} < \frac{1}{2}$$

The numerical answer to this inequality is found to be

$$n \geq 10$$

2-19. Poisson's Distribution. A random variable X is said to have a Poisson probability distribution if

$$P\{X = x\} = e^{-\lambda} \frac{\lambda^x}{x!} \tag{2-135}$$

where $\lambda > 0$, $x = 0, 1, 2, \ldots$, and $0! = 1$.

The corresponding cumulative distribution function (CDF) is

$$\begin{aligned} F(x) &= \sum_{k=0}^{[x]} e^{-\lambda} \frac{\lambda^k}{k!} \qquad x \geq 0 \\ F(x) &= 0 \qquad x < 0 \end{aligned} \tag{2-136}$$

It is to be noted that $F(x)$ satisfies the conditions required for a distribution function. In fact, $F(x)$ is monotonic, increasing, and, moreover,

$$F(0) = e^{-\lambda}$$

$$F(\infty) = \sum_{k=0}^{\infty} e^{-\lambda} \frac{\lambda^k}{k!} = e^{-\lambda}\left(1 + \frac{\lambda}{1!} + \frac{\lambda^2}{2!} + \cdots\right) = 1$$

It is of interest to note that the Poisson distribution is a certain type of limiting case of the binomial distribution, in which p is a specified function of n, namely p_n, where

$$\lim_{n \to \infty} np_n = \lambda > 0$$

Then

$$\lim_{n \to \infty} \binom{n}{x} p_n{}^x (1 - p_n)^{n-x} = e^{-\lambda} \frac{\lambda^x}{x!} \tag{2-137}$$

The validity of Eq. (2-137) can be checked through the following algebraic manipulations:

$$f(x) = \binom{n}{x} p_n{}^x (1 - p_n)^{n-x} = \frac{n(n-1) \cdots (n-x+1)}{n^x} \frac{\lambda^x}{x!} (1 - p_n)^{n-x} \tag{2-138}$$

$$f(x) = \left(1 - \frac{1}{n}\right)\left(1 - \frac{2}{n}\right) \cdots \left(1 - \frac{x-1}{n}\right)\frac{\lambda^x}{x!}(1 - p_n)^{n-x} \qquad (2\text{-}139)$$

$$f(x) = \frac{(1 - 1/n)(1 - 2/n) \cdots [1 - (x-1)/n]}{(1 - p_n)^x}\frac{\lambda^x}{x!}(1 - p_n)^n \qquad (2\text{-}140)$$

But
$$\lim_{n\to\infty} \frac{(1 - 1/n)(1 - 2/n) \cdots [1 - (x-1)/n]}{(1 - p_n)^x} = 1 \qquad (2\text{-}141)$$

Therefore,

$$\lim_{n\to\infty} (1 - p_n)^n = \lim_{n\to\infty} [(1 - p_n)^{-1/p_n}]^{-\lambda} = e^{-\lambda} \qquad (2\text{-}142)$$

Finally, for the limiting case we find

$$f(x) = e^{-\lambda}\frac{\lambda^x}{x!} \qquad (2\text{-}143)$$

Thus, in the binomial case, if the number of trials n becomes reasonably large and the probability of individual success p is relatively small, so that their product $np = \lambda$ is of moderate magnitude, the probability of the number of successes in n trials approaches the Poisson distribution. The following relative magnitudes illustrate a common range of application for Poisson's distribution:

$$n > 50 \qquad p < 0.1 \qquad \lambda < 10$$

In Chap. 6 it will be shown that λ is the "average" value for a random variable with a Poisson distribution.

Example 2-33. Assuming that, on an average, 3 per cent of the output of a factory making certain parts is defective and that 300 units are in a package, what is the probability that, at most, five defective parts may be found in a package?

Solution. The "average" number of defective parts in a package is $300 \times 0.03 = 9$. Assume a Poisson distribution with this average, i.e.,

$$\lambda = np = 9$$

According to Eq. (2-143), the probability of a box containing x defective parts is

$$\frac{9^x e^{-9}}{x!}$$

$$F(x) = P(X \leq x) = \sum_{k=0}^{[x]} \frac{e^{-9}9^k}{k!}$$

$$F(5) = e^{-9}\left(1 + \frac{9}{1!} + \frac{9^2}{2!} + \frac{9^3}{3!} + \frac{9^4}{4!} + \frac{9^5}{5!}\right)$$

Example 2-34. An industrial process has been running in control with 0.5 per cent defectives. Find the smallest integer k such that the probability of getting k or more defectives in a random sample of 100 is less than 0.10.

Solution. Assuming a Poisson distribution with $p = 0.005$ and $n = 100$, one finds $\lambda = np = 0.5$. Thus it is reasonable to use a Poisson distribution. In this case,

$$P(X \geq k) \leq 0.10$$
$$P(X < k - 1) \geq 0.90$$
$$\sum_{k=1}^{k} \frac{e^{-\lambda}\lambda^{k-1}}{(k-1)!} \geq 0.90$$
$$\sum_{k=1}^{k} \frac{e^{-0.5}0.5^{k-1}}{(k-1)!} \geq 0.90$$

From a Poisson distribution table one finds that

$$k - 1 = 1 \qquad k = 2$$

2-20. Expected Value of a Random Variable. Consider a discrete single-variate random variable X and its associated probability function:

$$[x_1, x_2, \ldots, x_n]$$
$$[p_1, p_2, \ldots, p_n]$$

If the random experiment under consideration is repeated a large number of times, the average or mean value of the numerical function X is found to be

$$\text{Average of } X = \bar{X} = \sum_{k=1}^{n} p_k x_k \tag{2-144}$$

For example, for the experiment of rolling an honest die, one finds

$$X = 1 \cdot \tfrac{1}{6} + 2 \cdot \tfrac{1}{6} + 3 \cdot \tfrac{1}{6} + 4 \cdot \tfrac{1}{6} + 5 \cdot \tfrac{1}{6} + 6 \cdot \tfrac{1}{6} = 3\tfrac{1}{2}$$

More generally, if $\psi(X)$ is a function of a random variable X (also called a *weighting function*), the mean value of $\psi(X)$ is defined as

$$\text{Mean of } \psi(X) = \overline{\psi(X)} = \sum_{k=1}^{n} p_k \psi(x_k) \tag{2-145}$$

In the literature of probability, the mean of a function is generally referred to as its *expected value*. An alternative notation for denoting the mean value of a random quantity is a capital E in front of that quantity, for instance, $E(X)$ or $E(X + Y)$ or $E(2X + X^3)$. When the function $\psi(X)$ is of the form $\psi(X) = X^j$, where j is a positive integer, its expected value is called the moment of the jth order of X. For example,

$$E(X) = \bar{X} = \text{first-order moment of } X = \sum_{k=1}^{n} p_k x_k$$

$$E(X^2) = \overline{X^2} = \text{second-order moment of } X = \sum_{k=1}^{n} p_k x_k^2 \qquad (2\text{-}146)$$

$$E(X^3) = \overline{X^3} = \text{third-order moment of } X = \sum_{k=1}^{n} p_k x_k^3$$

. . . .

The physical significance of moments is not discussed here. At present the reader is required only to acquaint himself with the concept of Eq. (2-145), that is, how the means of different weighting functions can be calculated. The concept of averaging is of considerable importance in engineering problems. For example, assume that X is a random voltage applied as the input to a device with an input-output relationship

$$Y = \psi(X)$$

Then $E(Y)$ is the d-c level for the output of the system. Similarly, if Y is applied across a unit resistor, the power consumed in the resistor, measured with respect to its d-c level, will have the same numerical value as the second moment of the random variable $(Y - \bar{Y})$, that is, the expectation of

$$\{\psi(X) - E[\psi(X)]\}^2 \qquad (2\text{-}147)$$

There are at least three special weighting functions of particular interest in probability and information theory. These are

X^j	$j = 1, 2, 3, \ldots$
e^X	e = base of natural logarithm
$\log X$	

Without discussing the details at this time, we merely point out the most important application feature of each of the above functions:

$E(X^j)$	This gives moments of different orders of X.
$E(e^X)$	When this mean is known, one can find the values of different moments without recourse to direct computation.
$E(-\log X)$	In the following chapter it will be shown that, when X is taken to be the probability function $f(x)$, the new random variable $[-\log f(x)]$ presents the amount of uncertainty associated with the occurrence of each outcome of the discrete experiment. Therefore, its mean value will stand for the average uncertainty of the system under consideration.

The concept of averaging can be generalized in a direct manner to weighting functions of n random variables associated with an n variate. For example, in the case of a bivariate random variable $[X,Y]$ and a weighting function $\psi(X,Y)$, we have

$$E[\psi(X,Y)] = \sum_j \sum_k \psi(x_k,y_j)p_{kj} \qquad (2\text{-}148)$$

PROBLEMS

2-1. Determine whether or not the following relations are correct (the primes denote the complements):

(a) $(A + B)(A + C) = A + BC$
(b) $(A + B) - B = AB'$
(c) $A'B = A + B$
(d) $(A - AB)C' = A(B + C)'$
(e) $(A + B)'C = A'B'C$
(f) $(A + B)(B + C)(C + A) = AB + AC + BC$
(g) $(A \cap B) \cap (B' \cap C) = \emptyset$

2-2. Let A, B, C be three arbitrary events of a sample space. Find the expressions for the following cases:

(a) At least one of the three events occurs.
(b) B occurs and either A or C occurs, but not both.
(c) Not more than two occur simultaneously.

2-3. Consider the set of points $S = \{(x,y)\}$ shown in Fig. P2-3.

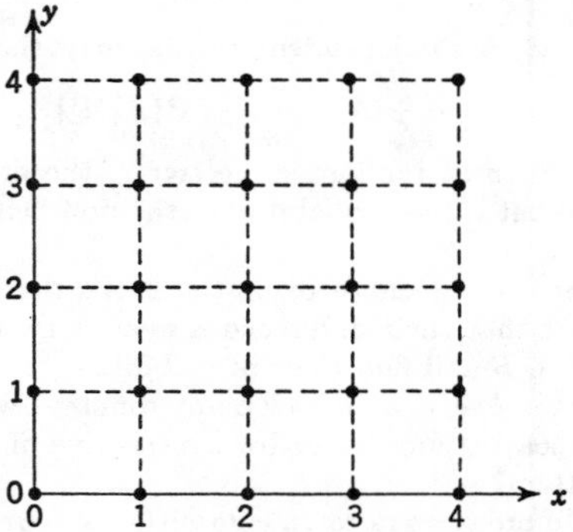

FIG. P2-3

(a) Find the subset $a = \{(x,y)|x^2 + y^2 \leq 4\}$.
(b) Describe the subset $b = \{(x,y)|y \leq x^2\}$.
(c) Describe the subset $c = \{(x,y)|x \leq y^2\}$.
(d) Describe the subset $b \cap c$.
(e) Describe the subset $(b \cup a)c'$.

2-4. Given a set $S = \{0,1,2,3,4,5,6,7,8,9,10\}$,

(a) Define the function $F_1(x) = x/2$ over S and draw its graph.
(b) Define the function $F_2(x) = x + 3$ over S and draw its graph.
(c) Determine the subset $a = \{x|(x/2)(x + 3) \leq 4\}$.

2-5. Show the following identities and draw the corresponding Sheffer-stroke diagrams.

(*a*) $(X|(Y|Y)) = X' \cup Y$.

(*b*) $(X|(X|X)) = U$.

(*c*) Verify the identity of the expression for the output as given in Fig. P2-5*a* and *b*.

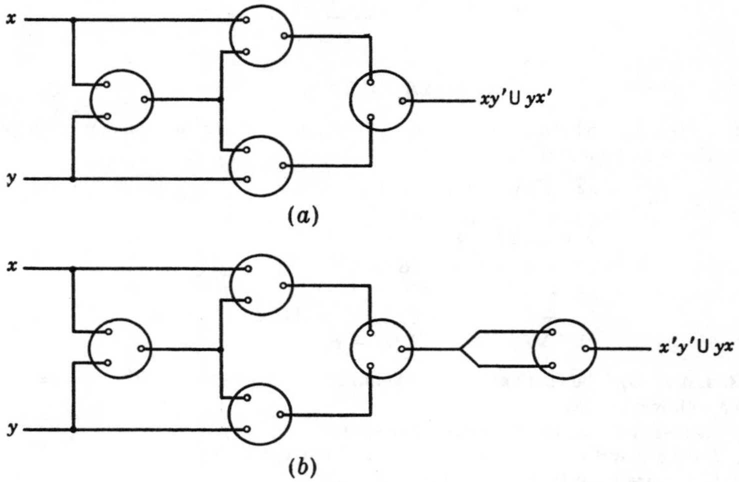

FIG. P2-5

2-6. If $A_1, A_2, \ldots, A_n$ are independent events, show that

$$P\{A_1 + A_2 + \cdots + A_n\} = 1 - P\{A'_1\}P\{A'_2\} \cdots P\{A'_n\}$$

2-7. Two cards are drawn at random successively, the first being replaced before the second is drawn. What is the probability of the first being a club and the second not a queen?

2-8. Two dice are thrown. Denote by A the event that the sum of the faces is even and by B the event that their difference is even. Describe the events $A + B$, AB, $A'B'$, AB', and $A' + B$ and find their probabilities.

2-9. Given five letters a, b, c, d, e, in how many different ways can one write three-letter words without repeating any letter (*a*) irrespective of their order and (*b*) considering the order of letters?

2-10. In how many different ways can a committee of four men and two women be selected from a total of 20 men and 10 women?

2-11. A survey of 1,000 people has indicated the following results: 714 listen to radio station A, 640 to station B, and 850 to station C. It also indicated that 530 listen to both A and B, 375 to both C and B, and 720 to A and C. Determine whether these data are not self-contradictory.

2-12. What is the probability of obtaining 8, 9, or 10 with two dice in one trial?

2-13. Two dice are thrown. Let A be the event that the sum of the faces is odd and B the event that at least one is a 1. Describe the events AB, $A + B$, AB' and find their probabilities.

2-14. What is the probability of drawing a club *or* a face card of any color in a single draw from an ordinary deck of cards?

2-15. Two events A and B associated with an experiment have respective probabilities of occurrence p and q. Show that in n trials the probability that AB occurs K_1 times; AB', K_2 times; $A'B$, K_3 times; and $A'B'$, K_4 times is

$$\frac{n!}{K_1!K_2!K_3!K_4!}p^{K_1+K_2}q^{K_1+K_3}(1-p)^{K_3+K_4}(1-q)^{K_2+K_4}$$

2-16. Urn A contains seven silver dollars and one \$10 gold coin. Urn B contains 10 silver dollars. Nine coins are taken from B and put in A; then eight coins are selected at random from the 17 coins in A and put back in urn B. If you were to select one of the two urns, which one should you select?

2-17. If the probability of a safe return from a certain trip is $P = 0.9$, what is the probability of exactly four safe returns out of six such trips?

2-18. A single card is removed from a regular deck of cards. From the remainder, we draw two cards and observe that they are both diamonds. What is the probability that the removed card was also a diamond?

2-19. Show that the two relay circuits of Fig. P2-19 are equivalent.

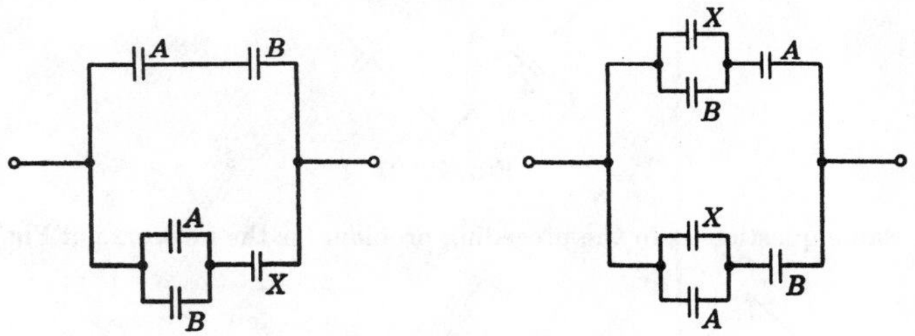

FIG. P2-19

2-20. Express the event of the functioning of the network in Fig. P2-20 in terms of the subevents $E_1, E_2, \ldots, E_6$, where E_k implies the functioning of the kth relay.

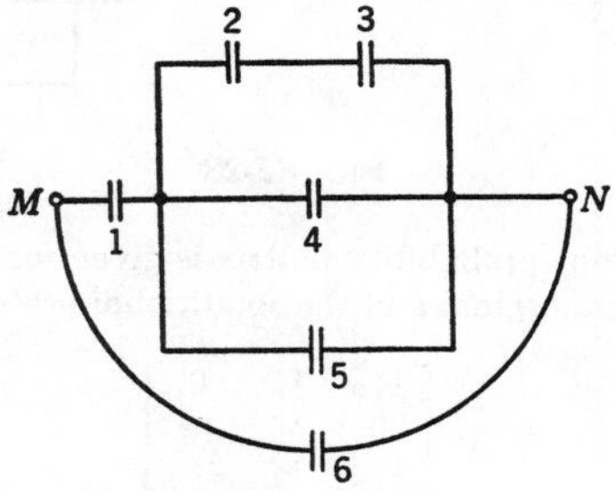

FIG. P2-20

2-21. Two persons toss a coin n times each. What is the probability that they score the same number of heads?

2-22. If a box contains 40 good and 10 defective objects, what is the probability that 10 objects selected at random from the box are all good?

2-23. What is the probability that in a bridge hand a player and his partner have a total of three aces?

2-24. Assuming that the ratio of male to female children is 1:2, find the probability that in a family of six children

(*a*) All children will be of the same sex.

(*b*) The four oldest children will be boys and the two youngest will be girls.

(*c*) Exactly half the children will be boys.

2-25. In a game of bridge, if a player has no ace, what is the probability that his partner has no ace either?

2-26. Find the probability that three, and only three, tails are obtained in a sequence of four tosses of a coin.

2-27. Assuming that the probability of each relay being closed is p, derive the probability for the flow of a current between nodes A and B of Fig. P2-27.

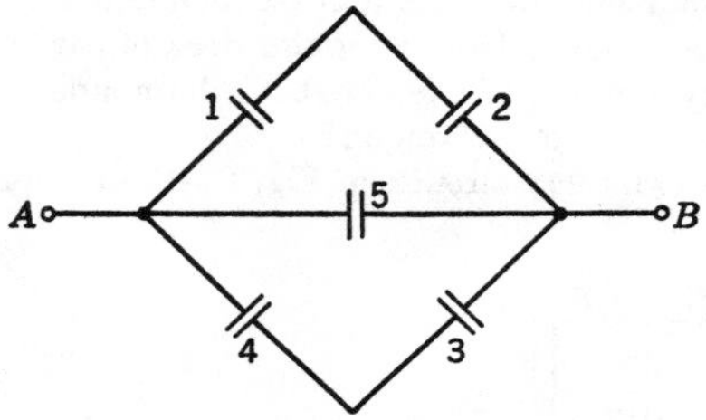

FIG. P2-27

2-28. Same question as in the preceding problem for the networks of Fig. P2-28.

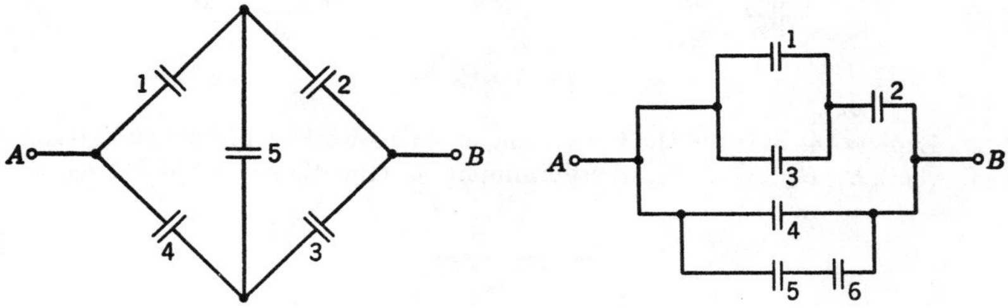

FIG. P2-28

2-29. The following joint probability matrix is given for discrete random variables X and Y. Evaluate the marginal and the conditional probability functions.

$$\begin{bmatrix} 1/12 & 1/6 & 0 \\ 0 & 1/9 & 1/5 \\ 1/18 & 1/4 & 2/15 \end{bmatrix}$$

2-30. The joint density function for two random variables X and Y is given below:

$$f(x,y) = k^2xy2^{-(x+y)} \qquad \text{for } x \text{ and } y \text{ positive integers}$$
$$f(x,y) = 0 \qquad \text{elsewhere}$$

Find $$P\{X < 4, Y < 4\}$$

2-31. Evaluate the probability of getting a four 0, 1, 2, 3, 4, and 5 consecutive times on five throws of a die.

2-32. If the probability of hitting a target is $\frac{1}{4}$ in each shot, independent of the number of shots fired,

(*a*) What is the probability of the target being hit twice in five shots?

(*b*) What is the probability of the target being hit at least twice in five shots?

2-33. A book of 200 pages contains 100 misprints. Assuming that these are distributed at random, estimate the chances that a page contains at least two misprints.

2-34. The random variable X assumes the values $[0,1,2]$ with respective probabilities $[\frac{1}{3},\frac{1}{4},\frac{5}{12}]$. The random variable Y assumes the values $[0,1]$ with probabilities $[\frac{1}{4},\frac{3}{4}]$. Assuming that the two variables are independent, determine their joint probability functions.

2-35. Study the different probability functions (joint, marginal, and conditional) associated with the following experiment. We draw five cards from an ordinary deck of cards and study the two random variables below:

X, number of aces drawn
Y, number of queens drawn

2-36. A random event E has the probability of occurrence $1/K$ in each experiment independently of the preceding outcome. Determine the following probabilities:

(*a*) E does not occur in n consecutive trials.

(*b*) E occurs in the nth experiment only but not in any of the previous ones.

(*c*) E occurs exactly twice in n experiments.

(*d*) Let $K = 4$, $n = 4$ and evaluate the results of parts (*a*), (*b*), and (*c*).

2-37. *Smith-Jones-Robinson Problem.* The following problem has appeared in the *Scientific American* (vol. 200, no. 2, p. 136, February, 1959) in an entertaining article entitled "Brain-teasers" That Involve Formal Logic.

	Engineer	Brakeman	Fireman
Smith			
Jones			
Robinson			

	Los Angeles	Omaha	Chicago
Mr. Smith			
Mr. Jones			
Mr. Robinson			

FIG. P2-37

1. Smith, Jones, and Robinson are the engineer, brakeman, and fireman on a train, but not necessarily in that order. Riding the train are three passengers with the same three surnames, to be identified in the following premises by "Mr." before their names.
2. Mr. Robinson lives in Los Angeles.
3. The brakeman lives in Omaha.
4. Mr. Jones long ago forgot all the algebra he learned in high school.
5. The passenger whose name is the same as the brakeman's lives in Chicago.
6. The brakeman and one of the passengers, a distinguished mathematical physicist, attend the same church.
7. Smith beat the fireman at billiards.

Who is the engineer?

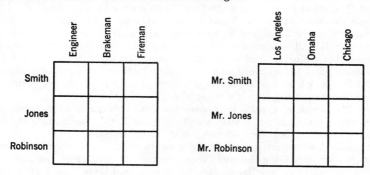

HINT: The solution by methods of set theory may become somewhat cumbersome. It is suggested in the above reference to use two matrices as notational aid. Each cell is the intersection of two sets, corresponding to the set of elements contained in the pertinent column and row. Put a 1 or a 0 in a cell indicating that such an intersection is a valid premise or not.

2-38. *Eddington's Controversy.* The following problem exemplifies the type of confusion that existed in probability prior to the introduction of set-theory considerations.

If A, B, C, D each speak the truth once in three times (independently), and A affirms that B denies that C declares that D is a liar, what is the probability that D was speaking the truth?

The following comments on Eddington's problem are given in an article entitled "Brain-Teasers" That Involve Formal Logic by M. Gardner (*op. cit.*).

"Eddington's answer of $25/71$ was greeted by howls of protest from his readers, touching off an amusing controversy that was never decisively resolved. The English astronomer Herbert Dingle, reviewing Eddington's book in *Nature* (Mar. 23, 1935), dismissed the problem as meaningless and symptomatic of Eddington's confused thinking about probability. Theodore Sterne, an American physicist, replied (*Nature*, June 29, 1935) that the problem was not meaningless but lacked sufficient data for a solution. Dingle responded (*Nature*, Sept. 14, 1935) by contending that, if one granted Sterne's approach, there were enough data to reach a solution of exactly $1/3$. Eddington then reentered the fray with a paper entitled The Problem of A, B, C and D (*Math. Gaz.*, October, 1935), in which he explained in detail how he had calculated his answer."

The difficulty lies chiefly in deciding exactly how to interpret Eddington's statement of the problem. If B is truthful in making his denial, are we justified in assuming that C said that D spoke the truth? Eddington thought not. Similarly, if A is lying, can we then be sure that B and C said anything at all? Fortunately we can side-step all these verbal difficulties by making (as Eddington did not) the following assumptions: (1) All four men made statements. (2) A, B, and C each made a statement that either affirmed or denied the statement that follows. (3) A lying affirmation is taken to be a denial, and a lying denial is taken to be an affirmation.

2-39.* If a stick is broken at random into three pieces, what is the probability that the pieces can be put together in a triangle?

HINT: The problem, despite its apparently clear statement, is ambiguous. It requires some additional information about the exact method of breaking the stick. The following two explanations are given in the cited reference.

"One method is to select, independently and at random, two points from the points that range uniformly along the stick, then break the stick at these two points. If this is the procedure to be followed, the answer is $1/4$, and there is an elegant way of demonstrating it with a geometrical diagram. . . .

"Suppose, however, that we interpret in a different way the statement 'break a stick at random into three pieces.' We break the stick at random, we select randomly one of the two pieces, and we break that piece at random. What are the chances that the three pieces will form a triangle? If after the first break we choose the smaller piece, no triangle is possible."

The latter interpretation of the problem gives $1/6$ for the required probability.

* This problem and its solution have appeared in the article Problems Involving Questions of Probability and Ambiguity by M. Gardner (*Sci. American*, October, 1959, pp. 174–176). Gardner's article contains several other examples of ambiguity which have puzzled even some well-known mathematicians.

2-40. The joint probability matrix of two variables is given below. Determine whether they are statistically independent.

$$\begin{matrix} 3 \\ 2 \\ 1 \\ y/x \end{matrix} \begin{matrix} \begin{bmatrix} 1/72 & 1/36 & 5/72 \\ 1/36 & 1/18 & 5/36 \\ 1/12 & 1/6 & 5/12 \end{bmatrix} \\ \begin{matrix} 1 & \quad 2 & \quad 3 \end{matrix} \end{matrix}$$

2-41. Two urns contain four white and three black balls and three white and seven black balls, respectively. One urn is selected at random and a ball is drawn from it. What is the probability that this ball is white?

2-42. A Markov chain has the transition probability matrix given below:

$$\begin{bmatrix} 0 & 2/3 & 1/3 \\ 0 & 7/12 & 5/12 \\ 1 & 0 & 0 \end{bmatrix}$$

The three states are initially selected with probabilities $1/2$, $1/6$, $1/3$.

(*a*) What is the probability of reaching state 2 via state 1 in one step?
(*b*) What is the probability of reaching state 2 via 1 in two steps?
(*c*) What is the probability of reaching state 3 in two steps?

2-43. Define the probability function for the number of boys in a family of six children, assuming that both sexes are equiprobable and no multiple birth occurs.

2-44. From the joint probability matrix below,

$$\begin{matrix} y_3 \\ y_2 \\ y_1 \\ \end{matrix} \begin{matrix} \begin{bmatrix} 0 & 5/36 & 1/3 \\ 1/12 & 1/9 & 1/18 \\ 1/36 & 1/4 & 0 \end{bmatrix} \\ \begin{matrix} x_1 & \quad x_2 & \quad x_3 \end{matrix} \end{matrix}$$

compute and tabulate:

(*a*) Marginal probability $P_1\{x_k\}$.
(*b*) Marginal probability $P_2\{y_j\}$.
(*c*) $P\{y_j|x_k\}$.
(*d*) $P\{x_k|y_j\}$.

CHAPTER 3

BASIC CONCEPTS OF INFORMATION THEORY: MEMORYLESS FINITE SCHEMES

The object of this chapter is to present the basic elements of information theory of discrete schemes in a manner parallel to the presentation of the elements of discrete probability theory. Our immediate aim is to develop a *measure* for *information content* of a discrete system. That measure will then be used for evaluating the rate of *transmission of information* in a *communication* system. No effort will be made to expound on the philosophical context of terms such as "information measure" or "communication." In order to grasp a basic understanding of this newly developed scientific field, it seems desirable to confine ourselves to an accurate abstract mathematical model rather than to deal with generalities of a semiphilosophical nature. The following approach is suggested:

We shall consider a discrete random experiment and its associated sample space Ω. Let X be a random variable (a real numerical function) associated with Ω; we know that, say, $E(X)$ has a particular physical meaning in regard to the random experiment. That is, if the experiment is repeated a large number of times, the values of X when averaged will approach $E(X)$. In summary, $E(X)$ has given a certain "physical" indication about the experiment. Similarly, $E(X^n)$ has a certain significance in our studies. Then the question arises, could we search for an indicative number associated with the random experiment such that it provides a "measure" of surprise or unexpectedness of occurrence of outcomes of the experiment? Shannon has suggested that the random variable $-\log P\{E_k\}$ is an indicative relative measure of the occurrence of the event E_k. In particular, he shows that the mean of this function is a good indication of the average uncertainty with respect to all the outcomes of the experiment.

The reader should note that the above terms in quotation marks are used here with their common meaning. Their more accurate technical meaning will be defined later.

3-1. A Measure of Uncertainty. Consider the sample space Ω of events pertaining to a random experiment. We partition the sample space in a finite number of mutually exclusive events E_k, whose probabilities p_k are assumed to be known (Fig. 3-1). The set of all events under consideration can be designated as a row matrix $[E]$ and the corresponding probabilities as another row matrix $[P]$.

$$[E] = [E_1, E_2, \ldots, E_n] \quad \text{with} \quad \bigcup_{k=1}^{u} E_k = U \qquad (3\text{-}1)$$

$$[P] = [p_1, p_2, \ldots, p_n] \quad \text{with} \quad \sum_{k=1}^{n} p_k = 1 \qquad (3\text{-}2)$$

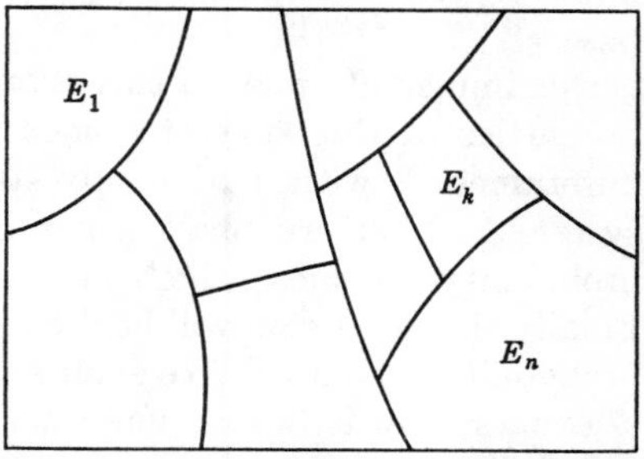

FIG. 3-1. A discrete probability space.

Equations (3-1) and (3-2) contain all the information that we have about the probability space which is called a *complete finite scheme.* For example, the following matrix represents such a situation:

$$\begin{bmatrix} E \\ P \end{bmatrix} = \begin{bmatrix} E_1 & E_2 & E_3 \\ 0.2 & 0.5 & 0.3 \end{bmatrix}$$

The fundamental problem of interest is to associate a measure of surprise or uncertainty, $H(p_1, p_2, \ldots, p_n)$, with such probability schemes. Of course at this point it is questionable what is meant by a measure of uncertainty. The clarification of this concept has to come gradually; it is, in essence, the central theme of information theory. The problem can be approached in either of two, not necessarily exclusive, ways:

1. First postulate the desired properties of such an uncertainty measure; then derive the functional form of $H(p_1, p_2, \ldots, p_n)$. The postulation of the desired properties can be based on some intuitive approach, such as physical motivation or "usefulness" for some purpose, but after such a postulate is adopted, mathematical discipline must prevail and no further intuitive approach may be employed.

2. Assume a known functional $H(p_1, p_2, \ldots, p_n)$ associated with a finite probability scheme and justify its "usefulness" for the physical problems under consideration.

Our present approach is primarily of type 2. The more mathematically inclined readers who prefer an axiomatic approach are referred to Sec. 3-19 or Feinstein (I).

Shannon and Wiener have suggested the following *measure of uncertainty* or *entropy* associated with the sample space of a complete finite scheme.

$$H(X) = -\sum_{i=1}^{n} p_i \log p_i \qquad (3\text{-}3)^*$$

* All logarithms are to base 2 unless otherwise specified.

where p_i is the probability of the occurrence of the event E_i as described in Eqs. (3-1) and (3-2). The base of the logarithm is rather arbitrary; however, for communication problems it is convenient to use the binary base.

Our immediate plan in this chapter is first to investigate the principal properties of this suggested measure of uncertainty and to justify its "usefulness" with respect to statistical problems of communication systems. Next we shall generalize this concept to two-dimensional probability schemes, which provide simple models for communications. Finally the discussion will be directed toward more general n-dimensional probability schemes. We shall always be restricted to *complete* systems of events; that is, we assume that Eqs. (3-1) and (3-2) are satisfied.

3-2. An Intuitive Justification. In this section we wish to justify the usefulness of the function suggested in Eq. (3-3) in connection with communication problems. In problems dealing with communication systems, it is often instructive to regard a finite exhaustive probability scheme as a mathematical model for a communication source. In this analogy, any elementary event or outcome, E_k, may be considered as a *letter of the alphabet* of the communication transmitter.

Now consider the random variable

$$X = -\log p \tag{3-4}$$

defined over the sample space of Fig. 3-1. To each event E_k there corresponds a value x_k of the random variable X, where by hypothesis

$$x_k = -\log P\{E_k\} = -\log p_k \tag{3-5}$$

The quantity $-\log p_k$ is frequently called the amount of *self-information* associated with the event E_k:

$$I(E_k) = -\log p_k \tag{3-6}$$

The unit of the amount of information is called a *bit*, where one bit is the amount of information associated with the selection of one of two equiprobable ($p_k = \frac{1}{2}$) events. In other words, if the sample space is partitioned into two equally likely events E_1 and E_2, then

$$I(E_1) = I(E_2) = -\log \tfrac{1}{2} = 1 \text{ bit} \tag{3-7}$$

A selection between two equally likely events requires one unit of information. If Ω were partitioned into 2^N equally probable events E_k ($k = 1, 2, \ldots, 2^N$), then the self-information associated with any event E_k would be

$$I(E_k) = -\log p_k = -\log 2^{-N} = N \qquad \text{bits} \tag{3-8}$$

The generalization from equiprobable events to the general case is straightforward. In fact, in order to evaluate the self-information associated with a particular event E_0, we divide the Ω space in two parts E_0 and E'_0; thus

$$I(E_0) = -\log p(E_0) = -\log p_0 \qquad \text{bits} \tag{3-9}$$

For instance, if $p_0 = 1/16$, the occurrence of E_0 in the average conveys to us 4 bits of information. The measure of self-information is essentially nonnegative:

$$I(E_k) = -\log p_k \geq 0 \tag{3-10}$$

The equality is only by the *certain* event; obviously, no information is conveyed by the knowledge of the occurrence of such an event.

The *average amount of information* or entropy of a finite complete probability scheme is defined by

$$H(X) = \overline{I(E_k)} = -\sum_{k=1}^{n} p_k \log p_k \tag{3-11}$$

where the random variable X is defined over the sample space of events Ω and the events satisfy Eqs. (3-1) and (3-2). $H(X)$ is the average amount of self-information per event, the average being taken over the entire sample space. In fact, if $-\log p_k$ indicates the measure of uncertainty associated with the event E_k, then $H(X)$ will clearly represent the mean or the expected value of the uncertainty associated with our probability scheme. As a simple example, let us consider the following three sets of complete events and compare their entropies.

(I) $E = [A_1, A_2] \qquad P = [1/256, 255/256]$
(II) $E = [B_1, B_2] \qquad P = [1/2, 1/2]$
(III) $E = [C_1, C_2] \qquad P = [7/16, 9/16]$

The average self-information associated with each of these schemes is given respectively by

(I) $\bar{I}_1 = -(1/256 \log 1/256 + 255/256 \log 255/256) = 0.0369$ bit
(II) $\bar{I}_2 = -(1/2 \log 1/2 + 1/2 \log 1/2) = 1$ bit
(III) $\bar{I}_3 = -(7/16 \log 7/16 + 9/16 \log 9/16) = 0.989$ bit

In system I it is relatively easy to guess whether A_1 or A_2 will occur. In system III this guess is much harder, and in II it is most difficult to predict the occurrence of one of the events B_1 or B_2. It is common sense to attribute a larger average uncertainty to system II than to system III and a larger average uncertainty to system III than to system I. This is in agreement with the results obtained by application of the chosen self-information function, that is,

$$\bar{I}_1 < \bar{I}_3 < \bar{I}_2$$

The average uncertainty associated with II is far more than that associated with I. For I, we are almost sure that A_2 generally occurs. For II, the average uncertainty is larger, as it is most difficult to say whether B_1 or B_2 occurs.

3-3. Formal Requirements for the Average Uncertainty. Shannon's approach, as well as several other authors', in suggesting a suitable H function has been to some extent directed toward an axiomatic description of such functions. The desired H function should have the following basic properties:

1. *Continuity.* That is, if the probabilities of the occurrence of events E_k are slightly changed, the measure of uncertainty associated with the system should vary accordingly in a continuous manner.

$$H(p_1,p_2, \ldots ,p_n) \text{ continuous in } p_k \qquad k = 1, 2, \ldots , n \qquad 0 \le p_k \le 1 \tag{3-12}$$

This requirement is obviously in conformity with our physical senses, since a slight change in the probability of the occurrence of an event should not provide us with a significantly large amount of information.

2. *Symmetry.* The H function must be functionally symmetric in every p_k. Indeed, the measure of uncertainty associated with a complete probability set $[E_k,E_k']$ must be exactly the same as the measure associated with the set $[E_k',E_k]$. Our measure must be invariant with respect to the order of these events.

$$H(p_1,p_2, \ldots ,p_n) = H(p_2,p_1, \ldots ,p_n) \tag{3-13}$$

3. *Extremal Property.* When all the events are equally likely, the average uncertainty must have its largest value. In this case, it is most uncertain which event is going to occur. Conversely, once we know which specific event among a number of n equally likely events has occurred, we have acquired the largest average amount of information relevant to the occurrence of events of a universe consisting of n complete events.

$$\text{Maximum of } H(p_1,p_2, \ldots ,p_n) = H\left(\frac{1}{n}, \frac{1}{n}, \ldots , \frac{1}{n}\right) \tag{3-14}$$

4. *Additivity.* Suppose that we have obtained a suitable measure of the average uncertainty $H(p_1,p_2, \ldots ,p_n)$ associated with a complete

set of events. Now, let us assume that the event E_n is divided into disjoint subsets (Fig. 3-2) such that

$$E_n = \bigcup_{k=1}^{m} F_k \tag{3-15}$$

$$p_n = \sum_{k=1}^{m} q_k \qquad P\{F_k\} = q_k \tag{3-16}$$

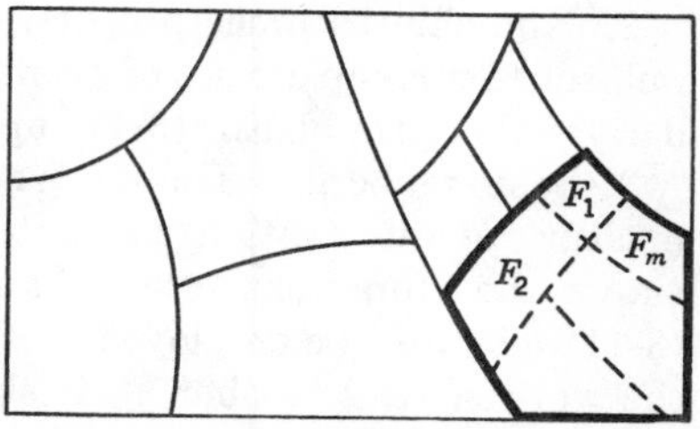

FIG. 3-2. A partitioning of the probability space illustrating the additive property of the information measure.

Evidently, the occurrence of the event E_n can be considered as another total sample space where the probabilities associated with events F_k can be normalized in the form

$$\frac{q_1}{p_n} + \frac{q_2}{p_n} + \cdots + \frac{q_m}{p_n} = 1 \tag{3-17}$$

[This recourse provides a rather convenient *relative* frame of reference. That is, we call the event E_n a sample space Ω_r, associated with the experiments of obtaining all events F_k ($k = 1, 2, \ldots, m$), when we know that E_n is bound to occur.] Therefore we have three probability spaces and hence the following three H functions:

$$\begin{aligned} &H_1(p_1,p_2, \ldots ,p_n) \\ &H_2(p_1,p_2, \ldots ,p_{n-1},q_1,q_2, \ldots ,q_m) \\ &H_3\left(\frac{q_1}{p_n}, \frac{q_2}{p_n}, \ldots , \frac{q_m}{p_n}\right) \end{aligned} \tag{3-18}$$

A suitable additive or linear measure which also satisfies our common sense is given by

$$H_2 = H_1 + p_n H_3 \tag{3-19}$$

The occurrence of the weighting factor p_n in this linear form is rather anticipated. However, the uninitiated reader will find the examples of the following section helpful in illustrating this point.

Complying with properties 1 to 4 given above, or with similar requirements, one should be able to derive a functional form for the desired uncertainty function. Such treatments have appeared in the work of Feinstein, Khinchin, Shannon, Schutzenberger, and others. Their findings are not too complicated, but for a detailed presentation much more space is required than is available in the present work. The following references to the literature are recommended for further reading.

1. Fadiev assumes properties 1, 2, and 4 and, subsequent to several lemmas, proves that H must be of the form suggested in Eq. (3-11) except for a multiplicative constant. (See Feinstein [I].)

2. Khinchin assumes properties 1, 3, and 4 and the fact that adding a null set to a complete set of events should not change its entropy, and he derives the form of Eq. (3-11) up to a positive constant multiplier.

3. Schutzenberger [I] aims for a more general axiomatic search for a measure of information associated with a complete set of events. He shows that functions other than the Shannon-Wiener entropy of Eq. (3-11) may also be employed. An example of such a function is given in the work of R. A. Fisher.* It should be pointed out, however, that the Shannon-Wiener suggested form is certainly the simplest of all such forms. The present richness and depth of the literature of information theory are to a great extent due to the simplicity of the form of Eq. (3-11).

3-4. ***H* Function as a Measure of Uncertainty.** In this section we shall present a treatment concerning the suggested measure of uncertainty. We have discussed that such a measure should obey the following requirements:

$$H(p_1,p_2, \ldots ,p_n) \text{ continuous in } p_k \text{ for all } 0 \leq p_k \leq 1 \tag{3-20}$$

$$H(p_k, 1 - p_k) = H(1 - p_k, p_k) \qquad k = 1, 2, \ldots , n \tag{3-21}$$

$$\text{maximum of } H(p_1,p_2, \ldots ,p_n) = H\left(\frac{1}{n}, \frac{1}{n}, \ldots , \frac{1}{n}\right) \tag{3-22}$$

$$H(p_1,p_2, \ldots ,p_{n-1},q_1,q_2, \ldots ,q_m) = H(p_1,p_2, \ldots ,p_{n-1},p_n) + p_nH\left(\frac{q_1}{p_n}, \frac{q_2}{p_n}, \ldots , \frac{q_m}{p_n}\right) \tag{3-23}$$

where

$$p_n = \sum_{k=1}^{m} q_k$$

* According to R. A. Fisher (*Proc. Cambridge Phil. Soc.*, vol. 22, pp. 700–725, 1925), the quantity of "information" in a sample from a distribution with density $f(x)$ and mean m is defined as

$$I = \int_{-\infty}^{\infty} \left[\frac{\partial \ln f(x)}{\partial m}\right]^2 f(x)\, dx$$

For example, for a normal distribution

$$f(x) = \frac{1}{\sqrt{2\pi}\,\sigma} \exp\left[-\left(\frac{x - m}{\sqrt{2}\,\sigma}\right)^2\right]$$

$$\log f(x) = -\frac{1}{2}\ln 2\pi - \ln \sigma - \left(\frac{x - m}{\sqrt{2}\,\sigma}\right)^2$$

$$\frac{\partial \ln f}{\partial m} = \frac{x - m}{\sigma^2}$$

$$I = \int_{-\infty}^{\infty} \left(\frac{x - m}{\sigma^2}\right)^2 \frac{1}{\sqrt{2\pi}\,\sigma} \exp\left[-\left(\frac{x - m}{\sqrt{2}\,\sigma}\right)^2\right] dx$$

$$I = \frac{1}{\sigma^2} \qquad \text{Fisher's "information" per observation}$$

In the following, we demonstrate that the function defined in Eq. (3-11) satisfies all these requirements.

Property 1: *Continuity.* The entropy function $H(p_1,p_2, \ldots ,p_n)$ is continuous in each and every independent variable p_k in the interval $]0, 1]$. The proof follows directly.

$$
\begin{aligned}
-H(p_1,p_2, \ldots ,p_n) &= p_1 \log p_1 + p_2 \log p_2 + \cdots + p_n \log p_n \\
&= p_1 \log p_1 + p_2 \log p_2 + \cdots + p_{n-1} \log p_{n-1} + (1 - p_1 - p_2 - \cdots \\
&\qquad - p_{n-1}) \log (1 - p_1 - p_2 - \cdots - p_{n-1})
\end{aligned} \tag{3-24}
$$

Note that all independent variables $p_1, p_2, \ldots , p_{n-1}$ and also $(1 - p_1 - p_2 - \cdots - p_{n-1})$ are continuous in $]0, 1]$ and that the logarithm of a continuous function is continuous itself.

Property 2: *Symmetry.* The entropy function is, obviously, a symmetrical function in all variables.

Property 3: *Extremal Value of the Entropy Function.* We should like to show that the entropy function has a maximum when all the individual probabilities are equal.

$$p_1 = p_2 = \cdots = p_n \tag{3-25}$$

This is in conformity with our intuitive feelings; i.e., in a system where all different states are equiprobable, our average uncertainty will be greatest (in other words, it is most difficult to predict which state is most likely to occur).

We may arbitrarily select p_n as a dependent variable depending on p_k $(k = 1, 2, \ldots , n - 1)$. In fact,

$$\frac{dH}{dp_k} = \sum_{i=1}^{n} \frac{\partial H}{\partial p_i}\frac{\partial p_i}{\partial p_k} = -\frac{d}{dp_k}(p_k \log p_k) - \frac{d}{dp_n}(p_n \log p_n)\frac{\partial p_n}{\partial p_k} \tag{3-26}$$

But $$p_n = 1 - (p_1 + p_2 + \cdots + p_k + \cdots + p_{n-1}) \tag{3-27}$$

Hence

$$\frac{dH}{dp_k} = -(\log_2 e + \log p_k) + (\log_2 e + \log p_n) \tag{3-28}$$

$$\frac{dH}{dp_k} = -\log \frac{p_k}{p_n} \tag{3-29}$$

$$\frac{dH}{dp_k} = 0 \qquad \text{yields} \qquad p_k = p_n \tag{3-30}$$

Since p_k was chosen arbitrarily, we come to the conclusion that, for an extremal point of the H function, we must have

$$p_1 = p_2 = \cdots = p_n = \frac{1}{n} \tag{3-31}$$

It remains to be shown if the latter relation makes the H function a maximum and not a minimum. For this we note that

$$H(1,0,0, \ldots ,0) = 0 \tag{3-32}$$

But

$$H\left(\frac{1}{n}, \frac{1}{n}, \ldots , \frac{1}{n}\right) = \log n > 0 \tag{3-33}$$

Thus when all the mutually exclusive events are equiprobable, the H function reaches its maximum value.

Property 4: *Additivity.* We prove the validity of this property by reducing the left member to a form identical with the right member of Eq. (3-23):

$$\begin{aligned} H(p_1,p_2, \ldots ,p_{n-1},q_1,q_2, \ldots ,q_m) &= -\sum_{k=1}^{n-1} p_k \log p_k - \sum_{k=1}^{m} q_k \log q_k \\ &= -\sum_{k=1}^{n} p_k \log p_k + p_n \log p_n - \sum_{k=1}^{m} q_k \log q_k \\ &= H(p_1,p_2, \ldots ,p_n) + p_n \log p_n - \sum_{k=1}^{m} q_k \log q_k \end{aligned} \tag{3-34}$$

But

$$\begin{aligned} p_n \log p_n - \sum_{k=1}^{m} q_k \log q_k &= p_n \sum_{k=1}^{m} \frac{q_k}{p_n} \log p_n - \sum_{k=1}^{m} q_k \log q_k \\ &= -p_n \sum_{k=1}^{m} \frac{q_k}{p_n} \log \frac{q_k}{p_n} \\ &= p_n H\left(\frac{q_1}{p_n}, \frac{q_2}{p_n}, \ldots , \frac{q_m}{p_n}\right) \end{aligned} \tag{3-35}$$

This proves the identity of the two sides of Eq. (3-23).

It is to be noted that, since H functions are essentially nonnegative, we have

$$H(p_1,p_2, \ldots ,p_{n-1},q_1,q_2, \ldots ,q_m) \geq H(p_1,p_2, \ldots ,p_{n-1},p_n) \tag{3-36}$$

That is, the partitioning of events into subevents cannot decrease the entropy of the system.

Example 3-1

(*a*) Evaluate the average uncertainty associated with the sample space of events shown in Fig. E3-1.

$$P\{\ \} = [\tfrac{1}{5},\tfrac{4}{15},\tfrac{8}{15}]$$

(*b*) Evaluate the average uncertainty pertaining to each of the following probability schemes.

$$[A, M = B \cup C], \quad [B \mid M, C \mid M]$$

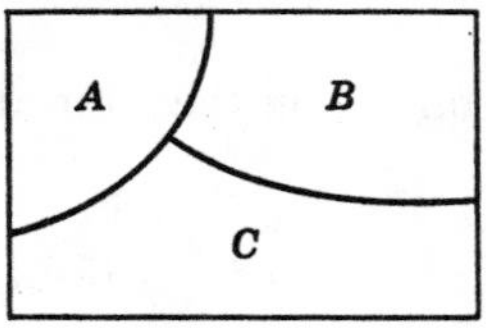

FIG. E3-1

(c) Verify the rule of the additivity of the entropies.

Solution

(a) $$H(\tfrac{1}{5},\tfrac{4}{15},\tfrac{8}{15}) = \tfrac{1}{15}(15 \log 5 + 12 \log 3 - 32) \text{ bits}$$

(b) $$H(\tfrac{1}{5},\tfrac{4}{5}) = \tfrac{1}{15}(15 \log 5 - 24) \text{ bits}$$
$$H(\tfrac{1}{3},\tfrac{2}{3}) = \tfrac{1}{15}(15 \log 3 - 10) \text{ bits}$$

(c) It is a matter of numerical computation to verify that

$$H(\tfrac{1}{5},\tfrac{4}{15},\tfrac{8}{15}) = H(\tfrac{1}{5},\tfrac{4}{5}) + \tfrac{4}{5}H(\tfrac{1}{3},\tfrac{2}{3})$$

Example 3-2. Verify the rule of additivity of entropies for the following probability schemes (Fig. E3-2*a*).

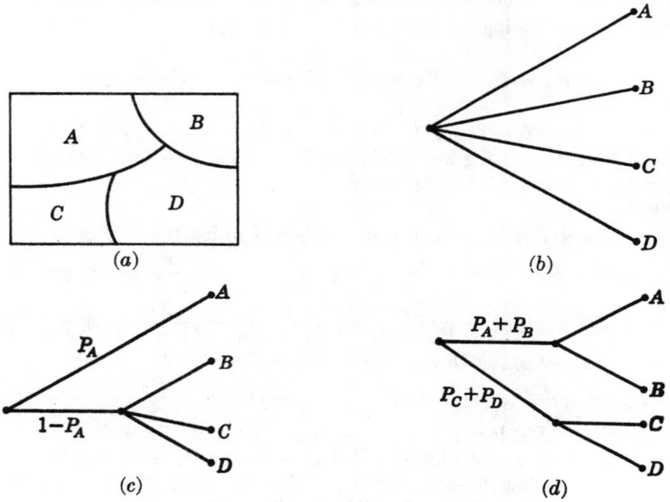

FIG. E3-2

(a) $[A,B,C,D]$ (Fig. E3-2*b*).
(b) $[A,A']$ $[B|A', C|A', D|A']$ (Fig. E3-2*c*).
(c) (Fig. E3-2*d*.)

Numerical example:

$$\left[\frac{\text{Event}}{\text{Probability}}\right] = \left[\frac{A \quad B \quad C \quad D}{\tfrac{1}{2} \quad \tfrac{1}{4} \quad \tfrac{1}{8} \quad \tfrac{1}{8}}\right]$$

Solution. The object of the problem is to demonstrate that the average uncertainty in a system is not affected by the arrangement of the events, as long as the probabilities of the individual events do not change.

(a) $$H = -P_A \log P_A - P_B \log P_B - P_C \log P_C - P_D \log P_D$$

where $$P_A = \tfrac{1}{2} \qquad P_B = \tfrac{1}{4} \qquad P_C = \tfrac{1}{8} \qquad P_D = \tfrac{1}{8}$$

$$\begin{aligned} H &= -\tfrac{1}{2} \log \tfrac{1}{2} - \tfrac{1}{4} \log \tfrac{1}{4} - \tfrac{1}{8} \log \tfrac{1}{8} - \tfrac{1}{8} \log \tfrac{1}{8} \\ &= \tfrac{1}{2} \log 2 + \tfrac{1}{4} \log 4 + \tfrac{1}{8} \log 8 + \tfrac{1}{8} \log 8 \\ &= \tfrac{1}{2} + \tfrac{1}{2} + \tfrac{3}{8} + \tfrac{3}{8} \\ &= 1\tfrac{3}{4} \text{ bits} \end{aligned}$$

(b) According to the additivity property [Eq. (3-19)] of the H functions,

$$
\begin{aligned}
H &= [-P_A \log P_A - (1 - P_A) \log (1 - P_A)] - (1 - P_A) \left(\frac{P_B}{1 - P_A} \log \frac{P_B}{1 - P_A} \right. \\
&\qquad \left. + \frac{P_C}{1 - P_A} \log \frac{P_C}{1 - P_A} + \frac{P_D}{1 - P_A} \log \frac{P_D}{1 - P_A} \right) \\
&= -P_A \log P_A - (1 - P_A) \log (1 - P_A) - P_B \log \frac{P_B}{1 - P_A} \\
&\qquad - P_C \log \frac{P_C}{1 - P_A} - P_D \log \frac{P_D}{1 - P_A} \\
&= -P_A \log P_A - (P_B + P_C + P_D) \log (1 - P_A) - P_B \log P_B + P_B \\
&\quad \log (1 - P_A) - P_C \log P_C + P_C \log (1 - P_A) - P_D \log P_D + P_D \log (1 - P_A) \\
&= -P_A \log P_A - P_B \log P_B - P_C \log P_C - P_D \log P_D
\end{aligned}
$$

where $\qquad P_A = \frac{1}{2} \qquad P_B = \frac{1}{4} \qquad P_C = \frac{1}{8} \qquad P_D = \frac{1}{8}$

$$
\begin{aligned}
H &= -\tfrac{1}{2} \log \tfrac{1}{2} - \tfrac{1}{2} \log \tfrac{1}{2} - \tfrac{1}{2}(\tfrac{1}{2} \log \tfrac{1}{2} + \tfrac{1}{4} \log \tfrac{1}{4} + \tfrac{1}{4} \log \tfrac{1}{4}) \\
&= \tfrac{1}{2} \log 2 + \tfrac{1}{2} \log 2 + \tfrac{1}{2}(\tfrac{1}{2} \log 2 + \tfrac{1}{4} \log 4 + \tfrac{1}{4} \log 4) \\
&= \tfrac{1}{2} + \tfrac{1}{2} + \tfrac{1}{2}(\tfrac{1}{2} + \tfrac{1}{2} + \tfrac{1}{2}) \\
&= 1\tfrac{3}{4} \text{ bits}
\end{aligned}
$$

(c)
$$
\begin{aligned}
H &= -(P_A + P_B) \log (P_A + P_B) - (P_C + P_D) \log (P_C + P_D) \\
&\quad + (P_A + P_B) \left(- \frac{P_A}{P_A + P_B} \log \frac{P_A}{P_A + P_B} - \frac{P_B}{P_A + P_B} \log \frac{P_B}{P_A + P_B} \right) \\
&\quad + (P_C + P_D) \left(- \frac{P_C}{P_C + P_D} \log \frac{P_C}{P_C + P_D} - \frac{P_D}{P_C + P_D} \log \frac{P_D}{P_C + P_D} \right) \\
&= -(P_A + P_B) \log (P_A + P_B) - (P_C + P_D) \log (P_C + P_D) \\
&\quad - P_A \log \frac{P_A}{P_A + P_B} - P_B \log \frac{P_B}{P_A + P_B} - P_C \log \frac{P_C}{P_C + P_D} - P_D \log \frac{P_D}{P_C + P_D} \\
&= -(P_A + P_B) \log (P_A + P_B) - (P_C + P_D) \log (P_C + P_D) \\
&\quad - P_A \log P_A + P_A \log (P_A + P_B) - P_B \log P_B + P_B \log (P_A + P_B) \\
&\qquad - P_C \log P_C + P_C \log (P_C + P_D) - P_D \log P_D + P_D \log (P_C + P_D) \\
&= -P_A \log P_A - P_B \log P_B - P_C \log P_C - P_D \log P_D
\end{aligned}
$$

where $\qquad P_A = \frac{1}{2} \qquad P_B = \frac{1}{4} \qquad P_C = \frac{1}{8} \qquad P_D = \frac{1}{8}$

$$
\begin{aligned}
H &= -(\tfrac{3}{4}) \log \tfrac{3}{4} - \tfrac{1}{4} \log \tfrac{1}{4} + \tfrac{3}{4}(-\tfrac{2}{3} \log \tfrac{2}{3} - \tfrac{1}{3} \log \tfrac{1}{3}) \\
&\qquad + \tfrac{1}{4}(-\tfrac{1}{2} \log \tfrac{1}{2} - \tfrac{1}{2} \log \tfrac{1}{2}) \\
&= -\tfrac{3}{4} \log 3 + \tfrac{3}{4} \log 4 + \tfrac{1}{4} \log 4 + \tfrac{3}{4}(-\tfrac{2}{3} \log 2 + \log 3) \\
&\qquad + \tfrac{1}{4}(\tfrac{1}{2} \log 2 + \tfrac{1}{2} \log 2) \\
&= -\tfrac{3}{4} \log 3 + \tfrac{3}{2} + \tfrac{1}{2} + \tfrac{3}{4}(-\tfrac{2}{3} + \log 3) + \tfrac{1}{4}(\tfrac{1}{2} + \tfrac{1}{2}) \\
&= -\tfrac{3}{4} \log 3 + \tfrac{4}{2} - \tfrac{1}{2} + \tfrac{3}{4} \log 3 + \tfrac{1}{4} \\
&= 1\tfrac{3}{4} \text{ bits}
\end{aligned}
$$

3-5. An Alternative Proof That the Entropy Function Possesses a Maximum. The Shannon-Wiener theory of information is strongly linked with the logarithmic function. Thus it is desirable to spend some time investigating some of the basic mathematical properties of the logarithmic function. Such mathematical presentations may seem distant from an immediate engineering application; however, they are of prime significance to those who are interested in basic research in the field.

First we shall prove a lemma on the convexity of the logarithmic function. Then the lemma will be employed in giving an alternative proof for property 3 of the previous section.

Lemma 1. The logarithmic function is a convex function.

The reader will recall that a function of the real variable $y = f(x)$ is said to be convex upward in a real interval if for any x_1 and x_2 in that interval one has

$$\tfrac{1}{2}[f(x_1) + f(x_2)] \leq f\left(\frac{x_1 + x_2}{2}\right) \tag{3-37}$$

Geometrically this relation can be simply interpreted by saying that the chord connecting points 1 and 2 lies below the curve. An equivalent definition can be given for a curve that is convex upward in an interval. That is,

$$af(x_1) + (1 - a)f(x_2) \leq f[ax_1 + (1 - a)x_2] \qquad 0 \leq a \leq 1 \tag{3-38}$$

The geometrical interpretation of Eq. (3-38) is that in the interval under consideration the chord lies everywhere below the curve (see Fig. 3-3a).

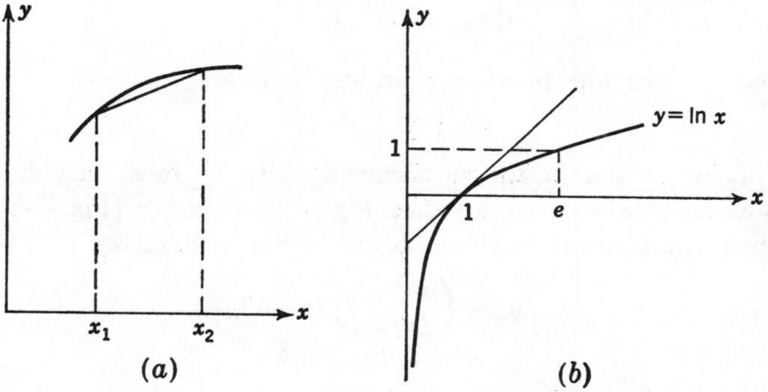

FIG. 3-3. (a) An upward convex function. (b) Logarithmic function is upward convex.

A necessary and sufficient condition for $y = f(x)$ to be convex on the real axis is that

$$\frac{d^2y}{dx^2} \leq 0 \tag{3-39}$$

for every point of the real axis, provided that the second derivative exists. This requirement is satisfied for the function

$$y = \ln x \tag{3-40}$$

In fact,

$$\frac{d^2y}{dx^2} = -\frac{1}{x^2} \tag{3-41}$$

$$\frac{d^2y}{dx^2} \leq 0 \qquad \text{for } 0 \leq x \leq \infty \tag{3-42}$$

Note that this property is independent of the base of the logarithm as long as the base is a number greater than unity:

$$\ln x = \ln 2 \cdot \log_2 x \tag{3-43}$$

Thus we have shown that for positive values of x_1 and x_2

$$\tfrac{1}{2}(\log x_1 + \log x_2) \leq \log \frac{x_1 + x_2}{2} \tag{3-44}$$

$$(x_1 x_2)^{1/2} \leq \frac{x_1 + x_2}{2} \tag{3-45}$$

The geometric mean of two positive numbers is smaller than their average.*

An alternative formulation of Eq. (3-38) can be given by using the following equivalent criterion for convex functions.† If $f(x)$ is convex on the real interval $a \leq x \leq b$, then for any three values of x, $a \leq x_1 \leq x_2 \leq x_3 \leq b$,

$$\begin{vmatrix} x_1 & f(x_1) & 1 \\ x_2 & f(x_2) & 1 \\ x_3 & f(x_3) & 1 \end{vmatrix} \leq 0 \tag{3-46}$$

Lemma 2. For any positive number we have

$$\ln x \leq x - 1 \tag{3-47}$$

This is a simple conclusion of the convexity of $\ln x$. Evidently, the tangent at point $x = 1$ is above the logarithmic curve (Fig. 3-3*b*). The equation of the tangent to the curve at $x = 1$ is given by

$$y_t = \left(\frac{dy}{dx}\bigg|_{x=1} \right)(x - 1) \tag{3-48}$$

$$y_t = x - 1 \tag{3-49}$$

$$\ln x \leq x - 1 \tag{3-50}$$

Again note that this property is equally true for the logarithmic function of the base 2, i.e.,

$$\log x = \ln x \log e \leq (x - 1) \log e$$

* This statement can be extended to the case of n positive numbers, that is,

$$\frac{1}{n} \sum_{k=1}^{n} x_k \geq \sqrt[n]{\prod_{k=1}^{n} x_k}$$

† A discussion on convex functions is generally included in books on advanced calculus. Those interested in further reading on the subject of convexity may refer to Hardy, Littlewood, and Polya. See also G. W. Medlin, On Limits of the Real Characteristic Roots of Matrices with Real Elements, *Proc. Am. Math. Soc.*, vol. 7, pp. 912–917, or G. Julia, "Les Principes géométriques d'analyse," Gauthier-Villars, Paris.

The above lemma will be of some use in our future work. At present, we may employ it to give an alternative proof for the fact that the average uncertainty is greatest when all the events are equiprobable. In order to show this, assume that the space of x contains m points, not necessarily with equal probabilities. It is required to show that $H(X)$ is smaller than the entropy of the equiprobable case, that is,

$$H(X) \le -m\left(\frac{1}{m}\log\frac{1}{m}\right) \tag{3-51}$$

or to prove

$$H(X) \le \log m \tag{3-52}$$

But by definition,

$$H(X) - \log m = \sum_{1}^{m} p_i \log\frac{1}{p_i} + \log\frac{1}{m} \tag{3-53}$$

Since we are dealing with exhaustive systems, $\log(1/m)$ can be replaced by

$$\left(\log\frac{1}{m}\right)\left(\sum_{1}^{m} p_i\right) \tag{3-54}$$

or

$$H(X) - \log m = \left(\sum_{1}^{m} p_i \log\frac{1}{p_i}\right) + \sum_{1}^{m} p_i \log\frac{1}{m} \tag{3-55}$$

$$H(X) - \log m = \sum_{1}^{m} p_i \log\frac{1}{p_i m} \tag{3-56}$$

Applying Lemma 2, we find

$$H(X) - \log m = \sum_{1}^{m} p_i \log\frac{1}{p_i m} \le \sum_{1}^{m} p_i\left(\frac{1}{p_i m} - 1\right)\log e \tag{3-57}$$

$$H(X) - \log m \le \log_2 e\left[\sum_{1}^{m}\left(\frac{1}{m} - p_i\right)\right] = 0 \tag{3-58}$$

$$H(X) \le \log m \tag{3-59}$$

The maximum entropy corresponds to the case when all m states have equal probabilities of occurrence $p_i = 1/m$.

3-6. Sources and Binary Sources. In the study of probability one usually employs concepts of sets but uses certain terminology which differs from that of set theory. Examples of such terminology were given in Sec. 2-6. Similarly, information theory uses certain specialized terms which need to be translated into a more universally understood mathematical form. For our immediate use the following terms are defined:

A source or transmitter is similar to the space of a random experiment. That is, a *source* is the assemblage of all possible events associated with the sample space of a complete random experiment. Each outcome of the experiment corresponds to an elementary output of the source and is called a *symbol* or a *character* or a *letter*.

The finite alphabet of a communication source consists of all its finite distinct characters, much in the same way that the sample space consists of all possible elementary outcomes of a discrete random experiment.

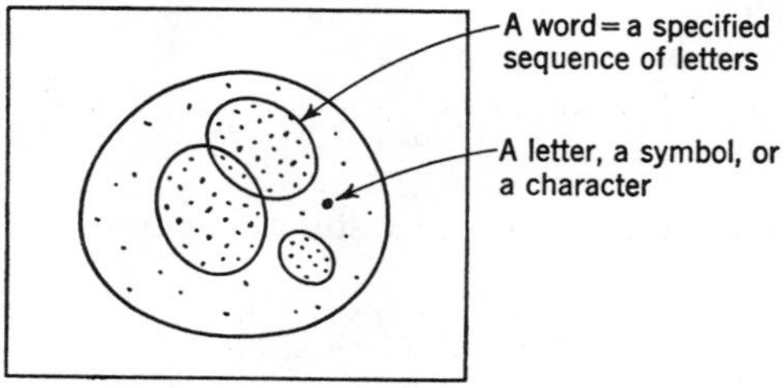

FIG. 3-4. A symbolic illustration of the message space of an independent source; words are specified as sequences of letters (with or without repetition).

A finite sequence of characters may be called a *word* or a *message* in the same way that the sequence of a number of outcomes associated with the repetition of an experiment may be designated as an event. This is schematically illustrated in Fig. 3-4. When the probabilities of the selection of successive letters are independent, we say that the source has no memory. This chapter is devoted to the study of discrete schemes without memory. The study of sources with memory will be deferred until Chap. 11.

A binary source is associated with the sample space of a random binary experiment when the experiment is repeated over and over. In lieu of saying that a random experiment has only two possible exclusive outcomes A and B, we adhere to communication terminology and say that a binary source has an alphabet of two letters A and B. The following three matrices summarize the information-theory performance of a binary source:

$$\begin{aligned}
&\text{Alphabet} = \{\text{letters}\} = [A,B] \\
&\text{Probability matrix } [P] = [p,\ 1-p] = [p,q] \\
&\text{Self-information matrix } [I] = [-\log p,\ -\log(1-p)] \qquad (3\text{-}60)\\
&\text{Average information per letter } H = \bar{I} = -p\log p \\
&\qquad\qquad - (1-p)\log(1-p)
\end{aligned}$$

The communication entropy for such a system will be

$$H(p) = -p \log p - q \log q = -p \log p - (1 - p) \log (1 - p) \quad (3\text{-}61)$$

A plot of the function $H(p)$ in terms of p is shown in Fig. 3-5. The maximum of this function, as anticipated, occurs at $p = \frac{1}{2}$, for which the entropy becomes 1 bit per letter. If a transmitter is sending the two letters A and B with equal probabilities, the average information per letter is a maximum of 1 bit per letter.

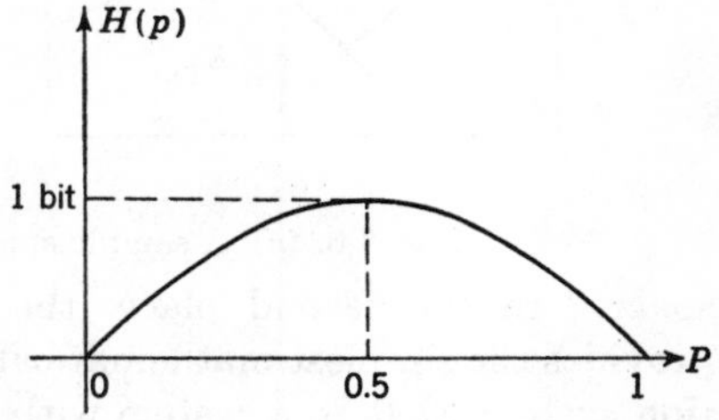

FIG. 3-5. Entropy of an independent binary source.

An interesting observation can be made here about the entropy of a binary source. That is, $H(p)$ of Eq. (3-61) is a function concave downward (or convex upward).

$$\tfrac{1}{2}[H(p_1) + H(p_2)] \leq H\left(\frac{p_1 + p_2}{2}\right) \quad (3\text{-}62)$$

Suppose that we have three specific binary sources for communication between two stations. If we assume the pertinent probabilities for the first letters of each source to be p_1, p_2, and $(p_1 + p_2)/2$, the above statement tells us that the average uncertainty of the third source is larger than the mean of the other two. Loosely speaking, it is *relatively* more difficult to predict the transmission of the letters of the third source.

For example, consider the following two binary sources s_1 and s_2.

$$p_{A1} = \tfrac{1}{3} \qquad p_{A2} = \tfrac{1}{4}$$
$$p_{B1} = \tfrac{2}{3} \qquad p_{B2} = \tfrac{3}{4}$$
$$H(s_1) = -\tfrac{1}{3} \log \tfrac{1}{3} - \tfrac{2}{3} \log \tfrac{2}{3} = -\tfrac{2}{3} + \log 3$$
$$H(s_2) = -\tfrac{1}{4} \log \tfrac{1}{4} - \tfrac{3}{4} \log \tfrac{3}{4} = 2 - \tfrac{3}{4} \log 3$$

A third binary source with an average probability $(p_{A1} + p_{A2})/2$ and $(p_{B1} + p_{B2})/2$ per letter will have an average entropy per letter of

$$p_A = \tfrac{1}{2}(\tfrac{1}{3} + \tfrac{1}{4}) = \tfrac{7}{24} \qquad p_B = \tfrac{1}{2}(\tfrac{2}{3} + \tfrac{3}{4}) = \tfrac{17}{24}$$
$$H(s) = -\tfrac{7}{24} \log \tfrac{7}{24} - \tfrac{17}{24} \log \tfrac{17}{24} = 3 + \log 3 - \tfrac{7}{24} \log 7 - \tfrac{17}{24} \log 17$$

The average information per letter for the third source is greater than the mean of the average information associated with letters of the first and the second source.

3-7. Measure of Information for Two-dimensional Discrete Finite Probability Schemes. In this section, we extend the definition of the measure of information from a one-dimensional to a two-dimensional probability scheme. The content of this section forms an important part of the basic concepts of information theory for several reasons. In the first place, the appropriate generalization from one-dimensional to two-dimensional can be considered as an induction rule for the derivation of the information measure of any finite-dimensional probability

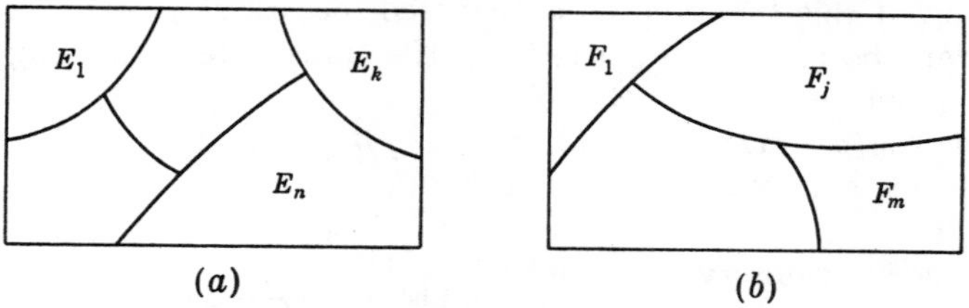

FIG. 3-6. (a) A sample space E. (b) A sample space F.

space. In the second place, the two-dimensional probability scheme provides the simplest mathematical model for an engineering communication system, that is, a system with a "transmitter" and a "receiver" or a transducer with in and out ports. Finally the concept of *mutual information* or *transinformation* which forms one of the fundamental concepts of information theory can be discussed in the light of this product space.

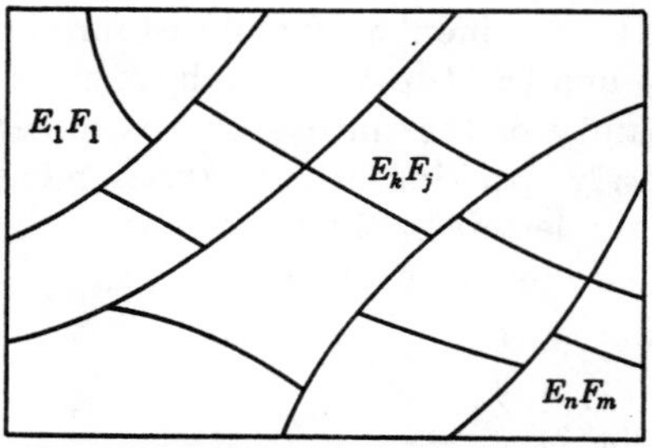

FIG. 3-7. Product space of $E \otimes F$.

Consider two finite discrete sample spaces Ω_1, Ω_2, and their product space Ω as illustrated in Figs. 3-6 and 3-7. In Ω_1 and Ω_2 we select complete sets of events in the sense of Eqs. (3-1) and (3-2).

$$\begin{aligned} \{E\} &= [E_1, E_2, \ldots, E_n] \\ \{F\} &= [F_1, F_2, \ldots, F_m] \end{aligned} \tag{3-63}$$

Each event E_k of Ω_1 may occur in conjunction with any event F_j of Ω_2; thus the following events form a complete set of events in the product space $\Omega_1\Omega_2$.

$$\{EF\} = \begin{bmatrix} E_1F_1 & E_1F_2 & \cdots & E_1F_m \\ E_2F_1 & E_2F_2 & \cdots & E_2F_m \\ \cdots & \cdots & \cdots & \cdots \\ E_nF_1 & E_nF_2 & \cdots & E_nF_m \end{bmatrix} \tag{3-64}$$

where E_kF_j stands for the simultaneous occurrence of the events E_k and F_j. In this fashion, we are confronted with the following three complete sets of probability schemes:

$$P\{E\} = [P\{E_k\}] \tag{3-65}$$
$$P\{F\} = [P\{F_j\}] \tag{3-66}$$
$$P\{EF\} = [P\{E_kF_j\}] \tag{3-67}$$

No stipulation is made about the independence or dependence of the events E_k and F_j. Of course, each one of the above three schemes is, by assumption, a finite complete probability scheme. The data pertaining to this fact can be conveniently obtained from the joint probability matrix below.

$$[P\{X,Y\}] = \begin{array}{c} \diagdown Y \\ X \end{array}\begin{bmatrix} p\{1,1\} & p\{1,2\} & \cdots & p\{1,m\} \\ p\{2,1\} & p\{2,2\} & \cdots & p\{2,m\} \\ \cdots & \cdots & \cdots & \cdots \\ p\{n,1\} & p\{n,2\} & \cdots & p\{n,m\} \end{bmatrix} \tag{3-68}$$

X and Y are random variables, associated with spaces Ω_1 and Ω_2, respectively, and (X,Y) with the product space. The marginal probabilities of the two-dimensional random variables (X,Y) yield the probabilities pertaining to each of the random variables X and Y. For example,

$$\begin{aligned} P\{x_1\} &= P\{E_1\} = P\{E_1F_1 \cup E_1F_2 \cup \cdots \cup E_1F_m\} \\ &= p\{1,1\} + p\{1,2\} + \cdots + p\{1,m\} \end{aligned} \tag{3-69}$$

$$\begin{aligned} P\{y_2\} &= P\{F_2\} = P\{F_2E_1 \cup F_2E_2 \cup \cdots \cup F_2E_n\} \\ &= p\{1,2\} + p\{2,2\} + \cdots + p\{n,2\} \end{aligned} \tag{3-70}$$

or

$$P\{x_k\} = \sum_{j=1}^{m} p\{x_k,y_j\} \tag{3-71}$$

$$P\{y_j\} = \sum_{k=1}^{n} p\{x_k,y_j\} \tag{3-72}$$

Thus we have three finite complete probability schemes, and naturally there are three corresponding entropies:

$$\begin{aligned} H(X,Y) = \; & -p\{1,1\} \log p\{1,1\} - p\{1,2\} \log p\{1,2\} - \cdots - p\{1,m\} \log p\{1,m\} \\ & -p\{2,1\} \log p\{2,1\} - p\{2,2\} \log p\{2,2\} - \cdots - p\{2,m\} \log p\{2,m\} \\ & \cdots\cdots\cdots\cdots\cdots\cdots\cdots\cdots\cdots\cdots \\ & -p\{n,1\} \log p\{n,1\} - p\{n,2\} \log p\{n,2\} - \cdots - p\{n,m\} \log p\{n,m\} \end{aligned} \tag{3-73}$$

$$\begin{aligned} H(X) = \; & -(p\{1,1\} + p\{1,2\} + \cdots + p\{1,m\}) \log (p\{1,1\} + p\{1,2\} + \cdots \\ & \qquad + p\{1,m\}) \\ & -(p\{2,1\} + p\{2,2\} + \cdots + p\{2,m\}) \log (p\{2,1\} + p\{2,2\} + \cdots \\ & \qquad + p\{2,m\}) \\ & \cdots\cdots\cdots\cdots\cdots\cdots\cdots\cdots\cdots\cdots \\ & -(p\{n,1\} + p\{n,2\} + \cdots + p\{n,m\}) \log (p\{n,1\} + p\{n,2\} + \cdots \\ & \qquad + p\{n,m\}) \end{aligned} \tag{3-74}$$

$$
\begin{aligned}
H(Y) = \\
&-(p\{1,1\} + p\{2,1\} + \cdots + p\{n,1\}) \log (p\{1,1\} + p\{2,1\} + \cdots \\
&\qquad\qquad + p\{n,1\}) \\
&-(p\{1,2\} + p\{2,2\} + \cdots + p\{n,2\}) \log (p\{1,2\} + p\{2,2\} + \cdots \\
&\qquad\qquad + p\{n,2\}) \\
&\cdots\cdots\cdots\cdots\cdots\cdots\cdots\cdots\cdots\cdots \\
&-(p\{1,m\} + p\{2,m\} + \cdots + p\{n,m\}) \log (p\{1,m\} + p\{2,m\} + \cdots \\
&\qquad\qquad + p\{n,m\})
\end{aligned}
\tag{3-75}
$$

The above three entropies can be expressed in a more condensed fashion by using directly the two-dimensional joint probability matrix of Eq. (3-68):

$$H(X,Y) = -\sum_{k=1}^{k=n} \sum_{j=1}^{j=m} p\{k,j\} \log p\{k,j\} \tag{3-76}$$

$$H(X) = -\sum_{k=1}^{k=n} \left[\left(\sum_{j=1}^{j=m} p\{k,j\} \right) \log \sum_{j=1}^{j=m} p\{k,j\} \right] \tag{3-77}$$

$$H(Y) = -\sum_{j=1}^{j=m} \left[\left(\sum_{k=1}^{k=n} p\{k,j\} \right) \log \sum_{k=1}^{k=n} p\{k,j\} \right] \tag{3-78}$$

$H(X,Y)$ represents the joint entropy, $H(X)$ the marginal entropy of X, and $H(Y)$ the marginal entropy of Y.

The marginal entropies can, of course, be directly expressed in terms of marginal probabilities $p\{x_k\}$ and $p\{y_j\}$, that is,

$$H(X) = -\sum_{k=1}^{k=n} p\{x_k\} \log p\{x_k\} \tag{3-79}$$

$$H(Y) = -\sum_{j=1}^{j=m} p\{y_j\} \log p\{y_j\} \tag{3-80}$$

The next section deals with conditional entropies associated with a discrete two-dimensional probability scheme.

3-8. Conditional Entropies. Reference is made to the matrix of Eq. (3-68) and Fig. 3-7; an event F_j, for example, may occur in conjunction with $E_1, E_2, \ldots,$ or E_n.

$$F_j = \bigcup_{k=1}^{n} E_k F_j \tag{3-81}$$

$$P\{X = x_k | Y = y_j\} = \frac{P\{X = x_k \cap Y = y_j\}}{P\{Y = y_j\}} \tag{3-82}$$

or

$$p\{x_k|y_j\} = \frac{p\{k,j\}}{p\{y_j\}} \tag{3-83}$$

Now consider the following probability scheme:

$$\{E|F_j\} = [E_1|F_j, E_2|F_j, \ldots, E_n|F_j] \tag{3-84}$$

$$P\{E|F_j\} = \left[\frac{p\{1,j\}}{p\{y_j\}}, \frac{p\{2,j\}}{p\{y_j\}}, \ldots, \frac{p\{n,j\}}{p\{y_j\}}\right] \tag{3-85}$$

The sum of the elements of this matrix is unity; that is, the probability scheme thus described in not only finite but also complete. Therefore an entropy may be directly associated with such a situation.

$$H(X|y_j) = -\sum_{k=1}^{n} \frac{p\{k,j\}}{p\{y_j\}} \log \frac{p\{k,j\}}{p\{y_j\}}$$

$$= -\sum_{k=1}^{n} p\{x_k|y_j\} \log p\{x_k|y_j\} \tag{3-86}$$

Now one may take the average of this conditional entropy for all admissible values of y_j, in order to obtain a measure of average conditional entropy of the system.

$$H(X|Y) = \overline{H(X|y_j)} = \sum_{j=1}^{m} p\{y_j\}[H(X|y_j)]$$

$$= -\sum_{j=1}^{m} p\{y_j\} \sum_{k=1}^{n} p\{x_k|y_j\} \log p\{x_k|y_j\} \tag{3-87}$$

$$H(X|Y) = -\sum_{j=1}^{m} \sum_{k=1}^{n} p\{y_j\}p\{x_k|y_j\} \log p\{x_k|y_j\} \tag{3-88}$$

Similarly, one can evaluate the average conditional entropy $H(Y|X)$:

$$H(Y|X) = -\sum_{k=1}^{n} \sum_{j=1}^{m} p\{x_k\}p\{y_j|x_k\} \log p\{y_j|x_k\} \tag{3-89}$$

The two conditional entropies (the word "average" will be omitted for briefness) can be written as

$$H(X|Y) = -\sum_{j=1}^{m} \sum_{k=1}^{n} p\{x_k,y_j\} \log p\{x_k|y_j\} \tag{3-90}$$

$$H(Y|X) = -\sum_{k=1}^{n} \sum_{j=1}^{m} p\{x_k,y_j\} \log p\{y_j|x_k\} \tag{3-91}$$

The conditional entropies along with marginals and the joint entropy compose the five principal entropies pertaining to a joint distribution. All logarithms are taken to the base 2 in order to obtain units in binary digits. Note that all entropies are essentially positive numbers as they are sums of positive numbers.

The physical interpretation of the different entropies will be discussed in the subsequent section.

Example 3-3. Determine five entropies pertaining to the joint probability matrix of Example 2-30.

Solution

$$H(X,Y) = -\sum_{1}^{6}\sum_{1}^{6} P_{ij} \log \tfrac{1}{36} = -\log \tfrac{1}{36} = 2(1 + \log 3)$$

$$H(X) = H(Y) = -\sum_{1}^{6} P_i \log \tfrac{1}{6} = -\log \tfrac{1}{6} = 1 + \log 3$$

$$H(X|Y) = H(Y|X) = -\sum_{1}^{6}\sum_{1}^{6} P_{ij} \log \tfrac{1}{6} = 1 + \log 3$$

3-9. A Sketch of a Communication Network. In this section, we wish to present an informal sketch of a model for a communication network. In contrast to the material of the previous sections, the content of this section is not presented in a strict mathematical frame. The words source, load, channel, transducer, transmitter, and receiver are used in their common engineering sense. Later on, we shall assign a strict mathematical description to some of these words, but for the present the reader is cautioned against any identification of these terms with similar terms defined in the professional literature.

In the study of physical systems from a systems engineering point of view, we generally focus our attention on a number of points of entry to the system. For example, in ordinary electric networks, we may be interested in the study of voltage-current relationships at the same port of entry in the network (Fig. 3-8*a*). This is generally known as a one-port system.

When the voltage-current relationships between two ports of entries are of interest, the situation is that of a two-port system. In a two-port system, a physical driving force is applied to one port and its effect observed at a second port. The second port may be connected to a "receiver" or "load" (Fig. 3-8*b*). Such a system is usually known as a two-port, or a loaded transducer. More generally, in many physical problems we may be interested in the study of an n-port network (Fig. 3-8*c*). From linear network theory, we know that a complete study of n-port systems requires a knowledge of transmission functions between different ports. For example, if we concentrate on different impedances of a network, the following matrices are considered for a general study of a one-port, two-port, and n-port, respectively.

$$[Z_{11}] \qquad \begin{bmatrix} Z_{11} & Z_{12} \\ Z_{21} & Z_{22} \end{bmatrix} \qquad \begin{bmatrix} Z_{11} & Z_{12} & \cdots & Z_{1n} \\ Z_{21} & Z_{22} & \cdots & Z_{2n} \\ \cdots & \cdots & \cdots & \cdots \\ Z_{n1} & Z_{n2} & \cdots & Z_{nn} \end{bmatrix} \tag{3-92}$$

(The impedances are used in the ordinary circuit sense, Z_{kj} being the transfer impedance between the kth and the jth port.)

An equivalent interpretation can be made for the study of probabilistic systems. In fact, the systems point of view does not rely on the deter-

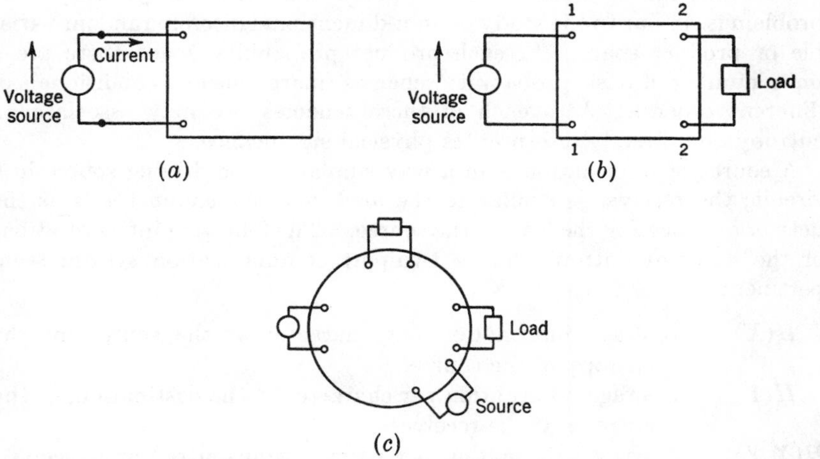

FIG. 3-8. (*a*) A one-port network. (*b*) A two-port analog of a channel connecting a source and a receiver. (*c*) An n-port analog of a communication system consisting of several sources, channels, and sinks.

ministic or probabilistic description of the performance. It is based on the *ports* of application of stimuli and observation of responses. For instance, consider a *source* of communication with a given *alphabet.* The source is linked to the *receiver* via a *channel.* The system may be described by a joint probability matrix, that is, by giving the probability of the joint occurrence of two symbols, one at the input and the other at the output. The joint probability matrix may be designated by

$$[P\{X,Y\}] = \begin{bmatrix} P\{x_1,y_1\} & P\{x_1,y_2\} & \cdots & P\{x_1,y_n\} \\ P\{x_2,y_1\} & P\{x_2,y_2\} & \cdots & P\{x_2,y_n\} \\ \cdots & \cdots & \cdots & \cdots \\ P\{x_m,y_1\} & P\{x_m,y_2\} & \cdots & P\{x_m,y_n\} \end{bmatrix} \tag{3-93}$$

But in a product space of the two random variables X and Y there are five basic probability schemes of interest. These are

$[P\{X,Y\}]$	joint probability matrix	(3-94)
$[P\{X\}]$	marginal probability matrix of X	(3-95)
$[P\{Y\}]$	marginal probability matrix of Y	(3-96)
$[P\{X\|Y\}]$	conditional probability matrix	(3-97)
$[P\{Y\|X\}]$	conditional probability matrix	

Thus we are naturally led to five distinct functions in the study of a simple communication model.

This idea can be generalized to n-port communication systems. The problem is similar to the study of an n-dimensional discrete random variable or product space. In each product probability space there are a finite number of basic probability schemes (marginals and conditionals of different orders). With each of these schemes, we may associate an entropy and directly interpret its physical significance.

A source of information is in a way similar to the driving source in a circuit; the receiver is similar to the load, and the channel acts as the network connecting the load to the source. The following interpretations of the different entropies for a two-port communication system seem pertinent.

$H(X)$ Average information per character at the source, or the entropy of the source.

$H(Y)$ Average information per character at the destination, or the entropy at the receiver.

$H(X,Y)$ Average information per pairs of transmitted and received characters, or the average uncertainty of the communication system as a whole.

$H(Y|X)$ A specific character x_i being transmitted; one of the permissible y_j may be received with a given probability. The entropy associated with this probability scheme when x_i covers sets of all transmitted symbols, that is, $\overline{H(Y|x_i)}$, is the conditional entropy $H(Y|X)$, a measure of information about the receiving port, where it is known that X is transmitted.

$H(X|Y)$ A specific character y_j being received; this may be a result of transmission of one of the x_i with a given probability. The entropy associated with this probability scheme when y_j covers all the received symbols, that is, $\overline{H(X|y_j)}$, is the entropy $H(X|Y)$ or equivocation, a measure of information about the source, where it is known that Y is received.

$H(X)$ and $H(Y)$ give indications of the probabilistic nature of the transmission and reception ports, respectively. $H(Y|X)$ gives an indication of the *noise* or *error* in the channel, and $H(X|Y)$ indicates a measure of equivocation, that is, how well one can recover the input content from the output.

All the probabilities encountered in the two-dimensional case can be derived from the joint probability matrix. Thus, a joint probability matrix specifies a communication channel, in much the same way that an impedance or admittance matrix specifies the performance of an ordinary linear two-port network with respect to its ports.

3-10. Derivation of the Noise Characteristics of a Channel. In communication problems in general, the joint probability matrix is not given. It is customary to specify the *noise characteristics* of a channel and the source alphabet probabilities. From these data we can directly derive the joint and the output probability matrices. For example, the joint probability matrix is

$$\begin{bmatrix} p\{x_1\}p\{y_1|x_1\} & p\{x_1\}p\{y_2|x_1\} & \cdots & p\{x_1\}p\{y_n|x_1\} \\ p\{x_2\}p\{y_1|x_2\} & p\{x_2\}p\{y_2|x_2\} & \cdots & p\{x_2\}p\{y_n|x_2\} \\ \cdots & \cdots & \cdots & \cdots \\ p\{x_m\}p\{y_1|x_m\} & p\{x_m\}p\{y_2|x_m\} & \cdots & p\{x_m\}p\{y_n|x_m\} \end{bmatrix}$$

which can be written as

$$[P\{X\}][P\{Y|X\}] = [P\{X,Y\}]$$

(In this form we assume that the marginal probability matrix is written in a diagonal form.)

Similarly, if for convenience $[P\{X\}]$ is written in the form of a row matrix, we have

$$[P\{X\}][P\{Y|X\}] = [P\{Y\}]$$

where $[P\{Y\}]$ will also be a row matrix designating the probabilities of the output alphabets.

This section offers for discussion two particularly simple communication channels:

1. Discrete noise-free channel
2. Discrete channel with independent input-output

Discrete Noise-free Channel. In such channels, as their name indicates, every letter of the input alphabet is in a one-to-one correspondence with a letter of the output alphabet. The joint probability matrix, as well as the channel probability matrix, is of the diagonal form:

$$[P\{X,Y\}] = \begin{bmatrix} p\{x_1,y_1\} & 0 & \cdots & 0 \\ 0 & p\{x_2,y_2\} & \cdots & 0 \\ \cdots & \cdots & \cdots & \cdots \\ 0 & 0 & \cdots & p\{x_n,y_n\} \end{bmatrix} \tag{3-98}$$

$$[P\{X|Y\}] = [P\{Y|X\}] = \begin{bmatrix} 1 & 0 & \cdots & 0 \\ 0 & 1 & \cdots & 0 \\ \cdot & \cdot & \cdots & \cdot \\ 0 & 0 & \cdots & 1 \end{bmatrix} \tag{3-99}$$

For a noise-free channel the entropies are

$$H(X,Y) = H(X) = H(Y) = -\sum_{i=1}^{n} p\{x_i,y_i\} \log p\{x_i,y_i\} \tag{3-100}$$

$$H(Y|X) = H(X|Y) = 0 \tag{3-101}$$

The interpretation of these formulas for a communication system is rather clear. To each transmitted symbol in a noise-free channel there corresponds one, and only one, received symbol. The average uncertainty at the receiving end is exactly the same as at the sending end. The individual conditional entropies are all equal to zero, a fact that reiterates a nonambiguous or noise-free transmission.

Discrete Channel with Independent Input-Output. In a similar fashion, one can visualize a channel in which there is no correlation between input and output symbols. That is, an input letter x_i can be received as any one of the symbols y_j of the receiving alphabet with equal probability. As will be shown, such a system is a degenerate one as it does not transmit any information. The joint probability matrix has n identical columns.

$$[P\{X,Y\}] = \overset{\diagdown Y}{X}\begin{bmatrix} p & p_1 & \cdots & p_1 \\ p_2 & p_2 & \cdots & p_2 \\ \cdot & \cdot & \cdots & \cdot \\ p_m & p_m & \cdots & p_m \end{bmatrix} \qquad \sum_{i}^{m} p_i = \frac{1}{n} \tag{3-102}$$

The input and output symbol probabilities are statistically independent of each other, that is,

$$p\{x_i,y_j\} = p_1\{x_i\}p_2\{y_j\} \tag{3-103}$$

This can be shown directly by calculation:

$$p_{ij} = np_i\left(\sum_{1}^{m} p_j\right) = np_i\frac{1}{n} = p_i \tag{3-104}$$

From this one concludes that

$$p\{x_i|y_j\} = p_1\{x_i\} = np_i \tag{3-105}$$

$$p\{y_j|x_i\} = p_2\{y_j\} = \frac{1}{n} \tag{3-106}$$

The different entropies can be computed directly:

$$H(X,Y) = -n\left(\sum_{i=1}^{m} p_i \log p_i\right) \tag{3-107}$$

$$H(X) = -\sum_{i=1}^{m} np_i \log np_i = -n\left(\sum_{i=1}^{m} p_i \log p_i\right) - \log n \tag{3-108}$$

$$H(Y) = -n\left(\frac{1}{n}\log\frac{1}{n}\right) = \log n \tag{3-109}$$

$$H(X|Y) = -\sum_{i=1}^{m} np_i \log np_i = H(X) \tag{3-110}$$

$$H(Y|X) = -\sum_{i=1}^{m} np_i \log \frac{1}{n} = \log n = H(Y) \tag{3-111}$$

The interpretation of the above formula is that a channel with independent input and output ports conveys no information whatsoever. To mention a network analogy, this channel seems to have the largest internal "loss," like a resistive network, in contrast to the noise-free channel which resembles a "lossless" network.

3-11. Some Basic Relationships among Different Entropies. In this section we should like first to investigate some of the fundamental mathematical relations that exist among different entropies in a simple two-port communication system and then point out their significance in communication theories. Our starting point is the evident fact that the different probabilities in a two-dimensional distribution (product space) are interrelated, plus the fact that the chosen logarithmic weighting function is a convex function on the positive real axis. We begin with the basic relationship that exists among the joint, marginal, and conditional probabilities, that is,

$$p\{x_k,y_j\} = p\{x_k|y_j\} \cdot p\{y_j\} = p\{y_j|x_k\} \cdot p\{x_k\} \tag{3-112}$$

$$\log p\{x_k,y_j\} = \log p\{x_k|y_j\} + \log p\{y_j\}$$
$$= \log p\{y_j|x_k\} + \log p\{x_k\} \tag{3-113}$$

The direct substitution of these relations in the defining equations of the entropies leads to the following basic identities:

$$H(X,Y) = H(X|Y) + H(Y) \tag{3-114}$$
$$H(X,Y) = H(Y|X) + H(X) \tag{3-115}$$

Next we should like to establish a fundamental inequality first shown by Shannon, namely,

$$H(X) \geq H(X|Y) \tag{3-116}$$

For the proof of this inequality, we employ once again Eq. (3-50) for $\log (p\{x_k\}/p\{x_k|y_j\})$.

$$H(X|Y) - H(X) = \sum_{j=1}^{m} \sum_{k=1}^{n} p\{x_k,y_j\} \log \frac{p\{x_k\}}{p\{x_k|y_j\}}$$

$$\leq \sum_{j=1}^{m} \sum_{k=1}^{n} p\{x_k,y_j\} \left(\frac{p\{x_k\}}{p\{x_k|y_j\}} - 1\right) \log e \qquad (3\text{-}117)$$

But the right side of this inequality is identically zero as

$$\sum_{j=1}^{m} \sum_{k=1}^{n} (p\{x_k\} \cdot p\{y_j\} - p\{x_k,y_j\}) \log e = \sum_{j=1}^{m} (p\{y_j\} - p\{y_j\}) \log e = 0 \qquad (3\text{-}118)$$

Hence,

$$H(X) \geq H(X|Y) \qquad (3\text{-}119)$$

and similarly one shows that

$$H(Y) \geq H(Y|X) \qquad (3\text{-}120)$$

The equality signs hold if, and only if, X and Y are statistically independent. It is only in such a case that our key inequality Eq. (3-50) becomes an equality (at point $x = 1$), that is,

$$\frac{p\{x_k\}}{p\{x_k|y_j\}} = 1 \qquad (3\text{-}121)$$

for all permissible values of k and j. This is the case of independence between X and Y.

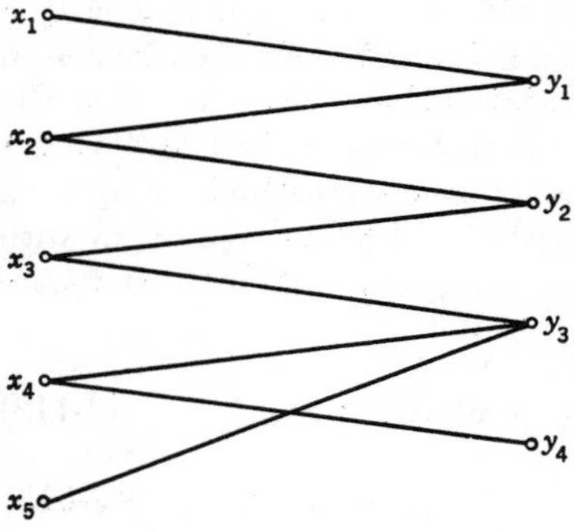

FIG. E3-4

Example 3-4. A transmitter has an alphabet consisting of five letters $\{x_1,x_2,x_3,x_4,x_5\}$ and the receiver has an alphabet of four letters $\{y_1,y_2,y_3,y_4\}$. The joint probabilities for the communication are given below. See Fig. E3-4.

	y_1	y_2	y_3	y_4
x_1	0.25	0	0	0
x_2	0.10	0.30	0	0
x_3	0	0.05	0.10	0
x_4	0	0	0.05	0.10
x_5	0	0	0.05	0

Determine the different entropies for this channel.

Solution

$$\begin{aligned} f_1(x_1) &= 0.25 & f_2(y_1) &= 0.25 + 0.10 = 0.35 \\ f_1(x_2) &= 0.10 + 0.30 = 0.40 & f_2(y_2) &= 0.30 + 0.05 = 0.35 \\ f_1(x_3) &= 0.05 + 0.10 = 0.15 & f_2(y_3) &= 0.10 + 0.05 + 0.05 = 0.20 \\ f_1(x_4) &= 0.05 + 0.10 = 0.15 & f_2(y_4) &= 0.10 \\ f_1(x_5) &= 0.05 \end{aligned}$$

$$\begin{aligned} f(x_1|y_1) &= \frac{f(x_1,y_1)}{f_2(y_1)} = \frac{0.25}{0.35} = \frac{5}{7} & f(y_1|x_1) &= \frac{f(x_1,y_1)}{f_1(x_1)} = \frac{0.25}{0.25} \\ f(x_2|y_2) &= \frac{0.30}{0.35} = \frac{6}{7} & f(y_2|x_2) &= \frac{0.30}{0.40} = \frac{3}{4} \\ f(x_3|y_3) &= \frac{0.10}{0.20} = \frac{1}{2} & f(y_3|x_3) &= \frac{0.10}{0.15} = \frac{2}{3} \\ f(x_4|y_4) &= \frac{0.10}{0.10} = 1 & f(y_4|x_4) &= \frac{0.10}{0.15} = \frac{2}{3} \\ f(x_2|y_1) &= \frac{0.10}{0.35} = \frac{2}{7} & f(y_1|x_2) &= \frac{0.10}{0.40} = \frac{1}{4} \\ f(x_3|y_2) &= \frac{0.05}{0.35} = \frac{1}{7} & f(y_2|x_3) &= \frac{0.05}{0.15} = \frac{1}{3} \\ f(x_4|y_3) &= \frac{0.05}{0.20} = \frac{1}{4} & f(y_3|x_4) &= \frac{0.05}{0.15} = \frac{1}{3} \\ f(x_5|y_3) &= \frac{0.05}{0.20} = \frac{1}{4} & f(y_3|x_5) &= \frac{0.05}{0.05} = 1 \end{aligned}$$

$$\begin{aligned} H(X,Y) &= -\sum_x \sum_y f(x,y) \log f(x,y) \\ &= -0.25 \log 0.25 - 0.10 \log 0.10 - 0.30 \log 0.30 - 0.05 \log 0.05 \\ &\quad - 0.10 \log 0.10 - 0.05 \log 0.05 - 0.10 \log 0.10 - 0.05 \log 0.05 \\ &= 2.665 \\ H(X) &= -\sum_x \sum_y f(x,y) \log f_1(x) \\ &= -0.25 \log 0.25 - 0.10 \log 0.40 - 0.30 \log 0.40 - 0.05 \log 0.15 \\ &\quad - 0.10 \log 0.15 - 0.05 \log 0.15 - 0.10 \log 0.15 - 0.05 \log 0.05 \\ &= 2.066 \\ H(Y) &= \sum_x \sum_y f(x,y) \log f_2(y) \\ &= -0.25 \log 0.35 - 0.10 \log 0.35 - 0.30 \log 0.35 - 0.05 \log 0.35 \\ &\quad - 0.10 \log 0.20 - 0.05 \log 0.20 - 0.05 \log 0.20 - 0.10 \log 0.10 \\ &= 1.856 \\ H(Y|X) &= -\sum_x \sum_y f(x,y) \log \frac{f(x,y)}{f_1(x)} \\ &= -0.10 \log \tfrac{1}{4} - 0.30 \log \tfrac{3}{4} - 0.05 \log \tfrac{1}{3} \\ &\quad - 0.10 \log \tfrac{2}{3} - 0.05 \log \tfrac{1}{3} - 0.10 \log \tfrac{2}{3} \\ &= 0.600 \\ H(X|Y) &= -\sum_x \sum_y f(x,y) \log \frac{f(x,y)}{f_2(y)} \\ &= -0.25 \log \tfrac{5}{7} - 0.10 \log \tfrac{2}{7} - 0.30 \log \tfrac{6}{7} - 0.05 \log \tfrac{1}{7} \\ &\quad - 0.10 \log \tfrac{1}{2} - 0.05 \log \tfrac{1}{4} - 0.05 \log \tfrac{1}{4} \\ &= 0.809 \end{aligned}$$

Note that

$$H(X,Y) < H(X) + H(Y)$$
$$2.665 < 2.066 + 1.856$$

and

$$H(X,Y) = H(Y) + H(X|Y) = H(X) + H(Y|X)$$
$$2.665 = 1.856 + 0.809 = 2.066 + 0.600$$

3-12. A Measure of Mutual Information. Consider a discrete communication system with given joint probabilities between its input and output terminals. Each transmitted symbol x_i while going through the channel has a certain probability $P\{y_j|x_i\}$ of being received as a particular symbol y_j. In the light of previous developments, one may look for a function relating a measure of mutual information between x_i and y_j. In other words, how many bits of information do we obtain in knowing that y_j corresponds to x_i when we know the over-all probability of x_i happening along with different y? In order to avoid a complex mathematical presentation, we follow a procedure similar to that of Sec. 3-3. We assume a definition for mutual information and justify its agreement with that of the previously adopted definition of the entropy. Finally, we shall investigate some of the properties of the suggested measure of mutual information. A measure for the mutual information contained in $(x_i|y_j)$ can be given as

$$I(x_i;y_j) = \log_2 \frac{p\{x_i|y_j\}}{p\{x_i\}} = \log \frac{p\{x_i,y_j\}}{p\{x_i\}p\{y_j\}} \tag{3-122}$$

This expression gives a reasonable measure of mutual information conveyed by a pair of symbols (x_i,y_j). For a moment, we concentrate on the received symbol y_j. Suppose that an observer is stationed at the receiver end at the position of the signal y_j. His a priori knowledge that a symbol x_i is being transmitted is the marginal probability $p\{x_i\}$, that is, the sum of the probabilities of x_i being transmitted and received as any one of the possible y_j. The a posteriori knowledge of our observer is based on the conditional probability of x_i being transmitted, given that a particular y_j is received, that is, $p\{x_i|y_j\}$. Therefore, loosely speaking, for this observer the gain of information is the logarithm of the ratio of his final and initial ignorance or uncertainties. However, the mathematically inclined reader may wish to forgo such justification and use (3-122) as a definition.

The following elementary properties can be derived for the mutual information function:

1. *Continuity.* $I(x_i;y_j)$ is a continuous function of $p\{x_i|y_j\}$.

2. *Symmetry or reciprocity.* The information conveyed by y_j about x_i is the same as the information conveyed by x_i about y_j, that is,

$$I(x_i;y_j) = I(y_j;x_i) \tag{3-123}$$

Obviously, Eq. (3-122) is symmetric with respect to x_i and y_j.

3. *Mutual and self-information.* The function $I(x_i;x_i)$ may be called the self-information of a symbol x_i. That is, if an observer is stationed at the position of the symbol x_i his a priori knowledge of the situation is that x_i will be transmitted with the probability $p\{x_i\}$ and his a posteriori knowledge is the certainty that x_i has been transmitted; thus

$$I(x_i) = I(x_i;x_i) = \log \frac{1}{p\{x_i\}} \tag{3-124}$$

Obviously,

$$I(x_i;y_j) \leq I(x_i;x_i) = I(x_i) \tag{3-125}$$
$$I(x_i;y_j) \leq I(y_j;y_j) = I(y_j) \tag{3-126}$$

An interesting interpretation of the concept of mutual information can be given by obtaining the average of the mutual information per symbol pairs, that is,

$$I(X;Y) = \overline{I(x_i;y_j)} = \sum_j \sum_i p\{x_i,y_j\} I(x_i;y_j) \tag{3-127}$$

$$I(X;Y) = \sum_j \sum_i p\{x_i,y_j\} \log \frac{p\{x_i|y_j\}}{p\{x_i\}} \tag{3-128}$$

It could be ascertained that this definition provides a proper measure for the mutual information of all the pairs of symbols. On the other hand, the definition ties in with our previously defined basic entropy formulas. Indeed, by direct application of the defining equations one can show that

$$I(X;Y) = H(X) + H(Y) - H(X,Y) \tag{3-129}$$
$$I(X;Y) = H(X) - H(X|Y) \tag{3-130}$$
$$I(X;Y) = H(Y) - H(Y|X) \tag{3-131}$$

The entropy corresponding to the mutual information, that is, $I(X;Y)$, indicates a measure of the information transmitted through the channel. For this reason it is referred to as transferred information or *transinformation* of the channel. Note that, based on the fundamental equation (3-116), the right side of Eq. (3-130) is a nonnegative number. Hence, the average mutual information is also nonnegative, while the individual mutual-information quantities may become negative for some symbol pairs. For a noise-free channel,

$$I(X;Y) = H(X) = H(Y) \tag{3-132}$$
$$I(X;Y) = H(X,Y) \tag{3-133}$$

For a channel where the output and the input symbols are independent,

$$\begin{aligned} I(X;Y) &= H(X) - H(X|Y) \\ &= H(X) - H(X) = 0 \end{aligned} \tag{3-134}$$

no information is transmitted through the channel.

Example 3-5. The joint probability matrix of a channel with binary input and output is given below:

$$\begin{array}{c} \\ x_1 \\ x_2 \end{array}\begin{array}{c} \begin{array}{cc} y_1 & y_2 \end{array} \\ \begin{bmatrix} 1/4 & 1/4 \\ 1/4 & 1/4 \end{bmatrix} \end{array}$$

Find the different entropies and the mutual information.

Solution. The marginal probabilities are

$$P\{x_1\} = P\{x_2\} = 1/2$$
$$P\{y_1\} = P\{y_2\} = 1/2$$

The entropies are

$$H(X) = H(Y) = 1$$
$$H(X,Y) = 2$$
$$I(X;Y) = H(X) + H(Y) - H(X,Y) = 0$$

The transinformation is zero, as the input and the output symbols are independent. In other words, there is no dependence or correlation between the symbols at the output and the input of the channel.

3-13. Set-theory Interpretation of Shannon's Fundamental Inequalities. A set-theory interpretation of Shannon's fundamental inequalities

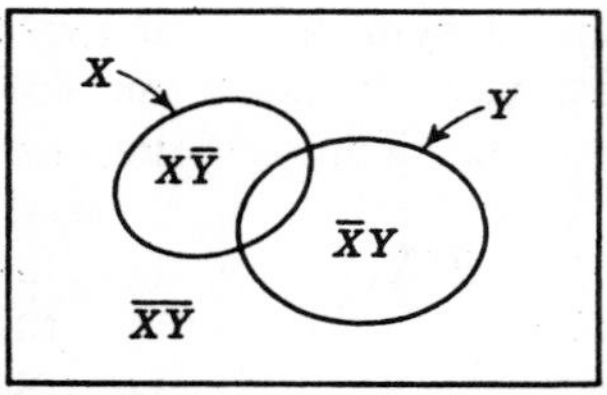

FIG. 3-9. A set-theory presentation of a simple communication system. (X,Y) represents the joint operation of the source and the channel.

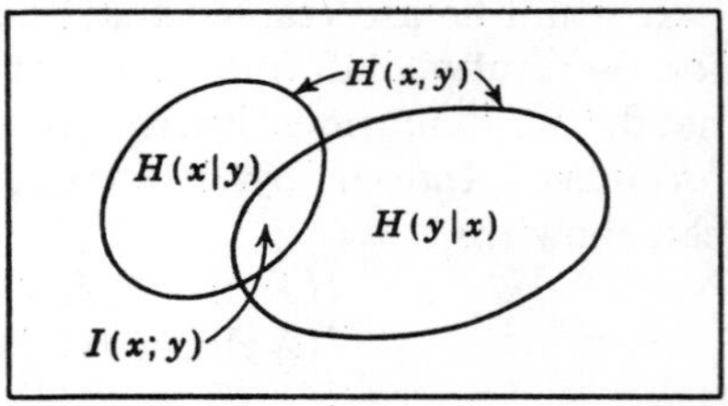

FIG. 3-10. A set-theory presentation of different entropies associated with a simple communication model.

along with the material discussed previously may be illuminating. Consider the variables A and B as sets. We may symbolically write $m(A)$ and $m(B)$ as some kind of measure (say the area) associated with sets X and Y. The entropies of discrete schemes are essentially nonnegative, and they possess the property of Eqs. (3-114) and (3-119). Thus one may observe that, in a sense, the law of "additivity" of the entropies holds for disjoint sets. Thus, the following symbolism may be useful in visualizing the interrelationships (see Figs. 3-9 and 3-10).

$m(A)$	$H(X)$	(3-135)
$m(B)$	$H(Y)$	(3-136)
$m(A \cup B)$	$H(X,Y)$	(3-137)
$m(AB')$	$H(X\|Y)$	(3-138)

$m(BA')$	$H(Y\|X)$	(3-139)
$m(A \cap B)$	$I(X;Y)$	(3-140)
$m(A \cup B) \leq m(A) + m(B)$	$H(X,Y) \leq H(X) + H(Y)$	(3-141)
$m(AB') \leq m(A)$	$H(X\|Y) \leq H(X)$	(3-142)
$m(BA') \leq m(B)$	$H(Y\|X) \leq H(Y)$	(3-143)
$m(A \cup B) = m(AB') + m(BA') + m(A \cap B)$	$H(X,Y) = H(X\|Y) + H(Y\|X) + I(X;Y)$	(3-144)

When the channel is noise-free, the two sets become "coincident" as follows:

$m(A) = m(B)$	$H(X) = H(Y)$	(3-145)
$m(A \cup B) = m(A) = m(B)$	$H(X,Y) = H(X) = H(Y)$	(3-146)
$m(AB') = 0$	$H(X\|Y) = 0$	(3-147)
$m(BA') = 0$	$H(Y\|X) = 0$	(3-148)
$m(A \cap B) = m(A) = m(B) = m(A \cup B)$	$I(X;Y) = H(X) = H(Y) = H(X,Y)$	(3-149)

When the channel is such that input and output symbols are independent, the two sets A and B are considered mutually exclusive:

$m(A \cup B) = m(A) + m(B)$	$H(X,Y) = H(X) + H(Y)$	(3-150)
$m(AB') = m(A)$	$H(X\|Y) = H(X)$	(3-151)
$m(A \cap B) = 0$	$I(X;Y) = 0$	(3-152)

This procedure may be extended to the case of channels with several ports. For example, for three random variables (X,Y,Z) one may write

$$H(X,Y,Z) \leq H(X) + H(Y) + H(Z) \tag{3-153}$$

$$H(Z|X,Y) \leq H(Z|Y) \tag{3-154}$$

See Fig. 3-11. For a formal proof of Eqs. (3-153) and (3-154) see Khinchin. Similarly, one may give formal proof for the following inter-

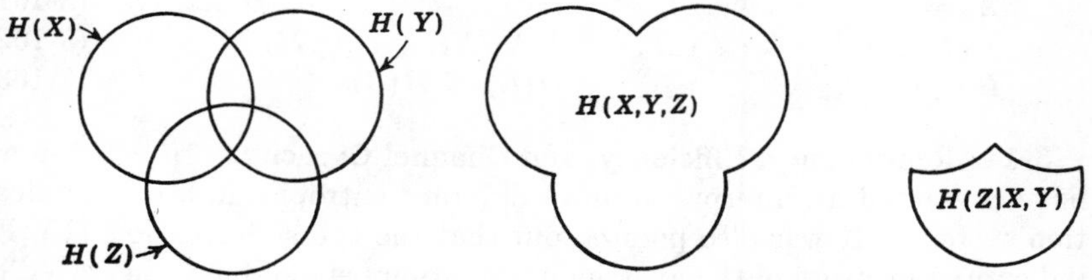

FIG. 3-11. A set-theory presentation of the entropies associated with (X,Y,Z) space.

esting equalities:

$$I(X;Y,Z) = I(X;Y) + I(X;Z|Y) \tag{3-155}$$
$$I(Y,Z;X) = I(Y;X) + I(Z;X|Y) \tag{3-156}$$

The set diagrams for these relations are given in Figs. 3-12 and 3-13.

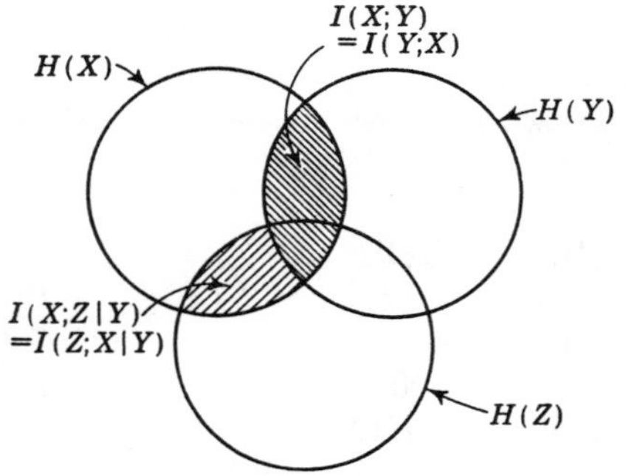

FIG. 3-12. A set-theory presentation of the entropies associated with (X,Y,Z) space.

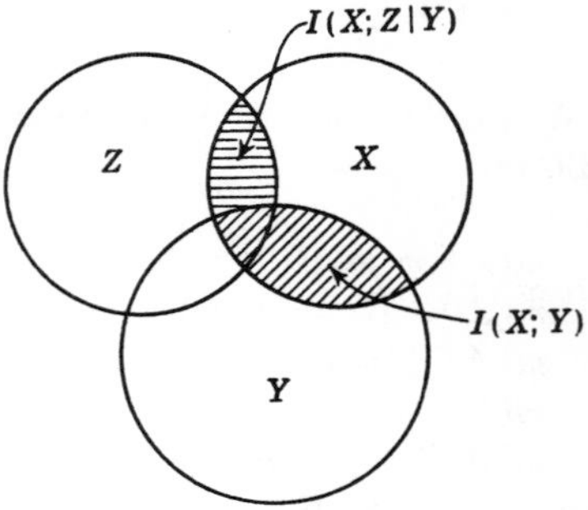

FIG. 3-13. A set-theory presentation of the transinformations $I(X;Y)$ and $I(X;Z|Y)$.

In conclusion, it seems worthwhile to present a simple set-theory rule for deriving relationships between different entropy functions of discrete schemes:

Draw a set A_k for each random variable X_k of the multidimensional random variable $(X_1,X_2, \ldots ,X_n)$. When two variables X_k and X_j are independent, their representative sets will be mutually exclusive. Two random variables X_i and X_h describing a noise-free channel (with a diagonal probability matrix) will have overlapping set representation. The following symbolic correspondences are suggested:

$A_kA'_j$	$H(X_k\|X_j)$	(3-157)
$A_k \cup A_j$	$H(X_k,X_j)$	(3-158)
$A_k \cap A_j$	$I(X_k;X_j)$	(3-159)
$A_kA_j = \emptyset$	$I(X_k;X_j) = 0$	(3-160)
$A_k = A_j$	$H(X_k) = H(X_j) = H(X_k,X_j)$	(3-161)
$A_1 \cap A_2 \cap \cdots \cap A_n$	$I(X_1;X_2; \ldots ;X_n)$	(3-162)
$B \subset C$	$H(B) \leq H(C)$	(3-163)

3-14. Redundancy, Efficiency, and Channel Capacity. In Sec. 3-9 we have presented an interpretation of different entropies in a communication system. It was also pointed out that the transinformation $I(X;Y)$ indicates a measure of the average information per symbol transmitted in the system. The significance of this statement is made clear by referring to Eq. (3-127). In this section, it is intended to introduce a suitable measure for efficiency of transmission of information by making a comparison between the actual rate and the upper bound of the rate of transmission of information for a given channel. In this respect, Shannon has introduced the significant concept of channel capacity. According to Shannon, in a discrete communication system the *channel capacity* is the maximum of transinformation.

$$C = \max I(X;Y) = \max [H(X) - H(X|Y)] \tag{3-164}$$

The maximization is with respect to all possible sets of probabilities that could be assigned to the source alphabet, that is, all discrete memoryless sources. Before proceeding with examples of application and computation of channel capacity, a somewhat analogous concept from linear network theory may be worth mentioning. Consider a linear, resistive, passive, two-port network connected to a linear resistor R at its output terminals. The power dissipated in R under a given regime depends on the network and the load. The maximum power dissipated in the load occurs when there is a matching between the load and the network, i.e., when the resistance of the network seen from the output terminals is identical with R. This situation can be further analyzed by observing that, for a given network, the power transferred to the load depends on the value of the load; the maximum power transfer occurs only when the load and the source are properly matched through a transducer. In a discrete communication channel, with prespecified noise characteristics, i.e., with a given transition probability matrix, the rate of information transmission depends on the source that drives the channel. Note that, in the network analogy, one could specify the load and determine the class of transducers that would match the given load to a specified class of sources. The maximum (or the upper bound) of the rate of information transmission corresponds to a proper matching of the source and the channel. This ideal characterization of the source depends in turn on the probability transition characteristics of the given channel.

Discrete Noiseless Channels. The following is an example of the evaluation of the channel capacity of the simplest type of sources.

Let $X = \{x_i\}$ be the alphabet of a source containing n symbols. Since the transition probability matrix is of the diagonal type, we have, according to Eq. (3-132),

$$C = \max I(X;Y) = \max [H(X)] = \max \Big[- \sum_{i=1}^{n} p\{x_i\} \log p\{x_i\} \Big] \tag{3-165}$$

According to Eq. (3-14), the maximum of $H(X)$ occurs when all symbols are equiprobable; thus the channel capacity is

$$C = \log n \qquad \text{bits per symbol} \tag{3-166}$$

The capacity of a channel, as well as the *rate of transmission of information* through the channel, can be equivalently expressed in bits per second instead of bits per symbol. For this, one has to introduce the concept of time required for the transmission of individual symbols. For instance, if the symbols have a common duration of t seconds, then the channel capacity per second C_t is given by

$$C_t = \frac{1}{t} C \qquad \text{bits per second} \tag{3-167}$$

For the simple noise-free communication system described above, we have

$$C_t = \frac{1}{t} C = \frac{1}{t} \log n \qquad \text{bits per second} \tag{3-168}$$

The difference between the actual rate of transmission of information $I(X;Y)$ and its maximum possible value is defined as the (*absolute*) *redundancy* of the communication system. The ratio of absolute redundancy to channel capacity is defined as the *relative redundancy*. For the afore-mentioned system,

$$\begin{aligned} \text{Absolute redundancy for noise-free channel} &= C - I(X;Y) \\ &= \log n - H(X) \end{aligned} \tag{3-169}$$

$$\begin{aligned} \text{Relative redundancy for noise-free channel} &= \frac{\log n - H(X)}{\log n} \\ &= 1 - \frac{H(X)}{\log n} \end{aligned} \tag{3-170}$$

The *efficiency* of the above system can be defined in an obvious fashion as

$$\begin{aligned} \text{Efficiency of noise-free channel} &= \frac{I(X;Y)}{\log n} = \frac{H(X)}{\log n} \\ &= 1 - \text{relative redundancy} \end{aligned} \tag{3-171}$$

When the time for the transmission of symbols is not necessarily equal, a similar procedure may be applied. Let t_i be the time associated with the symbol x_i; then the average transinformation of a noise-free channel per unit time is

$$R_t = \frac{-\sum_{i=1}^{n} p\{x_i\} \log p\{x_i\}}{\sum_{i=1}^{n} p\{x_i\} t_i} \tag{3-172}$$

R_t is known as the rate of transmission of information. For a given set of t_i ($i = 1, 2, \ldots, n$), one can evaluate the x_i leading to the maximum rate of transmission of information per second C_t. The computation will not be undertaken here.

Discrete Noisy Channel. The channel capacity is the maximum of the average mutual information when the noise characteristic $p\{y_j|x_i\}$ of the channel is prespecified.

$$C = \max\left(\sum_{i=1}^{n}\sum_{j=1}^{m} p_1\{x_i\}p\{y_j|x_i\} \log \frac{p\{y_j|x_i\}}{p_2\{y_j\}}\right) \tag{3-173}$$

where the maximization is with respect to $p_1\{x_i\}$. Note that the marginal probabilities $p_2\{y_j\}$ are related to the independent variables $p_1\{x_i\}$ through the familiar relation

$$p_2\{y_j\} = \sum_{i=1}^{n} p_1\{x_i\}p\{y_j|x_i\} \tag{3-174}$$

Furthermore, the variables are, of course, restricted by the following constraints:

$$p_1\{x_i\} \geq 0 \qquad i = 1, 2, \ldots, n \tag{3-175}$$

$$\sum_{i=1}^{n} p_1\{x_i\} = 1$$

The maximization of (3-173) with respect to the input probabilities does not necessarily lead to a set of admissible source-symbol probabilities.

From the physical point of view, the problem of channel capacity is a rather complex one. The communication channels are not generally of the aforesaid simplest types. When there is an interdependence between successive channel symbols, the statistical identification of the source and the maximization problem are more cumbersome. In these more general cases, the system will exhibit a stochastic nature. Therefore, more elaborate techniques need to be introduced for deriving the channel capacity of such systems. Because of this complexity, Shannon's fundamental channel-capacity theorems require adequate preliminary preparation. These will be considered in a later chapter.

3-15. Capacity of Channels with Symmetric Noise Structures. The computation of the channel capacity in general is a tedious mathematical problem, although its formulation is straightforward. The procedure of maximization requires some special mathematical techniques such as the method of Lagrangian multipliers. In the present section we should like to compute the capacity of some special channels with symmetric noise characteristics as considered by Shannon.

Consider a channel such that each input letter is transformed into a finite number of output letters with a similar set of probabilities for all the input letters. In this case the channel characteristic matrix $P\{y_j|x_i\}$ contains identical rows and identical columns but not necessarily in the same position. See Fig. E3-6, where we have

$$\begin{bmatrix} \frac{1}{3} & \frac{1}{3} & \frac{1}{6} & \frac{1}{6} \\ \frac{1}{6} & \frac{1}{6} & \frac{1}{3} & \frac{1}{3} \end{bmatrix}$$

For such channels the capacity can be computed without any difficulty. The key to the simplification is the fact that the conditional entropy $H(Y|X)$ is independent of the probability distribution at the input. Indeed, for a letter x_i with marginal probability a_i we may write

$$p\{y_j|x_i\} = \alpha_{ij} \tag{3-176}$$
$$p\{x_i,y_j\} = a_i\alpha_{ij} \tag{3-177}$$

The conditional entropy pertinent to the letter x_i will be

$$H(Y|x_i) = -\sum_{j=1}^{m} p\{y_j|x_i\} \log p\{y_j|x_i\} \tag{3-178}$$

Now let

$$H(Y|x_i) = \text{const} = h \qquad \text{for } i = 1, 2, \ldots, n$$

Thus, Eq. (3-89) yields

$$H(Y|X) = (a_1 + a_2 + \cdots + a_n)h = h \tag{3-179}$$

That is, the average conditional entropy is a constant number independent of the probabilities of the letters at the input of the channel.

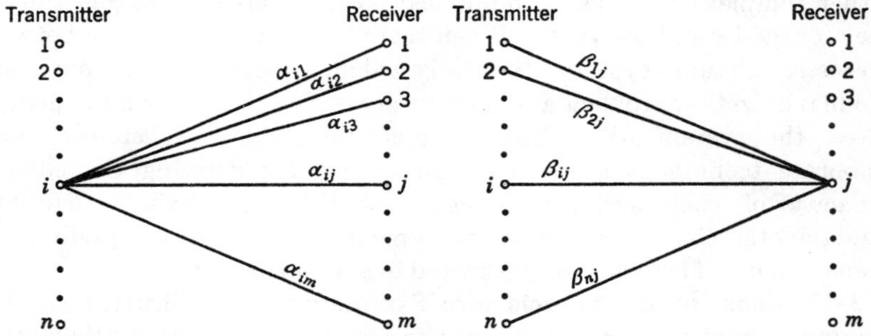

FIG. 3-14. A channel with a particularly symmetric structure.

Therefore, instead of maximizing the expression $H(Y) - H(Y|X)$, we may simply maximize the expression $H(Y) - h$, or $H(Y)$, as h is a constant. But the maximum of $H(Y)$ occurs when all the received letters have the same probabilities, that is,

$$C = \log m - h \tag{3-180}$$

We may wish to investigate further what restriction Eq. (3-180) imposes on the channel. For this, reference can be made to the channel probability matrix and the conditional probability matrix $P\{X|Y\}$ of the system. Let $p\{x_i|y_j\} = \beta_{ij}$ and note that

$$p\{x_i\}\alpha_{ij} = p\{y_j\}\beta_{ij}$$

It can be shown that when $I(X;Y) = C$, the β_{ij} matrix will also have identical rows, that is, the tree of all probabilities at the output of the channel assumes similar symmetry for all sets of the received letters (Fig. 3-14). Furthermore it can be shown that the probabilities of the transmitted letters $p\{x_i\}$ will have to be equal for $i = 1, 2, \ldots, n$.

Conversely, if the situation of Fig. 3-14 prevails, then

$$\begin{aligned} C &= \max [H(X) - H(X|Y)] \\ &= \max [H(X)] - h' = \log n - h' \end{aligned}$$

where h' is the conditional entropy $H(X|Y)$.

Example 3-6. Find the capacity of the channel illustrated in Fig. E3-6.

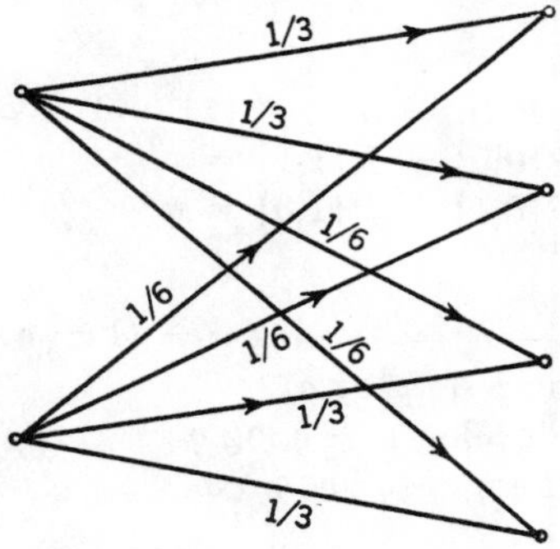

FIG. E3-6

Solution. Applying Eq. (3-180), one finds

$$\begin{aligned} C &= \log 4 - \tfrac{2}{3} \log 3 - \tfrac{1}{3} \log 6 \\ C &= \tfrac{5}{3} - \log 3 \qquad \text{bits} \end{aligned}$$

Example 3-7. A binary channel has the following noise characteristic:

$$\begin{matrix} & 0 & 1 \\ \begin{matrix} 0 \\ 1 \end{matrix} & \multicolumn{2}{c}{\begin{bmatrix} \frac{2}{3} & \frac{1}{3} \\ \frac{1}{3} & \frac{2}{3} \end{bmatrix}} \end{matrix}$$

(*a*) If the input symbols are transmitted with respective probabilities of $\frac{3}{4}$ and $\frac{1}{4}$, find

$$H(X),\ H(Y),\ H(X|Y),\ H(Y|X),\ I(X;Y)$$

(*b*) Find the channel capacity and the corresponding input probabilities.

Solution

(*a*)
$$\begin{aligned} H(X) &= 0.81 \qquad & H(Y) &= 0.98 \\ H(X|Y) &= 0.75 \qquad & H(Y|X) &= 0.92 \\ I(X;Y) &= 0.06 \end{aligned}$$

(*b*)
$$C = 1 + \tfrac{2}{3} \log \tfrac{2}{3} + \tfrac{1}{3} \log \tfrac{1}{3} = \tfrac{5}{3} - \log 3 = 0.08$$

$$p\{0\} = p\{1\} = \tfrac{1}{2}$$

3-16. BSC and BEC. The simplest type of source alphabet to be considered is binary $\{0,1\}$. In this section we assume that the output of such a source is transmitted via a binary symmetric (BSC) or a binary erasure (BEC) channel. Figure 3-15 shows a BSC and Fig. 3-16 a BEC.

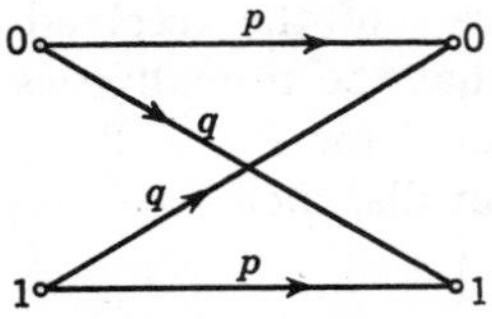

FIG. 3-15. A binary symmetric channel (BSC).

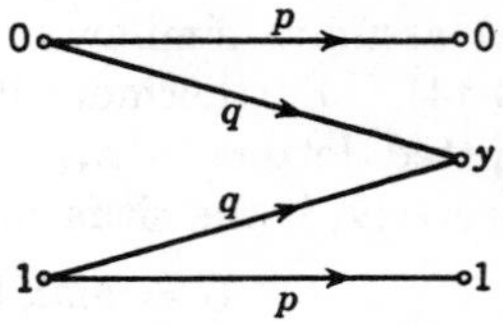

FIG. 3-16. A binary erasure channel (BEC).

The rate of transmission of information and the capacity will be derived for each case.

BSC. Let

$$P\{0\} = \alpha \qquad P\{1\} = 1 - \alpha$$
$$P\{0|0\} = P\{1|1\} = p$$
$$P\{0|1\} = P\{1|0\} = q$$

Then

$$H(X) = H(\alpha, 1-\alpha) = -\alpha \log \alpha - (1-\alpha) \log (1-\alpha)$$
$$H(Y|X) = -(p \log p + q \log q)$$
$$I(X;Y) = H(Y) + p \log p + q \log q$$
$$C = 1 + p \log p + q \log q \tag{3-181}$$

BEC. The channel has two input $\{0,1\}$ and three output symbols $\{0,y,1\}$. The letter y indicates the fact that the output is erased and no deterministic decision can be made as to whether the transmitted letter was 0 or 1. Let

$$P\{0\} = \alpha \qquad P\{1\} = 1 - \alpha$$
$$P\{0|0\} = P\{1|1\} = p$$
$$P\{y|0\} = P\{y|1\} = q$$

Then

$$H(X) = H(\alpha, 1-\alpha)$$
$$H(X|Y) = (1-p)H(X)$$
$$I(X;Y) = pH(X)$$
$$C = p \tag{3-182}$$

Equations (3-181) and (3-182), specifying the capacity of BSC and BEC, respectively, will be referred to frequently in the subsequent discussion.

Example 3-8. Consider the BSC shown in Fig. E3-8. Assume $P\{0\} = \alpha$ and that the successive symbols are transmitted independently. If the channel transmits, all possible binary words U of length 2 which are received as binary words V, derive

(a) The input entropy $H(U)$.

(b) The equivocation entropy $H(U|V)$.

(c) The capacity of the new channel (called the second-order extension of the first channel).

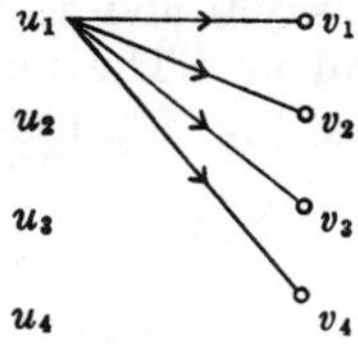

FIG. E3-8

(d) Generalize the results for the case of transmitting words each n binary digits long.

Solution. Let U be a random variable encompassing all the binary words 00, 01, 10, and 11 at the input. Let X_1 and X_2 be random variables referring to symbols in the first and the second position of each word, respectively. Similarly, let V, Y_1, and Y_2 correspond to the output. Symbolically we may write

$$U = X_1, X_2$$
$$V = Y_1, Y_2$$

Because of lack of memory, the probability distributions are given by

$$\begin{aligned} P\{U\} &= P\{X_1\}P\{X_2\} \\ P\{V\} &= P\{Y_1\}P\{Y_2\} \\ P\{V|U\} &= P\{Y_1|X_1\}P\{Y_2|X_2\} \\ P\{U|V\} &= P\{X_1|Y_1\}P\{X_2|Y_2\} \\ P\{U,V\} &= P\{U\}P\{V|U\} = P\{X_1,Y_1\}P\{X_2,Y_2\} \end{aligned}$$

(a) The source entropy $H(U)$ can be thought of as the entropy associated with the two independent random variables X_1 and X_2. Thus

$$H(U) = H(X_1,X_2) = H(X_1) + H(X_2) = -2[\alpha \log \alpha + (1 - \alpha) \log (1 - \alpha)]$$

since $H(X_1) = H(X_2)$.

(b)

$$\begin{aligned} H(U,V) &= H(X_1,Y_1) + H(X_2,Y_2) \\ H(V|U) &= H(Y_1|X_1) + H(Y_2|X_2) \\ H(U|V) &= H(X_1|Y_1) + H(X_2|Y_2) \end{aligned}$$

(c) The transinformation becomes

$$I(U;V) = H(U) - H(U|V) = 2I(X_1;Y_1)$$

The extended channel capacity is twice the capacity of the original channel.

(d) Similarly, one can show that the capacity of the nth-order extension of the channel equals nc, where c is the capacity of the original channel. Note that this statement is independent of the structure of the channel; that is, it holds for any memoryless channel.

3-17. Capacity of Binary Channels. Binary channels are of considerable interest in the transmission and storage of information. The vast field of digital computers offers many examples of such information channels. The problem undertaken in this section is the evaluation of the maximum rate of transmission of information of binary channels.

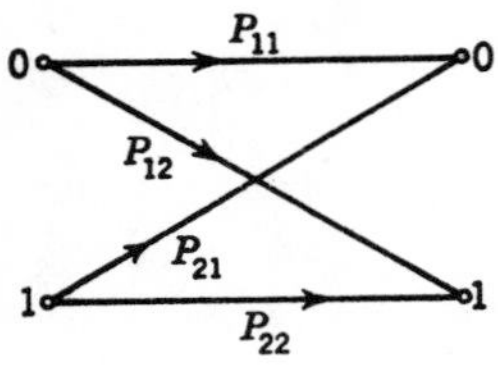

FIG. 3-17. BC.

The source transmits independently two symbols, say 0 and 1, with respective probabilities p_1 and p_2. The channel characteristic is known as (see Fig. 3-17)

$$\begin{bmatrix} p_{11} & p_{12} \\ p_{21} & p_{22} \end{bmatrix}$$

In order to evaluate the capacity of such a channel, when the entropy curve is available a simple geometric procedure can be devised (see Fig. 3-18).

The points A_1 and A_2 on the segment OM are selected so that

$$MA_1 = p_{11} \qquad OA_2 = p_{22}$$

The ordinates of the entropy curve at A_1 and A_2 are

$$\overline{B_1A_1} = H(p_{11}) \qquad \overline{B_2A_2} = H(p_{22})$$

Now, for any given channel output probabilities such as $OA = p$ and

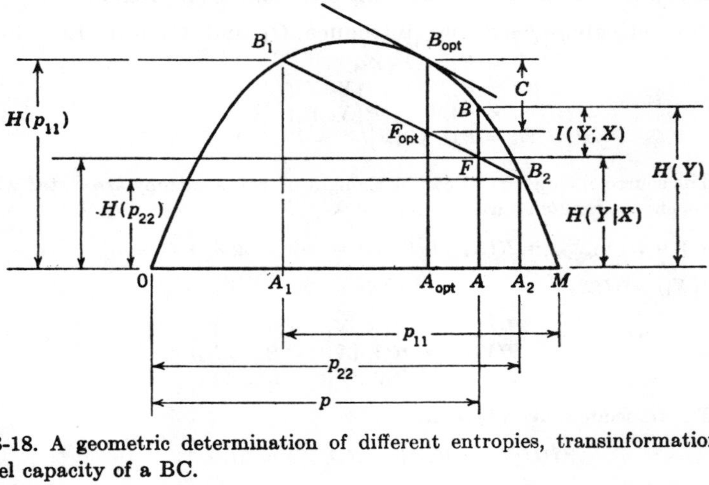

FIG. 3-18. A geometric determination of different entropies, transinformation, and channel capacity of a BC.

$MA = 1 - p$, where p is the probability of receiving 1, the transinformation can be geometrically identified. In fact,

$$I(X;Y) = H(Y) - H(Y|X)$$
$$I(X;Y) = H(p) - p_1H(p_{11}) - p_2H(p_{22})$$
$$I(X;Y) = \overline{BA} - \overline{FA}$$

Of course, the point A corresponding to the desired mode of operation is not known. A glance at Fig. 3-18 suggests that the largest value of transinformation is obtained when the probabilities at the receiving end are represented by point A_{opt} corresponding to point B_{opt}. The tangent of the entropy curve at point B_{opt} is parallel to B_1B_2. At B_{opt} the vertical segment representing the transinformation assumes its largest value. The corresponding source probabilities can be derived in a direct manner. C. E. Shannon has generalized this procedure to 3×3 and more complex channels.* His procedure is based on the use of a barycentric coordinate system. For complex channels, however, an analytic approach is often more desirable than a geometric procedure.

The following method for evaluation of the channel capacity has been suggested by S. Muroga. First one introduces auxiliary variables Q_1 and Q_2 which satisfy the following equations:

$$\begin{aligned} p_{11}Q_1 + p_{12}Q_2 &= +(p_{11} \log p_{11} + p_{12} \log p_{12}) \\ p_{21}Q_1 + p_{22}Q_2 &= +(p_{21} \log p_{21} + p_{22} \log p_{22}) \end{aligned} \tag{3-183}$$

The rate of transmission of information $I(X;Y)$ can be written as

$$\begin{aligned} I(X;Y) = H(Y) - H(Y|X) &= -(p_1' \log p_1' + p_2' \log p_2') \\ &+ p_1(p_{11} \log p_{11} + p_{12} \log p_{12}) + p_2(p_{21} \log p_{21} + p_{22} \log p_{22}) \end{aligned} \tag{3-184}$$

where p_1' and p_2' are the probabilities of receiving 0 and 1 at the output port, respectively. Next, we introduce Q_1 and Q_2 into Eq. (3-184), through Eq. (3-183):

$$I(X;Y) = -(p_1' \log p_1' + p_2' \log p_2') + (p_1p_{11} + p_2p_{21})Q_1 + (p_1p_{12} + p_2p_{22})Q_2$$

Thus, $$I(X;Y) = -(p_1' \log p_1' + p_2' \log p_2') + p_1'Q_1 + p_2'Q_2$$

The maximization of $I(X;Y)$ is now done with respect to p_1' and p_2', the probabilities at the output. In order to do this, we may use the method of Lagrangian multipliers. This method suggests maximizing the function

$$U = -(p_1' \log p_1' + p_2' \log p_2') + p_1'Q_1 + p_2'Q_2 + \mu(p_1' + p_2') \tag{3-185}$$

through a proper selection of the constant number μ. Therefore one requires

$$\begin{aligned} \frac{\partial U}{\partial p_1'} &= -(\log e + \log p_1') + Q_1 + \mu = 0 \\ \frac{\partial U}{\partial p_2'} &= -(\log e + \log p_2') + Q_2 + \mu = 0 \end{aligned} \tag{3-186}$$

The simultaneous validity of these equations requires that

$$\mu = -Q_1 + (\log e + \log p_1') = -Q_2 + (\log e + \log p_2') \tag{3-187}$$

* C. E. Shannon, Geometrische Deutung einiger Ergebnisse bei der Berechnung der Kanalkapazität, *NTZ-Nachrtech Z.*, vol. 10, no. 1, pp. 1–4, 1957.

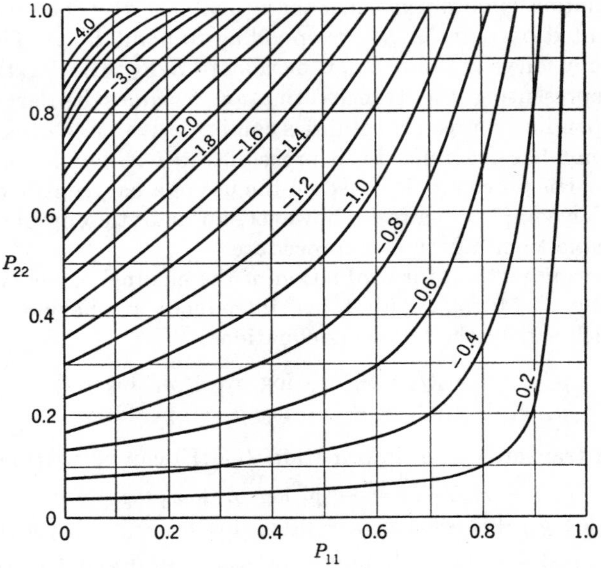

FIG. 3-19. A chart for determining values of Q_1 in terms of P_{11} and P_{22} for binary channels. The corresponding value of Q_2 is obtained by an interchange of P_{11} and P_{22}.

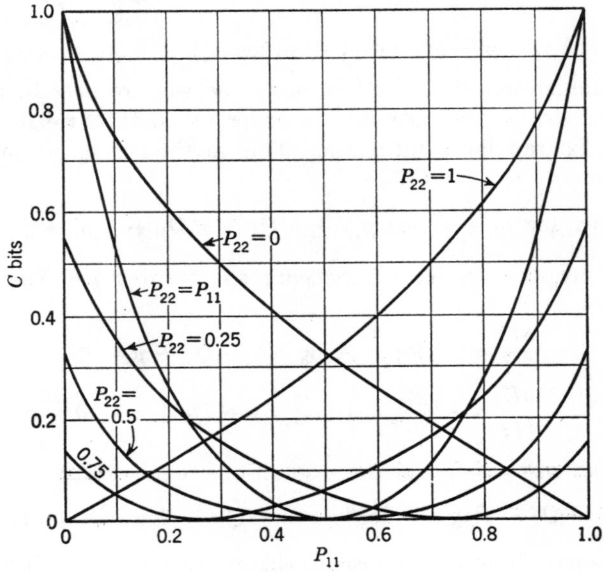

FIG. 3-20. Capacity of a binary channel in terms of P_{11} and P_{22}.

The channel capacity is found to be

$$C = \max [I(X;Y)] = Q_1 - \log p'_1 = Q_2 - \log p'_2$$

The values of Q_1 and Q_2 may be obtained from the set of Eqs. (3-183). But note that

$$p'_i = \exp (Q_i - C) \qquad i = 1, 2 \tag{3-188}$$

Thus

$$C = \log [\exp (Q_1) + \exp (Q_2)] \tag{3-189}$$

$$C = \log \sum_{i=1}^{i=2} \exp (Q_i) = \log (2^{Q_1} + 2^{Q_2})$$

A similar result was obtained earlier in a different way by Shannon. Later on Silverman and Chang derived further additional interesting results. The chart of Fig. 3-19 gives the value of Q_1 and Q_2 for a binary channel. The chart of Fig. 3-20 gives the corresponding capacities (*IRE Trans. on Inform. Theory*, vol. IT-4, p. 153, December, 1958). Note that the capacity of a binary channel is greater than zero except when

$$p_{11} + p_{22} = 1$$

Example 3-9. Find the capacity of the following three binary channels, first directly and then from the graph of Figs. 3-19 and 3-20, in each of the following three cases:

(*a*) $p_{11} = p_{22} = 1$

(*b*) $p_{11} = p_{12} = p_{21} = p_{22} = \frac{1}{2}$

(*c*) $p_{11} = p_{12} = \frac{1}{2} \qquad p_{21} = \frac{1}{4} \qquad p_{22} = \frac{3}{4}$

Solution

(*a*)

$$p_{11} = p_{22} = 1$$
$$p_{12} = p_{21} = 0$$

Direct computation yields

$$Q_1 = Q_2 = 0$$
$$C = \log (2^{Q_1} + 2^{Q_2}) = 1 \text{ bit}$$

This channel capacity is achieved when the input symbols are equiprobable.

(*b*)

$$p_{11} = p_{12} = p_{21} = p_{22} = \tfrac{1}{2}$$

The noise matrix is singular and leads to

$$Q_1 + Q_2 = -2$$
$$Q_1 = Q_2 = -1$$
$$C = \log (2^{-1} + 2^{-1}) = 0$$

Any input probability distribution will lead to zero transinformation as the input and the output are independent. This result can be verified by checking with Figs. 3-19 and 3-20.

(c)

$$p_{11} = p_{12} = \tfrac{1}{2}$$
$$p_{21} = \tfrac{1}{4} \qquad p_{22} = \tfrac{3}{4}$$
$$\begin{bmatrix} \tfrac{1}{2} & \tfrac{1}{2} \\ \tfrac{1}{4} & \tfrac{3}{4} \end{bmatrix} \begin{bmatrix} Q_1 \\ Q_2 \end{bmatrix} = \begin{bmatrix} \tfrac{1}{2}\log\tfrac{1}{2} + \tfrac{1}{2}\log\tfrac{1}{2} \\ \tfrac{1}{4}\log\tfrac{1}{4} + \tfrac{3}{4}\log\tfrac{3}{4} \end{bmatrix} = \begin{bmatrix} -1 \\ -2 + \tfrac{3}{4}\log 3 \end{bmatrix}$$
$$\begin{bmatrix} Q_1 \\ Q_2 \end{bmatrix} = \begin{bmatrix} 1 - \tfrac{3}{2}\log 3 \\ -3 + \tfrac{3}{2}\log 3 \end{bmatrix} = \begin{bmatrix} -1.378 \\ -0.622 \end{bmatrix}$$
$$C = \log\,(2^{-1.378} + 2^{-0.622}) = \log 1.0345 = 0.048 \text{ bit}$$

This answer can be verified from the graph of Fig. 3-20.

The generalization of the above method for a channel with an $m \times m$ noise matrix is straightforward. In fact, let

$$\begin{aligned} p_{11}Q_1 + \cdots + p_{1m}Q_m &= \sum_{j=1}^{m} p_{1j}\log p_{1j} \\ \cdots\cdots&\cdots\cdots \\ p_{m1}Q_1 + \cdots + p_{mm}Q_m &= \sum_{j=1}^{m} p_{mj}\log p_{mj} \end{aligned} \tag{3-190}$$

and assume that the solution to this set exists. Then the rate of transmission of information, as before, will become

$$-\sum_{i=1}^{m} p'_i \log p'_i + \sum_{i=1}^{m} p'_i Q_i \tag{3-191}$$

The use of the Lagrangian multiplier method will lead to

$$\begin{aligned} C &= Q_i - \log p'_i \\ C &= \log \sum_{i=1}^{m} 2^{Q_i} \end{aligned} \tag{3-192}$$

It is to be kept in mind that the values of C thus obtained may not necessarily correspond to a set of realizable input probabilities

$$\left(0 \le p_i \le 1, \sum_i p_i = 1\right)$$

In the latter case the calculation of the channel capacity is more complicated. Also, the solution to the set [Eq. (3-190)] may not exist, or the channel matrix may not even be a square matrix. In such cases some modifications of the above method are suggested in the afore-mentioned references. At any rate, although the formulation of the equations leading to the channel capacity is straightforward, computational difficulties exist and the present methods are not completely satisfactory.

The capacity of a general binary channel has been computed by R. A. Silverman, S. Chang, and J. Loeb. A straightforward computation leads to

$$C(\alpha,\beta) = \frac{-\beta H(\alpha) + \alpha H(\beta)}{\beta - \alpha} + \log\left[1 + \exp\frac{H(\alpha) - H(\beta)}{\beta - \alpha}\right]$$

where parameter $\alpha = p_{11}$, $\beta = p_{21}$, and H stands for the entropy of a binary source. Note that

$$C(\alpha,\beta) = C(\beta,\alpha) = C(1-\alpha, 1-\beta) = C(1-\beta, 1-\alpha)$$

The input probability $P\{0\}$ leading to the channel capacity is given by Silverman and Chang as

$$P\{0\} = P(\alpha,\beta) = \beta(\beta-\alpha)^{-1} - (\beta-\alpha)^{-1}\left[1 + \exp\frac{H(\beta) - H(\alpha)}{\beta - \alpha}\right]^{-1}$$

$$0.37 \cong \frac{1}{e} \le P\{0\} \le 1 - \frac{1}{e} \cong 0.63$$

The probability of receiving zeros when the channel capacity is achieved is

$$\left[1 + \exp\frac{H(\beta) - H(\alpha)}{\beta - \alpha}\right]^{-1}$$

Example 3-10. Find the capacity of the channel with the noise matrix as shown below:

$$\begin{bmatrix} \frac{1}{2} & \frac{1}{4} & 0 & \frac{1}{4} \\ 0 & 1 & 0 & 0 \\ 0 & 0 & 1 & 0 \\ \frac{1}{4} & 0 & \frac{1}{4} & \frac{1}{2} \end{bmatrix}$$

Solution

$$Q_1 = Q_4 = -2$$
$$Q_2 = Q_3 = 0$$
$$C = \log(2^{-2} + 2^0 + 2^0 + 2^{-2}) = \log 5 - 1 = 1.321 \text{ bits}$$

Example 3-11. Determine the capacity of a ternary channel with the stochastic matrix

$$[P] = \begin{bmatrix} \alpha & 1-\alpha & 0 \\ \frac{1}{2} & 0 & \frac{1}{2} \\ 0 & 1-\alpha & \alpha \end{bmatrix}$$
$$0 \le \alpha \le 1$$

Solution. Since the channel matrix is a square matrix, Eqs. (3-190) yield

$$[P][Q] = -[H]$$
$$[Q] = -[P]^{-1}[H]$$

$$\begin{bmatrix} Q_1 \\ Q_2 \\ Q_3 \end{bmatrix} = -\begin{bmatrix} \frac{1}{2\alpha} & 1 & -\frac{1}{2\alpha} \\ \frac{1}{2(1-\alpha)} & \frac{\alpha}{\alpha-1} & \frac{1}{2(1-\alpha)} \\ -\frac{1}{2\alpha} & 1 & \frac{1}{2\alpha} \end{bmatrix}\begin{bmatrix} h \\ 1 \\ h \end{bmatrix}$$

where $\qquad h = -\alpha \log \alpha - (1-\alpha)\log(1-\alpha)$

$$Q = \begin{bmatrix} -1 \\ -\frac{h-\alpha}{1-\alpha} \\ -1 \end{bmatrix}$$

$$C = \log(2^{Q_1} + 2^{Q_2} + 2^{Q_3}) = \log\left(1 + \exp\frac{\alpha - h}{1 - \alpha}\right)$$

According to the method applied by Muroga, Silverman, and Chang, a direct computation leads to the following values for the probability of the ith input symbol achieving channel capacity.

$$p_i = 2^{-C} \sum_{k=1}^{m} p_{ki}^{-1} 2^{Q_k} \qquad 1 < i \leq m$$

where p_{ki}^{-1} is the element of the inverse channel matrix $[P]^{-1}$.

In this example, one finds

$$p_1 = p_3 = 2^{-C} \cdot 2^{Q_1}$$
$$p_2 = \frac{1 + [\alpha/(\alpha - 1)] \exp [(\alpha - h)/(1 - \alpha)]}{1 + \exp [(\alpha - h)/(1 - \alpha)]}$$

Of course, if we desire to employ this method, the input probabilities must remain nonnegative. The condition $p_2 \geq 0$ yields

$$\frac{h - \alpha}{1 - \alpha} \geq \log \frac{\alpha}{1 - \alpha}$$

The equality is valid for $\alpha \approx 0.641$, and the channel capacity is achieved when $\alpha \geq 0.641$.

3-18. Binary Pulse Width Communication Channel. In many practical problems, there is a time (or cost) associated with the transmission of each letter of the alphabet. In such circumstances, it is desired to investigate the rate of transmission of information and the capacity of the channel in the absence of noise. To obtain the maximum rate of transmission of information, we assume independence of successive letters and we consider transmitting all words of duration T with equal probabilities. Therefore

$$R = \frac{\log N(T)}{T} \tag{3-193}$$

Shannon has defined the capacity of this noiseless channel as the limit of R when very long messages are considered, e.g.,

$$C = \lim_{T \to \infty} \frac{\log N(T)}{T} \tag{3-194}$$

Thus, the problem of calculation of the capacity for such communication channels, under the above assumptions, is reduced to a combinatorial problem, that is, computing $N(T)$.

Let the alphabet be $[a_1, a_2, \ldots, a_n]$ and the associated duration $[t_1, t_2, \ldots, t_n]$. The number of distinct words of duration T is given by

$$N(T) = \sum_{k=1}^{k=n} N(T - t_k) \tag{3-195}$$

This is a difference equation. The general solution of this equation is

$$N(T) = \sum_{k=1}^{n} A_k r_k^T \tag{3-196}$$

where the r_k's are roots of the characteristic equation

$$1 - \sum_{k=1}^{n} r^{-t_k} = 1 - f(r) = 0 \tag{3-197}$$

The constants $A_1, A_2, \ldots, A_k$ depend on the boundary conditions of the problem. At the moment, we are interested in an evaluation of $N(T)$ for very large values of T. From Eq. (3-197) it is clear that $f(r)$ is a monotonic decreasing function:

$$f(0) = \infty \qquad f(\infty) = 0$$

Therefore, the equation $f(r) = 1$ cannot have more than one positive real root. Hence for large values of T, the function $N(T)$ behaves as $A_i r_i^T$, where r_i is the positive root of the characteristic equation (3-197).

In the absence of additional constraint, the channel capacity becomes

$$C = \lim_{T\to\infty} \frac{T \log r_i + \log A_i}{T} \tag{3-198}$$
$$C = \log r_i$$

Example 3-12. Consider an alphabet consisting of two rectangular pulses of equal heights. The duration of the pulses are 2 and 4 time units. Find the capacity of a noiseless channel transmitting very long messages. (The messages are fed to the channel with equal probability and independently.)

Solution. The difference equation to be solved is

$$N(T) = N(T-2) + N(T-4)$$

Assuming $N(-3) = N(-2) = N(-1) = 0$, straightforward computation yields

T	$N(T)$	$\log N(T)$	$\frac{1}{T} \log N(T)$
2	1	0	0
3	1	0	0
4	2	1	0.250
5	2	1	0.200
6	3	1.585	0.264
7	3	1.585	0.226
8	5	2.322	0.290
9	5	2.322	0.258
10	8	3.000	0.300
15	21	4.392	0.292
19	55	5.781	0.304

Comparison between successive increments for values about $T = 19$ to $T = 23$ shows that $(1/T) \log N(T)$ approaches the value $C = 0.342$. We may alternatively employ Eq. (3-197):

$$1 - f(r) = 1 - r^{-2} - r^{-4} = 0$$

$$r_1 = -r_2 = \left(\frac{\sqrt{5} - 1}{2}\right)^{-\frac{1}{2}}$$

$$r_3 = -r_4 = j\left(\frac{\sqrt{5} + 1}{2}\right)^{-\frac{1}{2}}$$

The only positive root of the characteristic equation is $r_1 = (0.635)^{-\frac{1}{2}}$.

$$C = -\tfrac{1}{2} \log 0.635 = 0.328$$

3-19. Uniqueness of the Entropy Function. We have adopted the logarithmic form for the communication entropy as the most suitable form satisfying certain specified requirements. In this section we wish to prove formally that, if a few specified requirements are to be fulfilled, the logarithmic form is the unique function satisfying these constraints. The complexity of the proof depends on the type of constraints imposed. The following requirements for an entropy function seem to be reasonable.

1. Given a finite complete probability scheme $[p_1,p_2, \ldots ,p_n]$, the associated entropy function $H(p_1,p_2, \ldots ,p_n)$ must take on its largest value when all events are equiprobable.

2. For a joint finite complete scheme the associated entropies should satisfy the identity

$$H(X,Y) = H(X) + H(Y|X)$$

The average information conveyed by (X,Y) is the sum of the average information given by X and that provided by Y when X is given.

3. Adding an impossible event to a scheme should not change the entropy of the scheme.

$$H(p_1,p_2, \ldots ,p_n,0) = H(p_1,p_2, \ldots ,p_n)$$

4. The entropy function is continuous with respect to all its arguments.

Theorem. Let $H(p_1,p_2, \ldots ,p_n)$ be a function satisfying requirements 1, 2, 3, and 4 above for any values of p_k $(k = 1, 2, \ldots, n)$; then

$$H(p_1,p_2, \ldots ,p_n) = -\lambda \sum_{i=1}^{n} p_i \log p_i \qquad \lambda > 0$$

Proof. Let

$$H\left(\frac{1}{n},\frac{1}{n}, \ldots ,\frac{1}{n}\right) = f(n) \tag{3-199}$$

The first step in the proof is to show that $f(n) = \lambda \log n$. In fact,

$$f(n) = H\left(\frac{1}{n},\frac{1}{n}, \ldots ,\frac{1}{n},0\right) \leq H\left(\frac{1}{n+1},\frac{1}{n+1}, \ldots ,\frac{1}{n+1}\right) = f(n+1) \tag{3-200}$$

Thus, the desired $f(n)$ is a nondecreasing function of n. Note that according to requirement 2, for any complete probability scheme consisting of the sum of m mutually exclusive schemes, we can write

$$H(X_1,X_2, \ldots ,X_m) = \sum_{k=1}^{m} H(X_k) \tag{3-201}$$

If each scheme consists of r equally likely events, we have

$$H(X_1,X_2, \ldots ,X_m) = mf(r) = f(r^m) \tag{3-202}$$

m and r being any arbitrary positive integers. Now we choose integers t and n such that

$$r^m < t^n < r^{m+1}$$

or

$$m \log r < n \log t < (m+1) \log r \tag{3-203}$$

$$\frac{m}{n} < \frac{\log t}{\log r} < \frac{m+1}{n}$$

From the nondecreasing property of $f(n)$ we conclude that

$$\begin{gathered} f(r^m) \leq f(t^n) \leq f(r^{m+1}) \\ mf(r) \leq nf(t) \leq (m+1)f(r) \\ \frac{m}{n} \leq \frac{f(t)}{f(r)} \leq \frac{m+1}{n} \end{gathered} \tag{3-204}$$

Comparison between Eqs. (3-204) and (3-203) yields

$$\left| \frac{f(t)}{f(r)} - \frac{\log t}{\log r} \right| \leq \frac{1}{n} \tag{3-205}$$

Since n can be chosen arbitrarily large, for any positive integers r and t, we have

$$\frac{f(t)}{\log t} = \frac{f(r)}{\log r}$$

or

$$f(t) = \lambda \log t \tag{3-206}$$

The nondecreasing property of $f(t)$ requires that λ be a positive constant. Thus we have proved the uniqueness theorem for the special case when all events are equiprobable. Next, we consider the case where all the probabilities are required to be rational numbers but not necessarily all equal. Let α be a common denominator for the different rational p_k and let

$$p_k = \frac{\alpha_k}{\alpha} \qquad \sum_{k=1}^{n} \alpha_k = \alpha \qquad \alpha_k > 0 \quad k = 1, 2, \ldots , n \tag{3-207}$$

In order to define the entropy of this scheme (X), we shall transfer the problem to the previously discussed case. For this, consider a probability scheme Y depending on X. Let the scheme Z consist of α equally likely events: $[z_1, z_2, \ldots, z_\alpha]$. For convenience we partition these events into groups containing $\alpha_1, \alpha_2, \ldots,$ and α_n events, respectively. This partitioned scheme will be referred to as scheme Y. When the event X_k with probability p_k occurs, then in scheme Y all events partitioned in the kth group occur with equal probability. Therefore,

$$H\left(\frac{1}{\alpha_k}, \frac{1}{\alpha_k}, \ldots, \frac{1}{\alpha_k}\right) = \lambda \log \alpha_k \tag{3-208}$$

$$H(Y|X) = \sum_{k=1}^{n} p_k H\left(\frac{1}{\alpha_k}, \frac{1}{\alpha_k}, \ldots, \frac{1}{\alpha_k}\right)$$

$$= \lambda \sum_{k=1}^{n} p_k \log p_k + \lambda \log \alpha$$

The totality of events in Z forms the sum of the two schemes:

$$H(Z) = H(X,Y) = f(\alpha) = \lambda \log \alpha \tag{3-209}$$

But, according to the additivity requirement,

$$H(X) = H(X,Y) - H(Y|X)$$

$$= -\lambda \sum_{k=1}^{n} p_k \log p_k \tag{3-210}$$

Thus the uniqueness theorem is also valid when the p_k are rational numbers. Finally the postulate of continuity of the entropy function guarantees the validity of the theorem when the p_k are incommensurable. The proof given here is based on Khinchin's elegant presentation of Shannon's original idea. A more extensive proof based on less restrictive requirements has been very neatly derived by Fadiev [see Feinstein (I, Chap. 1)].

PROBLEMS

3-1. For a binary channel with

$$P\{B_1|A_1\} = p_1 \qquad P\{B_2|A_2\} = p_2$$

driven by a source

$$P\{A_1\} = \alpha \qquad P\{A_2\} = 1 - \alpha$$

find

(*a*) The average information rate of the input letters.
(*b*) The average information rate of the output letters.
(*c*) The average transinformation.

(*d*) The results of (*a*), (*b*), and (*c*) when $\alpha = \frac{1}{2}$, $p_1 = \frac{1}{4}$, $p_2 = \frac{1}{5}$.

(*e*) The input probabilities which make the transinformation a maximum for $p_1 = \frac{1}{4}$, $p_2 = \frac{1}{5}$.

(*f*) What is the capacity of the channel described in (*e*)?

3-2. Find the capacity of the channel illustrated in Fig. P3-2.

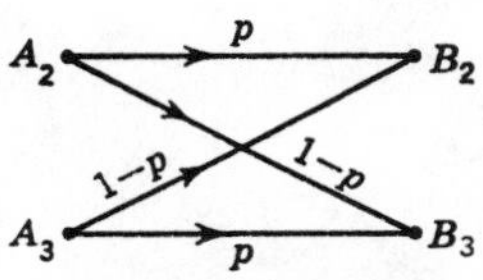

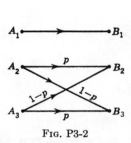

FIG. P3-2

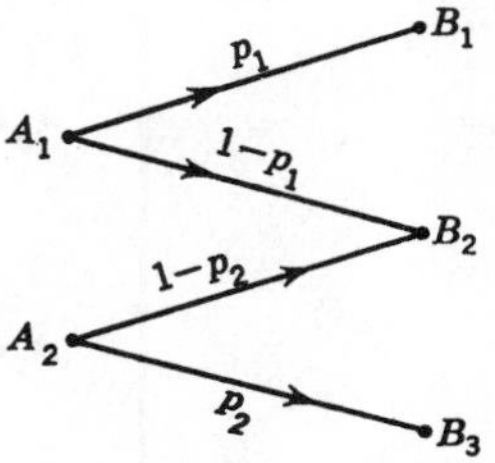

FIG. P3-3

3-3. (*a*) Compute the transinformation $I(X;Y)$ in the channel in Fig. P3-3, when symbols A_1 and A_2 are transmitted with respective probabilities α_1 and α_2 $(\alpha_1 + \alpha_2 = 1)$.

(*b*) Compute part (*a*) for $p_1 = \frac{2}{3}$, $p_2 = \frac{5}{6}$, $\alpha_1 = \alpha_2 = \frac{1}{2}$.

(*c*) In part (*b*) assume that at the receiving station, when B_2 is received, we "decide" that most likely A_1 was transmitted. This assumption provides us with a new transinformation $I(X;Y)$ which it is desired to calculate.

3-4. In the channel in Fig. P3-4 the messages $[x_1,x_2,x_3]$ are transmitted with respective probabilities $[\alpha_1,\alpha_2,\alpha_3]$. Find the channel capacity.

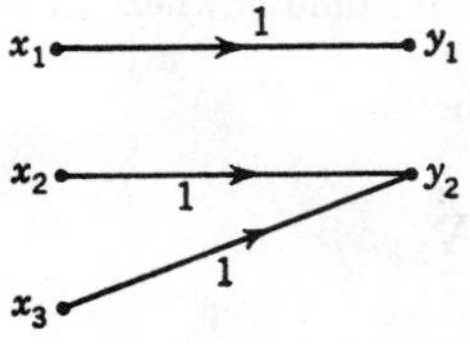

FIG. P3-4

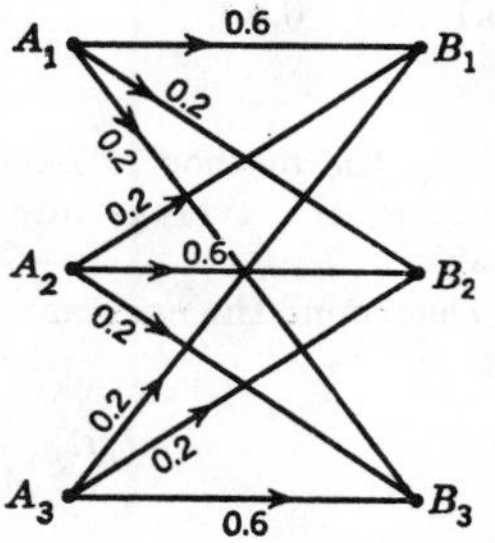

FIG. P3-5

3-5. A discrete source transmits messages $[A_1,A_2,A_3]$ with respective probabilities $[\frac{1}{2},\frac{1}{4},\frac{1}{4}]$. The source is connected to the channel given in Fig. P3-5. Determine

(*a*) $H(X)$.

(*b*) $H(Y)$.

(*c*) $I(X;Y)$.

(*d*) Channel capacity.

3-6. Same question as in Prob. 3-5 for the channel in Fig. P3-6:

$$P\{A_1\} = 0.6 \qquad P\{A_2\} = 0.3 \qquad P\{A_3\} = 0.1$$

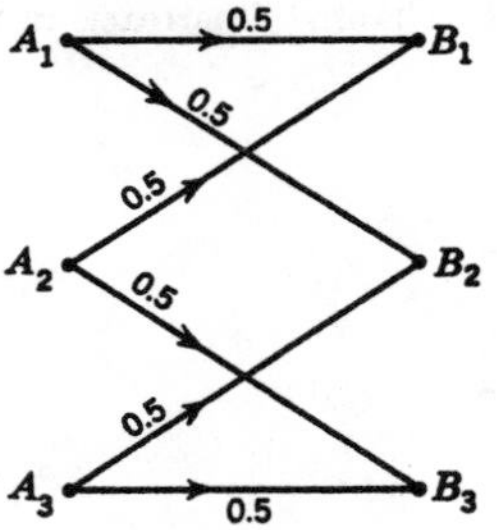

FIG. P3-6

3-7. Find the capacity of a binary channel with the channel matrices shown.

(a) $\begin{bmatrix} 9/10 & 1/10 \\ 1/10 & 9/10 \end{bmatrix}$ (b) $\begin{bmatrix} 9/10 & 1/10 \\ 2/10 & 8/10 \end{bmatrix}$

3-8. Compute the channel capacity when the channel is specified by each of the following matrices:

(a) $\begin{bmatrix} 3/4 & 1/4 & 0 \\ 1/8 & 3/4 & 1/8 \\ 1/4 & 0 & 3/4 \end{bmatrix}$ (b) $\begin{bmatrix} 2/3 & 1/3 & 0 \\ 3/4 & 0 & 1/4 \\ 0 & 3/4 & 1/4 \end{bmatrix}$

(c) $\begin{bmatrix} 0.8 & 0.1 & 0.1 \\ 0.1 & 0.8 & 0.1 \\ 0.1 & 0.1 & 0.8 \end{bmatrix}$ (d) $\begin{bmatrix} 3/4 & 1/8 & 1/8 & 0 \\ 1/8 & 3/4 & 0 & 1/8 \\ 1/8 & 1/8 & 3/4 & 0 \\ 0 & 0 & 1/4 & 3/4 \end{bmatrix}$

3-9. Using the method of Lagrange multipliers, give an alternative proof for Eq. (3-14), that is, the average information H has its maximum when all events are equiprobable.

HINT: Determine the constant λ such that the function

$$H(p_1,p_2, \ldots ,p_n) + \lambda \sum_{k=1}^{n} p_k$$

reaches its maximum value.

3-10. The following two finite probability schemes are given: $[p_1,p_2, \ldots ,p_n]$ and $[q_1,q_2, \ldots ,q_n]$. Show that

$$-\sum_{k=1}^{n} p_k \log q_k \geq -\sum_{k=1}^{n} p_k \log p_k$$

HINT: Let $q_k = p_k + r_k$ and express the above inequality in terms of p_k and r_k. Finally, apply Eq. (3-50) to the variable $x = 1 + r/p$.

3-11. It is possible to have the maximum of transinformation for more than one input probability distribution. The channel in Fig. P3-11 illustrates such a situation.

$$P\{X\} = \{p_1, p_2, p_3\}$$

Find $I(X;Y)$ and the condition for its maximum.

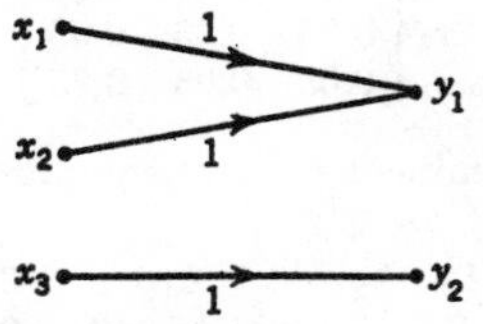

FIG. P3-11

3-12. Same problem as in Example 3-12, but with the duration of the pulses 2 and 5 time units.

3-13. Find the capacity of the memoryless channel specified by the matrix below:

$$\begin{bmatrix} \frac{1}{2} & \frac{1}{4} & \frac{1}{4} & 0 \\ \frac{1}{4} & \frac{1}{4} & \frac{1}{4} & \frac{1}{4} \\ 0 & 0 & 1 & 0 \\ \frac{1}{2} & 0 & 0 & \frac{1}{2} \end{bmatrix}$$

3-14. Under the hypothesis of Sec. 3-18, consider a telegraph channel where the symbols and their durations are

dot	2 time units
dash	4 time units
space	3 time units

(*a*) Derive the capacity of the channel for very long equiprobable messages by direct computation.

(*b*) Calculate the channel capacity by the described method of solving the characteristic equation $1 - f(r) = 0$.

3-15. In actual pulse-type communication like telegraphy, some additional constraint should be kept in mind. For example, in ordinary telegraphy we may consider the alphabet as consisting of four symbols: dot, dash, letter space, and word space. Two spaces may not be transmitted successively. A relevant diagram is given in Fig. P3-15 (dot 2 time units, letter space 3 time units, dash 4 time units, and word space 6 time units).

Extend the calculation of Sec. 3-18 to this case and derive a formula for the channel capacity under the hypothesis described in Sec. 3-18 [see S. Goldman (Appendix 1) and L. Brillouin (I, Chap. 4)].

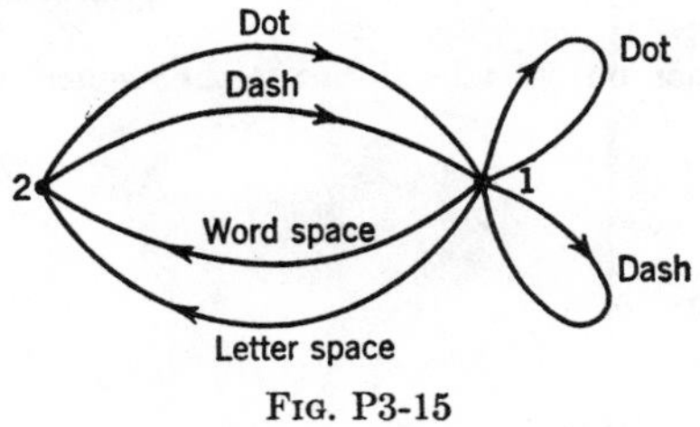

FIG. P3-15

3-16. An independent source transmits messages $[x_1,x_2,x_3]$ with probabilities [0.3, 0.35, 0.35]. The messages are transmitted over a noisy channel. Knowing the joint probability matrix for the transmitted-received pairs, find the channel matrix.

$$\begin{array}{c} \\ x_1 \\ x_2 \\ x_3 \end{array}\begin{array}{c} \begin{array}{ccc} y_1 & y_2 & y_3 \end{array} \\ \begin{bmatrix} 0.20 & 0.05 & 0.05 \\ 0.05 & 0.25 & 0.05 \\ 0.05 & 0.05 & 0.25 \end{bmatrix} \end{array}$$

3-17. Find the rate of transmission of three equiprobable messages over the following channel:

$$\begin{bmatrix} \frac{1}{2} & \frac{1}{2} & 0 & 0 & 0 & 0 \\ 0 & 0 & \frac{1}{2} & \frac{1}{2} & 0 & 0 \\ 0 & 0 & 0 & 0 & \frac{1}{2} & \frac{1}{2} \end{bmatrix}$$

3-18. Determine the capacity of the channel

$$\begin{bmatrix} \frac{1}{3} & \frac{2}{3} & 0 \\ \frac{2}{3} & \frac{1}{3} & 0 \\ 0 & 0 & 1 \end{bmatrix}$$

3-19. Let

$$N = \frac{n!}{n_1!n_2! \cdots n_k!}$$

where

$$n = n_1 + n_2 + \cdots + n_k$$

Using Stirling's formula, show that f r large values of $n_1, n_2, \ldots$ we have

$$\log N = -n \sum_{i=1}^{k} \frac{n_i}{n} \log \frac{n_i}{n}$$

Discuss the connection between this result and the definition of the entropy. [See Brillouin (I, pp. 7–8).]

3-20. (*Advanced Problem.*) A discrete random variable X with a specified first moment m may assume any one of a number n values with different probabilities. Using the Lagrange multiplier technique, find the probability distribution that gives the maximum entropy for the given m.

HINT: The problem is not a very simple one despite its appearance. A solution of this problem is given by B. S. Fleishman and G. B. Linkovskii, Maximum Entropy of an Unknown Discrete Distribution with Given First Moment (*Radiotekh. i Elektron.*, vol. 3, no. 4, pp. 554–556, 1958. English translation available).

3-21. (*Advanced Problem.*) Let k_1 and k_2 be the stochastic matrices of two given binary channels. Find the necessary and sufficient condition for their partial ordering, that is, $k_1 \supseteq k_2$.

HINT: Read the section on partial ordering in the Appendix; also see R. A. Silverman.

CHAPTER 4

ELEMENTS OF ENCODING

Thus far, coding offers a most significant application of information theory. The material presented here strongly relies on the content of Chap. 3. In this chapter, some of the fundamental theorems of information theory will be introduced. The noiseless encoding theorem and the fundamental theorem of discrete noiseless memoryless channels will be given in some detail, with several encoding procedures treated as applications of these theorems. A heuristic proof for the fundamental theorem of discrete memoryless channels in the presence of noise will be discussed. The formal proof for the latter theorem requires some further knowledge of probability theory beyond the contents of Chap. 3; consequently, it will be deferred until Chap. 12.

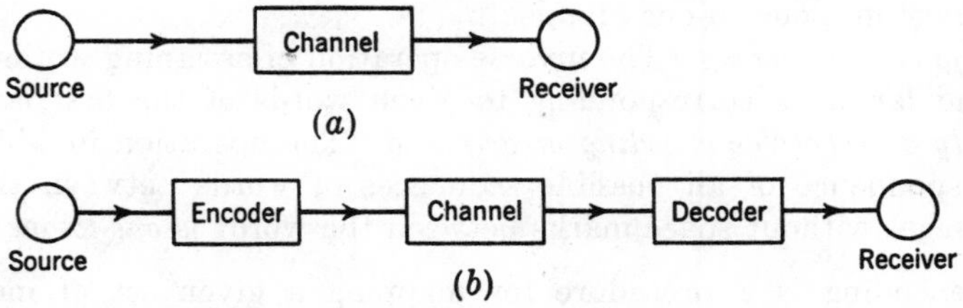

FIG. 4-1. A simplified model of a communication system (*a*) without encoder-decoder; (*b*) with encoder-decoder.

4-1. The Purpose of Encoding. The word *encoding*, like several other common terms of communication engineering, such as detection and modulation, is a descriptive word with a broad meaning. It is frequently used in a large variety of cases as a transformation procedure operating on the input signal prior to its entry into the communication channel, the main purpose of coding being, in general, to improve the "efficiency" of the communication link in some sense. This definition is, of course, unnecessarily broad and vague. In the present work we shall confine ourselves to a much more restricted definition which will be described later. Consider the basic elements of a communication setup as shown in Fig. 4-1*a*.

By an independent source we mean here a device that selects messages at random from a discrete message ensemble with prescribed probabilities.

$$\{m_1,m_2, \ldots ,m_N\}$$
$$p\{m_1\}, p\{m_2\}, \ldots , p\{m_N\}$$

In the development of this chapter we assume that successive messages are selected independently; that is, the source has no memory. Later on in Chap. 11 we shall discuss sources with memory where the selection of a message is affected by the selection of some of the previously transmitted messages.

The channel is assumed to be discrete and without memory. Its behavior is specified by a finite conditional probability matrix also referred to as a channel matrix. The channel of communication usually deals with symbols of some specified list. This list is generally referred to as the *alphabet* of the communication language. The following terminology is suggested for our subsequent work.

Letter, symbol, or character Any individual member of the alphabet set

Message or word A finite sequence of letters of the alphabet

Length of a word The number of letters in a word

Encoding or enciphering A procedure for associating words constructed from a finite alphabet of a language with given words of another language in a one-to-one manner

Decoding or deciphering The inverse operation of assigning words of the second language corresponding to given words of the first language

Uniquely decipherable encoding or decoding The operation in which the correspondence of all possible sequences of words between the two languages without space marks between the words is *one-to-one*

Thus, encoding is a procedure for mapping a given set of messages $[m_1,m_2, \ldots ,m_N]$ onto a new set of encoded messages $[c_1,c_2, \ldots ,c_N]$ so that the transformation is one-to-one. Also, generally by encoding we wish to improve the "efficiency" of the "transmission." It is, of course, possible to devise codes for a special purpose (such as secrecy) without relevance to the transmission efficiency in our adopted sense. It is also possible to resort to codes which do not have a one-to-one association. However, our present study will be confined strictly to one-to-one codes with an eye to improving some sort of "efficiency" of transmission.

If an alphabet set is denoted by

$$\{A\} = \{a_1,a_2, \ldots ,a_D\}$$

the sequences $a_1a_1a_2$, a_Da_1, and $a_2a_2a_2a_2$ will be referred to as words of this language. The lengths of these words are three, two, and four symbols, respectively. Similarly the set of letters $\{0,1\}$ constitutes what is commonly known as the binary alphabet; 001 is a word in the binary language.

By speaking of more efficient encoding, we agree to refer to encoding procedures that improve certain "cost functions." Perhaps the simplest cost function is obtained when we assign a constant cost figure t_i to each message m_i; t_i can be the duration or any other cost factor. Then the average cost per message becomes

$$R_t = \sum_{i=1}^{N} p\{m_i\} \cdot t_i \tag{4-1}$$

Obviously, the most efficient transmission is the one that minimizes the average cost R_t. In this chapter we confine ourselves to the simplest case when all symbols have identical cost. Thus, the average cost per message becomes proportional to the average of n_i, the number of symbols per message (or the average length of messages $\bar{L}$):

$$R_t = \bar{L} = \sum_{i=1}^{N} p\{m_i\}n_i \qquad t_i = n_i \quad i = 1, 2, \ldots, N \tag{4-1a}$$

An increase in transmission efficiency can be obtained by proper encoding of messages, that is, assigning new sequences of symbols to each message m_i so that the statistical distribution of the new symbols reduces the average word length $\bar{L}$.

The efficiency of the encoding procedure can be defined if, and only if, we know the lowest possible bound of $\bar{L}$. Of course, if such a lower bound does not exist, the term efficiency will be meaningless. Thus, an important question arises here. For a given set of messages and a given alphabet, what is the lowest possible $\bar{L}$ that can be obtained? In this chapter we show that, subject to certain restriction on the encoding rule, the lower bound for $\bar{L}$ is $H(X)/(\log D)$, where $H(X)$ is the entropy of the original message ensemble and D the number of symbols in the encoding alphabet. For the time being, we accept the following definition for efficiency of an encoding procedure: the ratio of the average information per symbol of encoded language to the maximum possible average information per symbol, that is,

$$\text{Efficiency} = \frac{H(X)}{\bar{L}} : \log D = \frac{H(X)}{\bar{L} \log D} \tag{4-2}$$

If a number of messages are encoded into new words taken from a D-symbol alphabet, the maximum possible information per symbol supplied by an independent source will be $\log D$. If the encoded words have an average length $\bar{L}$, then the entropy per symbol is $H(X)/\bar{L}$. Thus

$H(X)/\bar{L}:\log D$ gives a measure of the efficiency of the encoding as far as the entropy per symbol is concerned.

The redundancy of the code is defined as

$$\text{Redundancy} = 1 - \text{efficiency} = \frac{\bar{L}\log D - H(X)}{\bar{L}\log D} \tag{4-2a}$$

For example, let the transmitter have four messages as given below:

$$[M] = [m_1,m_2,m_3,m_4]$$
$$[P\{M\}] = [\tfrac{1}{2},\tfrac{1}{4},\tfrac{1}{8},\tfrac{1}{8}]$$

If we assume that $\{m_1,m_2,m_3,m_4\}$ constitutes our alphabet set, then according to Eq. (4-2) the entropy per symbol is

$$\text{Efficiency} = \frac{H(X)}{\log N} = \frac{-\tfrac{1}{2}\log\tfrac{1}{2} - \tfrac{1}{4}\log\tfrac{1}{4} - \tfrac{2}{8}\log\tfrac{1}{8}}{\log 4}$$
$$= \frac{7}{8} = 0.875 = 87.5 \text{ per cent}$$
$$\text{Redundancy} = 12.5 \text{ per cent}$$

We may wish to encode the messages into words selected from a binary alphabet with a one-to-one correspondence. The encoding may have been motivated by the need for improving the efficiency or simply by the necessity of using a binary language for the transmission of the data (as in computers and pulse-width communication). Then the following code may be suggested:

m_1	0	0
m_2	0	1
m_3	1	0
m_4	1	1

Thus we have a new transmitter transmitting 0's and 1's, with certain probabilities:

$$p\{0\} = \frac{\sum_{k=1}^{4} p\{m_k\}C_{k0}}{\sum_{k=1}^{4} p\{m_k\}n_k} = \frac{\tfrac{1}{2}\cdot 2 + \tfrac{1}{4}\cdot 1 + \tfrac{1}{8}\cdot 1}{\tfrac{1}{2}\cdot 2 + \tfrac{1}{4}\cdot 2 + \tfrac{1}{8}\cdot 2 + \tfrac{1}{8}\cdot 2} = \frac{11}{16}$$

$$p\{1\} = \frac{\sum_{k=1}^{4} p\{m_k\}C_{k1}}{\sum_{k=1}^{4} p\{m_k\}n_k} = \frac{5}{16}$$

where C_{k0} and C_{k1} are the number of 0's and 1's in the kth encoded message, respectively.

$$\bar{L}_1 = \sum_{k=1}^{4} p\{m_k\}n_k = 2$$

$$\text{Efficiency} = \frac{7/4}{2 \log 2} = \frac{7}{8} = 87.5 \text{ per cent}$$

Note that the encoding efficiency has not been improved.

Next the question arises whether a binary encoding procedure can be devised to provide 100 per cent efficiency. The answer to this question is in the affirmative. For instance, common sense suggests encoding a frequent message into a shorter code word. For example,

m_1	0
m_2	1 0
m_3	1 1 0
m_4	1 1 1

Now note that the frequency of occurrence of the new symbols is exactly the same:

$$p\{0\} = \frac{\frac{1}{2} \cdot 1 + \frac{1}{4} \cdot 1 + \frac{1}{8} \cdot 1}{\frac{1}{2} \cdot 1 + \frac{1}{4} \cdot 2 + \frac{1}{8} \cdot 3 + \frac{1}{8} \cdot 3} = \frac{1}{2}$$

$$p\{1\} = \tfrac{1}{2}$$

The average word length, coding efficiency, and redundancy are

$$\bar{L}_2 = \sum_{k=1}^{4} p\{m_k\}n_k = 1\tfrac{3}{4}$$

$$\text{Efficiency} = 1/1 = 100 \text{ per cent}$$

$$\text{Redundancy} = 0$$

Obviously, this latter encoding procedure is the best that one can obtain as far as the efficiency of the transmission of binary information is concerned. Furthermore, note that this encoding procedure establishes a one-to-one correspondence between the messages and their code words, without the necessity of having any space between successive messages. For example, if we write

0 0 0 0 0 1 0 0 1 0 0 0 1 1 0 1 0 0 0 1 1 0 1 0 0 1 0 0 1 0 0

then, according to the latter code, this message may be uniquely decoded into

$$m_1\, m_1\, m_1\, m_1\, m_1\, m_2\, m_1\, m_2\, m_1\, m_1\, m_3\, m_2\, m_1\, m_1\, m_3\, m_2\, m_1\, m_2\, m_1\, m_2\, m_1$$

This property of the code is referred to as *unique decipherability.*

We now are in a position to define accurately what is meant by encoding in the present work. Let a discrete source have a message ensemble

$\{M\} = [m_1, m_2, m_3, \ldots, m_N]$. Let a memoryless* finite channel be specified by its conditional probability matrix $[P\{y_j|x_i\}]$ (noise matrix). Let D be a finite alphabet $[a_1, a_2, \ldots, a_D]$. Then an encoding procedure is a technique for associating a code word c_k consisting of a sequence of letters from $[D]$ to every $m_k \in M$ in a one-to-one manner. We generally associate a transmission cost with symbols of the new language and search for encoding techniques that reduce the over-all transmission cost or achieve more reliable transmission through the channel. Therefore, encoding is in a sense a means of matching the source and the channel for a more "efficient" joint operation.

For our purpose, the most important kinds of codes are those that do not require spacing. That is, if m_k is encoded in C_k, then any string

$$[\ldots C_k C_j C_l \ldots]$$

is uniquely decoded as

$$[\ldots m_k m_j m_l \ldots]$$

Codes with this property are referred to as *separable* codes or *uniquely decipherable* codes. Obviously, the common English words are not separable. For example, if the three separate words "found," "at," and "ion" are transmitted without separation they form a different word, "foundation," which may not be implied by the three individual words. A sufficient condition for unique decipherability is that no encoded words can be obtained from each other by the addition of more letters. That is, the code should have the *prefix property* (also called *irreducibility*). It can be seen that irreducibility is a subproperty of unique decipherability. This is illustrated by examples below:

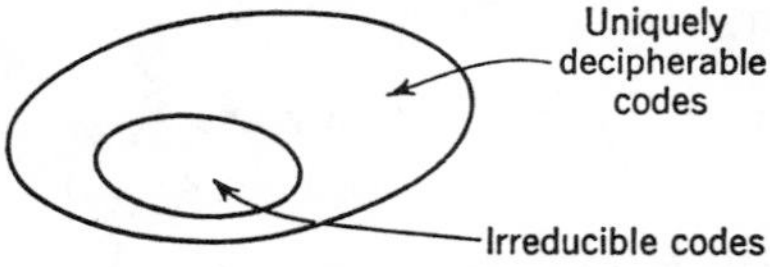

FIG. 4-2. The diagram points out that irreducible codes are a subclass of the uniquely decipherable codes.

$$C_1 = 1 \qquad C_2 = 10$$

The two words satisfy unique decipherability but not irreducibility. The same is true for the words $C_1 = 1$, $C_2 = 10$, $C_3 = 100$. In both these cases any string of words without spacing is uniquely decipherable, provided that an appropriate delay time for examining the string as a whole is allowed. The reader may wish to know when a uniquely decipherable encoding procedure can be devised. He would also be interested to know when several such schemes may exist and how one may find an encoding procedure with high efficiency. The answers to these questions will be presented in a systematic way. But first a few words on the

* A channel is said to be memoryless if the effect of the noise on the input letters is independent of the sequence of previously transmitted letters.

significance of encoding are in order. There are several justifications for including some encoding technique in the present discussion:

1. As yet, encoding techniques seem to provide a most direct application of information theory.

2. Binary encoding is of great importance in computing machines, telephones, and automata. Therefore, the particular case of $D = 2$ offers an interesting practical opportunity for encoding.

3. The two basic central theorems of information theory suggest that, in the presence or absence of noise, it is possible to approach the limit of efficiency, that is, to transmit at a rate less than or equal to the channel capacity with an arbitrarily small error probability. These theorems are essentially of a mathematical nature. Their proofs are rather tedious. Coding theory allows us to illustrate the significance of these theorems without undergoing an otherwise laborious mathematical development. The content of these theorems will be illustrated by several encoding techniques.

4. Finally, the mathematical tools required for research in the field of encoding seem to be commonly available. Thus it seems that further development in this area will be forthcoming. The reader may find it worthwhile to become acquainted with the elements of this new field.

4-2. Separable Binary Codes. When separability is the only constraint, the following simple encoding procedure may be employed. Divide the message set S into two arbitrary but nonempty subsets S_1 and S_2.

$$[\underbrace{m_1, m_2, \ . \ . \ . \ , m_k}_{S_1} | , \underbrace{. \ . \ . \ , m_N}_{S_2}]$$

Assign a 0 to all messages in S_1 and a 1 to messages in S_2. Now continue with the partitioning of S_1 into subsets S_{11} and S_{12}. All messages in S_{11} will have codes starting with 00, those in S_{12} will have codes starting with 01, and so on. The partitioning should continue as long as the subsets contain more than one message. The tree of Fig. 4-3 is an example of this partitioning process.

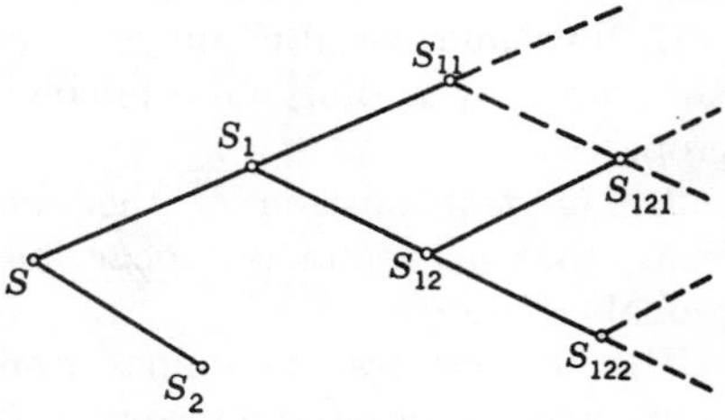

FIG. 4-3. A binary code tree.

If the subset, say S_{121112}, contains a single message, then the code 010001 is associated with that message. If, say, S_{21} contains a unique message, the associated code will be 10. It is easy to see that these codes have the prefix property as no path leading to a vertex can be a subset of a longer path leading to another vertex. Thus no word is derived by the addition of digits to shorter words. The partitioning of messages can be done in a variety of ways. For instance, we may partition one message at a time, such as

$S_1 = m_1$	0
$S_2 = [m_2, \ldots, m_N]$	codes starting with 1
$S_{21} = m_2$	10
$S_{22} = [m_3, \ldots, m_N]$	codes starting with 11
$S_{221} = m_3$	110
$S_{222} = [m_4, \ldots, m_N]$	codes starting with 111

Generally, the efficiency of a code is of special consideration. For this reason, the partitioning can be more conveniently achieved in the message probability space. For instance, if we wish the probability of the occurrence of 0 and 1 in the encoded messages to be not too unequal, it is logical successively to partition the messages into two more or less equiprobable subsets. Thus, the problems of separable encoding amount to devising appropriate partitioning schemes for the message space. This concept is put in focus in the succeeding sections.

4-3. Shannon-Fano Encoding. This method of encoding is directed toward constructing reasonably efficient separable binary codes for sources without memory. Let $[X]$ be the ensemble of the messages to be transmitted and $[P]$ their corresponding probabilities:

$$[X] = [x_1, x_2, \ldots, x_n]$$
$$[P] = [p_1, p_2, \ldots, p_n]$$

It is desired to associate a sequence C_k of binary numbers of unspecified length n_k to each message x_k such that:

1. No sequences of employed binary numbers C_k can be obtained from each other by adding more binary terms to the shorter sequence (prefix property).

2. The transmission of the encoded message is "reasonably" efficient, that is, 1 and 0 appear independently and with (almost) equal probabilities.

The first constraint eliminates any ambiguity in the receiving end and guarantees a one-to-one correspondence between any set of original messages and the corresponding set of encoded messages without the necessity of spacing between words. (This requirement is called *prefix* constraint.) The second constraint ensures the transmission of *almost* 1 bit of information per digit of the encoded messages. It will be shown that under favorable circumstances it is possible to have 1 bit of information per transmitted encoded digit, that is, 1 and 0 may appear with equal probability. The Shannon-Fano encoding procedure will be illustrated by the following example:

Messages	*Probabilities*	*Encoded messages*	*Length*
x_1	0.2500	0 0	2
x_2	0.2500	0 1	2
x_3	0.1250	1 0 0	3
x_4	0.1250	1 0 1	3
x_5	0.0625	1 1 0 0	4
x_6	0.0625	1 1 0 1	4
x_7	0.0625	1 1 1 0	4
x_8	0.0625	1 1 1 1	4
		Average length	2.75

The messages are first written in order of nonincreasing probabilities. Then the message set is partitioned into two most equiprobable subsets $\{X_1\}$ and $\{X_2\}$. A 0 is assigned to each message contained in one subset and a 1 to each of the remaining messages. The same procedure is repeated for subsets of $\{X_1\}$ and $\{X_2\}$; that is, $\{X_1\}$ will be partitioned into two subsets $\{X_{11}\}$ and $\{X_{12}\}$. Now the code word corresponding to a message contained in X_{11} will start with 00 and that corresponding to a message in X_{12} will begin with 01. This procedure is continued until each subset contains only one message. Note that each digit 1 or 0 in each partitioning of the probability space appears with more or less equal probability, independent of the previous or subsequent partitioning; therefore the second requirement is also fulfilled. The entropy of the original source and the average length of the encoded messages (average number of digits per message) are

$$H = -(\tfrac{1}{2} \log \tfrac{1}{4} + \tfrac{1}{4} \log \tfrac{1}{8} + \tfrac{1}{4} \log \tfrac{1}{16}) = 2\tfrac{3}{4} \text{ bits}$$
$$\bar{L} = \Sigma P\{x_i\} n_i = \tfrac{1}{2} \times 2 + \tfrac{1}{4} \times 3 + \tfrac{1}{4} \times 4 = 2\tfrac{3}{4}$$

Since each encoded message consists of sequences of independent binary digits, the entropy per digit for the encoded messages is 1 bit, that is, the efficiency of the transmission of information is 100 per cent. The encoding procedure is therefore said to be an optimum procedure minimizing the average length of messages. No other encoding procedure satisfying the above requirements can be found that leads to a smaller average number of digits per encoded message.

For deriving the most efficient code by this method, it is necessary that the message probability space Ω can be repeatedly partitioned into two equiprobable subspaces so that we finally reach the situation where each message corresponds to only one partitioned subspace. The probability of the occurrence of each message x_k must be of the form

$$P\{x_k\} = 2^{-n_k}$$

where n_k is a positive integer. The integers n_k satisfy the relation*

$$\sum_{k=1}^{N} 2^{-n_k} = 2^{-n_1} + 2^{-n_2} + \cdots + 2^{-n_N} = 1 \qquad (4\text{-}3)$$

A most particular case of this situation arises when

$$[P] = [2^{-1}, 2^{-2}, \ldots, 2^{-k}, \ldots, 2^{-N}, 2^{-N}] \qquad (4\text{-}4)$$

The encoding procedure is unambiguous (one-to-one), the prefix requirement is fulfilled, and the average number of digits in the encoded messages is

$$\bar{L} = \sum_{k=1}^{N} P\{x_k\} n_k = -\sum_{k=1}^{N} P\{x_k\} \log P\{x_k\} \qquad (4\text{-}5)$$

For such a message ensemble the average length of the encoded message is $\bar{L}$ bits, which is exactly the same as the entropy of the original message ensemble, that is,

$$H(X) = -\sum_{k=1}^{N} P\{x_k\} \log P\{x_k\} = \bar{L} \qquad \text{bits per message}$$

The efficiency of the transmission is 100 per cent, and the encoding is thus an optimum encoding procedure. The tree diagram of Fig. 4-4 illustrates this encoding procedure.

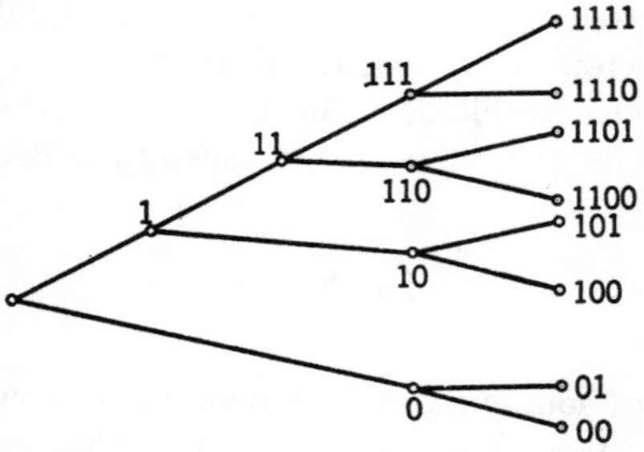

FIG. 4-4. Example of Shannon-Fano encoding procedure.

For practical purposes, it should be noted that, when the probability matrix of the message is such that the above successive equiprobable partitioning is not possible, the Shannon-Fano code may not be an optimum code. However, if the above requirement is "approximately" satisfied, then "reasonably" efficient encoding procedures can be expected. For instance, Fano (I) gives an example for which the efficiency remains very close to 100 per cent.

Example 4-1. Apply the Shannon-Fano encoding procedure to the following message ensemble:

$$[X] = [x_1, x_2, x_3, x_4, x_5, x_6, x_7, x_8, x_9]$$
$$[P] = [0.49, 0.14, 0.14, 0.07, 0.07, 0.04, 0.02, 0.02, 0.01]$$

* It is possible to have several equiprobable messages. In that case, n_k would be the same for all these messages; that is, if x_k and x_{k+1} are equiprobable and the Shannon-Fano method strictly applies, then $n_k = n_{k+1}$.

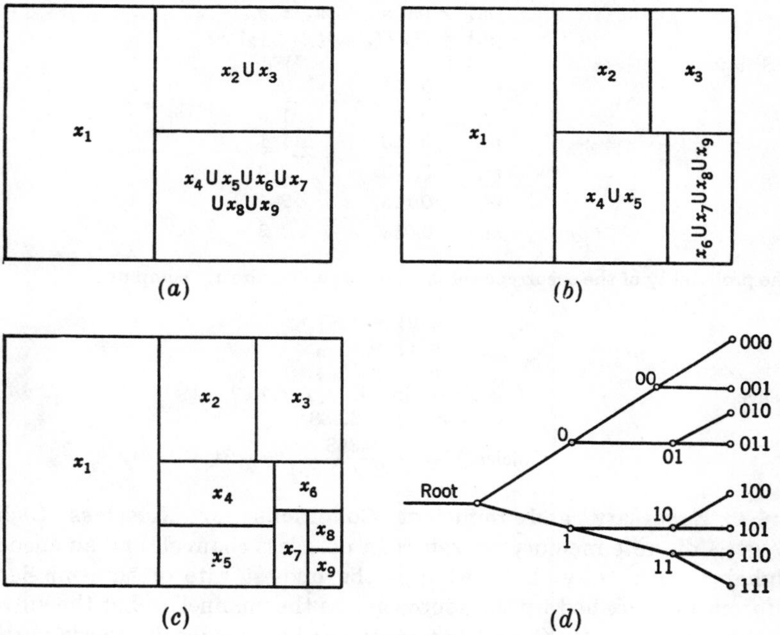

FIG. E4-1

Solution. Although the probability matrix is not identical with Eq. (4-4), successive partitioning can be efficiently achieved. See Figs. E4-1*a*, E4-1*b*, and E4-1*c*.

$$H(X) = -[0.49 \log 0.49 + 0.28 \log 0.14 + 0.14 \log 0.07 + 0.04 \log 0.04 + 0.04 \log 0.02 + 0.01 \log 0.01]$$

$$\bar{L} = 0.49 \times 1 + 0.28 \times 3 + 0.18 \times 4 + 0.02 \times 5 + 0.03 \times 6 = 2.33$$

$$\text{Efficiency} = \frac{H(X)}{2.33} = 0.993$$

Message	*Code*
x_1	0
x_2	100
x_3	101
x_4	1100
x_5	1101
x_6	1110
x_7	11110
x_8	111110
x_9	111111

The tree diagram of Fig. E4-1*d* exhibits the general procedure of the Shannon-Fano code.

Example 4-2. Apply a ternary partitioning technique (similar to the Shannon-Fano procedure) for encoding the following messages in codes using an alphabet [0,1,2]. Find $\bar{L}$ and the code efficiency.

$$\{m\} = [m_1, m_2, m_3, m_4, m_5, m_6]$$
$$p\{m\} = [3/8, 1/6, 1/8, 1/8, 1/8, 1/12]$$

Solution

m_1	0.375	0
m_2	0.167	1 0
m_3	0.125	1 1
m_4	0.125	2 0
m_5	0.125	2 1
m_6	0.083	2 2

The probability of the occurrence of 0, 1, and 2 can be directly computed.

$$p\{0\} = 16/39$$
$$p\{1\} = 13/39$$
$$p\{2\} = 10/39$$

Hence,

$$\bar{L} = \Sigma p\{m_i\} n_i = 39/24 = 1.62$$
$$H(X) = 2.388$$
$$\text{Efficiency} = \frac{2.388}{1.62 \times 1.584} = 0.94$$

4-4. Necessary and Sufficient Conditions for Noiseless Coding. Given a discrete memoryless source, a noiseless channel, and an encoding alphabet with D symbols, what is the highest rate of transmission of information supplied by the source across the channel? Let the entropy of the source be $H(X)$ and the messages be encoded in words with an average length $\bar{L}$; then the entropy of the encoded information is supplied at a rate

$$\frac{H(X)}{\bar{L}}$$

Since the maximum of this rate is log D and it occurs when all D symbols are equiprobable, it appears that the lowest value of $\bar{L}$ is

$$\frac{H(X)}{\log D}$$

However, it is not obvious that:

1. For a given set of message probabilities, it is possible to devise codes with preassigned word lengths.

2. There exist uniquely decipherable codes leading to

$$\bar{L} \geq \frac{H(X)}{\log D}$$

and no such codes exist with $\bar{L} < H(X)/(\log D)$.

The object of this section is to discuss part 1, that is, to derive the necessary and sufficient conditions for the existence of such *noiseless encoding* procedures. Part 2 will be discussed in a later section.

Let $\{X\}$ be the information source with N messages $[x_1,x_2, \ldots ,x_N]$ and $\{D\}$ the coding alphabet $[a_1,a_2, \ldots ,a_D]$. Our problem is to find a one-to-one correspondence between every element x_i of $\{X\}$ and a sequence of a's having, say, n_i digits, with the restriction that none of the encoded messages can be obtained from each other by adding a sequence of a's to the shorter encoded message. Thus, $[n_1,n_2, \ldots ,n_N]$, the length of the encoded words, cannot be arbitrary integers. They must satisfy certain realizability conditions if the prefix requirement is to be met. For instance, in the afore-mentioned partitioning procedure [Eq. (4-2)] we found that a code exists with

$$[n_1,n_2, \ldots ,n_N] = [1,2, \ldots ,N]$$

The following theorem has been derived in Szilard, Kraft, Fano (I), Mandelbrot (I), Sardinas and Patterson, McMillan (I), and Feinstein (I). The proof given here is based on the latter two references.

A Noiseless Coding Theorem. The necessary and sufficient condition for existence of an irreducible noiseless encoding procedure with specified word length $[n_1,n_2, \ldots ,n_N]$ is that a set of positive integers $[n_1,n_2, \ldots ,n_N]$ can be found such that

$$\sum_{i=1}^{N} D^{-n_i} \leq 1 \tag{4-6}$$

Proof. Clearly two encoded messages x_i and x_k can have the same length, that is, $n_i = n_k$. Let W_i be the number of encoded messages of length n_i and note that the number of encoded messages with only one letter cannot be larger than D.

$$W_1 \leq D \tag{4-7}$$

The number of encoded messages of length 2, because of our coding restriction, cannot be larger than

$$W_2 \leq (D - W_1)D = D^2 - W_1D \tag{4-8}$$

Similarly,

$$W_3 \leq [(D - W_1)D - W_2]D = D^3 - W_1D^2 - W_2D \tag{4-9}$$

Finally, if m is the maximum length of the encoded words, one concludes that

$$W_m \leq D^m - W_1D^{m-1} - W_2D^{m-2} - \cdots - W_{m-1}D \tag{4-10}$$

Dividing both sides of this inequality by D^m yields

$$0 \leq 1 - W_1D^{-1} - W_2D^{-2} - \cdots - W_{m-1}D^{-m+1} - W_mD^{-m}$$

or

$$\sum_{i=1}^{m} W_iD^{-i} \leq 1$$

It may not be obvious that this condition is identical with Eq. (4-6). But note that

$$m \geq n_i \qquad i = 1, 2, \ldots, N$$

and $\sum_{i=1}^{m} W_i D^{-i}$ means the sum of "the numbers of all sequences of length i multiplied by D^{-i}," where the summation extends from 1 to m.

Let us examine what is implied by the above inequality. We can rewrite it in the following way:

$$\sum_{j=1}^{m} W_j D^{-j} = \underbrace{\frac{1}{D} + \cdots + \frac{1}{D}}_{W_1} + \underbrace{\frac{1}{D^2} + \cdots + \frac{1}{D^2}}_{W_2} + \cdots + \underbrace{\frac{1}{D^m} + \cdots + \frac{1}{D^m}}_{W_m} \tag{4-11}$$

Each bracketed expression corresponds to a message x_i, and therefore the total number of terms is N.

$$[\underbrace{1, \ldots, 1}_{W_1}, \underbrace{2, \ldots, 2}_{W_2}, \ldots, \underbrace{m, \ldots, m}_{W_m}] \tag{4-12}$$
$$W_1 + W_2 + \cdots + W_m = N$$

The terms in W_k correspond to the encoded messages of length K. These latter terms can be considered as ΣD^{-n_i} when the summation takes place over all those terms with $n_i = k$. Therefore, by a simple reassignment of terms, we may equivalently write

$$\sum_{j=1}^{m} W_j D^{-j} = \sum_{i=1}^{N} D^{-n_i} \tag{4-13}$$

Thus

$$\sum_{j=1}^{m} W_j D^{-j} = \sum_{i=1}^{N} D^{-n_i} \leq 1$$

The desired set of positive integers $[n_1, n_2, \ldots, n_N]$ must satisfy the inequality of Eq. (4-6). This proves the necessity requirement of the theorem.

As an example, let

$$[X] = [X_1, X_2, X_3, X_4, X_5, X_6, X_7]$$

Assume that after encoding we get a set of messages with the following lengths:

$$n_1 = 2 \qquad n_2 = 2 \qquad n_3 = 3 \qquad n_4 = 3 \qquad n_5 = 3 \qquad n_6 = 4 \qquad n_7 = 5$$

Therefore

$$W_1 = 0 \quad W_2 = 2 \quad W_3 = 3 \quad W_4 = 1 \quad W_5 = 1 \quad W_6 = 0 \quad W_7 = 0$$

The sets of desired integers n_i and W_i are thus

$$[n_i] = [2,2,3,3,3,4,5] \qquad [W_i] = [0,2,3,1,1,0,0]$$

and
$$\sum_{j=1}^{m=5} W_j D^{-j} = 2 \cdot \frac{1}{D^2} + 3 \cdot \frac{1}{D^3} + 1 \cdot \frac{1}{D^4} + \frac{1}{D^5}$$

$$\sum_{i=1}^{7} D^{-n_i} = \frac{1}{D^2} + \frac{1}{D^2} + \frac{1}{D^3} + \frac{1}{D^3} + \frac{1}{D^3} + \frac{1}{D^4} + \frac{1}{D^5}$$

The two sums are obviously equal.

Now we show that the condition

$$\sum_{j=1}^{m} W_j D^{-j} = W_1 D^{-1} + W_2 D^{-2} + \cdots + W_m D^{-m} \leq 1$$

is sufficient for the existence of the desired codes. As all terms in Eq. (4-13) are positive, each term or the sum of a number of these terms must be positive and less than 1. Therefore we conclude that

$$W_1 D^{-1} \leq 1 \qquad \text{or} \qquad W_1 \leq D \tag{4-14}$$

and
$$W_1 D^{-1} + W_2 D^{-2} \leq 1 \qquad \text{or} \qquad W_2 \leq D(D - W_1) \tag{4-15}$$

and so on. But these are exactly the conditions that we have to satisfy in order to guarantee that no encoded message can be obtained from any other by the addition of a sequence of letters of the encoding alphabet. As an application of the foregoing theorem, let D be a binary set, that is, $A = [a_1,a_2]$; then the encoding theorem requires that

$$\sum_{i=1}^{N} 2^{-n_i} \leq 1 \tag{4-16}$$

As an application of the foregoing, consider the existence of a separable code book having N words of equal length n. The noiseless coding theorem suggests that such codes exist if

$$\sum_{k=1}^{N} D^{-n} = ND^{-n} \leq 1 \tag{4-17}$$

$$\log N \leq n \log D$$

This latter relation between N, n, and D guarantees the existence of the desired codes. In the particular case of $D = 2$, if we assume the further constraint that the words of the code book could be ordered in such a way that every two consecutive words differ by only one binit, the code is referred to as the Gray code. For instance, for $n = 2$ and $N = 4$, we have

$$00, 01, 11, 10$$

Gray codes are of some practical value in computing machines (for example, in analog-to-digital conversion as described in the references in the footnote).*

Example 4-3. Find the smallest number of letters in the alphabet (number D) for devising a code with a prefix property such that

$$[W] = [0,3,0,5]$$

Devise such a code.

Solution. The realizability condition is

$$3D^{-2} + 5D^{-4} \leq 1$$

The inequality is satisfied for

$$D^2 \geq \frac{3 + \sqrt{29}}{2}$$

The smallest permissible value is $D = 3$. That is, no binary code can be devised with the above constraint. To devise such a code, let the alphabet be [0.1.2]: then one of several encoding procedures is

$$\begin{aligned} m_1 &= 00 \\ m_2 &= 01 \\ m_3 &= 02 \\ m_4 &= 1000 \\ m_5 &= 1001 \\ m_6 &= 1002 \\ m_7 &= 2000 \\ m_8 &= 2222 \end{aligned}$$

Example 4-4. Show all possible sets of binary codes with the prefix property for encoding the message ensemble

$$[m_1, m_2, m_3]$$

in words not more than three digits long.

Solution. All possible desired sets can be obtained from the inequality

$$\begin{cases} w_1 2^{-1} + w_2 2^{-2} + w_3 2^{-3} \leq 1 \\ w_1 + w_2 + w_3 = 3 \qquad w_k \geq 0 \quad k = 1, 2, 3 \end{cases}$$

or

$$\begin{cases} 4w_1 + 2w_2 + w_3 \leq 8 \\ w_1 + w_2 + w_3 = 3 \end{cases}$$

* S. H. Caldwell, "Switching Circuits and Logical Design," John Wiley & Sons, Inc., New York, 1958; M. Phister, Jr., "Logical Design of Digital Computers," John Wiley & Sons, Inc., New York, 1959.

The possible sets of codes correspond to:

$$\begin{array}{lll} w_1 = 1 & w_2 = 1 & w_3 = 1 \\ w_1 = 1 & w_2 = 2 & w_3 = 0 \\ w_1 = 1 & w_2 = 0 & w_3 = 2 \\ w_1 = 0 & w_2 = 3 & w_3 = 0 \\ w_1 = 0 & w_2 = 2 & w_3 = 1 \\ w_1 = 0 & w_2 = 1 & w_3 = 2 \\ w_1 = 0 & w_2 = 0 & w_3 = 3 \end{array}$$

The corresponding codes can be found without difficulty.

4-5. A Theorem on Decodability. In Sec. 4-4, we have given the necessary and sufficient conditions for the existence of a set of irreducible code words of specified length. Actually a stronger theorem concerning the unique decipherability (not necessarily irreducible code words) has been derived by McMillan (I).*

McMillan's Theorem. Let $[m_1, m_2, \ldots, m_N]$ be a sequence of messages encoded in uniquely decipherable words of respective symbol length $[n_1, n_2, \ldots, n_N]$ taken from a finite alphabet $[a_1, a_2, \ldots, a_D]$. Then

$$\sum_{i=1}^{N} D^{-n_i} \leq 1 \tag{4-18}$$

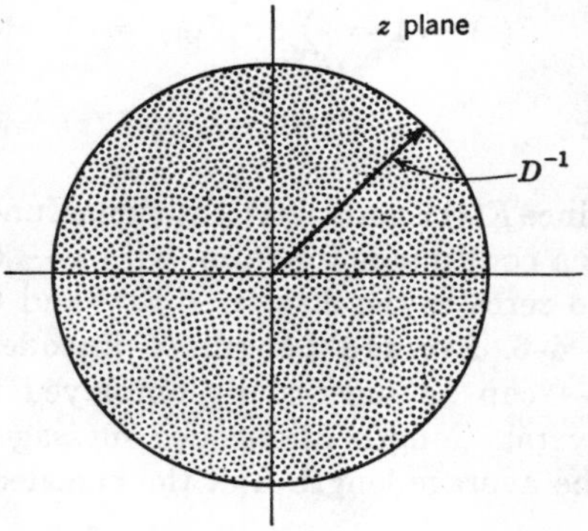

FIG. 4-5. The disk of unique decipherability in the complex plane.

Proof. Let l be the largest element of $[n_1, n_2, \ldots, n_N]$, and w_i, as before, the number of distinct words of length n_i. Then it is desired to prove that

$$\sum_{i=1}^{l} w_i D^{-i} \leq 1 \tag{4-18a}$$

McMillan employed an interesting method for proving Eq. (4-18a). Let

$$Q(z) = \sum_{1}^{l} w_i z^i \tag{4-19}$$

We prove that

$$Q(z) \leq 1$$

for

$$0 \leq z \leq D^{-1}$$

Let $N(k)$ be the number of distinct sequences of length k taken from the alphabet $[a]$. The decodability condition requires that

$$N(k) \leq D^k \tag{4-20}$$

* Also Mandelbrot (I) and L. Kraft in an unpublished 1949 MIT thesis.

Now consider the infinite series

$$F(z) = 1 + N(1)z + N(2)z^2 + \cdots \tag{4-21}$$

This series converges within the disk $|z| < D^{-1}$. (Why?)

Next we look into the property of unique decipherability which permits writing for a sequence of length k

$$N(k) = w_1N(k-1) + w_2N(k-2) + \cdots + w_lN(k-l) \tag{4-22}$$

It is easy to see that this recurrent formula holds if we let

$$N(0) = 1 \qquad \text{and} \qquad N(h) = 0 \qquad \text{for } h < 0$$

Note that

$$F(z) - 1 = \sum_{k=1}^{\infty} z^k N(k) = \sum_{k=1}^{\infty} z^k \sum_{i=1}^{l} w_i N(k-i) = F(z) \cdot Q(z) \tag{4-23}$$

or

$$F(z) = \frac{1}{1 - Q(z)}$$

Since $F(z)$ is an analytic rational function in $|z| < D^{-1}$ and the denominator is a continuous function, with $1 - Q(0) = 1$, it follows that $1 - Q(z)$ has no zeros in the disk $|z| < D^{-1}$ and $Q(z) \leq 1$ for $0 \leq z \leq D^{-1}$.

4-6. Average Length of Encoded Messages. The theorem of Sec. 4-4 can be successfully employed for obtaining a lower bound for the average length of encoded messages. In Sec. 4-3, we pointed out that the average length $\bar{L}$ of the encoded messages, that is

$$\bar{L} = \Sigma P\{x_i\} n_i \tag{4-24}$$

gives a measure of the efficiency of coding. In this section it will be shown that $\bar{L}$ cannot be decreased beyond a certain limit. More specifically:

Theorem. Let $\{X\}$ be a discrete message source, without memory, and x_i any message of this source with probability of transmission $P\{x_i\}$. If the $\{X\}$ ensemble is encoded in a sequence of uniquely decipherable characters taken from the alphabet $[a_1, a, \ldots, a_D]$ then

$$\bar{L} = \sum_{i=1}^{N} P\{x_i\} n_i \geq \frac{H(X)}{\log D} \tag{4-25}$$

Proof. The proof of this theorem lies in the following lemma:

Lemma. Consider two sets of nonnegative numbers $\{p_i\}$ and $\{q_i\}$, such that

$$\sum_{i=1}^{n} p_i = \sum_{i=1}^{n} q_i = 1$$

Then
$$- \sum_{i=1}^{n} q_i \log q_i \leq - \sum_{i=1}^{n} q_i \log p_i\text{*} \tag{4-26}$$

Now we shall apply this lemma to the set of nonnegative integers

$$q_i = \frac{D^{-n_i}}{\sum_{k=1}^{N} D^{-n_k}} \tag{4-27}$$

where D is a positive integer and $\{i\}$ the sequence of $[1,2, \ldots ,N]$. Indeed in Eq. (4-27) the denominator is a nonnegative number less than or equal to 1 and

$$\sum_{i=1}^{N} q_i = \sum_{i=1}^{N} \frac{D^{-n_i}}{\sum_{k=1}^{N} D^{-n_k}} = \frac{\sum_{i=1}^{N} D^{-n_i}}{\sum_{k=1}^{N} D^{-n_k}} = 1$$

Therefore we may write

$$H(X) = - \sum_{i=1}^{N} P\{x_i\} \log P\{x_i\} \leq - \sum_{i=1}^{N} P\{x_i\} \log \frac{D^{-n_i}}{\sum_{k=1}^{N} D^{-n_k}}$$

$$= \log \sum_{k=1}^{N} D^{-n_k} - \sum_{i=1}^{N} P\{x_i\} \log D^{-n_i} \tag{4-28}$$

or
$$H(X) \leq \log \sum_{k=1}^{N} D^{-n_k} + \sum_{i=1}^{N} P\{x_i\} n_i \log D$$

Applying the theorem of Sec. 4-4 yields

$$H(X) \leq \log 1 + \log D \sum_{i=1}^{N} P\{x_i\} n_i$$

or
$$\bar{L} \geq \frac{H(X)}{\log D} \tag{4-29}$$

* For example, let

$$[p_i] = [\tfrac{1}{2},\tfrac{1}{4},\tfrac{1}{4}]$$
$$[q_i] = [\tfrac{1}{8},\tfrac{5}{8},\tfrac{1}{4}]$$

Then
$$\tfrac{3}{8} - \tfrac{5}{8} \log \tfrac{5}{8} + \tfrac{1}{2} \leq \tfrac{1}{8} + \tfrac{5}{4} + \tfrac{1}{2}$$
or
$$\tfrac{18}{8} - \tfrac{5}{8} \log 5 \leq \tfrac{11}{8}$$
as, evidently,
$$7 \leq 5 \log 5$$

The proof is left for the reader. (Use convexity of $x \log x$. See also Prob. 3-10.)

This result is rather interesting as it clearly shows that even in the absence of noise no uniquely decipherable encoding procedure can be devised such that its average message length is less than a fixed number that is the ratio of the source entropy and log D. (Log D is the maximum possible entropy associated with the selected alphabet containing D characters or the capacity of the coding alphabet subject to the above constraint.) This lower bound is not generally achieved unless the n_j are all appropriately chosen integers. In such a case we obtain an optimum encoding procedure, that is,

$$L_o = \frac{H(X)}{\log D} = \frac{\text{source entropy}}{\text{capacity of coding alphabet}} \tag{4-30}$$

While L_o, the lower bound of $\bar{L}$, may not always be reached by an encoding procedure, it is always possible to give an obtainable bound for $\bar{L}$. In fact, if we let

$$-\frac{\log P\{x_k\}}{\log D} \le n_k < -\frac{\log P\{x_k\}}{\log D} + 1 \qquad k = 1, 2, \ldots, N \quad D \ge 2$$

we rest assured that

$$\frac{H(X)}{\log D} \le \bar{L} = \sum_{k=1}^{N} P\{x_k\}n_i < 1 + \frac{H(X)}{\log D} \tag{4-31}$$

Note that owing to the above inequality and its consequence

$$P\{x_k\} \ge D^{-n_k}$$

the noiseless coding theorem is satisfied. Thus uniquely decipherable codes exist. Shannon has developed a binary encoding scheme based on Eq. (4-31) which will be discussed in the next section.

The bound given by the theorem of this section is based on the assumption of decipherability of the encoded messages. If this strong requirement is removed, the average length may be reduced. As an example, consider the case

$$x_1 \rightarrow 1$$
$$x_2 \rightarrow 0$$
$$x_3 \rightarrow 100$$

You can show as an exercise that, with a proper selection of probabilities, one may violate the lower bound of Eq. (4-31).

Example 4-5. The output of a discrete source,

$$\{X\} = \{x_1,x_2,x_3,x_4,x_5,x_6\}$$
$$P\{X\} = \{2^{-1},2^{-2},2^{-4},2^{-4},2^{-4},2^{-4}\}$$

is encoded in the following six ways:

	C_1	C_2	C_3	C_4	C_5	C_6
x_1	0	1	0	111	1	0
x_2	10	011	10	110	01	01
x_3	110	010	110	101	0011	011
x_4	1110	001	1110	100	0010	0111
x_5	1011	000	11110	011	0001	01111
x_6	1101	110	111110	010	0000	011111

(*a*) Determine which of these codes are uniquely decipherable.
(*b*) Determine those which have the prefix property.
(*c*) Find the average length of each uniquely decipherable code.
(*d*) Does any one of the above codes give minimum average length?

Solution

(*a*) By direct inspection one finds that C_3, C_4, C_5, and C_6 are uniquely decipherable. While it is clear that C_1 and C_2 are not uniquely decipherable, we may wish to apply McMillan's realizability criterion.

For C_1: $$2^{-1} + 2^{-2} + 2^{-3} + 3 \cdot 2^{-4} > 1$$
For C_2: $$2^{-1} + 5 \cdot 2^{-3} > 1$$

Thus such uniquely decipherable codes cannot exist.

(*b*) C_3, C_4, and C_5 have the prefix property, but C_6 does not have such a property.

(*c*) $$\bar{L}_3 = \tfrac{1}{2} + \tfrac{2}{4} + \tfrac{3}{16} + \tfrac{4}{16} + \tfrac{5}{16} + \tfrac{6}{16} = 2\tfrac{1}{8}$$
$$\bar{L}_4 = 3 \qquad \bar{L}_5 = 2 \qquad \bar{L}_6 = 2\tfrac{1}{8}$$

(*d*) The entropy of the source is

$$H(X) = -\tfrac{1}{2}\log\tfrac{1}{2} - \tfrac{1}{4}\log\tfrac{1}{4} - \tfrac{4}{16}\log\tfrac{1}{16} = 2 \text{ bits per symbol}$$

According to Eq. (4-29), the average length of a uniquely decipherable code cannot be less than $H(X)$. Thus C_5 is a code that achieves minimum average length.

4-7. Shannon's Binary Encoding. Shannon has suggested a binary encoding procedure based on Eq. (4-31). First we must reassure ourselves whether such codes exist. For this, we employ the noiseless coding theorem. Equation (4-31) yields

$$2^{-n_k} \leq P\{x_k\} \qquad k = 1, 2, \ldots, N \tag{4-32}$$

If these inequalities are all satisfied, then

$$\sum_{k=1}^{N} 2^{-n_k} \leq 1$$

Thus we are sure that the desired code exists. Note that such codes will have the interesting property that their average length is constrained by

$$H(X) \leq \bar{L} < H(X) + 1 \tag{4-33}$$

The following steps describe this method:

1. Write down the message ensemble in the order of nonincreasing probabilities, say,

$$[x_1, x_2, \ldots, x_N]$$
$$P\{x_1\} \geq P\{x_2\} \geq \cdots \geq P\{x_N\} \tag{4-34}$$

2. Compute the sequences

$$\begin{aligned} \alpha_1 &= 0 \\ \alpha_2 &= P\{x_1\} \\ \alpha_3 &= P\{x_2\} + P\{x_1\} = P\{x_2\} + \alpha_2 \\ \alpha_4 &= P\{x_3\} + P\{x_2\} + P\{x_1\} = P\{x_3\} + \alpha_3 \\ &\cdots\cdots\cdots\cdots\cdots\cdots \end{aligned} \tag{4-35}$$

3. Determine the set of integers which is the smallest integer's solution of the inequalities

$$2^{n_i} P\{x_i\} \geq 1 \qquad i = 1, 2, \ldots \tag{4-36}$$

4. Expand the decimal numbers α_i in binary form to n_i places: that is, neglect the expansion beyond the n_i digits.*

Proof. To show the validity of the method, first we note that owing to the decodability theorem an encoding procedure with the prefix property must exist. In the second place, one observes that the numbers

* The solutions for the α of Eqs. (4-35) will be numbers expressed in decimal form. To express a decimal number N in binary form one must determine the set of τ_k such that

$$N = \cdots + \tau_3(2)^3 + \tau_2(2)^2 + \tau_1(2)^1 + \tau_0(2)^0 + \tau_{-1}(2)^{-1} + \tau_{-2}(2)^{-2} + \cdots$$

where τ_k ($k = 0, \pm 1, \pm 2, \ldots$) is either 0 or 1. The binary form is usually written in the following abbreviated manner, $\ldots \tau_3\tau_2\tau_1\tau_0 . \tau_{-1}\tau_{-2}\tau_{-3} \ldots$, where the "point" is called a *binary point*. Any single τ_k is called a *binary digit*. It has been suggested [Golay] that binary digit be abbreviated to *binit*. This convention would help to destroy the erroneous connection between a "bit" (a unit of binary information) and a "binary digit" (a term in a binary number). (Note the parallelism between the binary and decimal forms. By letting $\tau_k = 0, 1, 2, \ldots, 9$ and replacing the 2's in the parentheses with 10's we have a number expressed in decimal form. The "point" is then called the decimal point.) Consider the following illustrative examples.

Decimal form	*Binary form*
8.00	1000.00
7.00	0111.00
5.50	0101.10
0.25	0000.01
0.40	0000.0110011 · · ·

M. J. E. Golay, *Proc. IRE*, vol. 42, no. 9, p. 1452, September, 1954.

α_k correspond to ordinates of CDF as shown in Fig. 4-6:

$$0 = \alpha_1 < \alpha_2 < \alpha_3 < \cdots < \alpha_n < \alpha_{n+1} = 1 \tag{4-37}$$

Now assume that α_k has been expanded to n_k place as

$$\alpha_k \rightarrow .\tau_{-1}\tau_{-2} \cdots \tau_{-n_k}$$

and α_{k+1} to n_{k+1} place as

$$\alpha_{k+1} \rightarrow .\tau'_{-1}\tau'_{-2} \cdots \tau'_{-n_{k+1}} \qquad n_{k+1} \geq n_k$$

But

$$\alpha_{k+1} = \alpha_k + P\{x_k\} \tag{4-38}$$

This equality, written in the binary form, will become

$$.\tau'_{-1}\tau'_{-2} \cdots \tau'_{-n_{k+1}} = .\tau_{-1}\tau_{-2} \cdots \tau_{-n_k} + .\tau''_{-1}\tau''_{-2} \cdots \tau''_{-n_k} \tag{4-39}$$

Keeping Eq. (4-36) in mind, one can see that the suggested codes for α_k and α_{k+1} will be distinct binary numbers.

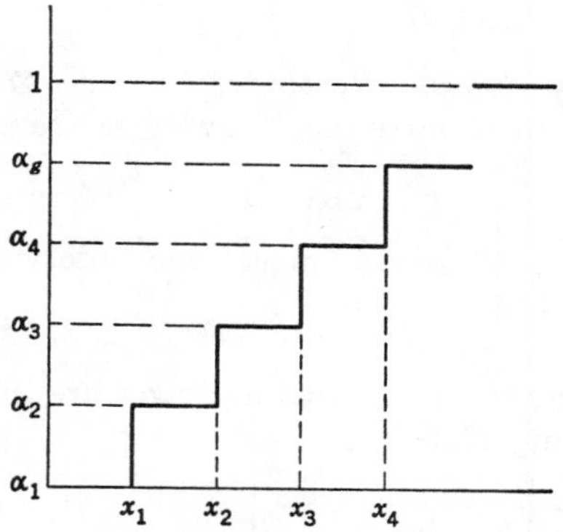

FIG. 4-6. A CDF associated with Shannon's binary encoding.

Example 4-6. Apply Shannon's encoding procedure to the following message ensemble:

$$[X] = [x_1,x_2,x_3,x_4]$$
$$[P] = [0.4,0.3,0.2,0.1]$$

Solution

Step 1 $\quad 0.4 > 0.3 > 0.2 > 0.1$

Step 2 $\quad \alpha_1 = 0 \quad \alpha_2 = 0.4 \quad \alpha_3 = 0.7 \quad \alpha_4 = 0.9$

Step 3

$0.4 \geq 2^{-2} \quad n_1 = 2$

$0.3 \geq 2^{-2} \quad n_2 = 2$

$0.2 \geq 2^{-3} \quad n_3 = 3$

$0.1 \geq 2^{-4} \quad n_4 = 4$

Step 4

$\alpha_1 = 00| \qquad x_1 \quad 00$

$\alpha_2 = 0.4 = 01|1 \qquad x_2 \quad 01$

$\alpha_3 = 0.7 = 101|1 \qquad x_3 \quad 101$

$\alpha_4 = 0.9 = 1110| \qquad x_4 \quad 1110$

4-8. Fundamental Theorem of Discrete Noiseless Coding. There is an important theorem due to C. E. Shannon which states:

Theorem. Let S be a discrete source without memory with a communication entropy $H(X)$ and a noiseless channel with capacity C bits per message. It is possible to encode the output of S so that, if the encoded messages are transmitted through the channel, the rate of transmission of information approaches C per symbol as closely as desired.

Proof. We have already seen that for a given source $[X]$ with N messages the length of each encoded message may be constrained by the following inequalities:

$$-\frac{\log P\{x_i\}}{\log D} \leq n_i < -\frac{\log P\{x_i\}}{\log D} + 1 \qquad i = 1, 2, \ldots, N \tag{4-40}$$

Furthermore it can be seen that the average length $\bar{L}_1$ satisfies the relation

$$\frac{H(X_1)}{\log D} \leq \bar{L}_1 < \frac{H(X_1)}{\log D} + 1 \tag{4-41}$$

Now suppose that we consider the source $X_2 = X_1 \otimes X_1$, that is, a source which transmits independently the following messages:

$$[X_2] = \begin{bmatrix} x_1x_1 & x_1x_2 & x_1x_3 & \cdots & x_1x_N \\ x_2x_1 & x_2x_2 & x_2x_3 & \cdots & x_2x_N \\ \cdots & \cdots & \cdots & \cdots & \cdots \\ x_Nx_1 & x_Nx_2 & x_Nx_3 & \cdots & x_Nx_N \end{bmatrix} \tag{4-42}$$

If it is assumed that the successive messages are independent, the corresponding probability matrix is

$$\begin{bmatrix} P\{x_1\}P\{x_1\} & P\{x_1\}P\{x_2\} & \cdots & P\{x_1\}P\{x_N\} \\ P\{x_2\}P\{x_1\} & P\{x_2\}P\{x_2\} & \cdots & P\{x_2\}P\{x_N\} \\ \cdots & \cdots & \cdots & \cdots \\ P\{x_N\}P\{x_1\} & P\{x_N\}P\{x_2\} & \cdots & P\{x_N\}P\{x_N\} \end{bmatrix} \tag{4-43}$$

This source will be referred to as a *second-order extension* of the original source. Now if this message ensemble is encoded, we expect that the average length $\bar{L}_2$ of the messages of the new source will satisfy the relation

$$\frac{H(X_2)}{\log D} \leq \bar{L}_2 < \frac{H(X_2)}{\log D} + 1 \tag{4-44}$$

As the successive messages are independent, one finds

$$H(X_2) = H(X_1) + H(X_1) = 2H(X_1) \tag{4-45}$$

Thus

$$\frac{2H(X_1)}{\log D} \leq \bar{L}_2 < \frac{2H(X_1)}{\log D} + 1 \tag{4-46}$$

In a similar fashion we consider the Mth-order extension of the original source. If the message ensemble of $X_M = \underbrace{X_1X_1X_1 \cdots X_1}_{M}$ is similarly encoded, we conclude that

$$\frac{H(X_M)}{\log D} \leq \bar{L}_M < \frac{H(X_M)}{\log D} + 1 \tag{4-47}$$

or

$$M\,\frac{H(X_1)}{\log D} \leq \bar{L}_M < M\,\frac{H(X_1)}{\log D} + 1$$

Finally,

$$\frac{H(X_1)}{\log D} \leq \frac{\bar{L}_M}{M} < \frac{H(X_1)}{\log D} + \frac{1}{M} \tag{4-48}$$

When M is made infinitely large, we obtain

$$\lim_{M\to\infty} \frac{\bar{L}_M}{M} = \frac{H(X_1)}{\log D} \tag{4-49}$$

This completes the proof of the so-called first fundamental coding theorem. It should be kept in mind that, while we asymptotically approach the above limit, the procedure does not necessarily yield a monotonically increasing improvement. That is, it is possible to have a situation where

$$\frac{1}{M}\bar{L}_M \geq \frac{1}{M-1}\bar{L}_{M-1}$$

4-9. Huffman's Minimum-redundancy Code. Huffman has suggested a simple method for constructing separable *codes with minimum redundancy* for a set of discrete messages [Huffman (I)]. The meaning of the latter term will be described shortly. Let $[X]$ be the message ensemble, $[P]$ the corresponding probability matrix, $[D]$ the encoding alphabet, and $L(x_k)$ the length of the encoded message x_k. Then

$$\bar{L} = E[L(x_k)] = \sum_{k=1}^{N} P\{x_k\}L(x_k) \tag{4-50}$$

A minimum redundancy or an *optimum code* is one that leads to the lowest possible value of $\bar{L}$ for a given D. This definition is accepted, having in mind the irreducibility requirements. That is, distinct messages must be encoded in uniquely decipherable words with the prefix property. To comply with these requirements, Huffman derives the following results:

1. For an optimum encoding, the longer code word should correspond to a message with lower probability; thus if for convenience the messages are numbered in order of nonincreasing probability,

$$P\{x_1\} \geq P\{x_2\} \geq P\{x_3\} \geq \cdots \geq P\{x_N\} \tag{4-51}$$

then

$$L(x_1) \leq L(x_2) \leq L(x_3) \leq \cdots \leq L(x_N) \tag{4-52}$$

Indeed, if Eq. (4-52) is not met for two messages x_k and x_j, one may interchange their corresponding codes and arrive at a lower value of $\bar{L}$. Thus such codes cannot be of the optimum type.

2. For an optimum code it is necessary that

$$L(x_{N-1}) = L(x_N) \tag{4-53}$$

If we assign similar code words to x_N and x_{N-1} except for the final digit, our purpose is served. Any additional digit for x_N and x_{N-1} unnecessarily increases $\bar{L}$. Therefore, at least two messages x_{N-1} and x_N should be encoded in words of identical length. However, not more than D such messages could have equal length. It can be shown that, for an optimum encoding, n_0, the number of least probable messages which should be encoded in words of equal length, is the integer satisfying the requirements

$$\frac{N - n_0}{D - 1} = \text{integer} \qquad 2 \leq n_0 \leq D$$

3. Each sequence of length $L(x_N) - 1$ digits either must be used as an encoded word or must have one of its prefixes used as an encoded word.

In the following we shall restrict ourselves to the binary case ($D = 2$). A similar procedure applies for the general case as shown in Example 4-8. Condition 2 now requires that the two least probable messages have the same length. Condition 2 specifies that the two encoded messages of length m are identical except for their last digits. We shall select these two messages to be the nth and $(n - 1)$st original messages. After such a selection we form a composite message out of these two messages with a probability equal to the sum of their probabilities. The set of messages X in which the composite message is replacing the afore-mentioned two messages will be referred to as an *auxiliary ensemble of order* 1 or simply $AE1$. Now we shall apply the rules for finding optimum codes to $AE1$; this will lead to $AE2$, $AE3$, and so on. The code words for each two least probable members of any ensemble AEK are identical except for their last digits, which are 0 for one and 1 for the other. The iteration cycle is continued up to the time that AEM has only two messages. A final digit 0 is assigned to one of the messages and 1 to the other. Now we shall trace back our path and remember each two messages which have to differ only in their last digits. The optimality of the procedure is a direct consequence of the previously described optimal steps. (For additional material see Fano [1].)

Huffman's method provides an optimum encoding in the described sense. The methods suggested earlier by Shannon and Fano do not necessarily lead to an optimum code.

Example 4-7. Given the following set of messages and their corresponding transmission probabilities

$$[m_1, m_2, m_3]$$
$$[\tfrac{1}{3}, \tfrac{1}{3}, \tfrac{1}{3}]$$

(*a*) Construct a binary code satisfying the prefix condition and having the minimum possible average length of encoded digits. Compute the efficiency of the code.

(*b*) Next consider a source transmitting messages

$$\begin{bmatrix} m_1m_1 & m_1m_2 & m_1m_3 \\ m_2m_1 & m_2m_2 & m_2m_3 \\ m_3m_1 & m_3m_2 & m_3m_3 \end{bmatrix}$$

Construct a binary code with the prefix property and minimum average length and compute its efficiency.

Solution

(*a*) If the binary code must have the prefix property, then we assign the following code:

$$m_1 \rightarrow 0$$
$$m_2 \rightarrow 1\ 0$$
$$m_3 \rightarrow 1\ 1$$

The average length of the code word is

$$\tfrac{1}{3} \cdot 1 + \tfrac{1}{3} \cdot 2 + \tfrac{1}{3} \cdot 2 = \tfrac{5}{3} \text{ binits}$$

0 and 1 appear with probabilities of $\tfrac{2}{5}$ and $\tfrac{3}{5}$, respectively.

$$\text{Efficiency} = \frac{\log 3}{\tfrac{5}{3} \log 2} = 0.95$$

Shannon's encoding also leads to the same result.

(*b*) We construct a Huffman code:

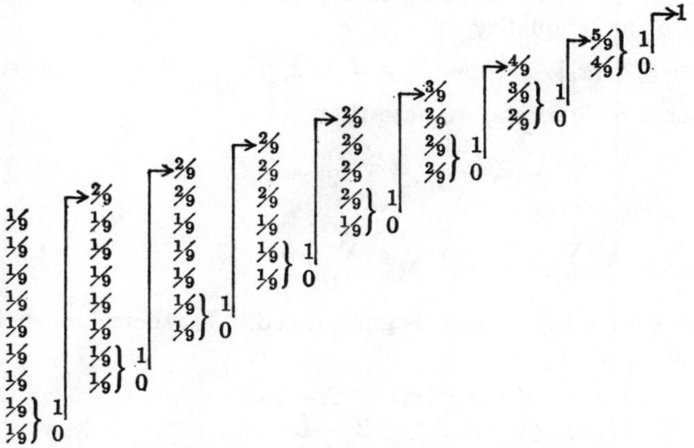

1	1	0	
1	0	1	
1	0	0	
0	1	1	
0	1	0	
0	0	1	
0	0	0	
1	1	1	1
1	1	1	0

$$p\{0\} = \tfrac{13}{29} \qquad p\{1\} = \tfrac{16}{29}$$

As all messages are equiprobable,

$$\text{Average length} = \frac{7 \times 3 + 2 \times 4}{9} = \frac{29}{9} = 3.22 \text{ binits}$$

$$\text{Efficiency} = \frac{2 \log 3}{3.22} = 0.98$$

Example 4-8. Apply Huffman's encoding procedure to the following message ensemble and determine the average length of the encoded message.

$$\{X\} = \{x_1,x_2,x_3,x_4,x_5,x_6,x_7,x_8,x_9,x_{10}\}$$
$$p\{X\} = \{0.18,0.17,0.16,0.15,0.10,0.08,0.05,0.05,0.04,0.02\}$$

The encoding alphabet is $\{A\} = \{0,1,2,3\}$.

Solution

						→1.00	
				→0.49	0		
x_1	0.18		0.18	0.18	1		1
x_2	0.17		0.17	0.17	2		2
			→0.16	0.16	3		
x_3	0.16		0.16	0			00
x_4	0.15		0.15	1			01
x_5	0.10		0.10	2			02
x_6	0.08		0.08	3			03
x_7	0.05	0					30
x_8	0.05	1					31
x_9	0.04	2					32
x_{10}	0.02	3					33

$$\begin{aligned}\bar{L} &= 0.18 \times 1 + 0.17 \times 1 + 0.16 \times 2 + 0.15 \times 2 + 0.18 \times 2 \\ &\quad + 0.05 \times 2 + 0.05 \times 2 + 0.04 \times 2 + 0.02 \times 2 \\ &= 1.65\end{aligned}$$

4-10. Gilbert-Moore Encoding. The Gilbert-Moore alphabetical encoding is an interesting and simple procedure. While Shannon's encoding discussed in Sec. 4-7 was based on Eq. (4-31), the Gilbert-Moore method is based on the inequality

$$2^{1-n_k} \leq P\{x_k\} < 2^{2-n_k} \qquad k = 1, 2, \ldots, N \tag{4-54}$$

To see that such a code exists, we note that

$$\sum_{k=1}^{N} 2^{1-n_k} \leq 1 < \sum_{k=1}^{N} 2^{2-n_k}$$

or

$$2 \sum_{k=1}^{N} 2^{-n_k} \leq 1 < 4 \sum_{k=1}^{N} 2^{-n_k} \tag{4-55}$$

Thus the existence of the desired code is guaranteed. Furthermore, such codes will have the property of

$$1 - n_k \leq \log P\{x_k\} < 2 - n_k$$
$$1 - \bar{L} \leq -H(X) < 2 - \bar{L}$$

or

$$1 + H(X) \leq \bar{L} < 2 + H(X) \tag{4-56}$$

The following steps summarize this method.

Step 1. Write down the messages in their specified order. (We assume that some "alphabetic order" has been specified for the symbols.)

Step 2. Let n_i be the length of the encoded symbol x_i; choose n_i such that

$$2^{1-n_i} \leq P\{x_i\} < 2^{2-n_i} \qquad i = 1, 2, \ldots, N$$

Step 3. Compute the nondecreasing sequence $\{\alpha_1,\alpha_2, \ldots\}$

$$\begin{aligned} \alpha_1 &= \tfrac{1}{2}P\{x_1\} \\ \alpha_2 &= P\{x_1\} + \tfrac{1}{2}P\{x_2\} \\ &\cdots\cdots\cdots\cdots \\ \alpha_i &= P\{x_1\} + P\{x_2\} + \cdots + P\{x_{i-1}\} + \tfrac{1}{2}P\{x_i\} \\ &\cdots\cdots\cdots\cdots\cdots\cdots\cdots\cdots \end{aligned} \tag{4-57}$$

(Note that $0 \leq \alpha_1 \leq \alpha_2 \leq \cdots \leq 1$.)

Step 4. The encoding of the message x_i is given by the binary expansion of the number α_i to the n_ith place.

To prove that the coding possesses the prefix property, note that either of the following two inequalities must be true for any two symbols ($i < j$).

(*a*) $P\{x_i\} \leq P\{x_j\}$

(*b*) $P\{x_i\} > P\{x_j\}$

If (*a*) is valid, then $n_i \geq n_j$, but since

$$\begin{aligned} \alpha_j &\geq \alpha_i + \tfrac{1}{2}P\{x_i\} + \tfrac{1}{2}P\{x_j\} \\ \alpha_j &\geq \alpha_i + 2^{-n_i} + 2^{-n_j} \end{aligned} \tag{4-58}$$

we find that the jth code word cannot be identical with the first n_j places of the ith code word. A similar conclusion can be reached if (*b*) is true. Thus the code has the prefix property.

As an example, consider the first four letters of the English alphabet and their corresponding probabilities:

[space,A,B,C]
[0.1859,0.0642,0.0127,0.0218]

The corresponding n_i and α_i are found to be

[4,5,8,7] and [0.09295,0.2180,0.25635,0.2736]

Thus the desired code is

Space: 0001
A: 00110
B: 01000001
C: 0100011

A similar encoding procedure has been suggested by replacing α_i with β_i:

$$\beta_i = \sum_{j=1}^{i-1} 2^{1-n_j} + 2^{-n_i}$$

and obtaining the first n_i digits of the binary expansion of β_i. This encoding procedure, which preserves the original message order in a binary numbering order and has also the prefix property, is referred to as an alphabetical encoding. The amount of computation for an alphabetical encoding is very little, but the existing method for finding the alphabetical encoding with the least average cost is rather complex.

One may wish to apply this latter procedure to the English alphabet in its ordinary alphabetical order. The Gilbert-Moore answer to this problem is given in Table 4-2. In the code listed in this table, word lengths have been shortened to a minimum without losing the prefix property. Such codes have been referred to as the best alphabetic codes. The average length of the best alphabetic code can be made reasonably close to the best possible average length obtained by Huffman's technique.

4-11. Fundamental Theorem of Discrete Encoding in Presence of Noise. In Sec. 4-8 we discussed the first fundamental theorem of information theory. It was shown there that, for a given discrete noiseless memoryless channel with capacity C and a given source (without memory) with an entropy H, it is possible to devise proper encoding procedures such that the encoded output of the source can be transmitted through the channel with a rate as close to C as desired. In this section we wish to extend the foregoing concept to cover the case of discrete channels when independent noise affects each symbol. It will be shown that the output of the source can be encoded in such a way that, when transmitted over a noisy channel, the rate of transmission may approach the channel capacity C with the probability of error as small as desired. This statement is referred to as the second fundamental theorem of information theory. Its full meaning will be minutely restated at the end of this section, where a more analytic statement is derived.

Second Fundamental Encoding Theorem. Let C be the capacity of a discrete channel without memory, R any desired rate of transmission of information ($R < C$), and S a discrete independent source with a specified entropy. It is possible to find an appropriate encoding procedure to encode the output of S so that the encoded output can be transmitted through the channel at the rate R and decoded with as small a probability of error or equivocation as desired. Conversely, such a reliable transmission for $R > C$ is not possible.

From a mathematical standpoint, the proof of this fundamental theorem and its converse is the central theme of information theory.

Subsequent to Shannon's original statement of this crucial theorem, much interest was stimulated toward producing a formal proof. After a number of years of research, formal proof is now available for channels without memory or with finite memory. It seems that further work will be forthcoming in the periphery of this basic theorem. Among those who have contributed considerably to the formalization of these basic theorems are Barnard, Elias, Fano, Feinstein, Khinchin, McMillan, Shannon, and Wolfowitz. The first complete proof for discrete noisy channels is due to Feinstein. This proof is quite complex and requires a number of preliminary mathematical lemmas. Feinstein's proof occupies more than two chapters of his book "Foundations of Information Theory" (McGraw-Hill Book Company, Inc., New York, 1958).

The presentation of such extensive proof is beyond the scope of this chapter. It is also questionable if the inclusion of such proof would be decisively helpful to the reader who may not have an advanced background in probability and a professional interest in information theory. However, a heuristic proof for binary symmetric channels will be given which will throw some light on the theorem while avoiding its complicated mathematical details. Those interested in a formal proof are referred to the original papers of the aforementioned contributors and the material presented in Chap. 12.

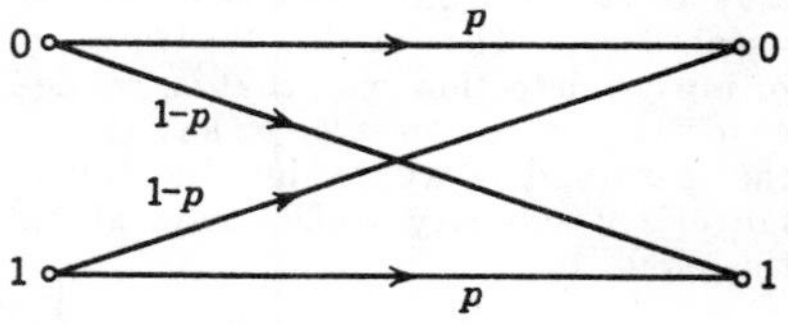

FIG. 4-7. A BSC.

A Heuristic Proof of the Fundamental Theorem for BSC. Consider a source S with a message ensemble $[A] = [a_1,a_2, \ldots ,a_N]$. The source is assumed to transmit any one of these N messages independently and with equal probability. In other words, you may think of it as a source which selects its signals completely "at random." The channel is specified to be a binary symmetric channel (BSC), as shown in Fig. 4-7.

$$P\{0|0\} = P\{1|1\} = p$$
$$P\{1|0\} = P\{0|1\} = 1 - p = q$$

The encoder must encode each of the N messages of $[A]$ into a string of 0's and 1's. We assume that the encoded messages all have the same-length n binary digits.

Finally, since the noise will affect the signals, we must devise an intelligent scheme for recognizing which input message was sent by inspecting the noise-altered message received. Consider the following: We have N source messages, each encoded into n binary digits. At the receiver, we have a catalog containing all possible 2^n n-symbol sequences. In the noisy transmission of an n-symbol sequence, the received sequence (there are at most 2^n of them) may not agree with any of the N sequences catalogued. In order to decide which of the N source sequences could have been sent, let us simply choose the catalogued sequence that differs from the received sequence by the least number of digits.

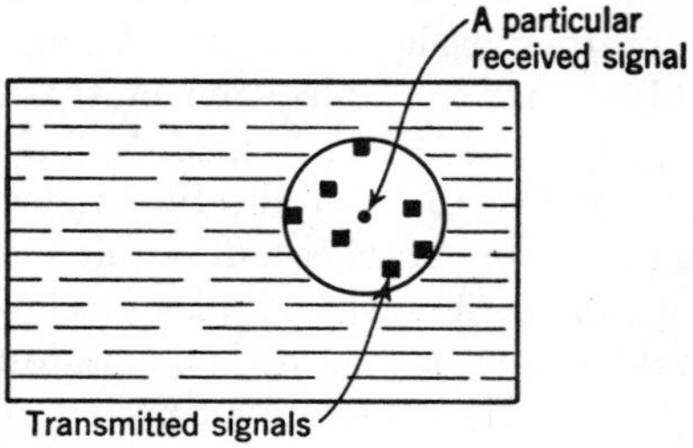

FIG. 4-8. Each square stands for a transmitted word. The small dot represents a received word not necessarily in the vocabulary. The circle illustrates a primitive detection rule, that is, we decode the received message as any one of the permissible words in the transmitter's vocabulary which may fall in this circle.

A geometric picture of the suggested code is given in Fig. 4-8. Each dot represents one of the 2^n possible received messages. The small squares indicate the N randomly selected source messages. For a given n, the number N should not be too large so that the message points can be somewhat evenly distributed in the message space with adequate distance among them to "overcome the effect of noise." Loosely speaking, for large values of n, one should be able to spread the message points so far apart that no other possible transmitted messages appear in their vicinity (that is, within the circle of the figure).

Now let us examine the reliability of the suggested encoding-decoding procedure when n is rather large. Consider the sequence of n 1's and 0's as a sequence of independent Bernoulli trials in which (referring to Fig. 4-7) noise-free transmission of each binary digit occurs with probability p and erroneous transmission with probability $1 - p = q$. Then the probability of receiving exactly $n - r$ correct digits (or, equivalently, exactly r erroneous digits) is

$$P\{r \text{ errors}\} = \binom{n}{n-r} p^{n-r} q^r \tag{4-59}$$

where

$$\binom{n}{n-r} = \frac{n!}{(n-r)!r!}$$

If we choose a random variable Z to denote the number of erroneous digits in a received message, then Z assumes values $k = 0, 1, 2, \ldots, n$ with probabilities

$$P\{Z = k\} = \binom{n}{n-k} p^{n-k} q^k$$

and its average value is

$$E(Z) = \bar{Z} = \sum_{k=0}^{n} kP\{Z = k\} = \sum_{k=1}^{n} k\binom{n}{n-k} p^{n-k} q^k = nq \tag{4-60}$$

In other words, in each sequence of length n binary digits, we can expect, *on an average*, nq digits to be altered by noise. Therefore, according to our decoding procedure, each catalogued sequence that differs from the received sequence by nq digits or less could, *on an average*, have been the sequence sent. The number of these sequences that can be considered as possible original messages, *on an average*, is

$$M = \sum_{k=0}^{nq} \binom{n}{k}$$

For $q \leq \frac{1}{2}$, the sum of the last nq terms on the right side of this equation is smaller than nq times the largest term, that is,

$$M \leqq 1 + nq \binom{n}{nq} \tag{4-61}$$

When n is large, the factorial can be approximated by Stirling's formula:

$$n! \approx \sqrt{2\pi}\, e^{-n} n^{n+\frac{1}{2}}$$

$$M \leqq 1 + nq \frac{(2\pi)^{\frac{1}{2}} e^{-n} n^{n+\frac{1}{2}}}{(2\pi)^{\frac{1}{2}} e^{-np} (np)^{np+\frac{1}{2}} (2\pi)^{\frac{1}{2}} e^{-nq} (nq)^{nq+\frac{1}{2}}} \tag{4-62}$$

Collecting terms yields

$$M \leqq 1 + nq \frac{e^{-n} n^{n+\frac{1}{2}}}{(2\pi)^{\frac{1}{2}} e^{-n} n^{(np+nq)} n (p)^{np} (q)^{nq} p^{\frac{1}{2}} q^{\frac{1}{2}}} \tag{4-63}$$

$$M \leqq \sqrt{\frac{nq}{2\pi p}}\, p^{-np} q^{-nq} + 1 \tag{4-64}$$

Of these M sequences which, according to our decoding scheme, can be considered as possible original messages, only one is correct, and $M - 1$ are potential misinterpretations of the received signal.

Now we use the further assumption that the binary encoding procedure is a random one. That is, for encoding any message, say a_k, we flip an honest coin n times. We obtain a sequence

$$a_k = \{HTTTHH \cdots T\}$$

Then each H is replaced by, say, a 0, and each T by a 1. Then

$$a_k = \{011100 \cdots 1\}$$

There are 2^n possible sequences that can be so constructed. However, there are only $N < 2^n$ messages in the message ensemble. Therefore, it is "intuitively" clear that the probability that an n-digit sequence, selected at random, corresponds to one of the N messages of $[A]$ is $N/2^n$. (See Shannon's proof in Chap. 12.)

Similarly, of the $M - 1$ potential misinterpretations of the received signal, on an average, only $N/2^n$ of them could correspond to one of the N messages of $[A]$. Thus, the number of messages of A, other than the correct original message, that could have been changed by noise into the received signal, *on an average*, is

$$\begin{aligned} M_A &\leq \frac{N}{2^n}(M - 1) \\ &\leq \frac{N}{2^n}\sqrt{\frac{nq}{2\pi p}}\, p^{-np}q^{-nq} \end{aligned} \tag{4-65}$$

The quantity M_A is indicative of the frequency of the occurrence of an error. To see the relation between the inequality (4-65) and the rate of transmission of information it is necessary to bring into consideration the rate of transmission of a binary symmetric channel and its capacity Note that

$$C = 1 + p \log p + q \log q \qquad \text{and} \qquad p^p = 2^{p \log p} \tag{4-66}$$

or

$$2^{-C} = \tfrac{1}{2}(2^{-p \log p})(2^{-q \log q}) = \tfrac{1}{2}p^{-p}q^{-q}$$

$$2^{-nC} = (\tfrac{1}{2})^n p^{-np}q^{-nq}$$

Substituting in (4-65),

$$M_A \leq \frac{N}{2^n}\sqrt{\frac{nq}{2\pi p}}\, 2^n 2^{-nC} \leq \frac{N}{2^{nC}}\sqrt{\frac{nq}{2\pi p}} \tag{4-67}$$

Now, by conveniently choosing N, the number of messages of A, to be equal to or less than $2^{Cn}/n$, the following simplification occurs:

$$M_A \leqq \sqrt{\frac{q}{2\pi pn}} \tag{4-68}$$

As the length of the encoded messages is increased, the number of the original sequences (not sent) which could have been erroneously decoded is diminished. When $n \to \infty$, then $M_A \to 0$ *irrespective* of the noise characteristic ($q \leq \tfrac{1}{2}$).

Finally, let us compute the entropy of the input to the channel when N is conveniently chosen to equal $2^{nC}/n$. (Values of 2^{nC} for different n and C are given in Fig. 4-9.)

$$H_n(X) = \frac{\log N}{n} = \frac{\log 2^{Cn} - \log n}{n} = C - \frac{1}{n}\log n \tag{4-69}$$

As n is made larger and larger, $H_n(X)$ approaches the channel capacity C, which also shows that the equivocation entropy approaches zero.

This completes the heuristic proof of the second fundamental theorem of information theory of a BSC. The foregoing proof demonstrates that, if the number of messages N is suitably selected, then by randomly encoding messages in binary digits, we obtain a family of random codes such that for at least one of these codes it is possible to approach the transmission rate C when n is made larger and larger. The closer we wish to get to this ideal rate, the more the length of the encoded messages must be increased. This necessarily causes a delay in transmission and reception. That is, a higher rate of transmission can be obtained at the expense of a longer delay. The above simple proof for the second fundamental theorem for BSC perhaps originated in several notes by scientists at Bell Telephone Laboratories (for instance, C. E. Shannon and E. N. Gilbert). Similar proofs were also given by G. A. Barnard and P. Elias. For a complete proof in the more general case, at present the mathematical machinery suggested by McMillan, Feinstein, and Khinchin

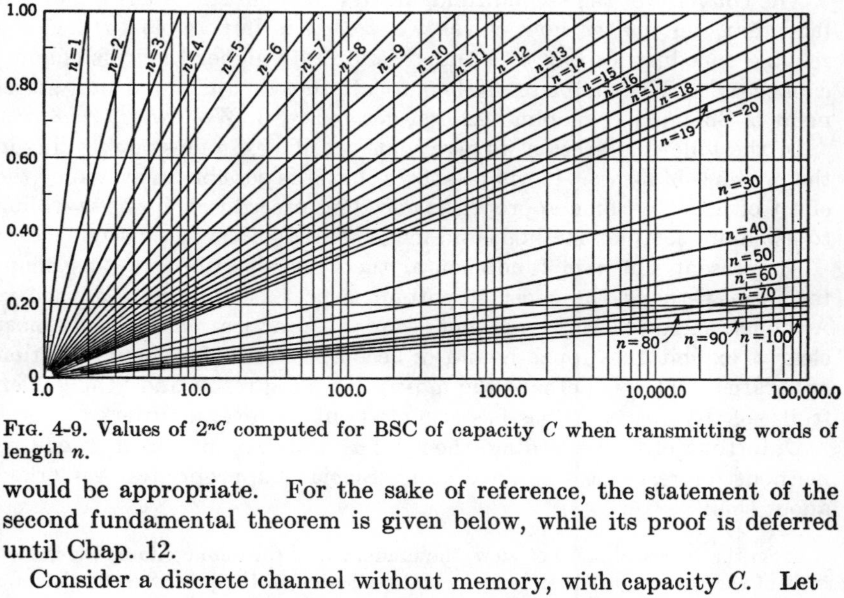

FIG. 4-9. Values of 2^{nC} computed for BSC of capacity C when transmitting words of length n.

would be appropriate. For the sake of reference, the statement of the second fundamental theorem is given below, while its proof is deferred until Chap. 12.

Consider a discrete channel without memory, with capacity C. Let

$$0 < H < C$$
$$\epsilon > 0$$

There exists a positive integer n depending on ϵ and H:

$$n = f(\epsilon, H)$$

such that if we consider the transmission of N words of length n

$$N \geq 2^{nH}$$

Then we are able to select the N transmitted symbols $u_1, u_2, \ldots, u_N$ such that at the receiver end we can associate with them N distinct categories of words $B_1, B_2, \ldots, B_N$ with

$$P\{v \in B_k|u_k\} \geq 1 - \epsilon \tag{4-70}$$

ϵ may be made as small as desired. Shannon's original statement of this most significant theorem in coding theory was given in two forms: one stating the possibility of a transmission rate close to the channel capacity with equivocation approaching zero and the other in terms of the vanishingly small probability of error in selecting input signals from the output signals. These two formulations, although not mathematically identical, lead to equivalent results from an engineering point of view. Those interested in mathematical developments are referred to Feinstein (I), Khinchin, and Chap. 12.

The converse of the second fundamental theorem states the important fact that, no matter how clever we are, it is impossible to devise a reliable encoding leading to a transmission rate higher than the channel capacity C. The complete mathematical proof for this important statement of Shannon's theorem was first derived by J. Wolfowitz (I).

In the following section we give some specific examples of encoding in the presence of noise. It will be shown that it is possible to overcome the effect of noise by some appropriate encoding procedures, if one is willing to use more complex methods and longer blocks of encoding sequences.

At present the two fundamental theorems described in this chapter form perhaps the most important aspect of information theory. Although these theorems may not seem of a practical nature, they most clearly exhibit the upper bound of accomplishment for communication apparatus. This is perhaps the most interesting result and "the golden fruit" of the theory, as has been pointed out by several writers.*

Unfortunately, the coding theory has not yet provided adequate methods for reaching this ideal aim. Slepian† appropriately remarked about this:

> From the practical point of view, the fundamental theorem contains the golden fruit of the theory. It promises us communication in the presence of noise of a sort that was never dreamed possible before: perfect transmission at a reasonable rate despite random perturbations completely outside our control. It is somewhat disheartening to realize that today, ten years after the first statement of this theorem, its content remains only a promise that we still do not know in detail how to achieve these results for even the most simple non-trivial channel.

4-12. Error-detecting and Error-correcting Codes. In the earlier sections of this chapter, we have discussed a number of basic encoding procedures for discrete independent sources connected to discrete

* For example, Robert Pierce, Frontispiece of PGIT, vol. IT-5, no. 2, June, 1959.
† D. Slepian, Coding Theory, *Nuovo cimento*, vol. 13, Suppl. 2, pp. 373–383, 1959.

memoryless channels in the absence of noise. From the practical point of view, it is of prime importance to devise encoding methods leading to a reliable transmission of information in the presence of noise. Examples of the need for such reliable transmission procedures are found in the operation of automatic telephone systems, large-scale digital computers, and in the new field of automata. In many applications of these types, the transmission of information must be kept error-free at quite a high level of reliability. Unfortunately, at present, there is no simple encoding method, analogous to Huffman's optimum coding, available for the transmission of information through noisy channels. The existing methods are generally complex and confined to binary channels with a relatively low rate of information transmission.

The first complete error-detecting and error-correcting encoding procedure was devised by Hamming in 1950. Hamming's method represents one of the simplest and most common encoding methods for the transmission of information in the presence of noise. We assume that the source transmits binary messages and that the channel is a binary symmetric channel (Fig. 4-7). In a message which is n digits long, a number of $m < n$ digits are directly employed to convey the information and the remaining $k = n - m$ digits are used for the detection and correction of error. The latter digits are called *parity checks*. Thus, in a certain sense, one may say that the relative redundancy of the procedure is $R \geq k/n$.

Hamming's single-error detecting code can be described as follows: The first $n - 1$ digits of the message are information digits; in the nth place we put either 0 or 1 so that the entire message has an even number of 1's. This is called an *even parity check* procedure. Evidently one can as well use an *odd parity check*, or one may wish to place the parity check at some other specified position. Examples of even and odd parity checks are given below:

$$\text{Messages}\quad\begin{bmatrix}100101\\010010\\101100\end{bmatrix}$$

Messages with even parity checks Messages with odd parity checks

$$\begin{bmatrix}1001011 & 1\\0100100 & 0\\1011001 & 1\end{bmatrix}_{\;P\text{ in last column}}\qquad\qquad\begin{bmatrix}1001010 & 0\\0100101 & 1\\1011000 & 0\end{bmatrix}_{\;P\text{ in last column}}$$

When a single parity check is used, if a single error occurs in a received message it will immediately be detected, although the position of the erroneous digit will not be determined. For example, with an even parity check, if we receive a message such as 110101$\underline{1}$, we detect an error, indicating that an odd number of digits has been transmitted in error. However, we have no specific knowledge of the position or the number of the errors. In the preceding we have tacitly assumed that the parity check was received without any error. This of course in itself may not be correct. Nonetheless the parity check improves the reliability of the transmission; that is, it increases the probability of the detection of error.

Hamming has also developed an error-correcting scheme which will not be presented here in detail, except for the case of a single error.

Single-error Detection and Correction. Our problem is to devise a method capable of:

1. Revealing the occurrence of a single error in any binary message block n digits long
2. Detecting the position of the erroneous digit

In an n-digit message, it is assumed, either no error or a single error occurs. But if the error occurs it may be in any one of the n possible positions. This procedure will be discussed in Sec. 4-13.

4-13. Geometry of the Binary Code Space. Consider all encoded messages having n digits and constructed as sequences of letters taken from an alphabet of D letters. Each encoded message can be considered as a point in an n-dimensional space. If for convenience some arbitrary numbers are associated with these D letters, then each point of the code space will have real coordinates. For example, when using a binary alphabet, we are led to points in the n-dimensional space with every coordinate being either 0 or 1. Such a geometric model has a certain natural appeal for discussing binary encoding problems. This model was initially employed by Hamming in 1950 and since then has found considerable use in connection with binary coding.

Let $U = [\alpha_1,\alpha_2, \ldots ,\alpha_n]$ be a binary word. That is, $\alpha_k = 0$ or 1. We may define the *distance* between two points U and $V = [\beta_1,\beta_2, \ldots ,\beta_n]$ as the number of coordinates for which all α_k and β_k are different.

$$D(U,V) = \sum_{k=1}^{n} (\alpha_k \oplus \beta_k) \tag{4-71}$$ *

To justify the use of the word "distance," the validity of certain mathematical properties of $D(U,V)$ should be examined: these are

* The notation $\oplus$ here implies

$$1 \oplus 0 = 1$$
$$0 \oplus 1 = 1$$
$$0 \oplus 0 = 0$$
$$1 \oplus 1 = 0$$

$$\begin{aligned} &D(U,V) = 0 && \text{if, and only if, } U = V \\ &D(U,V) = D(V,U) > 0 && \text{if } U \neq V \qquad (4\text{-}72) \\ &D(U,W) + D(V,W) \geq D(U,V) \end{aligned}$$

The validity of these properties is self-evident. As an example, note that

$$\begin{aligned} U &= 1001 & D(U,V) &= 4 \\ V &= 0110 & D(U,W) &= 1 \\ W &= 1000 & D(V,W) &= 3 \end{aligned}$$
$$1 + 3 = 4$$

Now that the *distance* has been defined, we are in a position to define a sphere of radius r and centered at point U as the locus of all points in the code space that are at a distance r from U. Thus, in the above example, point V is on a sphere with center at W and radius 3.

Now suppose that, at the input of the channel, all encoded words are at a distance of at least 2 from each other. If, because of an error in transmission, one single error occurs, then a word will be erroneously received as a meaningless word, that is, a word that does not exist in the transmission vocabulary. Thus, in such a setup any single error is detectable. If the minimum distance between points representing code words is taken to be 3 units and a single error occurs in the transmission of a word, then the point representing the received word will be 1 unit apart from the point representing the correct word. The correct word can be identified by finding the closest permissible word to the one received. Such schemes can be used for single-error detection and correction. The following data are given by Hamming:

Minimum required distance between every two coded words	*Description of the coding*
1	Error cannot be detected
2	A single error can be detected
3	A single error can be corrected
4	A single error can be corrected plus double error detected
5	Double-error correction

Having the concept of distance in mind, we now can raise the following question. How many code words at most can be included in a vocabulary containing only n-digit words subject to a single-error detection? Or alternatively, what is the largest number of vertices in a unit n-dimensional cube such that no two points are closer than 2 units from each other? A code book (n,d) containing the greatest number of words for a specified number of error detections and corrections is referred to as an optimal code. The Hamming codes (n,d) are generally referred to as *systematic codes.*

First consider the problem for $n = 2$. As shown in Fig. 4-10, there are four points in this space with the following coordinates:

$$\begin{bmatrix} U_1 \\ U_2 \\ U_3 \\ U_4 \end{bmatrix} = \begin{bmatrix} 0 & 1 \\ 1 & 1 \\ 0 & 0 \\ 1 & 0 \end{bmatrix}$$

Thus $\quad D(U_1,U_2) = D(U_1,U_3) = D(U_3,U_4) = D(U_4,U_2) = 1$

$$D(U_1,U_4) = D(U_2,U_3) = 2$$

There are two such sets of points with a distance of 2 units from each other. In other words, it is observed that a two-dimensional cube has 2^2 vertices, among which at most 2^1 points are 2 units apart.

For $n = 3$ we note (Fig. 4-11) that the points are the vertices of two distinct squares. Obviously, the two

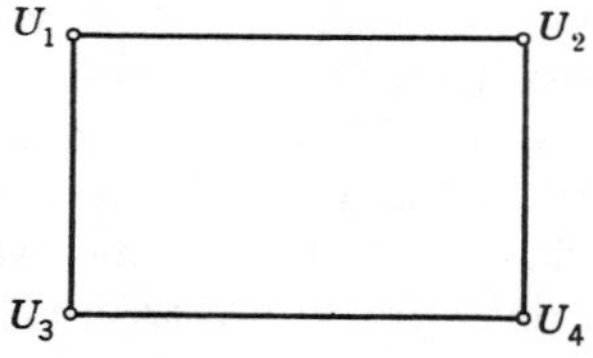

FIG. 4-10. Four points with a minimum mutual distance of 1 unit.

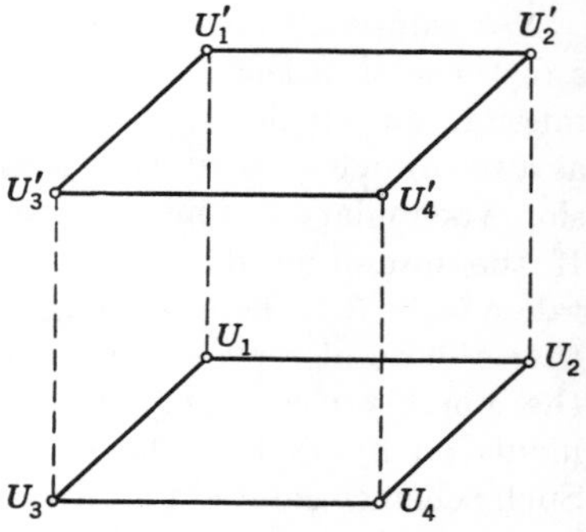

FIG. 4-11. Vertices of a unit cube in a three-dimensional space.

sets (U_1,U_4,U_2',U_3') and (U_2,U_3,U_1',U_4') satisfy the above distance requirements; that is, the distance among points of each set is at least 2 units.

$$\begin{bmatrix} U_1 \\ U_2 \\ U_3 \\ U_4 \end{bmatrix} = \begin{bmatrix} 010 \\ 110 \\ 000 \\ 100 \end{bmatrix} \qquad \begin{bmatrix} V_1 \\ V_2 \\ V_3 \\ V_4 \end{bmatrix} = \begin{bmatrix} 011 \\ 111 \\ 001 \\ 101 \end{bmatrix}$$

A three-dimensional cube has 2^3 vertices, among which at most 2^2 points are 2 units apart. This reasoning could be extended without any difficulty. We shall conclude that an n-dimensional cube has 2^n vertices, which can be considered the vertices of two distinct $(n - 1)$-dimensional cubes. Among all these vertices there are at most 2^{n-1} points which are 2 units apart. Therefore, for single-error detecting schemes with code words each n symbols long, we can have at most 2^{n-1} words. The result may be summarized by writing

$$B(n,d) = B(n,2) = 2^{n-1} \tag{4-73}$$

$B(n,d)$ being the upper bound for the number of code words of length n and a minimum mutual distance of 2 units.

The following interesting results have been obtained by Hamming and will be quoted without further proof. In the light of the above discussion, the reader may wish to prove them as an exercise.

$$\begin{aligned} B(n,1) &= 2^n \\ B(n,2) &= 2^{n-1} \\ B(n,3) &= 2^m \leq \frac{2^n}{n+1} \\ B(n,4) &= 2^m \leq \frac{2^{n-1}}{n} \\ B(n-1, 2K-1) &= B(n,2K) \end{aligned} \tag{4-74}$$

4-14. Hamming's Single-error Correcting Code. In this section we discuss Hamming's code for single-error detection as well as correction. The method demonstrates how one can improve the reliability of the transmission of information in the presence of noise. In order to correct a single error in any one of the n positions, we need $n + 1$ independent "pieces of information." (One piece of information is required to show that no error has occurred.) With n parity checks, it is possible to have at most 2^k distinct parity words. If a one-to-one correspondence among the parity words and error locations is to be established, we must require that

$$2^k \geq n + 1 \tag{4-75}$$

As for the transmission procedure, in lieu of transmitting a word $x_1x_2 \cdots x_m$, we compute its corresponding parity-check word $x_{m+1} \cdots x_n$ and transmit the word $x_1x_2 \cdots x_n$. (The parity checks must be such that distinct m words have distinct parity words.) Then, at the receiver, we must devise a technique for determining the position of any possible single error, or no error. The method can be illustrated in terms of an example.

Suppose that we wish to devise a single-error detecting and correcting code for blocks of four binary digits. The smallest number of the required parity checks k is given by

$$\begin{aligned} 2^k &\geq n + 1 = 5 \\ k &\geq \log 5 \end{aligned}$$

In order to be able to transmit four information digits we need to have at least three parity digits in each block. In fact, let x_i denote the digit in the ith position in a sequence of seven digits. The parity checks $[x_5,x_6,x_7]$ may be derived from the modular 2 equations:

$$\begin{aligned} x_1 + x_2 + x_3 + x_5 &= \text{even} \qquad s_1 \\ x_1 + x_2 + x_4 + x_6 &= \text{even} \qquad s_2 \\ x_1 + x_3 + x_4 + x_7 &= \text{even} \qquad s_3 \end{aligned} \tag{4-76}$$

For any received word, the truth set (validity) of these equations can be exhibited by the three sets of Fig. 4-12.

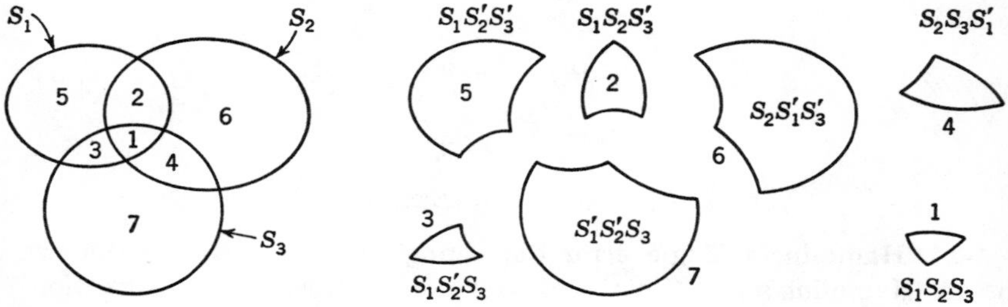

FIG. 4-12. A set-theoretic approach for deriving Hamming's single-error correcting equations among information and parity checks.

There are seven disjoint sets in this figure, and each is associated with only one variable x_i. The variable x_1 belongs to the common intersection $s_1s_2s_3$. The variable x_2 belongs to the set $s_1s_2s_3'$, and so on.

If only s_1 fails, then x_5 is incorrect.
If only s_2 fails, then x_6 is incorrect.
If only s_3 fails, then x_7 is incorrect.
If only s_1 and s_2 fail, then x_2 is incorrect.
If only s_1 and s_3 fail, then x_3 is incorrect.
If only s_2 and s_3 fail, then x_4 is incorrect.
If s_1, s_2, and s_3 fail, then x_1 is incorrect.
If s_1, s_2, and s_3 are valid, then there is no error.

Thus if we assume that not more than a single error may occur in blocks of seven digits, with this method we shall be able to correct all such possible errors. For example, if 1010110 is received, x_5 is in error.

The validity of the foregoing method is based on the fact that the suggested three sets embody seven disjoint subsets, each subset being in a one-to-one correspondence with a logical proposition concerning the validity or the failure of sets s_1, s_2, and s_3. Hence, a one-to-one correspondence between the seven variables and the corresponding seven logical propositions may be established.

Next, we may generalize the above procedure and suggest a logical method for writing the required number of modular 2 equations. For an

(n,k) code, the following steps in the selection of the appropriate terms of the basic k equations seem self-explanatory.

Step 1. Denote by s_i the logical proposition of the validity of the ith equation $(i = 1, 2, \ldots, k)$ and include x_{m+i} only in the ith equation. These are the totality of the parity checks.

Step 2. Include x_1 in all k equations.

Step 3. Include each of the next $\binom{k}{k-1} = k$ variables, that is, x_2, $x_3, \ldots, x_{k+1}$, in $k - 1$ equations [as they occur in a general set-theoretic diagram; that is, each equation contains $1 + \binom{k-1}{1} + \binom{k-1}{2} + \cdots + \binom{k-1}{k-1}$ terms].

Step 4. Include each of the next $\binom{k}{k-2} = \frac{1}{2}k(k-1)$ variables, that is, $x_{k+2}, \ldots$, in $k - 2$ equations (as they appear in a general set-theoretic diagram).

Step 5. Continue this method until the last information digit x_m is included.

The k equations obtained in this manner constitute the main rule for encoding and decoding. For encoding, one computes the parity checks to be transmitted along with any given information sequence. For decoding, one can consider the sequence $s = s_1 s_2 \cdots s_k$, where each individual term s_i assumes the value 0 if the ith equation is valid for the received message and the value 1 otherwise. There are 2^k distinct possibilities for s sequence, and as long as $2^k \geq n + 1$ we are able to make a one-to-one correspondence between every variable x_j and a distinct s sequence. As examples of the applications of the foregoing rules, we state the results for the following two cases:

Case A $\qquad m = 1 \qquad k = 2 \qquad n = 3$

$$x_1 + x_2 = 0$$
$$x_1 + x_3 = 0$$

Case B $\qquad m = 11 \qquad k = 4 \qquad n = 15$

$$x_1 + x_2 + x_3 + x_4 + x_6 + x_7 + x_8 + x_{12} = 0$$
$$x_1 + x_2 + x_3 + x_5 + x_6 + x_9 + x_{10} + x_{13} = 0$$
$$x_1 + x_2 + x_4 + x_5 + x_7 + x_9 + x_{11} + x_{14} = 0$$
$$x_1 + x_3 + x_4 + x_5 + x_8 + x_{10} + x_{11} + x_{15} = 0$$

In the practical application of single-error correcting codes, the following method of message numbering makes computation quite simple. This method is based on an appropriate bookkeeping procedure suggested by Hamming.

Step 1. Number the messages to be transmitted as 0, 1, 2, 3, . . . and choose for the message M_j the binary expression of number j. Since all messages must contain m information digits, add zeros to the left of the binary number j, if necessary.

Step 2. Number the positions in all n words from left to right.

Step 3. Assign $[x_1,x_2,x_4,x_8, \cdots]$ to check positions given by

$$\begin{aligned}
&x_1 + x_3 + x_5 + x_7 + x_9 + x_{11} + x_{13} + \cdots = 0 \qquad s_1\\
&x_2 + x_3 + x_6 + x_7 + x_{10} + x_{11} + x_{14} + \cdots = 0 \qquad s_2\\
&x_4 + x_5 + x_6 + x_7 + x_{12} + x_{13} + x_{14} + \cdots = 0 \qquad s_3\\
&\cdots\cdots\cdots\cdots\cdots\cdots\cdots\cdots\cdots\cdots\\
&\text{(Include only those position numbers containing}\\
&\qquad\qquad\qquad \text{a 1 in their } i\text{th digits)} = 0 \qquad s_i
\end{aligned}$$

Step 4. The selection of parities should be according to step 3. When a message is received, check the equations in step 3. When equation s_i is valid for the received messages, let $s_i = 0$; otherwise, $s_i = 1$. Next, compute the binary number

$$l = \cdots s_2s_1$$

The suggested method (Hamming) indicates that the digit in the lth position (step 2) must have been in error.

For example, when the method is applied to 16 messages ($n = 4, k = 3$), if 0101101 is received, we find $l = 100$; therefore x_4 is in error. The proof of the validity of the method is based on two facts: (1), the selection of independent equations or, what amounts to the same thing, the establishment of a one-to-one correspondence between $n + 1$ positions and distinct subsets of k sets, as described before; and (2), the proper ordering and assignment of binary numbers relevant to the chosen ordering system. The reader may wish to consult Hamming's paper or "Logical Design of Digital Computers," by M. Phister, Jr. (John Wiley & Sons, Inc., New York, 1959).

However, the constraint of a fixed number of errors in a block is not a practical one. In a BSC, let p be the probability of error in the transmission of a digit. Then the probability of receiving an incorrect word n digits long when $np \ll 1$ is

$$\begin{aligned}
\binom{n}{1}p(1-p)^{n-1} + \binom{n}{2}p^2(1-p)^{n-2} + \cdots + \binom{n}{n}p^n &= 1 - (1-p)^n\\
&= np - \tfrac{1}{2}n(n-1)p^2 + \cdots\\
&\approx np \qquad (4\text{-}77)
\end{aligned}$$

The probability of an incorrect word after applying a single-error correcting scheme is

$$\binom{n}{2}p^2(1-p)^{n-2} + \cdots + \binom{n}{n}p^n = 1 - (1-p)^n - np(1-p)^{n-1}$$
$$= \tfrac{1}{2}n(n-1)p^2 + \cdots \qquad (4\text{-}78)$$

Thus the probability of decoding an incorrect word will be reduced from np to approximately $\frac{1}{2}n(n-1)p^2$. (For example, if $p = \frac{1}{100}$, without any corrective measure, the probability of decoding an incorrect word of seven digits is 0.07.) When Hamming's single-error correcting code is

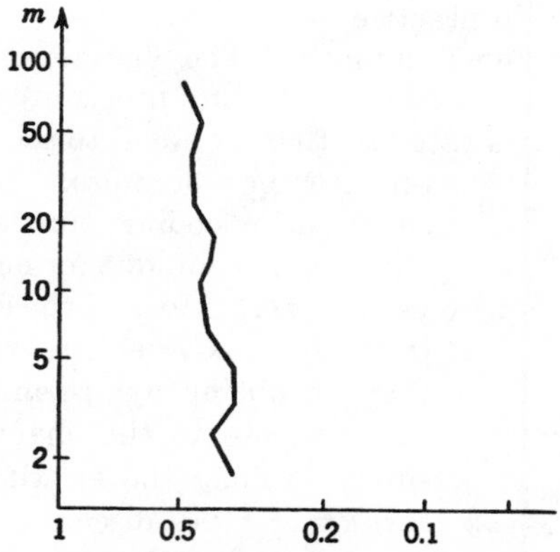

FIG. 4-13. To the left of the curve, the probability of erroneous decoding for Hamming's SEC code is larger than when no encoding is used, and, to the right, the probability is smaller.

applied, the probability of an incorrect word is often reduced. The ratio of the two probabilities gives an indication of the improvement brought about by the encoding scheme. This ratio is referred to as *the figure of merit* of the code. Thus the figure of merit for a single-error correcting code is

$$\frac{1-(1-p)^m}{1-(1-p)^n - np(1-p)^{n-1}} \qquad (4\text{-}79)$$

The probability of receiving an incorrect block after single-error correction can easily be computed for different n and $m = n - k$. Let N be the number of messages to be encoded. The number of information digits m is the smallest integer that is larger than $\log N$; note that

$$2^m \leq \frac{2^n}{n+1} \qquad (4\text{-}80)$$

The following table and the graph of Fig. 4-13 were obtained from the reference given in the footnote.*

N	4	8	16	32	64	128	256	512	1,024
m	2	3	4	5	6	7	8	9	10
n	5	6	7	9	10	11	12	13	14

The region to the right of the curve in Fig. 4-13 is the region where the probability of error becomes lower; thus Hamming's code can be successfully applied. Note that this region corresponds to the cases most frequently encountered in practice.

4-15. Elias's Iteration Technique. The Shannon-Feinstein fundamental theorem proves the existence of coding procedures allowing transmission of information at a rate less than or equal to the channel capacity in the presence of noise. However, no highly effective encoding procedures similar to Huffman's technique for noiseless channels are yet known. Most of the encoding techniques thus far discovered are rather complex, yet they do not always permit transmission at a rate as close to the channel capacity as desired. Among the existing codes, the error-correcting codes discussed earlier are perhaps the least complex ones. The iteration method suggested by P. Elias for binary symmetric erasure channels, as given below, is a good illustration of the application of error-correcting techniques (see Elias [II]).

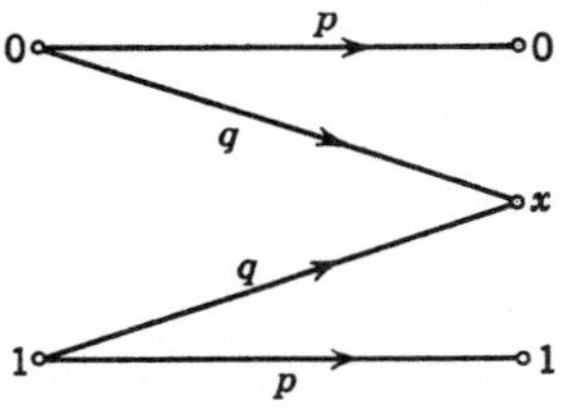

FIG. 4-14. A BEC.

Consider a BEC as illustrated in Fig. 4-14. The input sequence of 0's and 1's is divided into blocks each $N_1 - 1$ digits long. To each such block we add a parity check, say to the N_1th place. The parity check will be selected 0 or 1 so that the total number of 1's in each block becomes an even number (even parity check).

$$\begin{matrix} i_1, & i_2, & \ldots, & i_{N_1-1}, & C_1 \\ [0, & 0, & \ldots, & 1, & 1] \end{matrix}$$

The average number of erasure digits in a block is

$$E(z) = \bar{z} = \sum_{z=0}^{N_1} z \binom{N_1}{z} q^z p^{N_1-z} = N_1 q \tag{4-81}$$

Since with a single parity check any single error will be detected and cor-

* G. A. Shastova, *Radiotekh. i Elektron.*, vol. 3, no. 1, pp. 19–26, 1958.

rected, the average number of erasures in blocks of N_1 digits is

$$E(z) - E(z_1) \tag{4-82}$$

where $E(z_1)$ is the average number of single erasures in the block, i.e.,

$$E(z_1) = \binom{N_1}{1} q p^{N_1-1} \tag{4-83}$$

Equation (4-83) yields

$$E(z) - E(z_1) = N_1 q - N_1 q p^{N_1-1} = N_1 q(1 - p^{N_1-1}) \tag{4-84}$$

A simple upper bound for the average number of remaining erasures is

$$\begin{aligned} E(z) - E(z_1) &\le N_1 q[1 - (1 - q)^{N_1}] \\ &\le N_1 q(1 - 1 + N_1 q) = (N_1 q)^2 \end{aligned} \tag{4-85}$$

For example, assume that the following blocks ($N_1 = 10$) were received:

$$\begin{array}{ccccccccc|c} 1 & 0 & 0 & 1 & 0 & 1 & 0 & x & 0 & 1 \\ 0 & 1 & x & x & 0 & 1 & 0 & 1 & 0 & 1 \\ 1 & 1 & 1 & 0 & 0 & 1 & 1 & 1 & 0 & x \end{array}$$

In the first and the third block the erasure must have been a 0, while the originals of the double erasures in the second block remain unidentified.

By means of this technique the average number of erasures in each block is reduced from $N_1 q$ to not more than $(N_1 q)^2$. It must be observed, however, that the rate at which the information is supplied to the channel is in the meantime reduced by a factor of $(N_1 - 1)/N_1$. Elias's iteration technique suggests the transmission of $N_2 - 1$ blocks of the above type. The N_2th block is a parity block in which each digit is a parity check for the digits above it.

$$\left[\begin{array}{cccc|c} 0 & 1 & 0 & \cdots & 1 \\ 1 & 0 & 0 & \cdots & 0 \\ \cdot & \cdot & \cdot & \cdots & \cdot \\ 1 & 1 & 0 & \cdots & 1 \\ \hline 1 & 0 & 1 & \cdots & 0 \end{array}\right] \left[\begin{array}{c|c} \text{information digits} & \text{check digits} \\ \hline \text{check digits} & \text{check digit} \end{array}\right]$$

The matrix is partitioned into information digits and parity digits. The rows are transmitted in order. Thus, all rows with a single erasure are properly decoded. Since the information digits are statistically independent, when the last row of the matrix is received, we shall be able properly to decode each column that contains only a single erasure.

Define q_1 and q_2 as the average probability of erasure after correction of rows and columns, respectively. Then, the following relations hold:

$$N_1q_1 = N_1q(1 - p^{N_1-1}) \qquad \text{with } q_1 < N_1q^2 \tag{4-86}$$
$$N_2q_2 = N_2q_1(1 - p_1^{N_2-1}) \qquad \text{with } q_2 < N_2q_1^2 < N_2N_1^2q^4 \tag{4-87}$$

where
$$p_1 = 1 - q_1$$

For example, if $N_1 = N_2 = 10$ and $q = \frac{1}{8}$, the average erasure probability after matrix block correction is less than 2.5 per cent, roughly a fivefold improvement as far as the erasure probability is concerned. The iteration process can, of course, be continued. We may use N_3 matrix blocks in which the last block is a check. The new erasure probability for the remaining digits after the described corrections is

$$\begin{aligned} N_3q_3 &< N_3q_2(1 - p_2^{N_3-1}) \\ q_3 &< N_3q_2^2 < N_3N_2^2q_1^4 < N_3N_2^2N_1^4q^8 \end{aligned} \tag{4-88}$$

As the iteration process is continued, the remaining erasure probability becomes smaller and smaller. The rate of transmission in bits per symbol is meanwhile decreased. However, the decrease of the rate is very slow. For instance, for a particularly simple computation, let, with Elias, $N_k = 2N_{k-1}$ for $k = 1, 2, \ldots$; then

$$\begin{aligned} q_k &< 2N_{k-1}q_{k-1}^2 \\ q_k &< 2^{k-1}N_1q_{k-1}^2 \end{aligned} \tag{4-89}$$

For $q = \frac{1}{20}$ and $N_1 = 10$ we have

$$\begin{aligned} q_1 &< \tfrac{1}{10}\, 2^{-2} \\ q_2 &< \tfrac{1}{10}\, 2^{-4} \\ q_3 &< \tfrac{1}{10}\, 2^{-8} \end{aligned} \tag{4-90}$$

The rate at which the information is supplied to the channel, assuming equiprobable and statistically independent digits, is the ratio of the information to the total number of digits, that is,

$$\begin{aligned} R &= \frac{N_1 - 1}{N_1}\,\frac{N_2 - 1}{N_2}\,\frac{N_3 - 1}{N_3} \cdots \\ R &= \left(1 - \frac{1}{10}\right)\left(1 - \frac{2^{-1}}{10}\right)\left(1 - \frac{2^{-2}}{10}\right) \cdots \\ R &> 1 - \tfrac{1}{10}(1 + 2^{-1} + 2^{-2} + \cdots) \\ R &> 0.80 \end{aligned} \tag{4-91}$$

Note that with this procedure we have been able to reduce successfully the probability of erasure, while the rate of transmission of information has moderately decreased. In fact, as was pointed out in Chap. 3, the capacity of the BEC is

$$C = 1 - q = 0.95$$

We have not succeeded in reducing the error probability while approaching the maximum rate of transmission. Such an achievement would require much more elaborate encoding schemes. It is also to be noted that the lowering of the remaining erasure probability is accompanied by a greater delay required between the time of transmission and the time of decoding a digit. For a relation between the delay and the error probability see Elias (II).

Example 4-9. *Elias's Block Coding.* Consider a binary erasure channel with erasure probability $q = 1/16$. The source has a number of messages which are encoded in binary digits such that in a long sequence of messages 0 and 1 appear independently and with equal probabilities. Each information digit is transmitted twice in order to combat the effect of noise.

(*a*) Determine the rate of the input information.

(*b*) Determine the average fraction of information-digit erasure.

(*c*) Determine the rate of transmission of information in the channel.

(*d*) In order to improve the transinformation rate, we use Elias's block coding. More specifically, we transmit message matrices of the form

$$\begin{bmatrix} i_1 & i_2 & i_3 & c_1 \\ i_4 & i_5 & i_6 & c_2 \\ i_7 & i_8 & i_9 & c_3 \\ c_4 & c_5 & c_6 & c_7 \end{bmatrix}$$

c_1, c_2, and c_3 are parity checks on the first, second, and third rows, respectively; c_4, c_5, and c_6 are parity checks on the first, second, and third columns. c_7 is a parity check on the parity checks above it. The remaining digits are information digits. Repeat parts (*a*), (*b*), and (*c*) for this encoding procedure.

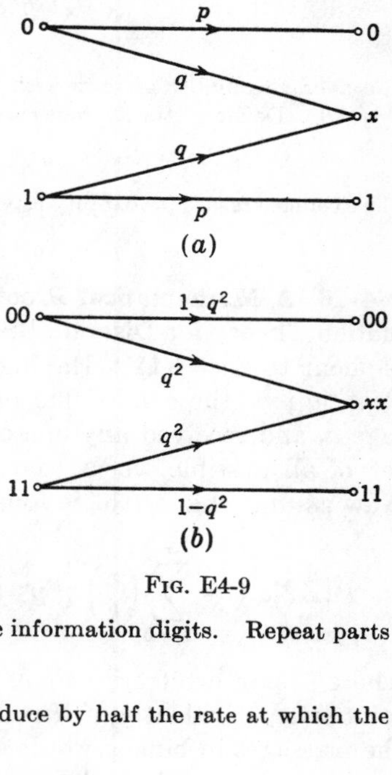

FIG. E4-9

Solution

(*a*) Repetition of each information digit will reduce by half the rate at which the information is supplied.

$$1 \times 1/2 = 1/2 \text{ bit per symbol}$$

(*b*) We assume that the input is supplied in groups of two digits; in each group one digit acts as a parity check on the other (information digit). Therefore, if we receive a pair $1x$ or $x0$, we know that the original pairs were 11 and 00, respectively. Thus, single errors are corrected. The probability of double erasure is $q \cdot q = 1/256$.

(*c*) The equivalent channel of Fig. E4-9*a* is self-explanatory. For simplicity, think of 00 and 11 as two equiprobable input messages to the channel; thus

$$H(X) = 1$$
$$2H(X|Y) = -q^2 \log 1/2 - q^2 \log 1/2 = 2q^2$$
$$R = I(X;Y) = H(X) - H(X|Y) = 1 - q^2$$

But this rate is accomplished for the equivalent channel. However, in our original scheme only one digit carried information. Thus the rate in question is

$$\tfrac{1}{2} - q^2 = \tfrac{1}{2} - \tfrac{1}{256} = \tfrac{127}{256}$$

(*d*) Each letter in the message matrix is equiprobable; thus the average input rate for each information digit is $\tfrac{9}{16}$ bit per symbol. The probability of the occurrence of more than a single erasure can be calculated as follows: The average number of erasures in blocks of N digits is Nq. The average number of a single erasure is

$$\binom{N}{1} qp^{N-1}$$

Thus the desired average is

$$\begin{aligned} Nq - Nqp^{N-1} &= Nq(1 - p^{N-1}) \\ &= \tfrac{4}{16}[1 - (\tfrac{15}{16})^3] \end{aligned}$$

The average number of rows with an erasure (after a single correction) is $\tfrac{1}{4}[1 - (\tfrac{15}{16})^3]$. Define q_1, the average erasure probability after correction of the rows, as

$$q_1 = \tfrac{1}{16}[1 - (\tfrac{15}{16})^3]$$

The average erasure probability after checking by rows and columns is

$$q_2 = Nq_1 - Nq_1p_1^{N-1} = \tfrac{1}{4}[1 - (\tfrac{15}{16})^3](1 - p_1^3)$$

4-16. A Mathematical Proof of the Fundamental Theorem of Information Theory for Discrete BSC.* The material of this section is a supplement to Sec. 4-11. Having the notation of that section in mind, let $P\{A_i^r|a_i\}$ be the conditional probability of having transmitted the message a_i and received any one of the messages in the set A_i^r, that is, the set of all possible words that differ from a_i in not more than r digits. Now assume that we have found an encoding procedure for which

$$P\{A_i^r|a_i\} = \sum_{K=0}^{r} \binom{n}{k} q^K p^{n-K} > 1 - \epsilon \qquad i = 1, 2, \ldots, N \qquad (4\text{-}92)$$

where ϵ is an arbitrarily small positive number and A_i are mutually disjoint sets. Evidently, if N is not too large, it would be possible to encode the messages in binary words so that the above conditional probabilities are as large as desired. In fact, the conditional probability can be increased by increasing n, the length of the words, when N and ϵ are specified. The disjointness of the A_i is not of much concern as long as N is quite small. For example, if only two messages a_1 and a_2 are to be transmitted, by assigning an adequately long sequence of 0's and 1's to a_1 and a_2, respectively, the requirements of Eq. (4-92) are met. But it is

* The proof given here is a condensed version of a proof derived by D. D. Joshi. For further information see D. D. Joshi, L'Information en statistique mathématique et dans la théorie des communications, *Publ. inst. statistique univ. Paris*, vol. 8, no. 2, pp. 95–99, 1959.

not at all evident that such an encoding procedure would exist when N is comparatively large and ϵ arbitrarily small. To this end, we compute some bounds on the number of N_0, the upper bound of N.

Based on distance considerations, it becomes clear that the number of disjoint regions A_i^r is less than the maximum number of disjoint spheres with radius r packed in an n cube. A direct estimate of the upper bound of this number can be made (Hamming's) by computing the ratio of the total number of points 2^n to the number of points in each sphere. That is,

$$N_0 = B(n,r) = \frac{2^n}{\binom{n}{0} + \binom{n}{1} + \cdots + \binom{n}{r}}$$

or for $r < n/2$

$$\frac{2^n}{\frac{n-r}{n-2r}\binom{n}{r}} < N_0 < \frac{2^n}{\binom{n}{r}} \tag{4-93}$$

Let $\alpha = r/n$ and $\beta = 1 - \alpha$; then for sufficiently large n, say $n \geq n_0$, Eq. (4-93) gives

$$\sum_{n=0}^{n\alpha} \binom{n}{n\alpha} p^{n\alpha} q^{n\beta} > 1 - \epsilon \qquad n \geq n_0 \tag{4-94}$$

Thus

$$\frac{2^n}{\frac{\beta}{\beta - \alpha}\binom{n}{n\alpha}} < N_0 < \frac{2^n}{\binom{n}{n\alpha}}$$

$$n - \log \frac{\beta}{\beta - \alpha} - \log \binom{n}{n\alpha} < \log N_0 < n - \log \binom{n}{n\alpha} \tag{4-95}$$

For a moment, consider a hypothetical BSC with parameters

$$q_0 = \alpha \geq \frac{r}{n} \qquad p_0 = \beta$$

$$C_0 = 1 + \alpha \log \alpha + \beta \log \beta < C$$

It will be shown that the nth-order extension of this channel will lead to an ideal transmission, that is, when n approaches infinity, N_0 and the detection error remain bounded by Eq. (4-95) and Eq. (4-93), respectively. Subsequently, it will be demonstrated that the fundamental theorem remains valid when the parameters of the hypothetical channel approach those of the specified channel (p,q,C). We employ an approximate form of Stirling's formula, that is,

$$(2\pi)^{1/2} n^{n+1/2} e^{-n} < n! < (2\pi)^{1/2} (n + \tfrac{1}{2})^{n+1/2} e^{-(n+1/2)} \tag{4-96}$$

or

$$\begin{aligned} \log n! &< \log \sqrt{\frac{2\pi}{e}} + (n + \tfrac{1}{2}) \log (n + \tfrac{1}{2}) - n \log e \\ \log n! &> \log \sqrt{2\pi} + (n + \tfrac{1}{2}) \log n - n \log e \end{aligned} \tag{4-97}$$

Proper application of these bounds yields

$$\log \binom{n}{n\alpha} > \log \frac{e}{\sqrt{2\pi n\alpha\beta}} - n(\alpha \log \alpha + \beta \log \beta) - (n\alpha + \tfrac{1}{2}) \log \left(1 + \frac{1}{2n\alpha}\right) - (n\beta + \tfrac{1}{2}) \log \left(1 + \frac{1}{2n\beta}\right)$$

$$\log \binom{n}{n\alpha} < \log \frac{1}{\sqrt{2\pi en\alpha\beta}} - n(\alpha \log \alpha + \beta \log \beta) + (n + \tfrac{1}{2}) \log \left(1 + \frac{1}{2n}\right) \tag{4-98}$$

The completion of the proof of the theorem requires the introduction of the channel's capacity into our calculations.

$$\log N_0 > nc_0 + \log \sqrt{2\pi en\alpha\beta} - (n + \tfrac{1}{2}) \log \left(1 + \frac{1}{2n}\right) - \log \frac{\beta}{\beta - \alpha}$$

$$\log N_0 < nc_0 + \log \sqrt{2\pi n\alpha\beta}\, e + (n\alpha + \tfrac{1}{2}) \log \left(1 + \frac{1}{2n\alpha}\right) + (n\beta + \tfrac{1}{2}) \log \left(1 + \frac{1}{2n\beta}\right)$$

The entropy per transmitted symbol satisfies the inequalities

$$\frac{1}{n} \log N_0 - c_0 > \frac{1}{2n} \log \frac{2\pi en\alpha(\beta - \alpha)^2}{\beta} - \left(1 + \frac{1}{2n}\right) \log \left(1 + \frac{1}{2n}\right)$$

$$\frac{1}{n} \log N_0 - c_0 < \frac{1}{2n} \log 2\pi n\alpha\beta e^2 + \left(\alpha + \frac{1}{2n}\right) \log \left(1 + \frac{1}{2n\alpha}\right) + \left(\beta + \frac{1}{2n}\right) \log \left(1 + \frac{1}{2n\beta}\right) \tag{4-99}$$

By applying the above detection scheme and greatly increasing the word length, one finds

$$\lim_{n \to \infty} \left(\frac{1}{n} \log N_0\right) - C_0 = 0 \tag{4-100}$$

The upper bound of the number of words in the transmitter's vocabulary approaches 2^{nC_0} as n is made larger and larger. Meanwhile, the error in decoding each message a_i is kept under control. If the message a_i is transmitted with a probability p_i for $i = 1, 2, \ldots, N$, the *average error* remains bounded, that is,

$$\text{Average error} < \sum_{i=1}^{N_0} p_i\epsilon = \epsilon \tag{4-101}$$

Finally, one may choose α arbitrarily close to q and follow the same reasoning to conclude that

$$\lim_{n \to \infty} \frac{1}{n} \log N = C$$

This completes the proof of the fundamental theorem for discrete memoryless BSC. Similar proofs were also given earlier by G. A. Barnard, P. Elias, E. N. Gilbert, and D. Slepian.

4-17. Encoding the English Alphabet. The redundancy of the English language has been estimated by Shannon and several other authors. The word "estimate" is used here, as the problem in itself is not mathematically well defined. In a rough estimate we may assume that the alphabet consists of 26 letters with mutually independent probabilities. The maximum entropy of such a system is log 26 = 4.64 bits per letter when all letters are equiprobable. Of course such an estimate is unrealistic. In an approximation, we may compute the desired entropy based on the frequency of letters as shown in Table 4-1. The corresponding entropy is found to be 4.3 bits per letter (D. A. Bell, p. 164).

TABLE 4-1. FREQUENCY OF LETTERS IN ENGLISH LANGUAGE

A	7.81	1111	*N*	7.28	1100
B	1.28	101000	*O*	8.21	1110
C	2.93	01010	*P*	2.15	110111
D	4.11	11010	*Q*	0.14	1101100101
E	13.05	100	*R*	6.64	1011
F	2.88	01011	*S*	6.46	0110
G	1.39	00001	*T*	9.02	001
H	5.85	0001	*U*	2.77	01000
I	6.77	0111	*V*	1.00	1101101
J	0.23	1101100110	*W*	1.49	101001
K	0.42	11011000	*X*	0.30	1101100111
L	3.60	10101	*Y*	1.51	00000
M	2.62	01001	*Z*	0.09	1101100100

The main shortcoming of this calculation is that the successive letters are supposed to be transmitted independently of each other. This is, of course, not true, as the transmission of English letters in ordinary meaningful text is more of a stochastic nature of the Markov type. That is, the probability of the transmission of a letter is strongly affected by the probability of the transmission of the preceding letters. For example, the letter T is almost never followed by X but is very often followed by H. Therefore, in a better approximation we should compute the entropy based on the frequency of the occurrence of two successive letters. This will lead to the computation of the entropy of a discrete source with 26^2 symbols. Similarly, one could compute the frequency of the occurrence of any possible three-letter combinations and find the corresponding entropy. Several authors have estimated the entropy of English text. Their estimates indicate that the redundancy of the English language is somewhere between 0.50 and 0. 80. (It should be kept in mind that redundancy is not always undesirable. For example, in the presence of noise redundancy contributes to improvement in the intelligibility of the text.)

TABLE 4-2. CODES FOR ENGLISH ALPHABET

Probability	Letter	Huffman code	Alphabetical code
0.1859	space	000	00
0.0642	*A*	0100	0100
0.0127	*B*	0111111	010100
0.0218	*C*	11111	010101
0.0317	*D*	01011	01011
0.1031	*E*	101	0110
0.0208	*F*	001100	011100
0.0152	*G*	011101	011101
0.0467	*H*	1110	01111
0.0575	*I*	1000	1000
0.0008	*J*	0111001110	1001000
0.0049	*K*	01110010	1001001
0.0321	*L*	01010	100101
0.0198	*M*	001101	10011
0.0574	*N*	1001	1010
0.0632	*O*	0110	1011
0.0152	*P*	011110	110000
0.0008	*Q*	0111001101	110001
0.0484	*R*	1101	11001
0.0514	*S*	1100	1101
0.0796	*T*	0010	1110
0.0228	*U*	11110	111100
0.0083	*V*	0111000	111101
0.0175	*W*	001110	111110
0.0013	*X*	0111001100	1111110
0.0164	*Y*	001111	11111110
0.0005	*Z*	0111001111	11111111
	Cost	4.1195	4.1978

According to Shannon's* statistical study, printed English texts (27 letters including a space) are approximately 75 per cent redundant. If all letters were equiprobable, the same "information" could be encoded in texts roughly one-fourth the size of the noncoded texts. This estimate has been supported by Burton and Licklider.† Although interesting

* C. E. Shannon, Prediction and Entropy of Printed English, *Bell System Tech. J.*, vol. 29, pp. 147–160, 1951.

† N. G. Burton and J. C. R. Licklider, Long-range Constraints in the Statistical Structure of Printed English, *Am. J. Psychol.*, vol. 68, pp. 650–653, 1955.

work on this subject has been done more recently at Harvard University, it remains outside the scope of this study. The interested reader is referred to the following article and references given there: George A. Miller and Elizabeth A. Friedman, The Reconstruction of Mutilated English Texts, *Inform. and Control*, vol. 1, pp. 38–55, 1957.

Historically speaking, the first significant encoding of a language structure was the common Morse code. In this code, *dot, dash,* and *space* are used. The dash occupies a time equal to the time of transmission of three dots with no significant time space between them. The space occupies a time equal to that of a dot. If these durations are taken as cost units, then the average cost of the Morse-encoded English text is found to be

$$\sum_{\substack{\text{English}\\\text{alphabet}}} P(x_i)L(x_i) = 6.0 \text{ bits per letter}$$

The Morse code is based on a compromise between two objectives (Bell, p. 169), to assign the easiest symbols to the most frequent letters and also to assign shorter codes to the more frequent letters. For example, the letter E, which is the most frequently used letter in the English language, is represented by a single dot. If the letters were encoded strictly in accordance with the criterion of the shortest symbol for the most frequent letter, the average cost per letter of English would be reduced to 5.55 bits. The Shannon-Fano encoding procedure has been used (Bell, p. 64) in deriving Table 4-1.

Huffman's optimum encoding procedure has been applied in a direct manner to obtain the code given in Table 4-2. E. N. Gilbert and E. F. Moore have derived other types of encoding for English texts. They have obtained some interesting binary codes with the prefix property (alphabetical codes). The cost of the best of these codes is close to the optimum cost given by Huffman's method. An article by Gilbert and Moore contains an alphabetical code with a cost of 4.1978 compared with 4.1195 of Huffman's code. According to these authors an alphabetical encoding might be used as a means of saving memory space, in a data-processing machine in general and in a language-translating machine in particular, if it were desired to preserve the conventional alphabetical order of dictionaries.

PROBLEMS

4-1. Verify if the following sets of word lengths may correspond to a uniquely decipherable binary code.

(*a*) $[W] = [0,2,3,2]$

(*b*) $[W] = [0,2,2,2,5]$

4-2. In Fig. P4-2 each box represents a message output of an independent source. The probability of the transmission of each message is known to be $2^{-k\alpha}$, where k is a parameter for each message as indicated in the figure.

(*a*) Devise a binary encoding for this message ensemble.

(*b*) Devise a binary encoding with the lowest average length.

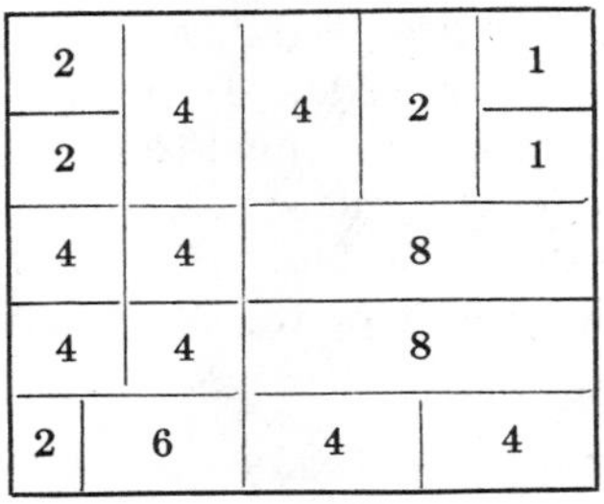

FIG. P4-2

4-3. See whether it is possible to encode 195 messages in separable words with

(*a*) $$D = 3$$
(*b*) $$D = 4$$

4-4. A source without memory has six characters with the following associated probabilities:

$$\begin{bmatrix} A, & B, & C, & D, & E, & F \\ \frac{1}{3}, & \frac{1}{4}, & \frac{1}{8}, & \frac{1}{8}, & \frac{1}{12}, & \frac{1}{12} \end{bmatrix}$$

(*a*) What is the entropy of this source?

(*b*) Devise an encoding procedure with the prefix property giving minimum possible average length for the transmission over a binary noiseless channel. What is the average length of the encoded messages?

4-5. Consider a BSC with $P\{1|0\} = P\{0|1\} = \frac{1}{8}$. The input to the channel consists of four equiprobable words

$$\begin{matrix} m_1 & 1 & 1 & 1 \\ m_2 & 1 & 0 & 0 \\ m_3 & 0 & 1 & 0 \\ m_4 & 0 & 0 & 1 \end{matrix}$$

(*a*) Compute $P\{1\}$ and $P\{0\}$ at the input.

(*b*) Compute the efficiency of the code.

(*c*) Compute the channel capacity.

4-6. (*a*) Apply Shannon's encoding procedure to the following set of messages:

$$\begin{bmatrix} m_1, & m_2, & m_3, & m_4 \\ 0.1, & 0.2, & 0.3, & 0.4 \end{bmatrix}$$

(*b*) Determine the efficiency of the code in each case.

(*c*) If the same technique is applied to the second-order extension of this source, how much will the efficiency be improved?

4-7. Same questions as in Prob. 4-6 for the following set of messages:

$$\begin{matrix} [m_1, & m_2, & m_3, & m_4, & m_5] \\ [\frac{1}{8}, & \frac{1}{16}, & \frac{3}{16}, & \frac{1}{4}, & \frac{3}{8}] \end{matrix}$$

4-8. Apply the Gilbert-Moore techniques for encoding the messages listed in Prob. 4-7.

4-9. Answer all parts of Prob. 4-6 for the case where the Gilbert-Moore technique is employed.

4-10. Given a discrete source with the following messages:

$$\{m\} = [m_1, \ m_2]$$
$$p\{m\} = [0.9, 0.1]$$

(a) Derive a Shannon code for the above messages.
(b) Find $\bar{L}$ and the code efficiency.
(c) Do parts (a) and (b) for the second-order extension of the source.
(d) Do parts (a) and (b) for the third-order extension of the source.

4-11. For the binary Huffman code, prove that

$$H(X) \leq \bar{L} \leq H(X) + 1 - 2p_{\min}$$

where $p_{\min}$ is the smallest probability in the message probability set.

4-12. Find the figure of merit of a Hamming's single-error correcting code for a BSC with $p = 0.01$ in the following cases:
(a) Number of information digits is 4.
(b) Number of information digits is 11.
(c) Number of information digits is 26.

4-13. Find the figure of merit of a Hamming's double-error correcting code for a BSC.

4-14. (a) Find an optimum binary encoding for the following messages:

$$\begin{matrix} [x_1, & x_2, & x_3] \\ [\frac{1}{2}, & \frac{1}{5}, & \frac{3}{10}] \end{matrix}$$

(b) Encode the output of the second-order extension of the source to the channel in an optimum binary code.
(c) Determine the coding efficiency in (a) and (b).
(d) What is the smallest order of the extension of the channel if we desire to reach an efficiency of $1 - 10^{-3}$ and $1 - 10^{-4}$, respectively?

4-15. A pulse-code communication channel has eight distinct amplitude levels $[x_1, x_2, \ldots, x_8]$. The respective probabilities of these levels are $[p_1, p_2, \ldots, p_8]$. The messages are encoded in sequences of three binary pulses (that is, the third-order extension of the source). The encoded messages are transmitted over a binary channel (p,q).
(a) Compute $H(X)$.
(b) Compute $H(X|Y)$.
(c) Compute $I(X;Y)$.
(d) Calculate (a), (b), and (c) for the numerical case, where

(1) $p_1 = p_2 = p_3 = \frac{1}{8}$
$p_4 = p_5 = p_6 = \frac{1}{16}$
$p_7 = \frac{1}{4}$
$p_8 = \frac{3}{16}$
$p = 0.9$

(2) $p_1 = p_2 = p_3 = \frac{1}{8}$
$p_4 = p_5 = \frac{1}{16}$
$p_7 = \frac{1}{4}$
$p_8 = \frac{3}{16}$
$p = 0.99$

4-16. We wish to transmit eight blocks of binary digits over a BEC. The first three positions are used for the information and the rest for parity checks. The following equations indicate the relations between information and parity digits.

$$\begin{bmatrix} x_4 \\ x_5 \\ x_6 \end{bmatrix} = \begin{bmatrix} 1 & 1 & 0 \\ 0 & 1 & 1 \\ 1 & 1 & 1 \end{bmatrix} \begin{bmatrix} x_1 \\ x_2 \\ x_3 \end{bmatrix}$$

(*a*) Determine how many combinations of single and double erasures may be corrected.

(*b*) Find the average erasure per block after correcting all possible single and double errors.

(*c*) What is the average rate of information over the channel?

4-17. Apply Hamming's single-error correcting in the following cases:

(*a*)	$m = 2$	$k = 3$
(*b*)	$m = 4$	$k = 3$
(*c*)	$m = 4$	$k = 4$
(*d*)	$m = 6$	$k = 4$
(*e*)	$m = 11$	$k = 5$

4-18. Show that for a Huffman binary code

$$H \leq \bar{L} \leq H + 1$$

4-19. Prove that Huffman's encoding for a given alphabet has a cost which is less than or equal to that of any uniquely decipherable encoding for that alphabet (see Gilbert-Moore, theorem 11).

PART 2

CONTINUUM WITHOUT MEMORY

The science of physics does not only give us [mathematicians] an opportunity to solve problems, but helps us also to discover the means of solving them, and it does this in two ways: it leads us to anticipate the solution and suggests suitable lines of argument.

Henri Poincaré
La valeur de la science

CHAPTER 5

CONTINUOUS PROBABILITY DISTRIBUTION AND DENSITY

5-1. Continuous Sample Space. In Sec. 2-15 we presented the concept of a discrete sample space and its associated discrete random variable. In this section we should like to introduce the idea of a random variable assuming a continuum of values.

Consider, for instance, X to be a random noise voltage which can assume *any* value between zero and 1 volt. Since by assumption the outcomes of this experiment are points on the real line interval [0,1], clearly X assumes a continuum of values. Furthermore, if we make this assumption, then we may state that X is a random variable taking a continuum of values.

The preceding intuitive approach in defining a random variable is unavoidable in an introductory treatment of the subject. On the other hand, a mathematically rigorous treatment of this more or less familiar concept requires extensive preparation in the professional field of measure theory. Such a presentation is beyond the scope of this book; for a complete coverage see Halmos and Loève. For the time being, the reader may satisfy himself with the following.

As in the case of a discrete sample space, an event is interpreted as a subset of a continuous sample space. In the former case we have already given methods for calculating probabilities of events. For the continuous case, however, it is not possible to give a probability measure satisfying all four requirements of Eqs. (2-36) to (2-39) such that every subset has a probability. The proof of this statement is involved with a number of mathematical complexities among which is the so-called "continuum hypothesis." Because of these difficulties in the study of continuous sample space, one has to confine oneself to a family of subsets of the sample space which does not contain all the subsets but which has enough subsets so that set algebra can be worked out within the members of that family (for example, union and intersection of subsets, etc.). Such a family of subsets of the sample space Ω will be denoted by $\mathfrak{F}$ ($\mathfrak{F}$ stands for the mathematical term field). More specifically, the events of $\mathfrak{F}$ must satisfy the following two requirements:

1. If $A_1, A_2, \ldots \in \mathfrak{F}$, then

$$\bigcup_{i=1}^{\infty} A_i \in \mathfrak{F} \tag{5-1}$$

2. If $A \in \mathfrak{F}$, then

$$U - A \in \mathfrak{F} \tag{5-2}$$

The first property simply implies that the union of a denumerable sequence of events must also be an event. The second property requires that the complement of an event also be an event.

With such a family in mind, the next step will be to define a probability measure $P\{A\}$ for every event A of that family. This can be done in a way similar to the definition of a probability measure over the discrete sample space, namely,

1. For each $A \in \mathfrak{F}$,

$$0 \leq P\{A\} \tag{5-3}$$

2. For all denumerable unions of disjoint events of $\mathfrak{F}$ family,

$$P\{\bigcup_{i=1}^{\infty} A_i\} = \sum_{i=1}^{\infty} P\{A_i\} \tag{5-4}$$

3.
$$P\{U\} = 1 \tag{5-5}$$

We assume the validity of these axioms and then proceed with defining the probability distribution and density of a continuous random variable.

It is to be noted that, in the strict sense, a random variable need not be real-valued. One can directly define a complex-valued random variable through two real-valued variables:

$$X + \sqrt{-1}\, Y$$

This simply requires the measure space to be a complex two-dimensional space rather than an ordinary real space.

5-2. Probability Distribution Functions. For simplicity, consider first the case of a random variable taking values in a one-dimensional real coordinate space. The probability that the random variable X assumes values such that

$$E = \{a < X \leq b\} \tag{5-6}$$

is shown by

$$P\{a < X \leq b\} = P\{E\}$$

The event E in Eq. (5-6) consists of the set of all subevents such that their corresponding values of X satisfy the above inequality. In particular, consider the event E_1 defined by

$$E_1 = \{X \leq a\} \qquad -\infty < a < +\infty$$

Note that

$$E_1E = \emptyset$$
$$E \cup E_1 = \{X \leq b\}$$

Therefore, according to Eq. (2-69),

$$P\{E_1 \cup E\} = P\{E_1\} + P\{E\}$$

That is,

$$P\{E\} = P\{X \leq b\} - P\{X \leq a\} \tag{5-7}$$

In general, if x is any real number, we may write

$$P\{X \leq x\} = F(x) \tag{5-8}$$

$F(x)$ is called the *probability distribution* function, or *cumulative distribution function* (CDF), of the random variable X. Note that the distribution function is defined for all real values of x. It is a *monotonic nondecreasing* function of x continuous on the right for every x. The following two properties of the distribution function are evident in the light of Eqs. (2-38) and (2-39):

$$\begin{aligned} \lim_{x \to -\infty} F(x) &= 0 \\ \lim_{x \to \infty} F(x) &= 1 \end{aligned} \tag{5-9}$$

Any monotonic nondecreasing function continuous on the right for every x satisfying Eq. (5-9) *can be regarded as a distribution function.*

There are two important classes of CDF. Although they do not cover all possible cases of CDF, they are the most important ones: (1) discrete and (2) continuous. A random variable and its CDF are said to be discrete if the only values that the variable can assume with positive probability are at most denumerable. In other words, if there exists a denumerable sequence of distinct numbers a_j ($j = 1, 2, \ldots$) such that $\sum_j P\{X = a_j\} = 1$, then the CDF is defined as

$$F(x) = \sum_{a_j \leq x} P\{X = a_j\} \qquad a_1 < a_2 < a_3 < \cdots \tag{5-10}$$

The binomial and Poisson's distributions discussed in Secs. 2-18 and 2-19 are the most common examples of the discrete case.

When the random variable and its CDF admit a continuum of values, they are said to be of the continuous type. The normal distribution of Sec. 5-4 is the most common example of the continuous case.

Example 5-1. Suppose that in Example 2-28 the random variable X takes on any one of the values 1, 2, 3, 4, 5, 6 with equal probability $\frac{1}{6}$. Then for $x \leq 6$

$$F(x) = P\{X \leq x\} = \sum_{i=1}^{[x]} P\{X = i\} = \tfrac{1}{6}[x]$$

where $[x]$ denotes the greatest integer smaller than or equal to a chosen x. Figure E5-1 shows the graph of the discontinuous function $F(x)$. Note that the probability distribution function rises by jumps in the case of discrete variables.

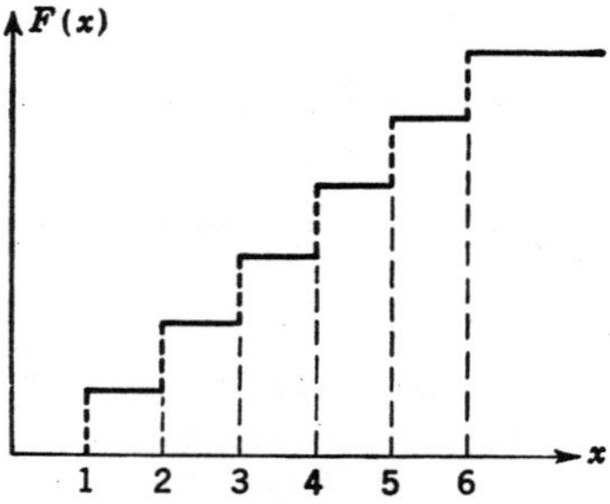

FIG. E5-1

Example 5-2. Assume that we have a circular disk with a circumference of unit length. A completely symmetrical pointer at the center of the disk is whirled. The pointer stops at a point on the periphery. Let X be the reading of the pointer from the point zero. If perfect symmetry is assumed, the probability of X being in any interval is proportional to the length of that interval, i.e.,

$$P\{a < X \leq b\} = K(b - a) \qquad 0 \leq a \leq b \leq 1$$

K being a constant of proportionality. Accordingly,

$$\begin{aligned} P\{-\infty < X \leq 0\} &= 0 \\ P\{1 < X \leq \infty\} &= 0 \\ P\{0 < X \leq 1\} &= K \end{aligned}$$

Hence, $K = 1$ and

$$F(x) = P\{0 < X \leq x\} = x \qquad 0 \leq x \leq 1$$

5-3. Probability Density Function. If $F(x)$ is such that

$$F(x) = \int_{-\infty}^{x} f(t)\, dt \tag{5-11}$$

where $f(t)$ is a real-valued integrable function, then $F(x)$ is said to be *absolutely continuous*. It is known that almost everywhere

$$f(x) = \frac{dF(x)}{dx} \tag{5-11a}$$

In case $F(x)$ is an absolutely continuous CDF, we have

$$P\{a < X \leq b\} = \int_{a}^{b} f(x)\, dx \tag{5-12}$$

$f(x)$ is known as the *probability density function* (PDF).

Since the probability distribution function is a nondecreasing monotonic function, the density function will be nonnegative over the real axis:

$$f(x) \geq 0 \tag{5-13}$$

Furthermore,

$$\int_{-\infty}^{\infty} f(x)\,dx = F(+\infty) - F(-\infty) = 1 \qquad (5\text{-}14)$$

If $F(x)$ is a continuous function about $x = a$, the probability of X assuming the value $x = a$ is zero. In fact,

$$P\{X = a\} = \lim_{\epsilon \to 0} P\{a - \epsilon < X \le a\} \qquad (5\text{-}15)$$

$$P\{X = a\} = \lim_{\epsilon \to 0} [F(a) - F(a - \epsilon)] = 0 \qquad (5\text{-}16)$$

For continuous random variables the probability of the random variable being in an interval decreases with the length of that interval and in the limit becomes zero.

For a discrete random variable, if $X = a$ is a possible value for the random variable, then

$$P\{X = a\} = \lim_{\epsilon \to 0} [F(a) - F(a - \epsilon)] \neq 0 \qquad (5\text{-}17)^*$$

The mathematical implication of this equation is rather clear. However, the engineering-minded reader may find it convenient for his own

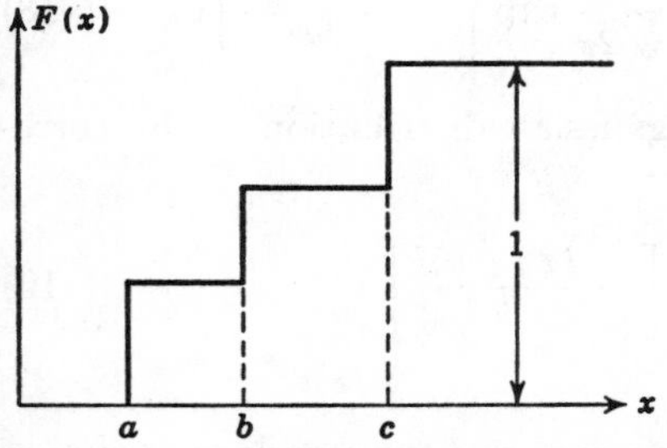

FIG. 5-1. Example of a discrete CDF.

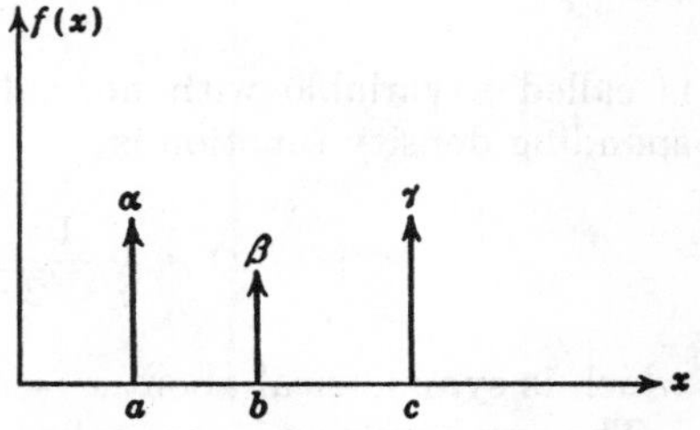

FIG. 5-2. Example of the discrete distribution corresponding to Fig. 5-1 in terms of impulse functions. $P(X = a) = \alpha$. $P(X = b) = \beta$. $P(X = c) = \gamma$. $\alpha + \beta + \gamma = 1$.

use to illustrate the density distribution function in the discrete case with the help of Dirac or impulse functions.† A unit impulse effective at a point $x = a$ will be denoted by $u_0(x - a)$.

The discrete probability distribution function of Fig. 5-1 leads to the density distribution function of Fig. 5-2.

* When dealing with continuous probabilities due to Eq. (5-15), we understand that the expressions $P\{X < a\}$ and $P\{X \le a\}$ are equivalent.

† While the rigorous use of Dirac "functions" requires special mathematical consideration, their employment is frequent and commonplace in electrical engineering literature. In this respect, Fig. 5-2 may be of interest to the electrical engineer.

Example 5-3. A random process gives measurements x between 0 and 1 with a probability density function

$$\begin{aligned} f(x) &= 12x^3 - 21x^2 + 10x \qquad 0 \le x \le 1 \\ f(x) &= 0 \qquad \text{elsewhere} \end{aligned}$$

(a) Find $P\{X \le \frac{1}{2}\}$ and $P\{X > \frac{1}{2}\}$.
(b) Find a number K such that $P\{X \le K\} = \frac{1}{2}$.
Solution

(a)
$$P\{X \le \tfrac{1}{2}\} = \int_0^{1/2} (12x^3 - 21x^2 + 10x)\,dx = \tfrac{9}{16}$$
$$P\{X > \tfrac{1}{2}\} = 1 - \tfrac{9}{16} = \tfrac{7}{16}$$

(b)
$$\int_0^K (12x^3 - 21x^2 + 10x)\,dx = \tfrac{1}{2}$$
$$3K^4 - 7K^3 + 5K^2 = \tfrac{1}{2}$$

The permissible answer is the root of this equation between 0 and 1; this is found to be

$$K = 0.452$$

5-4. Normal Distribution. A random variable X with a cumulative distribution function given by

$$F(x) = P\{X < x\} = \int_{-\infty}^{x} \frac{1}{\sigma\sqrt{2\pi}} \exp\left[-\frac{(t-a)^2}{2\sigma^2}\right] dt \tag{5-18}$$

is called a variable with normal or gaussian distribution. The corresponding density function is

$$f(x) = \frac{1}{\sigma\sqrt{2\pi}} \exp\left[-\frac{(x-a)^2}{2\sigma^2}\right] \tag{5-19}$$

which is symmetrical about $x = a$.

The numbers a and σ are called the average and the standard deviation of the random variable, respectively; their significance is discussed in Chap. 6.

One may be interested in checking the suitability of the function $f(x)$ of Eq. (5-19) as a density function. In other words, one has to show that

$$\int_{-\infty}^{\infty} f(x)\,dx = 1$$

For this purpose, we may shift the density curve to the left by a units. Next, we consider the double integral

$$\begin{aligned} \left(\int_{-\infty}^{\infty} \frac{1}{\sigma\sqrt{2\pi}} \epsilon^{-x^2/2\sigma^2}\,dx\right)^2 &= \frac{1}{2\pi\sigma^2} \int_{-\infty}^{\infty} e^{-x^2/2\sigma^2}\,dx \int_{-\infty}^{\infty} e^{-y^2/2\sigma^2}\,dy \\ &= \frac{1}{2\pi\sigma^2} \int_{-\infty}^{\infty}\int_{-\infty}^{\infty} e^{-(x^2+y^2)/2\sigma^2}\,dx\,dy \end{aligned} \tag{5-20}$$

Finally, introducing the familiar polar coordinates yields

$$\left(\int_{-\infty}^{\infty} \frac{1}{\sigma\sqrt{2\pi}} e^{-x^2/2\sigma^2}\,dx\right)^2 = \frac{1}{2\pi}\int_0^{2\pi}\int_0^{\infty} e^{-r^2/2\sigma^2}\frac{r}{\sigma^2}\,dr\,d\theta = 1 \qquad (5\text{-}21)$$

In the next chapter, it will be shown that the letter a in Eq. (5-18) denotes the "average" value of the random variable with a normal distribution.

When the parameters of the normal distribution have values

$$a = 0 \qquad \text{and} \qquad \sigma = 1$$

the distribution function is called *standard normal distribution.*

$$F(x) = P\{X < x\} = \int_{-\infty}^{x} \frac{1}{\sqrt{2\pi}} e^{-(t^2/2)}\,dt \qquad (5\text{-}22)$$

Tables of standard normal distribution are commonly available. To use such tables one first employs a transformation of variable of the type

$$t = \frac{x - a}{\sigma}$$

in Eq. (5-19) in order to transform the normal curve into a standard normal curve. Standard normal distributions are given in Table T-2 of the Appendix. The use of this table for evaluating the probability of a random variable being in an interval is self-explanatory. All one has to remember is that Eq. (5-12) suggests the equivalence between probability and the area under the density curve between points of interest. For example, if X has a standard normal probability density distribution, then

$$\begin{aligned}
&P\{0 < X < 2\} = 0.47725\\
&P\{-2 < X < 2\} = 0.95450\\
&P\{(X < -2) \cup (X > 2)\} = 1 - 0.95450 = 0.04550\\
&P\{X < 2\} = 0.97725\\
&P\{X > 2\} = 1 - 0.97725 = 0.02275
\end{aligned}$$

More detailed information is given in Table T-3 of the Appendix.

Example 5-4. The average life of a certain type of electric bulb is 1,200 hours. What percentage of this type of bulb is expected to fail in the first 800 working hours? What percentage is expected to fail between 800 and 1,000 hours? Assume *a normal distribution with* σ = 200 *hours.*

Solution. Referring to Sec. 2-8, one notes that in a large number of samples *the frequency of the failures* is approximately equal to *the probability of failure.* In this connection the word *percentage* is used synonymously with *frequency.* Using the average life of 1,200 hours for a, we make a change of variable $y = (x - a)/\sigma$ which allows the normal curve in y to be symmetrical about $y = 0$, hence permitting the use of a table of normal probability.

$$y_0 = \frac{x - a}{\sigma} = \frac{800 - 1{,}200}{200} = -2$$

$$\int_{x=800}^{x=1{,}200} \frac{1}{200\sqrt{2\pi}} \exp\left[-\frac{(x - 1{,}200)^2}{80{,}000}\right] dx = \int_0^2 \frac{1}{\sqrt{2\pi}} e^{-y^2/2}\, dy = 0.477$$

The area under the whole normal curve being unity, the desired probability is

$$0.500 - 0.477 = 0.023$$

For the second part of the problem, let

$$y_0' = \frac{1{,}000 - 1{,}200}{200} = -1$$

$$\int_0^1 f(y)\, dy = 0.341$$

$$0.500 - 0.341 = 0.159$$
$$0.159 - 0.023 = 0.136$$

The reader should note that, in view of our assumption of normal distribution, there is a fraction of bulbs with negative life expectancy ($-\infty$ to zero). This fraction is included in the above calculation; that is, the number $0.023 = P\{-\infty < X < 800\}$ will be somewhat larger than

$$P\{0 < X < 800\}$$

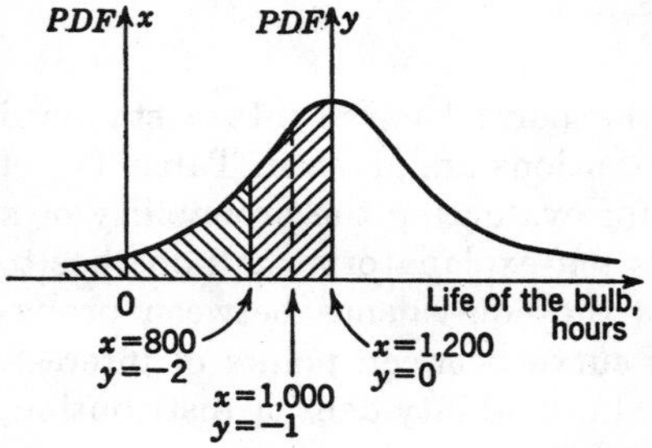

FIG. E5-4

Of course, if we had used a density distribution bounded between 0 and infinity, we should not be confronted with the problem of negative life expectancy. On the other hand, tables of such distributions are not readily available. The calculation of $P\{800 < X < 1{,}200\}$ in lieu of $P\{0 < X < 800\}$ was a simple matter of using Table T-2 of normal distributions.

5-5. Cauchy's Distribution. A random variable X is said to have a Cauchy distribution if

$$F(x) = P\{X < x\} = \int_{-\infty}^{x} \frac{dt}{\pi(1 + t^2)} \tag{5-23}$$

The corresponding density function is

$$f(x) = \frac{1}{\pi(1 + x^2)} \tag{5-24}$$

Note that $$\int_{-\infty}^{\infty} f(x)\, dx = \frac{1}{\pi}\,[\tan^{-1} x]_{-\infty}^{\infty} = 1$$

The graphs of the corresponding density and CDF are shown in Fig. E5-5.

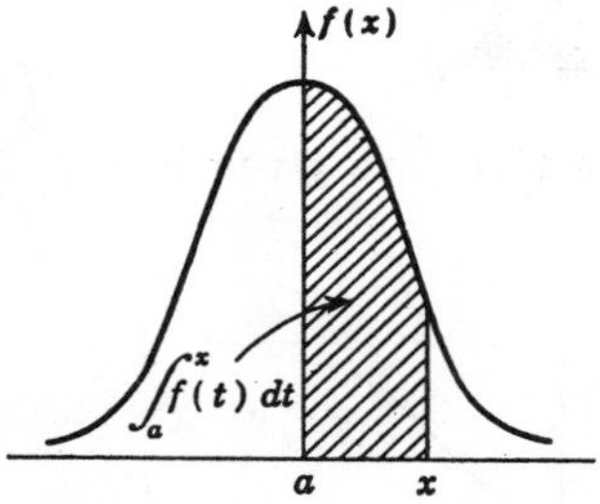

FIG. 5-3. A normal PDF.

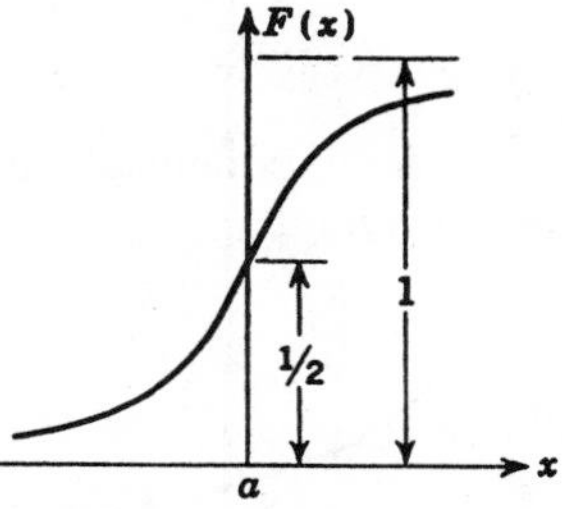

FIG. 5-4. A normal CDF.

Example 5-5. Consider a point M on the vertical axis of a two-dimensional rectangular system with $OM = 1$. A straight line MN is drawn at a random angle θ (Fig. E5-5). What is the probability distribution of the random variable $ON = X$?

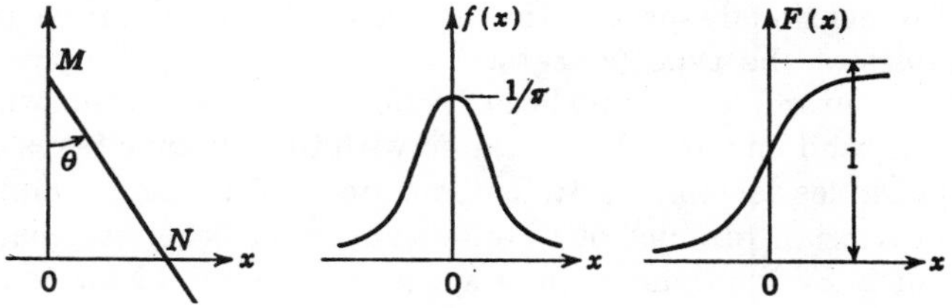

FIG. E5-5

Solution. The problem suggests that the angle θ in Fig. E5-5 has a uniform probability distribution. Thus the probability of drawing a line in a particular $d\theta$ is $d\theta/\pi$. The random variable of interest is $ON = X = \tan\theta$. Accordingly,

$$d\theta = \frac{dx}{1 + x^2}$$

$$P\{x - dx \leq X \leq x\} = P\{\theta - d\theta < \text{angle} \leq \theta\} = \frac{d\theta}{\pi} = \frac{dx}{\pi(1 + x^2)}$$

$$f(x) = \frac{1}{\pi(1 + x^2)}$$

$$F(x) = \int_{-\infty}^{x} f(x)\,dx = \int_{-\infty}^{x} \frac{dx}{\pi(1 + x^2)}$$

$$F(x) = \frac{1}{\pi}[\tan^{-1} x]_{-\infty}^{x} = \frac{1}{\pi}\left(\tan^{-1} x + \frac{\pi}{2}\right)$$

5-6. Exponential Distribution. A probability density distribution of the type

$$\begin{aligned} f(x) &= ae^{-ax}\,dx \qquad a > 0 \qquad x > 0 \\ f(x) &= 0 \qquad \text{elsewhere} \end{aligned} \tag{5-25}$$

is referred to as an exponential distribution. The corresponding CDF is given by

$$P\{X < x\} = \int_0^x ae^{-at}\,dt = [-e^{-at}]_0^x$$
$$F(x) = 1 - e^{-ax} \tag{5-26}$$

A graph of the exponential density and its CDF are given in Fig. 5-5*a* and *b*, respectively.

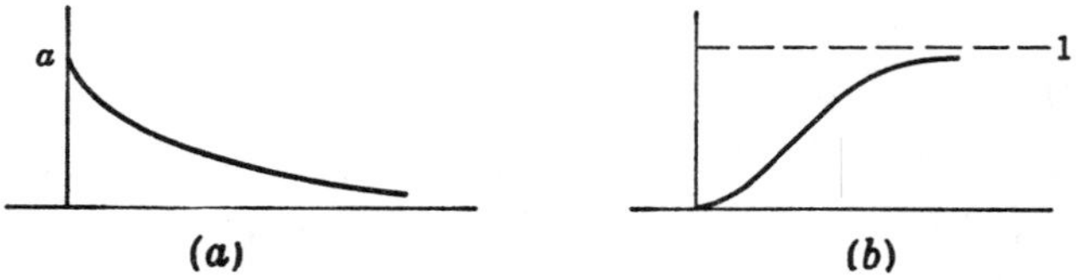

FIG. 5-5. (*a*) Example of an exponential PDF. (*b*) CDF of the density illustrated in Fig. 5-5*a*.

5-7. Multidimensional Random Variables. The coordinate space can be a multidimensional space. In this case the random variable X assumes values of the type $(x_1,x_2,x_3, \ldots ,x_n)$, that is, n-tuples of real numbers. For example, if four dice of different colors are thrown simultaneously, the random variable associated with the outcome takes certain-number quadruples as values. In fact, we are considering a sample space that is the cartesian product of a finite number of other sample spaces. If the outcome E_k is in the sample space Ω_k, the n-fold outcome E is defined as a point in the cartesian product space Ω, that is,

$$E_k \in \Omega_k \qquad k = 1, 2, \ldots, n \tag{5-27}$$
$$E = \{E_1,E_2, \ldots ,E_n\}$$
$$\Omega = \Omega_1 \otimes \Omega_2 \otimes \cdots \otimes \Omega_n$$

Then

$$E \in \Omega$$

If the outcome E is a permissible point of the product sample space Ω and if the events E_k are mutually independent (this is not always the case), then the probability measure associated with E equals the product of the individual probability measures, i.e.,

$$m(E) = m(E_1)m(E_2) \cdots m(E_n) \tag{5-28}$$

It is to be noted that by the event $E_k \in \Omega_k$ we understand the set of all events in the Ω space where the kth random variable assumes a specified value but variables other than that are arbitrary. Such a set of events is usually called a *cylinder set.* The probability measure associated with this cylinder is defined as the probability of the *event* E_k.

By analogy with the one-dimensional case, we define the cumulative probability distribution function (CDF) of the n-dimensional random variable $(X_1,X_2, \ldots ,X_n)$ as

$$F(x_1,x_2, \ldots ,x_n) = P\{-\infty < X_1 \leq x_1, -\infty < X_2 \leq x_2, \ldots , -\infty < X_n \leq x_n\}$$
$$= P\{X_1 \leq x_1, X_2 \leq x_2, \ldots , X_n \leq x_n\} \tag{5-29}$$

Now we explore this defining equation for the two most important cases of continuous and discrete variables.

Continuous Case. The CDF is also defined by

$$F(x_1,x_2, \ldots ,x_n) = \int_{-\infty}^{x_1} \int_{-\infty}^{x_2} \cdots \int_{-\infty}^{x_n} f(t_1,t_2, \ldots ,t_n) \, dt_1 \, dt_2 \cdots dt_n \tag{5-30}$$

where $f(x_1,x_2, \ldots ,x_n)$ is the probability density function. Note that

$$\int_{-\infty}^{\infty} \int_{-\infty}^{\infty} \cdots \int_{-\infty}^{\infty} f(x_1,x_2, \ldots ,x_n) \, dx_1 \, dx_2 \cdots dx_n = 1 \tag{5-31}$$

The study of n different random variables $X_1, X_2, \ldots , X_n$ is equivalent to the consideration of one n-dimensional random variable

$$X = (X_1,X_2, \ldots ,X_n)$$

The one-dimensional variables $X_1, X_2, \ldots , X_n$ are said to be independent if for all permissible values of the variables and all joint CDF's we have

$$F(x_1,x_2, \ldots ,x_n) = F_1(x_1)F_2(x_2) \cdots F_n(x_n) \tag{5-32}$$

where $F_i(x_i)$ denotes the cumulative distribution function of the one-dimensional random variable X_i, that is,

$$F_i(x_i) = \int_{-\infty}^{\infty} \cdots \int_{-\infty}^{x_i} \cdots \int_{-\infty}^{\infty} f(x_1, \ldots ,x_i, \ldots ,x_n) \, dx_1 \, dx_i \cdots dx_n$$
$$= P\{-\infty < X_1 < \infty, \ldots , -\infty_1 < X_i < x_i, \ldots , -\infty < X_n < \infty\} \tag{5-33}$$

$F_i(x_i)$ is called the ***marginal probability distribution function*** of X_i, in other words, the cumulative distribution of X_i irrespective of the values assumed by other variables.

When the density functions (continuous type) exist, the condition of independence [Eq. (5-30)] can be written in the equivalent form:

$$f(x_1,x_2, \ldots ,x_n) = f_1(x_1)f_2(x_2) \cdots f_n(x_n) \tag{5-34}$$

where $f_i(x_i)$ is the density function of the random variable X_i, without regard to other variables. This is the so-called "marginal density function" of the variable X_i. The condition must be satisfied for all permissible values of the variables.

Discrete Case. In this case the random variable

$$X = (X_1, X_2, \ldots, X_n)$$

takes on only a denumerable number of n-tuples as values such that the total probability is concentrated in only a denumerable number of points of the n-dimensional space. It is obvious that in this case each of the component random variables X_i can take only denumerably many values and that the marginal distribution of X_i,

$$F_i(x_i) = P\{X_i \le x_i\} \tag{5-35}$$

is also discrete. The definition of independence can now be given as before in terms of the marginal distribution.

As is easy to see, it is much more convenient in the discrete case to specify probabilities of the type $P\{X_1 = A_1, \ldots, X_n = A_n\}$ rather than to give the analytical form of the CDF.

5-8. Joint Distribution of Two Variables: Marginal Distribution

Continuous Case. The case of a two-dimensional random variable (X,Y) is of considerable interest and will be treated in some detail. Let $f(x,y)$ be the corresponding probability density function; then by a reasoning similar to that of Sec. 5-2 we find that

$$P\{a_1 < X \le b_1,\ a_2 < Y \le b_2\} = \int_{a_1}^{b_1} \int_{a_2}^{b_2} f(x,y)\,dx\,dy$$

$$P\{-\infty < X \le x,\ -\infty < Y \le y\} = F(x,y) = \int_{-\infty}^{x} \int_{-\infty}^{y} f(x,y)\,dx\,dy \tag{5-36}$$

with the understanding that $f(x,y) \ge 0$

and
$$\int_{-\infty}^{\infty} \int_{-\infty}^{\infty} f(x,y)\,dx\,dy = 1$$

Differentiating Eq. (5-36) partially first with respect to x and then with respect to y gives

$$\begin{aligned} \frac{\partial}{\partial x} F(x,y) &= \int_{-\infty}^{y} f(x,t)\,dt \\ \frac{\partial}{\partial x}\frac{\partial}{\partial y} F(x,y) &= f(x,y) \end{aligned} \tag{5-37}$$

The latter equation is in a sense a "dual" expression for Eq. (5-36).

The probability of the variable (X,Y) assuming values in the rectangle $(a_1 < x \le b_1,\ a_2 < Y \le b_2)$ can be directly computed. Consider the following sets of events in relation to Fig. 5-6.

$$\begin{aligned} E_1 &= \{X \le b_1,\ Y \le b_2\} \qquad & E_2 &= \{X \le a_1,\ Y \le a_2\} \\ E_3 &= \{X \le b_1,\ Y \le a_2\} \qquad & E_4 &= \{X \le a_1,\ Y \le b_2\} \end{aligned} \tag{5-38}$$

The event of interest can be written as

$$E = \{a_1 < X \leq b_1,\ a_2 < Y \leq b_2\} \tag{5-39}$$

Now keeping in mind that the probability $P\{E\}$ would correspond to the

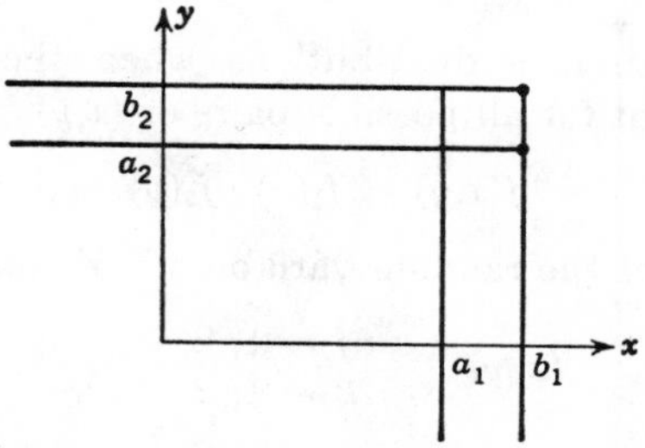

FIG. 5-6. Certain events associated with the sample space of a random variable.

double integral of the density over the above region, from Fig. 5-6 one concludes that

$$P\{E\} = P\{a_1 < X \leq b_1,\ a_2 < Y \leq b_2\} = F(b_1,b_2) - F(b_1,a_2) + F(a_1,a_2) - F(a_1,b_2) \tag{5-40}$$

The probability $P\{X \leq x\}$, irrespective of the values assumed by the second component Y, can be written as

$$F_1(x) = P\{X \leq x\} = \int_{-\infty}^{x} dx \int_{-\infty}^{\infty} f(x,y)\, dy \tag{5-41}$$

$F_1(x)$ is defined as the *marginal distribution* of the variable X. The *marginal density function* can be obtained in the following manner:

$$f_1(x) = \frac{dF_1(x)}{dx} = \int_{-\infty}^{\infty} f(x,y)\, dy \tag{5-42}$$

The *marginal probability distribution* and *density function* for the variable Y can be given in a similar manner:

$$F_2(y) = P\{Y \leq y\} = \int_{-\infty}^{y} dy \int_{-\infty}^{+\infty} f(x,y)\, dx \tag{5-43}$$

$$f_2(y) = \frac{dF_2(y)}{dy} = \int_{-\infty}^{\infty} f(x,y)\, dx \tag{5-44}$$

Note that the marginal distributions can be alternatively written as

$$\begin{aligned} F_1(x) &= F(x,\infty) \\ F_2(y) &= F(\infty,y) \end{aligned} \tag{5-45}$$

When the two variables X and Y are independent, then, for every pair of (a_1,b_1) and (a_2,b_2), we have

$$P\{a_1 < X \leq b_1, a_2 < Y \leq b_2\} = P\{a_1 < X \leq b_1\} P\{a_2 < Y \leq b_2\} \tag{5-46}$$

$$F(x,y) = F_1(x) \cdot F_2(y) \tag{5-47}$$

Note that for continuous distributions when the two variables are mutually independent for all possible pairs of (x,y)

$$f(x,y) = f_1(x) \cdot f_2(y) \tag{5-48}$$

Discrete Case. Let the random variable (X,Y) take the values

$$(x_j,y_k) \qquad \begin{matrix} j = 1, 2, \ldots \\ k = 1, 2, \ldots \end{matrix}$$

and let $P\{X = x_j, Y = y_k\} = P\{j,k\}$. Then the CDF of the random variable is

$$F(x,y) = \sum_{\substack{x_jy_k \\ x_j \leq x \\ y_k \leq y}} P\{j,k\} \tag{5-49}$$

and one can also calculate the marginal probabilities as before. Note that if one or both of j and k are finite in number, the calculations become much easier.

Example 5-6. The density function of a two-dimensional continuous distribution is given as

$$f(x,y) = e^{-(x+y)} \qquad \text{for } x \geq 0, y \geq 0$$
$$f(x,y) = 0 \qquad \text{elsewhere}$$

Find the probability of $(\frac{1}{2} < X < 2; 0 < Y < 4)$.

Solution. First we observe whether the given density function is a permissible one. In fact, $f(x,y)$ is nonnegative in the specified range and

$$\int_0^\infty \int_0^\infty e^{-(x+y)}\, dx\, dy = \int_0^\infty e^{-x}[-e^{-y}]_0^\infty\, dx = 1$$

The desired answer is given by

$$\int_{1/2}^2 \int_0^4 e^{-x}e^{-y}\, dx\, dy = (e^{-1/2} - e^{-2})(1 - e^{-4})$$

5-9. Conditional Probability Distribution and Density

Continuous Case. The conditional probability distribution for two-dimensional random variables can be derived in a way basically similar to that of the conditional probability defined in Sec. 2-8. However, the required mathematical care is beyond the level of an introductory presentation. An accurate account of the conditional probability distribution and density is given in Loève (p. 359). The following presentation, often given in introductory texts, lacks adequate mathematical rigor.

The conditional probability of $P\{Y \leq y\}$ subject to $(x - \Delta x < X \leq x)$ can be written as

$$P\{Y \leq y | x - \Delta x < X \leq x\} = \frac{P\{x - \Delta x < X < x,\ Y < y\}}{P\{x - \Delta x < X < x\}} \tag{5-50}$$

The difficulty with this elementary presentation stems from the fact that the denominator in general might be zero. In the above relationships one may now introduce the appropriate probability density functions, when such functions exist.

$$P\{Y \leq y | x - \Delta x < X \leq x\} = \frac{\int_{x-\Delta x}^{x} \int_{-\infty}^{y} f(x,y)\, dx\, dy}{\int_{x-\Delta x}^{x} f_1(x)\, dx} \tag{5-51}$$

Taking the limit as Δx approaches zero yields

$$\lim_{\Delta x \to 0} P\{Y \leq y | x - \Delta x < X \leq x\} = \frac{\int_{-\infty}^{y} f(x,y)\, dy}{f_1(x)} \tag{5-52}$$

If we denote the left-hand member of this equation by $P\{Y \leq y|x\}$ and if the derivatives of both sides are taken with respect to the variable quantity y, the following results:

$$\frac{dP\{Y \leq y|x\}}{dy} = \frac{f(x,y)}{f_1(x)} \tag{5-53}$$

We define $f_x(y|x) = dP\{Y \leq y|x\}/dy$ as the conditional probability density function of Y, given X. Similarly,

$$f_y(x|y) = \frac{f(x,y)}{f_2(y)} \tag{5-54}$$

The conditional density $f_x(y|x)$ is a function of one variable y and the parameter x which assumes a given value in each case. If the conditional density $f_x(y|x)$ does not depend on the parameter x, X and Y are mutually independent.

Discrete Case. In the discrete case we do not have the conditional densities. The familiar form discussed in Sec. 2-8 is quite satisfactory for the calculation of the various conditional probabilities.

Example 5-7. Consider the two-dimensional density function

$$f(x,y) = 2 \quad \text{for } \begin{cases} 0 < x < 1 \\ 0 < y < x \end{cases}$$
$$f(x,y) = 0 \quad \text{outside}$$

(*a*) Find the marginal density functions.
(*b*) Find the conditional density functions.

Solution

(*a*) $$f_1(x) = \int_{-\infty}^{\infty} f(x,y)\,dy = \int_0^x 2\,dy = 2x \qquad 0 < x < 1$$
$$= 0 \qquad \text{outside}$$
$$f_2(y) = \int_{-\infty}^{\infty} f(x,y)\,dx = \int_y^1 2\,dx = 2(1-y) \qquad 0 < y < 1$$
$$= 0 \qquad \text{outside}$$

(*b*) $$f_x(y|x) = \frac{f(x,y)}{f_1(x)} = \frac{1}{x} \qquad 0 < x < 1$$
$$f_y(x|y) = \frac{f(x,y)}{f_2(y)} = \frac{1}{1-y} \qquad 0 < y < 1$$

A pictorial interpretation of conditional probability is given in Fig. 5-7. Let R be a region within the range of definition of the density function of a two-dimensional random variable (also called *bivariate*). We ask ourselves what is the probability of the variable Y being in the interval

$$y_0 < Y \leq y_0 + \Delta y_0$$

given that $x_0 < X \leq x_0 + \Delta x_0$. Assuming that all densities exist, the required probability is

$$\frac{f(x_0,y_0)\,\Delta x_0\,\Delta y_0}{f_1(x_0)\,\Delta x_0} \tag{5-55}$$

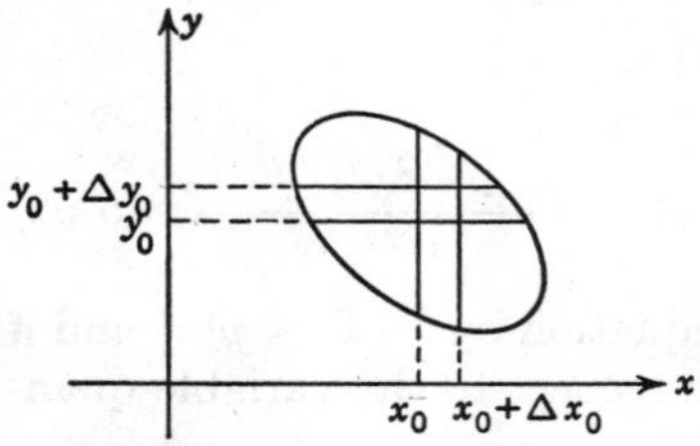

FIG. 5-7. Illustration of different distributions associated with a two-dimensional random variable.

For arbitrary points of the region R, this ratio divided by Δy_0 is the conditional density for y, given x, and is designated as

$$f_x(y|x) = \frac{f(x,y)}{f_1(x)} \tag{5-56}$$

A further justification may be desirable for showing that the function in Eq. (5-56) satisfies the requirements for a probability density function. In fact, since the numerator and the denominator are essentially nonnegative, the above ratio is a finite nonnegative number [assuming $f_1(x) \neq 0$]. Furthermore,

$$\int_{-\infty}^{\infty} f_x(y|x_0)\,dy = \int_{-\infty}^{\infty} \frac{f(x_0,y)}{f_1(x_0)}\,dy$$
$$= \frac{1}{f_1(x_0)} \int_{-\infty}^{\infty} f(x_0,y)\,dy = \frac{f_1(x_0)}{f_1(x_0)} = 1 \tag{5-57}$$

5-10. Bivariate Normal Distribution. In this section, we shall consider in some detail the most frequently used two-dimensional distribution, namely, the bivariate normal distribution

$$f(x,y) = C \exp\left[-\left(\frac{x^2}{a^2} - 2\frac{xy}{k} + \frac{y^2}{b^2}\right)\right] \tag{5-58}$$

a, b, and k are given positive constants. The analysis of the bivariate normal distribution will be divided into the following three parts:

1. Determining the value of the constant C
2. Determining the marginal densities
3. Determining the conditional densities

1. In order that $f(x,y)$ be a density function one must have

$$C = \frac{1}{\int_{-\infty}^{\infty}\int_{-\infty}^{\infty} \exp\{-[(x^2/a^2) - 2(xy/k) + (y^2/b^2)]\}\, dx\, dy} = \frac{1}{I} = 1 \tag{5-59}$$

Note that

$$\frac{x^2}{a^2} - 2\frac{xy}{k} + \frac{y^2}{b^2} = \left(\frac{1}{a^2} - \frac{b^2}{k^2}\right)x^2 + \left(\frac{y}{b} - \frac{bx}{k}\right)^2 \tag{5-60}$$

$$I = \int_{-\infty}^{\infty} \exp\left[-\left(\frac{1}{a^2} - \frac{b^2}{k^2}\right)x^2\right] dx \int_{-\infty}^{\infty} \exp\left[-\left(\frac{y}{b} - \frac{bx}{k}\right)^2\right] dy \tag{5-61}$$

A change of variable is in order. Let

$$\mu = \sqrt{2}\left(\frac{y}{b} - \frac{bx}{k}\right) \tag{5-62}$$

Then,

$$I = \int_{-\infty}^{\infty} \exp\left[-\left(\frac{1}{a^2} - \frac{b^2}{k^2}\right)x^2\right] dx \int_{-\infty}^{\infty} \frac{b}{\sqrt{2}} e^{-(\mu^2/2)}\, d\mu$$

$$I = b\sqrt{\pi}\int_{-\infty}^{\infty} \exp\left[-\left(\frac{1}{a^2} - \frac{b^2}{k^2}\right)x^2\right] dx \tag{5-63}$$

$$I = b\sqrt{\pi} \div \left(\sqrt{\frac{1}{a^2} - \frac{b^2}{k^2}}\sqrt{\frac{1}{\pi}}\right)$$

Finally,

$$C = \frac{1}{\pi abk}\sqrt{k^2 - a^2b^2} \tag{5-64}$$

2. For marginal densities,

$$f_1(x) = \int_{-\infty}^{\infty} f(x,y)\, dy = \int_{-\infty}^{\infty} C \exp\left[-\left(\frac{x^2}{a^2} - 2\frac{xy}{k} + \frac{y^2}{b^2}\right)\right] dy \tag{5-65}$$

A change of variable yields the following results:

$$f_1(x) = \frac{1}{\sigma_x \sqrt{2\pi}} e^{-(x^2/2\sigma_x^2)} \tag{5-66}$$

$$f_2(y) = \frac{1}{\sigma_y \sqrt{2\pi}} e^{-(y^2/2\sigma_y^2)} \tag{5-67}$$

with

$$\sigma_x^2 = \frac{1}{2}\frac{a^2k^2}{k^2 - a^2b^2} \qquad \sigma_y^2 = \frac{1}{2}\frac{b^2k^2}{k^2 - a^2b^2} \tag{5-68}$$

The marginal density distributions are also normal distributions.

3. The conditional probability densities, as obtained from the joint and marginal densities, are

$$f_x(x|y) = \frac{f(x,y)}{f_2(y)} = \frac{k}{\sigma_x \sqrt{2\pi(k^2 - a^2b^2)}} \exp\left[-\frac{k^2}{2(k^2 - a^2b^2)}\left(\frac{x}{\sigma_x} - \frac{aby}{k\sigma_y}\right)^2\right] \tag{5-69}$$

$$f_y(y|x) = \frac{f(x,y)}{f_1(x)} = \frac{k}{\sigma_y \sqrt{2\pi(k^2 - a^2b^2)}} \exp\left[-\frac{k^2}{2(k^2 - a^2b^2)}\left(\frac{y}{\sigma_y} - \frac{abx}{k\sigma_x}\right)^2\right] \tag{5-70}$$

For any given value of x or y, the associated conditional densities as well as the marginal densities are normally distributed.

5-11. Functions of Random Variables. One of the most fundamental problems in mathematics and physics is the problem of transforming a set of given data from one coordinate frame to another. For instance, we may have some information concerning a variate $X = (X_1,X_2, \ldots ,X_n)$, and, knowing a function of this variate, say $Y = g(X)$, we wish to obtain comparable information on function Y. The simplest examples of such functions are given by ordinary mathematical functions of one or more variables. In the field of probability, knowing the probability density of a random variable X, we desire to find the density distribution of a new random variable $Y = g(X)$. A moment of reflection is sufficient to realize the significance of such queries in physical problems. In almost any physical problem, we express the result of a complex observation or experiment in terms of certain of its basic constituents. We express the current in a system in terms of certain parameters, say resistances, voltages, etc. Thus, the problem generally requires computing the value of an assumed function, knowing the value of its arguments. The computation of interest may be of a deterministic or a probabilistic origin. Our present interest in the problem is, of course, in the latter direction.

First we shall consider the case of a real single-valued continuous strictly increasing function. Then the procedure will be extended to cover the more general cases.

One-dimensional Case. Let X be a random variable with CDF $F(x)$ and let $g(x)$ be a real single-valued continuous strictly *increasing* function. The CDF of the new random variable $Y = g(X)$ can be easily calculated as

$$\begin{aligned} G(y) &= P\{g(X) \leq y\} = 1 && \text{if } y \geq g(+\infty) \\ G(y) &= P\{g(X) \leq y\} = F[g^{-1}(y)] && \text{if } g(-\infty) < y < g(+\infty) \\ G(y) &= 0 && \text{if } y \leq g(-\infty) \end{aligned} \tag{5-71}$$

$g^{-1}(y)$ being the inverse of $g(y)$.

Thus, $G(y)$, the CDF of the new variable, is completely determined in terms of $F(x)$ and the transformation $g(x)$. If X has a density $f(x)$, the density of $Y = g(X)$ can easily be found as

$$\begin{aligned} F[g^{-1}(y)] &= \int_{-\infty}^{g^{-1}(y)} f(t)\, dt \qquad t = g^{-1}(\mu) \\ &= \int_{g(-\infty)}^{x} f[g^{-1}(\mu)] \frac{d}{d\mu} [g^{-1}(\mu)]\, d\mu \end{aligned} \tag{5-72}$$

When this latter integral exists, the density function of Y is

$$\begin{aligned} \rho(y) &= 0 && \text{for} \quad y \leq g(-\infty) \\ \rho(y) &= \frac{f(x)}{|dy/dx|} && g(-\infty) < y < g(+\infty) \\ \rho(y) &= 0 && y \geq g(+\infty) \end{aligned} \tag{5-73}$$

Because of the diversity of the cases encountered in practice, it is advisable not to use any ready-made formula for obtaining the density function of the new variable. For this reason the examples given below have been worked out directly from the definitions. However, in most cases the proper application of Eqs. (5-72) and (5-73) is adequate.

As a particular application of this problem, consider the case of a linear half-wave rectifier followed by a hypothetical amplifier. The output signal is simply

$$\begin{aligned} Y &= AX && X \geq 0 \\ Y &= 0 && X < 0 \end{aligned}$$

For the positive values of X we have

$$\rho(y) = \frac{1}{|A|} f\left(\frac{y}{A}\right)$$

For example, if X has a normal distribution with zero mean and σ standard deviation, then

$$\rho(y) = \frac{1}{|A|\sqrt{2\pi}\,\sigma} \exp\left(-\frac{y^2}{2A^2\sigma^2}\right) \qquad y \geq 0$$

It should be noted that

$$P\{-\infty < X \leq 0\} = P\{Y = 0\} = \tfrac{1}{2}$$

The density of Y consists of the above continuous distribution and a discrete probability of $\tfrac{1}{2}$ applied at $y = 0$.

Example 5-8. Find (*a*) the distribution and (*b*) the density functions for

$$Y = aX + b \qquad a \neq 0,\ b \text{ real}$$

assuming that $F(x)$ and $f(x)$, the distribution and the density of X, are known.

Solution

(*a*) Distribution function:

$$G(y) = P\{aX + b \leq y\} \qquad a \neq 0$$

$$G(y) = P\left\{X \leq \frac{y-b}{a}\right\} = F\left(\frac{y-b}{a}\right) \qquad \text{if } a > 0$$

$$G(y) = P\left\{X \geq \frac{y-b}{a}\right\} = 1 - F\left(\frac{y-b}{a}\right) \qquad \text{if } a < 0$$

(*b*) Density function:

$$G(y) = \int_{-\infty}^{(y-b)/a} f(t)\,dt \qquad \text{if } a > 0$$

$$= \int_{-\infty}^{y} \frac{1}{a} f\left(\frac{\mu-b}{a}\right) d\mu \qquad t = \frac{\mu - b}{a}$$

$$G(y) = \int_{(y-b)/a}^{\infty} f(t)\,dt \qquad \text{if } a < 0$$

$$G(y) = \int_{y}^{-\infty} f\left(\frac{\mu-b}{a}\right)\frac{d\mu}{a} = \frac{1}{|a|}\int_{-\infty}^{y} f\left(\frac{\mu-b}{a}\right) d\mu$$

The density function in both cases is given by

$$\frac{1}{|a|} f\left(\frac{y-b}{a}\right)$$

Example 5-9. Find (*a*) the distribution and (*b*) the density functions for

$$Y = e^X$$

Solution

(*a*) Distribution function:

$$G(y) = P\{e^X \leq y\} = 0 \qquad y \leq 0$$

$$G(y) = F(\ln y) \qquad y > 0$$

(*b*) Density function. If x has a density $f(x)$, then

$$G(y) = \int_{-\infty}^{\ln y} f(t)\,dt \qquad y > 0 \quad t = \ln \mu$$

$$dt = \frac{d\mu}{\mu}$$

$$G(y) = \int_0^y f(\ln \mu)\,\frac{d\mu}{\mu}$$

$$\rho(y) = 0 \qquad y \leq 0$$

$$\rho(y) = \frac{f(\ln y)}{y} \qquad y > 0$$

Example 5-10. Find the transformation of the variable which changes any given density function into a rectangular distribution, i.e.,

$$\rho(y) = 0 \qquad y \le 0$$
$$\rho(y) = 1 \qquad 0 < y < 1$$
$$\rho(y) = 0 \qquad y \ge 1$$

Solution. We confine ourselves to the case when $F(x)$, the CDF of X, is strictly increasing and continuous. Then consider $Y = F(X)$ as the new variable. The density distribution of Y is

$$\rho(y) = 0 \qquad \text{for} \qquad y \le F(-\infty) = 0$$
$$\rho(y) = \frac{f(x)}{(d/dx)F(x)} = 1 \qquad F(-\infty) < y < F(+\infty)$$
$$(0 < y < 1)$$
$$\rho(y) = 0 \qquad y \ge F(+\infty) = 1$$

Example of a Simple Nonlinear Device. Consider the nonlinear device of Fig. 5-8 with an input-output relationship

$$Y = aX^2$$

Knowing the probability density function of the input signal X, we wish

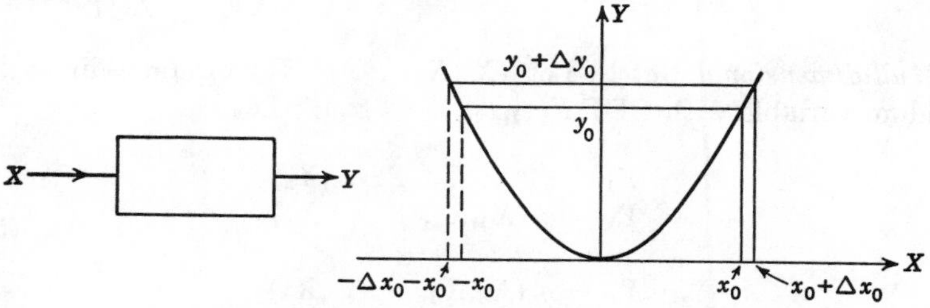

FIG. 5-8. A nonlinear transformation of a random variable.

to derive the corresponding function for the output signal. The curve $Y = aX^2$ can be divided into two increasing parts.

$$P\{y_0 < Y \le y_0 + \Delta y_0\} = P\{x_0 < X \le x_0 + \Delta x_0\} + P\{-x_0 - \Delta x_0 \le X \le -x_0\}$$

In the limit when Δx_0 and Δy_0 tend to zero,

$$\rho(y_0) = \frac{f(x_0) + f(-x_0)}{|dy/dx|_{x=x_0}} = \frac{f(x_0) + f(-x_0)}{|2ax_0|}$$

Note that the probability density of Y is not generally equal to twice the density of X, unless $f(x) = f(-x)$. As an example, suppose that X is a random signal with normal distribution

$$f(x) = \frac{1}{\sqrt{2\pi}} \exp\left(-\frac{1}{2}x^2\right) \qquad -\infty \le x \le \infty$$

In this case

$$\begin{aligned} \rho(y) &= \frac{2}{\sqrt{2\pi}} \frac{1}{2ax} \exp\left(-\frac{1}{2}x^2\right) \qquad 0 \le x \le \infty \\ &= \frac{1}{\sqrt{2\pi a y}} \exp\left(-\frac{1}{2a}y\right) \qquad y \ge 0 \end{aligned}$$

When the distribution of X is, say,

$$f(x) = \frac{1}{\sqrt{2\pi}} \exp\left[-\frac{1}{2}(x-\alpha)^2\right] \qquad -\infty \le x \le \infty$$

then we have

$$\rho(y) = \frac{1}{\sqrt{2\pi}} \frac{1}{2ax} \left\{ \exp\left[-\frac{1}{2}(x-\alpha)^2\right] + \exp\left[-\frac{1}{2}(-x-\alpha)^2\right] \right\}$$

$$\rho(y) = \frac{1}{2\sqrt{2\pi a y}} \left\{ \exp\left[-\frac{1}{2}\left(\sqrt{\frac{y}{a}}-\alpha\right)^2\right] + \exp\left[-\frac{1}{2}\left(-\sqrt{\frac{y}{a}}-\alpha\right)^2\right] \right\} \qquad y \ge 0$$

Multidimensional Case. Let $(X_1,X_2, \ldots ,X_n)$ be an n-dimensional random variable with CDF $F(x_1,x_2, \ldots ,x_n)$. Let

$$\begin{aligned} Y_1 &= g_1(X_1,X_2, \ldots ,X_n) \\ Y_2 &= g_2(X_1,X_2, \ldots ,X_n) \\ &\cdots\cdots\cdots\cdots \\ Y_n &= g_n(X_1,X_2, \ldots ,X_n) \end{aligned} \tag{5-74}$$

It can be shown without difficulty that a generalization of the relation (5-73) is valid under certain appropriate circumstances. First of all, the transformation of Eqs. (5-74) must be one-to-one, and all the g_k must be differentiable and have continuous first partial derivative. Furthermore,

$$\Delta = \begin{bmatrix} \frac{\partial Y_1}{\partial X_1} & \frac{\partial Y_1}{\partial X_2} & \cdots & \frac{\partial Y_1}{\partial X_n} \\ \frac{\partial Y_2}{\partial X_1} & \frac{\partial Y_2}{\partial X_2} & \cdots & \frac{\partial Y_2}{\partial X_n} \\ \cdots & \cdots & \cdots & \cdots \\ \frac{\partial Y_n}{\partial X_1} & \frac{\partial Y_n}{\partial X_2} & \cdots & \frac{\partial Y_n}{\partial X_n} \end{bmatrix} \neq 0 \tag{5-75}$$

In the second place, the variable $(X_1,X_2, \ldots ,X_n)$ must possess a density function $f(x_1,x_2, \ldots ,x_n)$. Under such assumptions, the density function of $(Y_1, Y_2, \ldots, Y_n)$ can be derived as a logical extension of Eqs. (5-73).

$$\rho(y_1, y_2, \ldots, y_n) = \frac{1}{|\Delta|} f(x_1, x_2, \ldots, x_n) \tag{5-76}$$

Example 5-11. Let the density function for the random variable (X_1, X_2) be

$$f(x_1, x_2) = \frac{1}{2\pi} e^{-(x_1^2 + x_2^2)/2}$$

Find the density function for the variable (Y_1, Y_2), where

$$\begin{bmatrix} Y_1 \\ Y_2 \end{bmatrix} = \begin{bmatrix} 1 & 1 \\ 1 & -1 \end{bmatrix} \begin{bmatrix} X_1 \\ X_2 \end{bmatrix}$$

Solution. According to Eq. (5-75),

$$|\Delta| = 2$$

Also note that

$$\begin{bmatrix} X_1 \\ X_2 \end{bmatrix} = \tfrac{1}{2} \begin{bmatrix} 1 & 1 \\ 1 & -1 \end{bmatrix} \begin{bmatrix} Y_1 \\ Y_2 \end{bmatrix}$$

Hence,

$$g(y_1, y_2) = \frac{1}{4\pi} \exp\left[-\frac{1}{2}\left(\frac{y_1^2}{2} + \frac{y_2^2}{2} \right) \right]$$

The simplicity of this problem is due to the linear transformation of the variable. Otherwise, Δ would not remain invariant for all values of the variable.

Multivalued Transformations. Let $Y = g(X)$ be any real continuous transformation defined for all possible values of the random variable X. The probability distribution for the random variable Y is

$$G(y) = P\{g(X) \leq y\} \tag{5-77}$$

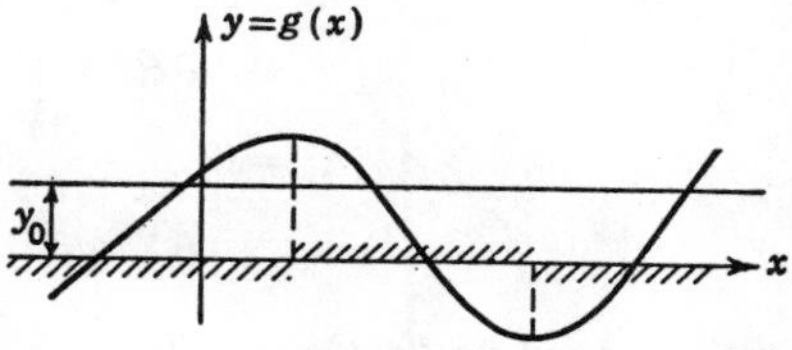

FIG. 5-9. A multivalued transformation of the random variable.

For convenience of analysis, the range of the values of x may be divided into pieces in which $y = g(x)$ is monotonic either nonincreasing or nondecreasing. Without loss of generality, we assume that the functional relationship $y = g(x)$ is such that in each of the above intervals y is differentiable, single-valued, and has a nonzero derivative. Thus, the equation $y_0 = g(x_0)$ may have zero, one, or several real roots $x_0 = g^{-1}(y_0)$. The desired CDF is

$$\begin{aligned} G(y_0) &= P\{g(X) \leq y_0\} = P\{X \leq g^{-1}(y_0)\} \\ &= F(g^{-1}(y_0)) \qquad \text{for } g'(x) > 0 \\ G(y_0) &= P\{X \geq g^{-1}(y_0)\} = 1 - F(g^{-1}(y_0)) \\ &\qquad + P\{X = g^{-1}(y_0)\} \qquad \text{for } g'(x) < 0 \end{aligned} \tag{5-78}$$

If $F(x)$ has a density, then $P\{X = g^{-1}(y_0)\} = 0$ for any arbitrary y_0, and by differentiating $G(y)$ we find the density distribution $\rho(y)$:

$$\rho(y) = \frac{dG(y)}{dy} = \frac{dF(g^{-1}(y))}{dg^{-1}(y)} \frac{dg^{-1}(y)}{dy}$$

$$= f(g^{-1}(y)) \frac{1}{g'(g^{-1}(y))} \qquad \text{for } g'(x) > 0$$

$$\rho(y) = -f(g^{-1}(y)) \frac{1}{g'(g^{-1}(y))} \qquad \text{for } g'(x) < 0$$

The above two cases can be combined in

$$\rho(y) = f(g^{-1}(y)) \frac{1}{|g'(g^{-1}(y))|} \tag{5-79}$$

5-12. Transformation from Cartesian to Polar Coordinate System. A frequent application of the material of the previous section occurs in the transformation of polar and rectangular coordinate systems. Suppose that we are studying the position of a random point M of the two-dimensional plane with reference to the rectangular coordinates X and Y. The position of the point $M(X,Y)$ can be considered as a two-dimensional random variable. We assume that the joint density distribution $f(x,y)$ is known. The problem is to change from the cartesian to the polar coordinate system $M(R,\phi)$ and to determine the corresponding density function $\rho(r,\varphi)$. The following equations are self-explanatory:

$$R = (X^2 + Y^2)^{1/2} \qquad X = R\cos\phi$$
$$\phi = \tan^{-1}\frac{Y}{X} \qquad Y = R\sin\phi \tag{5-80}$$

$$|\Delta| = \begin{vmatrix} \dfrac{\partial R}{\partial X} & \dfrac{\partial R}{\partial Y} \\ \dfrac{\partial \phi}{\partial X} & \dfrac{\partial \phi}{\partial Y} \end{vmatrix} = (X^2 + Y^2)^{-1/2} \tag{5-81}$$

Thus, according to Eq. (5-76),

$$\rho(r,\varphi) = (x^2 + y^2)^{1/2} f(x,y) \tag{5-82}$$

For instance, if X and Y are normally distributed independent random variables with densities

$$\frac{1}{\sigma_x\sqrt{2\pi}} \exp\left(-\frac{x^2}{2\sigma_x^2}\right) \qquad \frac{1}{\sigma_y\sqrt{2\pi}} \exp\left(-\frac{y^2}{2\sigma_y^2}\right) \tag{5-83}$$

then their joint density distribution will be

$$f(x,y) = \frac{1}{2\pi\sigma_x\sigma_y} \exp\left(-\frac{x^2}{2\sigma_x^2} - \frac{y^2}{2\sigma_y^2}\right) \tag{5-84}$$

The density distribution in the polar system is

$$\rho(r,\varphi) = \frac{r}{2\pi\sigma_x\sigma_y} \exp\left[-\frac{r^2}{2}\left(\frac{\cos^2\varphi}{\sigma_x{}^2} + \frac{\sin^2\varphi}{\sigma_y{}^2}\right)\right] \tag{5-85}$$

In engineering literature, this distribution is sometimes called a *Rayleigh distribution.* In the particular case when $\sigma_x = \sigma_y = \sigma$, the Rayleigh distribution becomes

$$\rho(r,\varphi) = \frac{r}{2\pi\sigma^2} \exp\left(-\frac{r^2}{2\sigma^2}\right) \tag{5-86}$$

This probability distribution function is independent of the direction of the point in the plane with respect to the reference coordinate system.

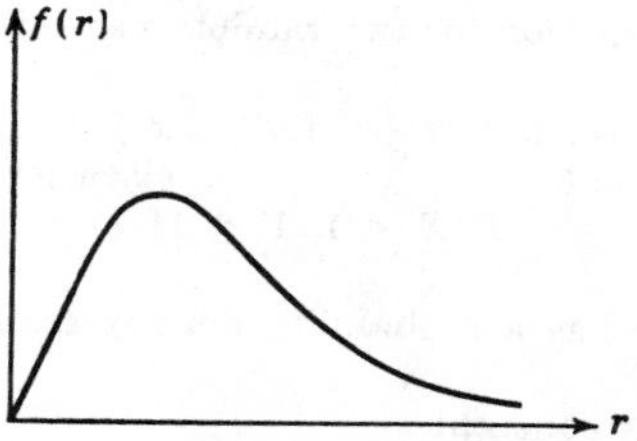

FIG. 5-10. The Rayleigh density function.

In other words, ϕ has a uniform distribution between zero and 2π. In this case the probability of the point M not being closer than a distance d to the origin is given by

$$\int_d^\infty \int_0^{2\pi} \frac{r}{2\pi\sigma^2} \exp\left(-\frac{r^2}{2\sigma^2}\right) dr\, d\varphi = \int_d^\infty \frac{r}{\sigma^2} \exp\left(-\frac{r^2}{2\sigma^2}\right) dr$$
$$= \exp\left(-\frac{d^2}{2\sigma^2}\right) \tag{5-87}$$

The probability of the point M being in the region

$$r_1 < R \leq r_2 \qquad \rho_1 < \phi \leq \rho_2$$

is given by

$$\int_{r_1}^{r_2} \int_{\rho_1}^{\rho_2} \frac{r}{2\pi\sigma_x\sigma_y} \exp\left[-\frac{r^2}{2}\left(\frac{\cos^2\varphi}{\sigma_x{}^2} + \frac{\sin^2\varphi}{\sigma_y{}^2}\right)\right] dr\, d\varphi \tag{5-88}$$

Note that the joint density distribution of Eq. (5-86) is the product of the individual marginal probabilities $1/2\pi$ and

$$\frac{r}{\sigma^2} \exp\left(-\frac{r^2}{2\sigma^2}\right) \tag{5-89}$$

The latter distribution is illustrated in Fig. 5-10.

PROBLEMS

5-1. For what value of K is the function

$$F(x) = 1 - Kx^{-\alpha} \qquad x \geq x_0 \quad \alpha > 0$$
$$F(x) = 0 \qquad x < x_0$$

a CDF? Find the corresponding density function.

5-2. Let X be a random variable varying in the range $0 < x < 1$. For what value of K is the function

$$f(x) = Kx(1 - x) \qquad 0 < x < 1$$
$$f(x) = 0 \qquad \text{otherwise}$$

a probability density function? For this value of K determine the CDF.

5-3. The joint density function for two random variables X and Y is given below:

$$f(x,y) = xye^{-(x+y)} \qquad \text{for} \qquad x \geq 0 \quad y \geq 0$$
$$f(x,y) = 0 \qquad \text{elsewhere}$$

Find
$$P\{X < 1, Y < 1\}$$

5-4. A random variable has a probability density distribution as shown in Fig. P5-4.

(a) Find the value of the constant k.
(b) Find the CDF.
(c) Determine $P\{-\frac{1}{2} < x < \frac{1}{2}\}$.

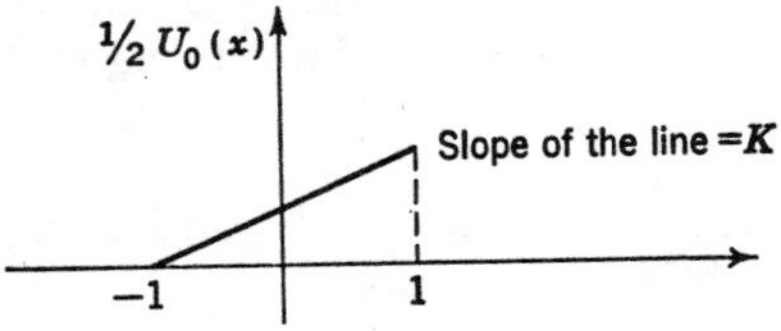

FIG. P5-4

5-5. Evaluate the parameter K which will make the function

$$f(x,y) = Ke^{-(4x+3y)} \qquad x > 0 \quad y > 0$$
$$f(x,y) = 0 \qquad \text{elsewhere}$$

a permissible probability density function. Find

$$P\{1 < x < 2, 1 < y < 2\}$$
$$P\{1 < x < 2\}, P\{1 < y < 2\}$$
$$F(x,y), f_1(x), f_2(y)$$
$$f(x|y), f(y|x)$$

5-6. The current I in a certain electric circuit is assumed to be a random variable normally distributed about its average value of $a = 10$ amperes with $\sigma = 1$. Determine the following:

(*a*) The probability of the current I being less than 11.5 amperes.
(*b*) The probability of the current I being larger than 20 amperes.
(*c*) The probability of I being between 10 and 20 amperes.
(*d*) The probability of I being between 9 and 11 amperes.

5-7. In a lot consisting of 200 items 10 are defective. Find the probability that a random sample of $n = 10$ of this lot yields exactly one defective.

5-8. If the probability that any person 25 years old will die within one year is $p = 0.01$, find the probability that out of a group of 100 such persons (*a*) none, (*b*) only one person, (*c*) not more than one person, (*d*) more than one person, and (*e*) at least one person will die in a year.

5-9. A random voltage V has a normal distribution

$$\frac{1}{\sqrt{2\pi}\,\sigma} \exp\left(-\frac{v^2}{\sigma^2}\right)$$

This voltage is applied to an ideal full-wave rectifier; that is, the output Y is given by the equation

$$Y = |X|$$

Determine the density distribution of the output.

5-10. A certain fluctuating electric current can be considered as a random variable whose value is I amperes. I is uniformly distributed between 9 and 11 amperes. Assuming that this current flows in a 2-ohm resistor, what is the density distribution of the power, that is, $2I^2$?

5-11. The amplitude of a random noise variable has a normal density distribution

$$\rho(x) = \frac{1}{\sqrt{2\pi}\,\sigma} \exp\left(-\frac{x^2}{2\sigma^2}\right)$$

Find the density distribution for its power spectrum, that is, $Y = X^2$.

5-12. A random voltage X has a uniform probability density in the interval $-K \leq x \leq K$ $(K > 0)$. X is the input to a nonlinear device with the characteristic shown in Fig. P5-12. Find the density distribution of the output Y in all three following cases:
(*a*) $K < a$.
(*b*) $a < K < x_0$.
(*c*) $x_0 < K$.

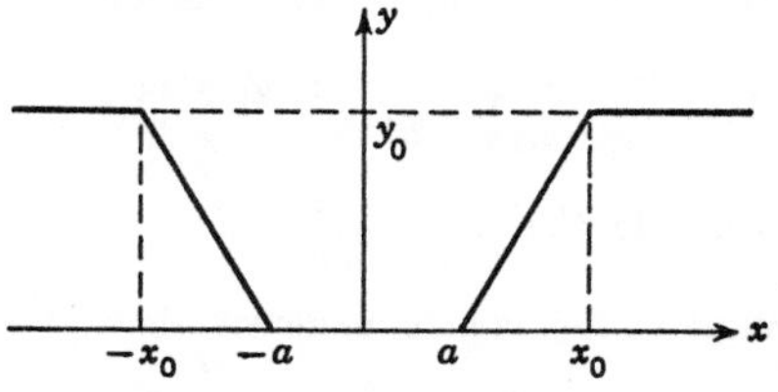

FIG. P5-12

5-13. The CDF of a random variable X is given by Fig. P5-13.
(*a*) Give the density function $f(x)$ graphically with all pertinent values.
(*b*) Determine $E(X)$.

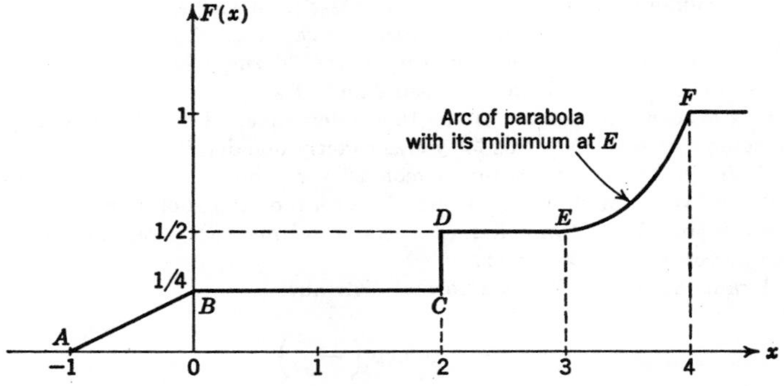

FIG. P5-13

5-14. (*a*) Determine the value of the constant m which makes the function

$$f(x) = mxe^{-x} \qquad x \geq 0$$

a permissible density function.

(*b*) Determine $P\{X \leq 1\}$.

(*c*) Determine $P\{1 < X \leq 2\}$.

5-15. Let I be a random current having a normal distribution with means of 10 amperes and $\sigma = 1$. This current is applied across a ½-ohm resistor. Find the probability distribution of the power dissipated in the resistor.

5-16. Check if the two variables X and Y with joint density

$$f(x,y) = \frac{1}{4\pi^2}[1 - \sin(x + y)] \qquad \text{for} \quad -\pi \leq x \text{ and } y \leq \pi$$
$$= 0 \qquad \text{elsewhere}$$

are independent.

5-17. Same question for

$$f(x,y) = 8xy \qquad 0 \leq x \leq y, 0 \leq y \leq 1$$

5-18. A number is chosen at random on a semicircle and projected onto the diameter. Find the density function for the point of projection.

5-19. Two independent random variables have the following distributions:

$$f_1(x) = 1 \quad 0 < x < 1 \qquad f_2(y) = ae^{-ay} \quad y > 0$$
$$= 0 \quad \text{elsewhere} \qquad = 0 \quad \text{elsewhere}$$

(*a*) Find the density distribution of $Z = X + Y$.

(*b*) Find $P\{z > 1\}$.

5-20. The random variable X has an exponential density distribution

$$f(x) = 1 - e^{-x} \qquad 0 \leq x \leq \infty$$

Find the density of the variables

(*a*) $Y = 3X + 2$

(*b*) $Z = 2X - 3$

(*c*) $U = 1 - e^{-X}$

5-21. A random variable (X,Y) has a uniform probability distribution of $\frac{1}{4}$ in the region $(|X| \leq 1, |Y| \leq 1)$. Find $P\{X^2 + Y^2 \leq 1/\pi^2\}$.

5-22. If A, B, and C are uniformly distributed between 0 and 1, what is the probability that the equation $Ax^2 + Bx + C = 0$ will have real roots?

5-23. The random variable X has a normal distribution

$$f(x) = \frac{1}{\sqrt{2\pi}\,\sigma} \exp\left(-\frac{x^2}{2\sigma^2}\right)$$

Find the distribution of $Y = X^3$.

5-24. The density of a two-dimensional random variable (X,Y) is given below:

$$f(x,y) = e^{-(x+y)} \qquad x > 0 \quad y > 0$$

Make the transformation of the random variable to (U,V),

$$U = X + Y$$
$$V = \frac{X}{Y}$$

and derive the new density function.

5-25. Let X and Y be random variables with exponential distribution on the positive real axis. Make the transformation of variable

$$U = X + Y$$
$$V = \frac{X}{X + Y}$$

(*a*) Find the probability density $\rho(u,v)$.
(*b*) Are the variables U and V independent?

CHAPTER 6

STATISTICAL AVERAGES

In Chaps. 2 and 5 we introduced the elements of discrete and continuous probability theory. This chapter is devoted to an integrated discussion of the concepts of averages (or expected values), moments, and related generating functions. In most probability problems we have a number of random variables and a set of functions associated with them. For example, in the simplest case, we may have a weighting function $f(X)$ (also called utility function, cost function, or loss function) associated with a random variable X. Then the general nature of questions of interest is to obtain what may be called the *statistical average* or the *expected value* of the weighting function $f(X)$, that is, the average value of $f(X)$ in the long run when X assumes all its possible values with their specified probabilities.

The computation of average values of random variables and the corresponding cost functions is of considerable interest in physical problems. The major part of this chapter will be devoted to the study of such expected values. Section 6-2 treats the expected value of the sum and the product of a number of functions of several independent random variables. Sections 6-3 to 6-8 concentrate on a particular form of weighting function, namely, X^r, which is of great practical significance as it leads to different moments. Later sections of the chapter will be devoted to relating these topics to the familiar theory of Fourier and Laplace transforms. Finally we shall apply this material in deriving a simple form of the central-limit theorem in a subsequent chapter.

The concept of statistical averaging applies to discrete as well as to continuous random variables. At the beginning of this chapter there may be a tendency, in proving certain theorems, to employ a discrete probability scheme. The results are equally valid for continuous schemes.

6-1. Expected Values; Discrete Case. Let X be a discrete random variable assuming the values $x_1, x_2, \ldots, x_n, \ldots$, with respective probabilities $p_1, p_2, \ldots, p_n, \ldots$, and let $g(X)$ be a real single-valued function. The mathematical expectation of $g(X)$ is defined as

$$E[g(X)] = \sum_{i=1}^{\infty} p_i g(x_i) \tag{6-1}$$

This definition is contingent upon the convergence of the above series; however, in most physical problems the convergence restriction is only of theoretical significance. The mathematical expectation of a function is, in a way, the "average" of that function over all possible values that the function assumes. For example, if the outcome of a random experiment assumes numerical values $g(x_1)$, $g(x_2)$, . . . , with frequencies n_1, n_2, . . . , and the experiment is repeated N times, then the arithmetic average of the function $g(x)$ is

$$g(X) \text{ average } = \frac{n_1 g(x_1) + n_2 g(x_2) + n_3 g(x_3) + \cdots}{N} \tag{6-2}$$

When the number of trials N is made very large, it can be intuitively inferred that the average value of $g(X)$ in Eq. (6-2) approaches the expectation of $g(X)$ as defined in Eq. (6-1). The mathematical expectation is also called the *average, statistical average*, and the *mean value* of the random variable.

The reason for using the term "statistical average" will be apparent when one has proved the theorem which is generally known as the law of large numbers.

Note that, in the simple case of $g(X) = X$, Eq. (6-1) gives the average value of the random variable X itself. Thus,

$$E(X) = X \text{ average } = \sum_{i=1}^{\infty} p_1 x_1 \tag{6-3}$$

Also note that when $X = \text{const} = K$, then $E(X) = E(K) = K$.

The above definition can be extended to the case of multidimensional discrete random variables. For example, suppose that a function $g(X,Y)$ is associated with a two-dimensional discrete random variable (X,Y) having a joint probability distribution $P\{i,j\}$. The concept of mean value as outlined in Eqs. (6-1) and (6-2) leads to the following definition for the mathematical expectation of the function $g(X,Y)$:

$$E[g(X,Y)] = \sum_i \sum_j P\{i,j\} g(x_i,y_j) \tag{6-4}$$

The above defining equations can be directly extended to cover the case of a random variable assuming a continuum of values. Let $f(x)$ be the density function and $g(x)$ a real single-valued integrable function; then we define

$$E[g(X)] = \int_{-\infty}^{+\infty} g(x)f(x)\,dx \tag{6-5}$$

For instance,

$$E[X] = \int_{-\infty}^{+\infty} xf(x)\,dx$$

Similarly, for a two-dimensional random variable (X,Y) and the weighting function $g(x,y)$, we define

$$E[g(X,Y)] = \int_{-\infty}^{\infty} \int_{-\infty}^{\infty} g(x,y) f(x,y)\, dx\, dy \tag{6-6}$$

All integrals in the defining equations must exist.

Example 6-1. An urn contains three white and two black balls. A and B agree to the following game. Each person draws two balls at a single drawing, the balls being replaced after each drawing. B will pay A the amount of \$5 for each white ball and \$2 for each black ball.

(*a*) What is the mathematical expectation of the player A?

(*b*) How much should A pay B for the drawing of a white ball and a black ball so that their expectations are the same?

Solution

(*a*) There are three possible cases: WW, WB or BW, and BB.

$$\begin{aligned} P(WW) &= 3/5 \cdot 2/4 = 3/10 \\ P(WB) &= 3/5 \cdot 2/4 = 3/10 \\ P(BW) &= 2/5 \cdot 3/4 = 3/10 \\ P(BB) &= 2/5 \cdot 1/4 = 1/10 \end{aligned}$$

Let X be the gain of A in a drawing; then X will assume the following three values at random:

$$\begin{aligned} x_1 &= 10 \qquad & x_2 &= 7 \qquad & x_3 &= 4 \\ p_1 &= 3/10 \qquad & p_2 &= 6/10 \qquad & p_3 &= 1/10 \end{aligned}$$

Now one may apply Eq. (6-1).

$$E(X) = 3/10 \cdot 10 + 6/10 \cdot 7 + 1/10 \cdot 4 = 7.6$$

In the long run, A can expect an average gain of \$7.60 in each drawing.

(*b*) Let B's gain be x and y dollars for the drawing of a white and a black ball, respectively. Then if Y is the gain of B,

$$E(Y) = 3/10(2x) + 6/10(x + y) + 1/10(2y)$$

The answer to the question is given by any values of x and y satisfying

$$3x + 2y - 19 = 0$$

6-2. Expectation of Sums and Products of a Finite Number of Independent Discrete Random Variables. In this section we employ Eq. (6-4) in obtaining the expectation of sums and products of a number of discrete random variables.

Sum of Random Variables. Let

$$g(X,Y) = X + Y \tag{6-7}$$

Then in the discrete case

$$E(X+Y) = \sum_i \sum_j P\{i,j\}(x_i + y_j)$$
$$= \sum_i \sum_j P\{i,j\}x_i + \sum_i \sum_j P\{i,j\}y_j$$
$$= \sum_i x_i \sum_j P\{i,j\} + \sum_j y_j \sum_i P\{i,j\} \tag{6-8}$$

Finally
$$E(X+Y) = E(X) + E(Y) \tag{6-9}$$

This relation is also valid when X and Y are random variables of a continuous type.

More generally, for the expectation of the sum of a finite number of discrete random variables (not necessarily independent), one obtains

$$E(X_1 + X_2 + \cdots + X_n) = E(X_1) + E(X_2) + \cdots + E(X_n) \tag{6-10}$$

provided that the expectation of the individual variables has a finite value.

Product of Two Independent Random Variables. Let

$$g(X,Y) = XY$$

For independent discrete variables we have

$$P\{i,j\} = p_1\{i\} \cdot p_2\{j\}$$

Hence,

$$E(XY) = \sum_i \sum_j x_i y_j P\{i,j\} = \sum_i \sum_j [x_i p_1(i)][y_j p_2(j)]$$
$$= \sum_i p_1(i) x_i \sum_j p_2(j) y_j \tag{6-11}$$
$$E(XY) = E(X) \cdot E(Y) \tag{6-12}$$

This result also holds for independent random variables with continuous distributions:

$$E(XY) = \int_{-\infty}^{\infty} \int_{-\infty}^{\infty} xyf(x,y)\,dx\,dy$$
$$= \int_{-\infty}^{\infty} \int_{-\infty}^{\infty} [xf_1(x)\,dx][yf_2(y)\,dy] = E(X)E(Y) \tag{6-12a}$$

When one of the variables has a constant value K, we have

$$E(KX) = E(K) \cdot E(X) = KE(X)$$

By induction one arrives at the result that the expectation of the product of a finite number of discrete *independent* random variables is the product of their expectations.

$$E(X_1 X_2 \cdots X_n) = E(X_1) \cdot E(X_2) \cdots E(X_n) \tag{6-13}$$

Note that the independence of the variables is required for Eq. (6-13) but not for Eq. (6-9).

Example 6-2. Two dice are thrown; find the expected value for the sum and the product of their face numbers.

Solution. The joint probability $P\{i,j\}$ and the marginal probabilities $p_1(i)$ and $p_2(j)$ are

$$P\{i,j\} = \tfrac{1}{36}$$
$$p_1(i) = \tfrac{1}{6} \qquad p_2(j) = \tfrac{1}{6}$$
$$E(X) = E(Y) = \tfrac{1}{6}(1 + 2 + 3 + 4 + 5 + 6) = \tfrac{7}{2}$$
$$\begin{aligned} E(X + Y) = \tfrac{1}{36}(&2 + 3 + 4 + 5 + 6 + 7 \\ &+ 3 + 4 + 5 + 6 + 7 + 8 \\ &+ 4 + 5 + 6 + 7 + 8 + 9 \\ &+ 5 + 6 + 7 + 8 + 9 + 10 \\ &+ 6 + 7 + 8 + 9 + 10 + 11 \\ &+ 7 + 8 + 9 + 10 + 11 + 12) \\ = \tfrac{1}{36}(&27 + 33 + 39 + 45 + 51 + 57) = \tfrac{1}{36} \cdot 252 = 7 \end{aligned}$$

This direct calculation checks with the result $E(X + Y) = \tfrac{7}{2} + \tfrac{7}{2} = 7$.

The product of the two face numbers assumes the following values, with their corresponding probabilities.

1	2	3	4	5	6	8	9	10	12	15	16	18	20	24	25	30	36
$\tfrac{1}{36}$	$\tfrac{2}{36}$	$\tfrac{2}{36}$	$\tfrac{3}{36}$	$\tfrac{2}{36}$	$\tfrac{4}{36}$	$\tfrac{2}{36}$	$\tfrac{1}{36}$	$\tfrac{2}{36}$	$\tfrac{4}{36}$	$\tfrac{2}{36}$	$\tfrac{1}{36}$	$\tfrac{2}{36}$	$\tfrac{2}{36}$	$\tfrac{2}{36}$	$\tfrac{1}{36}$	$\tfrac{2}{36}$	$\tfrac{1}{36}$

$$\begin{aligned} E(XY) = \tfrac{1}{36}(&1 + 4 + 6 + 12 + 10 + 24 + 16 + 9 \\ &+ 20 + 48 + 30 + 16 + 36 + 40 \\ &+ 48 + 25 + 60 + 36) \end{aligned}$$
$$E(XY) = \tfrac{1}{36} \cdot 441 = \tfrac{49}{4}$$
$$E(X) \cdot E(Y) = \tfrac{7}{2}\,\tfrac{7}{2} = \tfrac{49}{4}$$

6-3. Moments of a Univariate Random Variable. Equations (6-1) and (6-5) describe the mathematical expectation associated with a general function of a random variable $g(X)$. The particularly simple function

$$g(X) = X^r$$

plays an important role in the theory of probability and application problems. The expectation of X^r, that is,

$$m_r = E[X^r] = \sum_{i=1}^{\infty} p_i x_i^r$$

or

$$m_r = E[X^r] = \int_{-\infty}^{+\infty} x^r f(x)\, dx \tag{6-14}$$

is called the rth moment about the origin of the random variable X. The definition is contingent on the convergence of the series or the existence of the integral, i.e., on the finiteness of the rth-order moment. For $r = 0$ one has the zero-order moment about the origin.

$$E(X^0) = 1$$

For $r = 1$ the first-order moment about the origin or the *mean value* of the random variable is

$$E(X^1) = E(X) = m_1 = \sum_{i=1}^{\infty} p_i x_i \tag{6-15}$$

or

$$m_1 = \int_{-\infty}^{+\infty} x f(x)\, dx$$

For $r = 2$ the second-order moment about the origin is

$$E(X^2) = p_1 x_1^2 + p_2 x_2^2 + \cdots + p_n x_n^2 + \cdots \tag{6-16}$$

or

$$E(X^2) = \int_{-\infty}^{+\infty} x^2 f(x)\, dx$$

The rth-order moment about a point c is defined by

$$E[(X - c)^r] \tag{6-17}$$

A very useful and familiar case is the moment of the variable centered about the point m_1, the mean value of the variable. Such moments are called *central moments.*

$$\mu_r = \text{central moment of order } r = E[X - E(X)]^r = E(X - m_1)^r \tag{6-18}$$

The values of central moments of first and second order in terms of ordinary moments are discussed below.

First-order Central Moment

$$\begin{aligned} \mu_1 &= \text{first-order central moment} = \text{expectation of deviation} \\ &\quad \text{of random variable from its mean value } m_1 \\ &= E(X - m_1) \\ \mu_1 &= E(X - m_1) = E(X) - E(m_1) = m_1 - m_1 = 0 \end{aligned} \tag{6-19}$$

Second-order Central Moment

$$\begin{aligned} \mu_2 &= \text{second-order central moment} = E[(X - m_1)^2] \\ \mu_2 &= E(X^2) - 2E(Xm_1) + E(m_1^2) = E(X^2) - m_1^2 \end{aligned} \tag{6-20}$$

The second-order central moment is also called the *variance* of the random variable X.

$$\mu_2 = \text{var}(X) = E[(X - m)^2] = E(X^2) - m_1^2 = m_2 - m_1^2 \tag{6-21}$$

The nonnegative square root of the variance is called the *standard deviation* of the random variable X.

$$\text{Standard deviation} = \sigma_X = \sqrt{\mu_2} = \sqrt{m_2 - m_1^2} \tag{6-22}$$

The physical interpretation of the first and second moments in engineering problems is self-evident. The first moment m_1 is the ordinary mean or the average value of the quantity under consideration, and m_2 is the average of the square of that quantity (mean square). For instance, if X is an electric current, m_1 and m_2 are the average (or d-c level) of the current and the power dissipated in the unit resistance, respectively. Similarly, the standard deviation is the root mean square (rms) of the current, about its mean value.

Example 6-3. In part (*a*) of Example 6-1, find m_1, m_2, μ_2, and σ.
Solution

$$
\begin{aligned}
E(X) &= m_1 = 7.6 \\
E(X^2) &= m_2 = \tfrac{3}{10} \cdot 100 + \tfrac{6}{10} \cdot 49 + \tfrac{1}{10} \cdot 16 = 61.0 \\
\mu_2 &= m_2 - m_1^2 = 61.00 - 57.76 = 3.24 \\
\sigma &= \sqrt{\mu_2} = 1.8
\end{aligned}
$$

The above definition can be extended to the sum of two or more random variables. The pertinent algebraic operations will be simplified by the use of some familiar mathematical formalism. Let $E(X)$ and $\sigma(X)$ stand for the expectation and the standard deviation of a random variable X, respectively; then for the sum of two random variables we write

$$
\begin{aligned}
\operatorname{var}(X+Y) &= E[(X+Y)^2] - [E(X+Y)]^2 \\
&= E(X^2) + E(Y^2) + 2E(XY) \\
&\qquad - [E^2(X) + E^2(Y) + 2E(X)\cdot E(Y)] \qquad (6\text{-}23)
\end{aligned}
$$

If the two variables are independent, then $E(XY) = E(X) \cdot E(Y)$, and Eq. (6-23) yields

$$\sigma^2_{x+y} = \sigma_x^2 + \sigma_y^2 \qquad (6\text{-}23a)$$

This result can be extended to obtain the standard deviation of the sum of a finite number of independent random variables.

$$\sigma^2_{x_1+x_2+\cdots+x_n} = \sigma_{x_1}^2 + \sigma_{x_2}^2 + \cdots + \sigma_{x_n}^2 \qquad (6\text{-}24)$$

Example 6-4. X is a random electric current normally distributed, with $\bar{X} = 0$ and $\sigma = 1$. If this current is passed through a full-wave rectifier, what is the expected value of the output?

Solution. Let the output of the rectifier be Y; then

$$Y = |X|$$

Applying Eqs. (5-73), the probability density of Y is

$$\rho(y) = 2\frac{f(x)}{1} = 2\frac{1}{\sqrt{2\pi}}e^{-x^2/2} = \sqrt{\frac{2}{\pi}}\,e^{-y^2/2} \qquad y \geq 0$$

The density distribution has the shape of a normal curve for $y \geq 0$ and is zero elsewhere.

$$E(Y) = \int_0^\infty \sqrt{\frac{2}{\pi}}\,ye^{-y^2/2}\,dy = \sqrt{\frac{2}{\pi}}\,[-e^{-y^2/2}]_0^\infty = \sqrt{\frac{2}{\pi}} \qquad \text{amperes}$$

Example 6-5. The internal noise of an amplifier has an rms (root mean square) value of 2 volts. When the signal is added, the rms output is 5 volts. What would be the rms value of the output when the signal is tripled?

Solution. This simple example is rather familiar to electrical engineers. While the solution may look obvious, we shall give the hypotheses under which the familiar solution is obtained. Let S and N be *independent* random variables (signal and noise) with zero means and given standard deviations σ_x and σ_n.

$$\text{Root mean square of } S = \sqrt{E(S^2)} = \sqrt{\sigma_s^2} = \sigma_s$$
$$\text{Root mean square of } N = \sqrt{E(N^2)} = \sqrt{\sigma_n^2} = \sigma_n = 2$$

As the random variables are assumed to be independent, by Eq. (6-23*a*) we find

$$E[(S+N)^2] = E(S^2) + E(N^2) = \sigma_s^2 + \sigma_n^2 = 5^2$$
$$\sigma_s^2 = 5^2 - 2^2 = 21$$

The problem asks for

$$\sqrt{E(3S+N)^2}$$

Now
$$E(3S+N)^2 = E(9S^2) + E(N^2) = 9 \times 21 + 2^2$$
$$\sqrt{E(3S+N)^2} = \sqrt{193} = 13.9 \text{ volts}$$

Note that no additional assumption is necessary as to the nature of the distribution functions of the signal and the noise.

6-4. Two Inequalities

An Inequality for Second-order Moments. We should like to compare the second-order moment of a random variable about a point c with the second-order central moment μ_2 of the same variable.

$$E[(X-c)^2] = E[(X - m_1 + m_1 - c)^2]$$
$$E[(X-c)^2] = E[(X-m_1)^2] + 2E[(X-m_1)(m_1-c)] + E[(m_1-c)^2]$$
$$E[(X-c)^2] = \mu_2 + (m_1-c)^2$$

The second-order moment about any point $c \neq m_1$ is larger than the central second moment:

$$E[(X-c)^2] > E[(X-m_1)^2] \qquad \text{for } c \neq m_1$$

The fact that the second-order central moment of a random variable is the smallest of all second-order moments is of basic significance in the theory of error. In the analogy with electrical engineering, note that the smallest possible root mean square of a fluctuating current or voltage $f(t)$ is obtained when that quantity is measured with respect to its mean value (d-c level) rather than any other level.

Chebyshev Inequality. The Chebyshev inequality suggests an interesting relation that exists between the variance and the spreading out of the probability density. Let X be a random variable with a probability density distribution $f(x)$, first moment m_1, and standard deviation σ. Chebyshev's inequality states that the probability of $X - m_1$ assuming values larger than $k\sigma$ is less than $1/k^2$.

$$P\{|X - m_1| \geq k\sigma\} \leq \frac{1}{k^2} \qquad K > 0 \tag{6-25}$$

To prove the validity of this inequality, let $Y = |X - m_1|$, and refer to Fig. 6-1. The desired probability is

$$P\{Y \geq k\sigma\} = \int_{-\infty}^{m_1 - k\sigma} f(x)\,dx + \int_{m_1+k\sigma}^{\infty} f(x)\,dx$$

Multiplication by $k^2\sigma^2$ yields

$$k^2\sigma^2 P\{|X - m_1| \geq k\sigma\} = \int_{-\infty}^{m_1-k\sigma} k^2\sigma^2 f(x)\,dx + \int_{m_1+k\sigma}^{\infty} k^2\sigma^2 f(x)\,dx \quad (6\text{-}26)$$

Note that in each of the ranges of integration

$$k\sigma \leq |x - m_1|$$

Thus

$$k^2\sigma^2 P\{|X - m_1| \geq k\sigma\} \leq \int_{-\infty}^{m_1-k\sigma} (x - m_1)^2 f(x)\,dx + \int_{m_1+k\sigma}^{\infty} (x - m_1)^2 f(x)\,dx \quad (6\text{-}27)$$

But the right side is certainly not greater than the second-order central moment of X; therefore,

$$k^2\sigma^2 P\{|X - m_1| \geq k\sigma\} \leq \int_{-\infty}^{\infty} (x - m_1)^2 f(x)\,dx = \sigma^2$$

Dividing both sides of this inequality by $k^2\sigma^2$ gives the desired inequality.

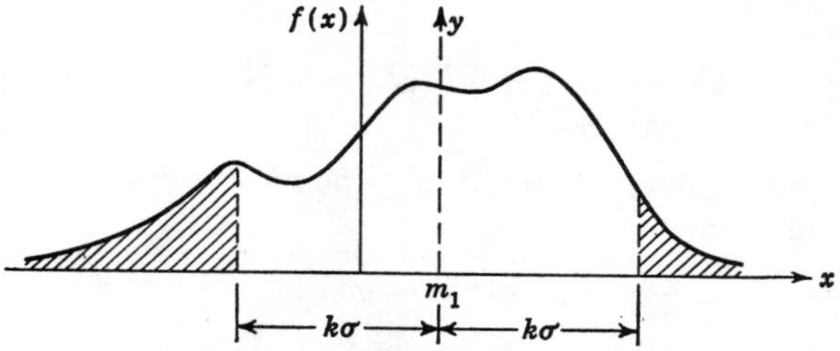

FIG. 6-1. Illustration of Chebyshev's inequality.

The Chebyshev inequality expresses interesting bounds on the probability of the centralized random variable exceeding any units of standard deviation. For example,

$$\begin{aligned} P\{|X - m_1| \geq 2\sigma\} &\leq 0.250 \\ P\{|X - m_1| \geq 3\sigma\} &\leq 0.111 \end{aligned} \quad (6\text{-}28)$$

This result may be applied to a specific known density distribution such as a normal distribution. For normal distributions, from tables we derive

$$\begin{aligned} P\{|X - m_1| \geq 2\sigma\} &\approx 0.045 \\ P\{|X - m_1| \geq 3\sigma\} &\approx 0.0026 \end{aligned} \quad (6\text{-}29)$$

A comparison of Eqs. (6-28) and (6-29) shows that Chebyshev's results are weak and that they give only a rough estimate of the spreading of the distribution. Similar results are valid for a discrete random variable.

Example 6-6. Let X be the fraction of number of heads obtained in throwing an honest coin 10^8 times. Show that

$$P\{|X - \tfrac{1}{2}| > 0.001\} < 0.01$$

Solution. Let X_i be a random variable denoting the number of heads in the ith throwing,

$$X = \frac{1}{n}(X_1 + X_2 + \cdots + X_n) \qquad n = 10^8$$

$$\bar{X}_i = \tfrac{1}{2} \qquad \overline{X_i^2} = \tfrac{1}{2} \qquad \sigma_{x_i} = \tfrac{1}{2} \qquad \sigma_x = \frac{1}{2\sqrt{n}}$$

Applying the Chebyshev inequality to X, we find

$$P\{|X - m| > K\} \le \frac{\sigma_x^2}{K^2} \qquad m = \tfrac{1}{2} \quad K = 10^{-3}$$

$$P\{|X - \tfrac{1}{2}| > 0.001\} \le \tfrac{1}{400} < 0.01$$

6-5. Moments of Bivariate Random Variables. Equations (6-3) to (6-6) give expressions for the expected value of a function of a one-dimensional and a two-dimensional random variable. In Sec. 6-3 we discussed different moments of a one-dimensional random variable by letting $g(X) = X^r$, $r = 1, 2, 3, \ldots$. The object of this section is to derive similar formulations for moments of a two-dimensional random variable by letting

$$g(X,Y) = X^iY^j \qquad i, j = 1, 2, 3, \ldots \tag{6-30}$$

The moment of order i, j of a two-dimensional discrete random variable is defined by

$$\alpha_{ij} = E(X^iY^j) \tag{6-31}$$

For a continuous random variable we have

$$\alpha_{ij} = E(X^iY^j) = \int_{-\infty}^{\infty}\int_{-\infty}^{\infty} x^iy^jf(x,y)\,dx\,dy \tag{6-31a}$$

The central moments correspond to the variable being centered at the point representing its two-dimensional first moment, i.e., the point $(\bar{X},\bar{Y})$:

$$\mu_{ij} = E[(X - \bar{X})^i(Y - \bar{Y})^j] \tag{6-32}$$

The central moments of second order are of considerable interest. They can easily be computed in terms of the central moments of X and Y. Following the above notation,

$$\begin{aligned} \alpha_{10} &= \bar{X} = E(X) \\ \alpha_{01} &= \bar{Y} = E(Y) \\ \alpha_{11} &= \overline{XY} = E(XY) \\ \alpha_{20} &= \overline{X^2} = E(X^2) \\ \alpha_{02} &= \overline{Y^2} = E(Y^2) \end{aligned} \tag{6-33}$$

The three central moments of second order are

$$\begin{aligned} \mu_{20} &= E[(X - \bar{X})^2] = \sigma_x^2 = \alpha_{20} - \alpha_{10}^2 \\ \mu_{11} &= E[(X - \bar{X})(Y - \bar{Y})] = E(XY) - \bar{X}\bar{Y} = \alpha_{11} - \alpha_{10} \cdot \alpha_{01} \\ \mu_{02} &= E[(Y - \bar{Y})^2] = \sigma_y^2 = \alpha_{02} - \alpha_{01}^2 \end{aligned} \tag{6-34}$$

In the next section we shall show that the ratio of $\mu_{11}/\sqrt{\mu_{20}\mu_{02}}$ gives an indication of the degree of linear dependence between the two variables.

6-6. Correlation Coefficient. In everyday problems of the physical sciences, we are often confronted with the study of two variable quantities, X and Y, with a functional relation $y = f(x)$ between them. Sometimes no specific knowledge of any such relationship between the two variables is available, but a set of several of their paired values is known. In such situations, as a rule, we do not search for an analytic relation among the variables but wish to find out if the values of one are influenced by the values of the other. For example, one variable may represent the heights of students in a university and the other variable the heights of their fathers. While it is hopeless to establish a functional relationship between these two variables, it is quite reasonable to expect a certain degree of dependence between the two. A measure of the degree of this common influence is given by what is referred to as the correlation coefficient* ρ defined by

$$\rho = \frac{\mu_{11}}{\sqrt{\mu_{20}\mu_{02}}} = \frac{\mu_{11}}{\sigma_x\sigma_y} \tag{6-35}$$

It can be shown that the permissible values for the correlation coefficient are confined to the interval $[-1,+1]$. One way of showing this is by considering a new random variable

$$Z = a(X - \bar{X}) + b(Y - \bar{Y}) \tag{6-36}$$

where a and b are real parameters. The second moment of the random variable Z must, of course, remain a nonnegative number for all values of a and b.

$$E(Z^2) = E[a(X - \bar{X}) + b(Y - \bar{Y})]^2 = a^2\mu_{20} + 2ab\mu_{11} + b^2\mu_{02} \tag{6-37}$$

* The coefficient of correlation appeared in the work of Karl Pearson in 1891. An interesting historical and technical account of this topic can be found in the article On the Mathematics of Simple Correlation by C. D. Smith (*Math. Mag.*, vol. 32, no. 2, pp. 57–69, November–December, 1958).

The leading coefficient of the quadratic form

$$\mu_{20}\left(\frac{a}{b}\right)^2 + 2\mu_{11}\left(\frac{a}{b}\right) + \mu_{02} \tag{6-38}$$

is nonnegative. Hence the discriminant should not become positive:

$$\mu_{11}^2 - \mu_{20}\cdot\mu_{02} \leq 0 \tag{6-39}$$

or, in terms of the correlation coefficient,

$$|\rho| = \left|\frac{\mu_{11}}{\sqrt{\mu_{20}\cdot\mu_{02}}}\right| \leq 1 \tag{6-40}$$

In the extreme case, when $\rho = \pm 1$, there is complete dependence between the variables X and Y. When the two variables are independent, $(X - \bar{X})$ and $(Y - \bar{Y})$ are also independent variables. This fact implies

$$\mu_{11} = E(X - \bar{X})\cdot E(Y - \bar{Y}) = 0 \tag{6-41}$$

Thus, when X and Y are independent, their correlation coefficient is zero; the converse, however, is not true.* The correlation coefficient ρ indicates a measure of the linear interdependence between the two variables.

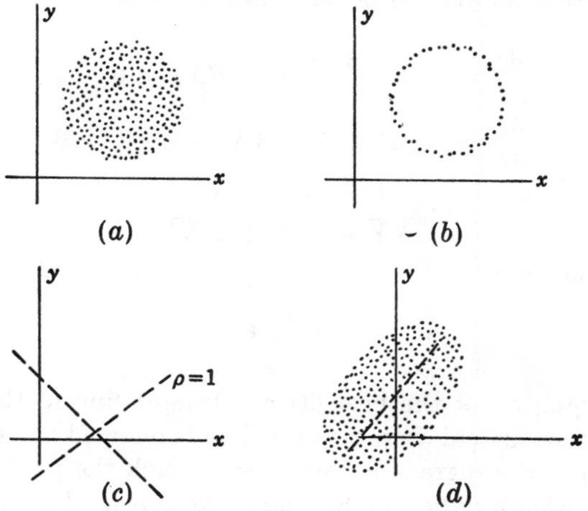

FIG. 6-2. (a) No correlation; (b) no correlation but strong dependence; (c) linear dependence; (d) scatter diagram.

When points of a rectangular coordinate system are used to show ordered pairs of values corresponding to pairs of associated discrete random variables, we obtain a diagram commonly known as a *scatter diagram.* A scatter diagram generally gives an indication of the degree of linear

* It is important to note that two random variables may be dependent but not linearly correlated (see, for example, Prob. 6-16).

relationship between the two variables. For example, if the two variables are uncorrelated, i.e., if $\rho = 0$, we may find the somewhat uniform scatter diagram of Fig. 6-2*a* or *b*. In these cases the variables are linearly independent, although they may be bound by some nonlinear relationships (Fig. 6-2*b*). Conversely, when the variables are linearly related ($\rho = \pm 1$) the scatter diagram takes the general form of Fig. 6-2*c*.

In problems of engineering statistics, frequently several pairs of values of two related discrete random variables are known. The problem of interest is to find an estimate of this linear relationship such that we have the *best straight-line fitting.* The word "best" is taken in the sense of least mean square; that is, if the desired line has a cartesian equation of the type

$$Y = A + BX$$

then A and B must be such that the sum

$$g(A,B) = E(Y - A - BX)^2 \tag{6-42}$$

has its smallest possible value. This condition requires that the first partial derivatives of $g(A,B)$ be simultaneously zero.

$$\begin{aligned} \frac{\partial g}{\partial A} &= -2E(Y - A - BX) = 0 \\ \frac{\partial g}{\partial B} &= -2E(XY - AX - BX^2) = 0 \end{aligned} \tag{6-43}$$

or

$$\begin{aligned} \bar{Y} &= A + B\bar{X} \\ \overline{XY} &= A\bar{X} + B\overline{X^2} \end{aligned} \tag{6-44}$$

These equations give

$$Y = \bar{Y} + \frac{\mu_{11}}{\sigma_x^2}(X - \bar{X}) \tag{6-45}$$

This is the equation of the best-fitting straight line in the least-square sense. In the statistical literature this line is referred to as the *regression line.* Note that the regression line goes through the point $(\bar{X},\bar{Y})$.

6-7. Linear Combination of Random Variables. Let X be the linear combination of a finite number of random variables.

$$X = a_1X_1 + a_2X_2 + \cdots + a_nX_n$$

We assume that the coefficients $a_1, a_2, \ldots, a_n$ have fixed values. The mathematical expectation of X is

$$\begin{aligned} E(X) &= E(a_1X_1 + a_2X_2 + \cdots + a_nX_n) \\ \bar{X} &= a_1\bar{X}_1 + a_2\bar{X}_2 + \cdots + a_n\bar{X}_n \end{aligned} \tag{6-46}$$

The variance of X is

$$\sigma_x^2 = E[(X - \bar{X})^2] = E[a_1(X_1 - \bar{X}_1) + a_2(X_2 - \bar{X}_2) + \cdots + a_n(X_n - \bar{X}_n)]^2 \quad (6\text{-}47)$$

Note that Eqs. (6-46) and (6-47) are valid for dependent as well as independent variables. But only when all the variables are independent does Eq. (6-47) lead to the following simple expression:

$$\sigma_x^2 = a_1^2\sigma_1^2 + a_2^2\sigma_2^2 + \cdots + a_n^2\sigma_n^2 \quad (6\text{-}48)$$

For example, consider X to be the number of heads in a sequence of N throws of a biased coin with the probability of a head in each throw being equal to p. Then

$$X = X_1 + X_2 + \cdots + X_n$$

where $X_i = 1$ if the ith throw results in a head and $X_i = 0$ otherwise. Now

$$P\{X_i = 1\} = p$$
$$P\{X_i = 0\} = 1 - p$$
$$E(X_i) = p$$
$$\sigma_{X_i}^2 = E(X_i^2) - [E(X_i)]^2 = p(1 - p)$$

As the X_i are independent, we have, by application of Eqs. (6-46) and (6-48),

$$E(X) = E(X_1) + E(X_2) + \cdots + E(X_n) = Np$$
$$\sigma_X^2 = \sigma_{X_1}^2 + \sigma_{X_2}^2 + \cdots + \sigma_{x_n}^2 = Np(1 - p)$$

Example 6-7. Let

$$Y = \frac{X - m}{\sigma}$$

Find the mean and the standard deviation of Y when

$$\bar{X} = m \qquad \text{and} \qquad \sigma_x = \sigma$$

Solution

$$\bar{Y} = \frac{1}{\sigma}(\bar{X} - \bar{m}) = \frac{1}{\sigma}(m - m) = 0$$

$$\sigma_y^2 = \frac{1}{\sigma^2}\sigma_x^2 + \text{variance of } \frac{m}{\sigma} = 1 + 0 = 1$$

Example 6-8. The joint density function for two random variables X and Y is given as

$$f(x,y) = x + y \qquad \text{for } \begin{cases} 0 \le x \le 1 \\ 0 \le y \le 1 \end{cases}$$
$$f(x,y) = 0 \qquad \text{elsewhere}$$

Find the correlation coefficient.

Solution

$$\bar{X} = \int_0^1 \int_0^1 x(x + y)\, dx\, dy = 7/12$$
$$\bar{Y} = \int_0^1 \int_0^1 y(x + y)\, dx\, dy = 7/12$$
$$\overline{X^2} = \int_0^1 \int_0^1 x^2(x + y)\, dx\, dy = 5/12$$
$$\overline{Y^2} = 5/12$$
$$\sigma_x^2 = \sigma_y^2 = 5/12 - (7/12)^2 = 11/144$$
$$\mu_{11} = E(X - \bar{X})(Y - \bar{Y}) = E(XY) - \bar{X}\bar{Y} = 1/3 - 49/144 = -1/144$$
$$\rho = \frac{\mu_{11}}{\sigma_x \sigma_y} = \frac{-1/144}{11/144} = -\frac{1}{11}$$

6-8. Moments of Some Common Distribution Functions. In this section we should like to apply some of the material of the preceding sections to some familiar distributions.

Binomial Distribution. The random variable X assumes the values

$$x_k = k \qquad k = 0, 1, 2, \ldots, n$$

with the probability

$$p_k = \binom{n}{k} p^k (1 - p)^{n-k}$$

The first and second moments are

$$m_1 = E(X) = \sum_{k=0}^{n} k \frac{n!}{k!(n - k)!} p^{k-1} p (1 - p)^{n-k}$$
$$= np \sum_{k=1}^{n} \frac{(n - 1)!}{(k - 1)!(n - k)!} p^{k-1} (1 - p)^{n-k} \tag{6-49}$$
$$m_1 = np[p + (1 - p)]^{n-1} = np$$

$$m_2 = \sum_{k=0}^{n} k^2 \binom{n}{k} p^k (1 - p)^{n-k}$$
$$= np \sum_{k=1}^{n} k \binom{n - 1}{k - 1} p^{k-1} (1 - p)^{n-k}$$
$$m_2 = np \left[\sum_{k=1}^{n} (k - 1) \binom{n - 1}{k - 1} p^{k-1} (1 - p)^{n-k} \right. \tag{6-50}$$
$$\left. + \sum_{k=1}^{n} \binom{n - 1}{k - 1} p^{k-1} (1 - p)^{n-k} \right]$$
$$m_2 = np[(n - 1)p + 1]$$
$$\sigma^2 = m_2 - m_1^2 = np[(n - 1)p + 1] - (np)^2 = np(1 - p) \tag{6-51}$$

Poisson's Distribution. The random variable X assumes the values

$$x_k = n \qquad k = 0, 1, 2, \ldots, n, \ldots$$

with the probability

$$P_n = \frac{\lambda^n}{n!} e^{-\lambda}$$

$$m_1 = E(X) = e^{-\lambda}\left[0 + \lambda + \frac{\lambda^2}{1!} + \frac{\lambda^3}{2!} + \cdots + \frac{\lambda^n}{(n-1)!} + \cdots\right]$$

$$m_1 = e^{-\lambda} \cdot e^{\lambda} \cdot \lambda = \lambda \tag{6-52}$$

$$m_2 = E(X^2) = e^{-\lambda}\left(0 + 1^2\frac{\lambda}{1} + 2^2\frac{\lambda^2}{2!} + \cdots n^2\frac{\lambda^n}{n!} + \cdots\right) \tag{6-53}$$

In order to compute m_2 in a closed form, note that

$$\frac{n^2}{n!} = \frac{n(n-1) + n}{n!} = \frac{1}{(n-1)!} + \frac{1}{(n-2)!}$$

Thus

$$m_2 = e^{-\lambda}(e^{\lambda}\lambda + e^{\lambda}\lambda^2) = \lambda + \lambda^2 \tag{6-54}$$

The variance and standard deviation are, respectively,

$$\mu_2 = m_2 - m_1^2 = \lambda$$

$$\sigma = \sqrt{\lambda} \tag{6-55}$$

Normal Distribution. The random variable X has a density distribution function

$$f(x) = \frac{1}{\sigma\sqrt{2\pi}} e^{-(x^2/2\sigma^2)}$$

$$m_1 = E(X) = \frac{1}{\sigma\sqrt{2\pi}}\int_{-\infty}^{\infty} xe^{-(x^2/2\sigma^2)}\,dx = -\frac{\sigma}{\sqrt{2\pi}}\left[e^{-(x^2/2\sigma^2)}\right]_{-\infty}^{\infty} = 0 \tag{6-56}$$

$$m_2 = E(X^2) = \frac{1}{\sigma\sqrt{2\pi}}\int_{-\infty}^{\infty} x^2e^{-(x^2/2\sigma^2)}\,dx = \sigma^2 \tag{6-57}$$

$$\mu_2 = m_2 - m_1^2 = \sigma^2$$

$$\sqrt{\mu_2} = \sigma$$

When the distribution is centered at a point with abscissa m, the mean value of the variable is m and its standard deviation is σ. A normal distribution is completely determined by its two parameters m and σ:

$$\frac{1}{\sigma\sqrt{2\pi}} e^{-[(x-m)^2/2\sigma^2]}$$

Cauchy's Distribution. Cauchy's distribution provides an example where a moment is not defined; in fact,

$$f(x) = \frac{1}{\pi(1 + x^2)}$$

$$E(X) = \int_{-\infty}^{\infty} \frac{x}{\pi(1 + x^2)}\, dx = \frac{1}{2\pi} [\log (1 + x^2)]_{-\infty}^{\infty}$$

This integral is not convergent; hence the first moment is not defined.

Bivariate Normal Distribution. The normal bivariate density function was discussed in Sec. 5-11. For convenience of interpretation, here we shall give the normal bivariate in terms of its statistical parameters.

$$f(x,y) = \frac{1}{2\pi\sigma_1\sigma_2(1 - \rho^2)^{1/2}} \exp \left\{ \frac{-1}{2(1 - \rho^2)} \left[\left(\frac{x - m_1}{\sigma_1} \right)^2 - 2\rho \left(\frac{x - m_1}{\sigma_1} \right) \left(\frac{y - m_2}{\sigma_2} \right) + \left(\frac{y - m_2}{\sigma_2} \right)^2 \right] \right\}$$

In this equation the parameters m_1, m_2, σ_1, σ_2, and ρ have direct statistical interpretations; for example,

$$\begin{aligned}
\alpha_{10} &= \bar{X} = \int_{-\infty}^{\infty} \int_{-\infty}^{\infty} xf(x,y)\, dx\, dy = m_1 \\
\alpha_{01} &= \bar{Y} = \int_{-\infty}^{\infty} \int_{-\infty}^{\infty} yf(x,y)\, dx\, dy = m_2 \\
\alpha_{20} &= \overline{X^2} = \int_{-\infty}^{\infty} \int_{-\infty}^{\infty} x^2 f(x,y)\, dx\, dy = \sigma_1^2 + m_1^2 \\
\alpha_{02} &= \overline{Y^2} = \int_{-\infty}^{\infty} \int_{-\infty}^{\infty} y^2 f(x,y)\, dx\, dy = \sigma_2^2 + m_2^2 \qquad (6\text{-}58) \\
\mu_{20} &= \int_{-\infty}^{\infty} \int_{-\infty}^{\infty} (x - m_1)^2 f(x,y)\, dx\, dy = \sigma_1^2 \\
\mu_{02} &= \int_{-\infty}^{\infty} \int_{-\infty}^{\infty} (y - m_2)^2 f(x,y)\, dx\, dy = \sigma_2^2 \\
\mu_{11} &= \int_{-\infty}^{\infty} \int_{-\infty}^{\infty} (x - m_1)(y - m_2) f(x,y)\, dx\, dy = \rho\sigma_1\sigma_2
\end{aligned}$$

These relations can be verified by direct computation. m_1 and m_2 are average values of X and Y; σ_1 and σ_2, their respective standard deviations; and ρ, the correlation coefficient.

Binomial Distribution in Two Dimensions. Consider an experiment involving the joint occurrence of two specific events E_1 and E_2.

$$P\{E_1\} = p_1$$
$$P\{E_2\} = p_2$$
$$P\{E_1E_2\} = 0$$

If the experiment is repeated k times, the corresponding probability distribution, that is, the probability of having x times event E_1 and y

times event E_2, is

$$f(x,y) = \frac{k!}{x!y!(k-x-y)!} p_1^x p_2^y (1 - p_1 - p_2)^{k-y-x} \tag{6-59}$$

The different moments and correlation coefficients are found to be*

$$\mu_{11} = -kp_1p_2 \qquad \mu_2 = kp_1(1-p_1) \qquad \mu_{02} = kp_2(1-p_2)$$
$$\rho = -\sqrt{\frac{p_1p_2}{(1-p_1)(1-p_2)}} \tag{6-60}$$

Poisson's Distribution in Two Dimensions. Using the same notation as in the case of binomial distribution in two dimensions, let p_1 and p_2 tend to zero and k tend to infinity while

$$kp_1 \to a$$
$$kp_2 \to b$$

The two-dimensional Poisson distribution is the limit of the two-dimensional binomial distribution.

$$f(x,y) = \frac{a^x b^y}{x!y!} e^{-a-b} \tag{6-61}$$

Example 6-9. Let θ be a random variable uniformly distributed in the interval $-(\pi/2)$ to $+(\pi/2)$. Find the first and the second moment of the function

$$X = g(\theta) = A \sin \theta$$

Solution. The first step is to find the density function for the variable X. According to Eqs. (5-73),

$$f(x) = \frac{\varphi(\theta)}{|d\theta/dx|} = \frac{1/\pi}{|A(1 - x^2/A^2)^{1/2}|} = \frac{1}{\pi(A^2 - x^2)^{1/2}}$$

This density function is shown in Fig. E6-9.

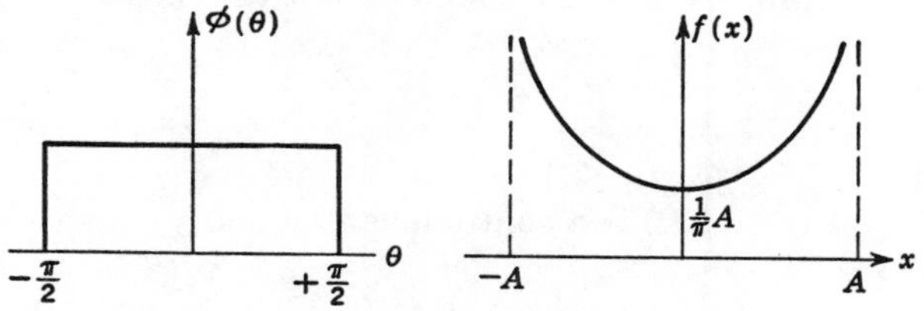

FIG. E6-9

* See, for instance, A. Guldberg, Sur les lois de probabilités et la corrélation, *Ann. Inst. Henri Poincaré*, fascicule II, vol. 5, pp. 159–176, 1935.

The moments of first and second orders are

$$m_1 = E(X) = \int_{-A}^{A} \frac{x\,dx}{\pi(A^2 - x^2)^{1/2}} = 0$$

$$m_2 = E(X^2) = \int_{-A}^{A} \frac{x^2\,dx}{\pi(A^2 - x^2)^{1/2}} = \frac{A^2}{2}$$

$$\sigma = \frac{A}{\sqrt{2}}$$

6-9. Characteristic Function of a Random Variable. The characteristic function of a random variable X is defined as the mathematical expectation of e^{jtX}, where t is a real variable, e the base of the natural logarithm, and $j = \sqrt{-1}$. When the distribution of the random variable is absolutely continuous, one has

$$\phi_x(t) = E(e^{jtX}) = \int_{-\infty}^{\infty} e^{jtx} f(x)\,dx \tag{6-62}$$

$f(x)$ being the probability density function of the variable X. When the random variable X is of the discrete type, the characteristic function is defined by

$$\phi_x(t) = E(e^{jtX}) = \sum_i p_i e^{jtx_i} \tag{6-62a}$$

The characteristic function is always a well-defined function, that is, the integral of Eq. (6-62) always converges. Moreover, it can be proved that the characteristic function uniquely determines a distribution function, in the discrete, the continuous, and other possible cases. The proof is not given here. However, the reader can see that $\phi_x(t)$ is the inverse Fourier transform of $f(x)$, that is, the two functions are interrelated by Fourier integrals.

$$f(x) = \frac{1}{2\pi} \int_{-\infty}^{\infty} e^{-jtx} \phi_x(t)\,dt \tag{6-63}$$

The following properties of the characteristic function are of immediate interest.

$$\begin{array}{ll} \text{(I)} & \phi_x(0) = 1 \\ \text{(II)} & |\phi_x(t)| \le 1 \\ \text{(III)} & \phi_x(t) \text{ is a continuous function} \end{array} \tag{6-64}$$

Property I is self-evident from Eq. (6-62). In order to show the validity of property II, note that

$$\left| \int_{-\infty}^{\infty} e^{jtx} f(x)\,dx \right| \le \int_{-\infty}^{\infty} f(x)\,dx = 1 \tag{6-65}$$

While the inequality (6-65) is self-evident, one may alternatively use the following novel physical reasoning.

Consider a unit circle in a complex plane. We select a number of points on the unit circle with coordinates

$$\cos tx_k \qquad j \sin tx_k \qquad k = 1, 2, \ldots$$

To each point we assign a mass p_k equal to the probability associated with the value x_k of our discrete random variable X,

$$\begin{array}{l} x_1, x_2, \ldots, x_k, \ldots \\ p_1, p_2, \ldots, p_k, \ldots \end{array}$$

Obviously, the center of gravity of these masses located on a convex curve (unit circle) must remain within the convex curve for all values of the real parameter t. But this center of gravity coincides with the point $\phi_x(t)$; thus

$$\phi_x(t) = E(e^{jtX}) = E(\cos tX + j \sin tX) \tag{6-66}$$

$$\phi_x(t) = \sum_{k=1}^{\infty} p_k(\cos tx_k + j \sin tx_k) = \sum_{k=1}^{\infty} p_k \cos tx_k + j \sum_{k=1}^{\infty} p_k \sin tx_k$$

$$|\phi_x(t)| \leq 1 \tag{6-67}$$

Property III requires more detailed mathematical consideration and so is omitted here.

6-10. Characteristic Function and Moment-generating Function of Random Variables. For two independent random variables X, Y and their sum $Z = X + Y$ the following relation is immediately evident:

$$\phi_z(t) = E[e^{jt(X+Y)}] = E(e^{jtX}e^{jtY}) = E(e^{jtX})E(e^{jtY}) \tag{6-68}$$

$$\phi_z(t) = \phi_x(t)\phi_y(t) \tag{6-69}$$

By induction one concludes that the characteristic function of the sum of a number of independent random variables is the product of their characteristic functions.

Next let us compute the characteristic function $\phi_y(t)$ of a random variable Y which is a linear function of another random variable X:

$$Y = aX + b \qquad a, b \text{ real numbers} \tag{6-70}$$

$$\phi_y(t) = E(e^{jtY}) = E[e^{jt(aX+b)}] = E(e^{jtaX}e^{jtb})$$

$$\phi_y(t) = E(e^{jtaX})E(e^{jtb})$$

Finally

$$\phi_y(t) = e^{jbt}\phi_x(at) \tag{6-71}$$

The characteristic functions of some of the common distribution functions will now be derived.

Binomial. The characteristic function of the binomial distribution is

$$\phi_x(t) = \sum_{0}^{n} e^{jkt} \binom{n}{k} p^k (1-p)^{n-k}$$
$$= \sum_{0} \binom{n}{k} (pe^{jt})^k (1-p)^{n-k} \tag{6-72}$$

This obviously is the binomial expansion of

$$[(pe^{jt}) + (1-p)]^n = \phi_x(t) \tag{6-73}$$

Poisson. The characteristic function of the Poisson distribution is

$$\phi_x(t) = \sum_{k=0}^{\infty} e^{jkt} e^{-\lambda} \frac{\lambda^k}{k!}$$
$$\phi_x(t) = e^{-\lambda} \sum_{k=0}^{\infty} \frac{(\lambda e^{jt})^k}{k!} \tag{6-74}$$
$$\phi_x(t) = e^{-\lambda} e^{\lambda e^{jt}} = e^{\lambda(e^{jt}-1)} \tag{6-75}$$

Normal. The characteristic function of the standardized normal distribution is

$$\phi_x(t) = \int_{-\infty}^{\infty} \frac{e^{jtx}}{\sqrt{2\pi}} e^{-(x^2/2)}\, dx \tag{6-76}$$
$$\phi_x(t) = \frac{1}{\sqrt{2\pi}} \int_{-\infty}^{\infty} e^{jtx - x^2/2}\, dx$$
$$= \frac{1}{\sqrt{2\pi}} \int_{-\infty}^{\infty} e^{-\frac{1}{2}(x^2 - 2jxt - t^2 + t^2)}\, dx$$
$$= \frac{1}{\sqrt{2\pi}} \exp\left(-\frac{t^2}{2}\right) \int_{-\infty}^{\infty} e^{-\frac{1}{2}(x-jt)^2}\, dx$$
$$\phi_x(t) = e^{-(t^2/2)} \tag{6-77}$$

Example 6-10. Find the characteristic function of a standardized random variable

$$Y = \frac{X - m}{\sigma}$$

where $m = \bar{X}$ and $\sigma = \sigma_x$.

Solution. Applying Eq. (6-71), one finds

$$\phi_y(t) = e^{-j(mt/\sigma)} \phi_x\left(\frac{t}{\sigma}\right)$$

The different-order moments of a random variable can be obtained directly from the expansion of its characteristic function.

$$e^{jtx} = 1 + j\frac{tx}{1!} + j^2 \frac{t^2x^2}{2!} + \cdots \tag{6-78}$$

Applying this equation to (6-62), one obtains

$$\phi_x(t) = \int_{-\infty}^{\infty} f(x)\,dx + \frac{jt}{1!}\int_{-\infty}^{\infty} xf(x)\,dx + \frac{j^2t^2}{2!}\int_{-\infty}^{\infty} x^2f(x)\,dx + \cdots \tag{6-79}$$

While the characteristic function $\phi_x(t)$ always exists, the above expansion is not always possible. Also it is to be noted that, even if all the moments exist, the above expansion is valid only in its region of convergence. Subject to these restrictions, the rth-order moment is

$$m_r = \left[\frac{1}{j^r}\frac{d^r\phi_x(t)}{dt^r}\right]_{t=0} \tag{6-80}$$

The expansion of the characteristic function gives

$$\phi_x(t) = \phi_x(0) + \phi_x'(0)\frac{t}{1!} + \phi_x''(0)\frac{t^2}{2!} + \cdots + \phi_x^{(n)}\frac{t^n}{n!} + \cdots$$

Moment-generating functions. While the different moments have been derived directly from the characteristic function, they could alternatively be obtained from the moment-generating function. The latter function is defined for a discrete and a continuous random variable, respectively, as

$$\psi_x(t) = E(e^{tX}) = \sum_i p_i e^{tx_i} \tag{6-81}$$

$$\psi_x(t) = E(e^{tX}) = \int_{-\infty}^{\infty} e^{tx}f(x)\,dx \tag{6-81a}$$

Note that $\psi(jt) = \phi(t)$.

In a manner similar to the derivation of Eq. (6-80), one can see that

$$\left[\frac{d^r\psi_x(t)}{dt^r}\right] = E\left(\frac{\partial^r}{\partial t^r}e^{tx}\right) = E(x^re^{tx})$$

$$m_r = E(x^r) = \left[\frac{d^r\psi_x(t)}{dt^r}\right]_{t=0} \tag{6-82}$$

The defining equation of the characteristic function can be directly extended to cover multivariate distributions. For instance, for a bivariate distribution we have

$$\phi(u,v) = E[e^{j(uX+vY)}] = \int_{-\infty}^{\infty}\int_{-\infty}^{\infty} e^{j(ux+vy)}f(x,y)\,dx\,dy \tag{6-83}$$

The power expansion of this function will lead to the calculation of different-order moments. For example,

$$\alpha_{rl} = E(X^rY^l) = \left[\frac{\partial^{r+l}}{\partial u^r\,\partial v^l}\phi(u,v)\right]_{u=v=0}\frac{1}{j^{r+l}} \tag{6-84}$$

Example 6-11. Find the first and second moments of a Poisson distribution.
Solution

$$\psi_x(t) = E(e^{tX}) = \sum_{x=0}^{\infty} e^{tx} \frac{e^{-\lambda}\lambda^x}{x!} \qquad x = 0, 1, 2, \ldots$$

$$= e^{-\lambda} \sum_{0}^{\infty} \frac{(\lambda e^t)^x}{x!} = e^{-\lambda} e^{\lambda e^t}$$

$$\frac{d\psi_x(t)}{dt} = e^{-\lambda}\lambda e^t e^{\lambda e^t}$$

$$\frac{d^2\psi_x(t)}{dt^2} = e^{-\lambda}\lambda e^t e^{\lambda e^t}(1 + \lambda e^t)$$

$$m_1 = \left[\frac{d\psi_x(t)}{dt}\right]_{t=0} = \lambda$$

$$m_2 = \left[\frac{d^2\psi_x(t)}{dt^2}\right]_{t=0} = \lambda(1 + \lambda)$$

$$\sigma^2 = \lambda(1 + \lambda) - \lambda^2 = \lambda$$

6-11. Density Functions of the Sum of Two Random Variables. In a number of problems which occur in practice one is interested in finding the density function for the sum of two independent random variables X and Y when the individual density functions, say $f_1(x)$ and $f_2(y)$, are known. From the fact that the two variables are independent, we conclude that the density of the two-dimensional random variables (X,Y) is given by

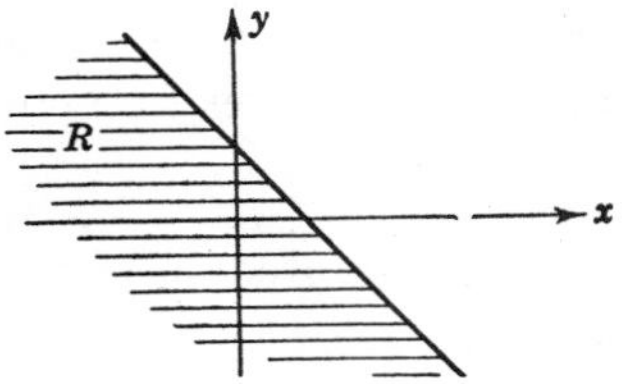

FIG. 6-3. Determination of the probability density for the sum of two independent variables.

$$f(x,y) = f_1(x)f_2(y) \tag{6-85}$$

Our problem is first to find the probability distribution function for the variable $X + Y$, that is,

$$P\{X + Y \leq t\}$$

This is done by first drawing the line

$$x + y = t \tag{6-86}$$

in the (X,Y) plane with t as a parameter (Fig. 6-3). Now the probability under consideration can be obtained by integrating the density function [Eq. (6-85)] over the shaded region R.

$$P\{X + Y \leq t\} = \iint_R f(x,y)\,dx\,dy = \int_{-\infty}^{\infty}\int_{-\infty}^{t-y} f_1(x)f_2(y)\,dx\,dy \tag{6-87}$$

$$P\{X + Y \leq t\} = \int_{-\infty}^{\infty} F_1(t - y)f_2(y)\,dy \tag{6-88}$$

where $F_1(x)$ is assumed to be the CDF of the variable X.

The desired density function is obtained by taking the derivative of the integral with respect to t:

$$g(t) = \int_{-\infty}^{\infty} f_1(t - y) f_2(y)\, dy \tag{6-89}$$

An alternative method of obtaining the density function for the sum of two independent random variables involves the use of the concept of the characteristic function and its relation with the Fourier transform. In fact,

$$\text{Density for } X = f_1(x) = \frac{1}{2\pi} \int_{-\infty}^{\infty} e^{-jtx} \phi_x(t)\, dt \tag{6-90}$$

$$\text{Density for } Y = f_2(y) = \frac{1}{2\pi} \int_{-\infty}^{\infty} e^{-jty} \phi_y(t)\, dt \tag{6-91}$$

$$\text{Density for } Z = f(z) = \frac{1}{2\pi} \int_{-\infty}^{\infty} e^{-jtz} \phi_z(t)\, dt$$

$$= \frac{1}{2\pi} \int_{-\infty}^{\infty} e^{-jtz} \phi_x(t) \phi_y(t)\, dt \tag{6-92}$$

It is to be noted that the relationship between a density function and its associated characteristic function is that of the familiar Fourier integral. Having this in mind, we recall the convolution theorem of the Fourier (or Laplace) transforms. According to the theorem for Fourier transforms, the inverse Fourier (or Laplace) transform of the product of two functions corresponds to the convolution integrals of the two inverse functions. That is, if

$$\begin{aligned} f_1(x) &= \mathfrak{F}[\phi_x(t)] \\ f_2(y) &= \mathfrak{F}[\phi_y(t)] \end{aligned} \tag{6-93}$$

then

$$f(z) = \mathfrak{F}[\phi_x(t)\phi_y(t)] = f_1(x) * f_2(y) \tag{6-94}$$

where

$$f_1(x) * f_2(y) = \int_{-\infty}^{\infty} f_1(z - y) f_2(y)\, dy \tag{6-95}$$

The result of Eq. (6-95) can be generalized in order to obtain the probability density function of the sum of a finite number of independent random variables. Let

$$X = \sum_1^n X_k$$

where the X_k are all independent. The probability density of X is

$$f(x) = f_1(x_1) * f_2(x_2) * \cdots * f_n(x_n) \tag{6-96}$$

If every variable is normally distributed, then their sum will also be normally distributed. The same is true when all variables have distributions of the Poisson or binomial type (see Probs. 6-14 and 6-7).

Example 6-12. X and Y are two independent random variables with standard normal distributions. Derive the distribution of their sum.

Solution. According to Eq. (6-94), the probability density of $Z = X + Y$ is

$$g(z) = f_1(x) * f_2(x)$$

$$\begin{aligned} g(z) &= \int_{-\infty}^{\infty} \frac{1}{\sqrt{2\pi}} \left\{ \exp\left[\frac{-(z-x)^2}{2}\right]\right\} \frac{1}{\sqrt{2\pi}} \left[\exp\left(-\frac{x^2}{2}\right)\right] dx \\ &= \frac{1}{2\pi} \int_{-\infty}^{\infty} \exp\left[-\left(x - \frac{z}{2}\right)^2 - \frac{z^2}{4}\right] dx \\ &= \frac{1}{2\sqrt{\pi}} e^{-z^2/4} \frac{1}{\sqrt{\pi}} \int_{-\infty}^{\infty} \exp\left[-\left(x - \frac{z}{2}\right)^2\right] dx \end{aligned}$$

Make the change of variable:

$$x - \frac{z}{2} = \frac{v}{\sqrt{2}}$$

$$\frac{1}{\sqrt{\pi}} \int_{-\infty}^{\infty} \exp\left(-\frac{v^2}{2}\right) \frac{dv}{\sqrt{2}} = 1$$

Thus

$$g(z) = \frac{1}{2\sqrt{\pi}} e^{-(z^2/4)}$$

The density distribution of Z is also a normal distribution with zero mean but a standard deviation of $\sqrt{2}$. Note that the mean and standard deviation could have been predicted by applying Eq. (6-23*a*).

$$E(Z) = E(X) + E(Y) = 0$$
$$E(Z^2) = E(X^2) + E(Y^2) = 2$$

Example 6-13. A gun fires at a target point which is assumed to be the center of the rectangular coordinate system. Let (x,y) be the coordinate of the point at which the bullet hits, and assume that the associated random variables X and Y are independent. We furthermore assume that X and Y are normally distributed with 0 mean and standard deviation 1. Find the probability distribution of the random variable

$$R^2 = X^2 + Y^2$$

Solution. It was shown in Sec. 5-11 that the density of X^2 is

$$f_1(x^2) = \frac{1}{x} \frac{e^{-x^2/2}}{\sqrt{2\pi}}$$

Application of the rule of convolution yields

$$\begin{aligned} f(r^2) &= \int_0^r \frac{1}{\sqrt{2\pi z}} e^{-z/2} \frac{1}{\sqrt{2\pi(r-z)}} e^{-(r-z)/2}\, dz \\ &= \frac{1}{2\pi} e^{-r/2} \int_0^r \frac{dz}{\sqrt{z(r-z)}} \end{aligned}$$

From an integral table one finds

$$\int_0^r \frac{1}{\sqrt{z(r-z)}}\, dx = \sin^{-1} \frac{2z - r}{r}\Big]_0^r = \pi$$

Finally we find

$$f(r^2) = \tfrac{1}{2} e^{-(r/2)} \qquad r \geq 0$$

Example 6-14. The random variables X and Y are uniformly distributed over the real interval zero to a ($a > 0$). Find the density distribution of

(a) $$Z = X + Y$$
(b) $$Z = X - Y$$

Solution

(a) Since it is likely that most readers are more familiar with the ordinary (one-sided) Laplace transform than with the Fourier transform, in this example we shall change the parameter t to s, which is the symbol commonly used in engineering texts in conjunction with Laplace transforms. We also assume that the readers are familiar

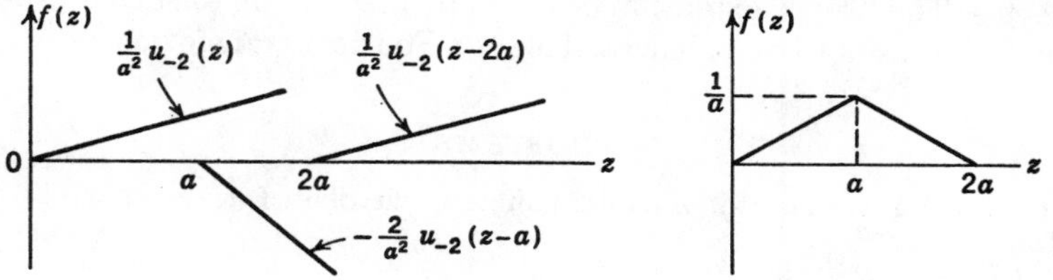

FIG. E6-14

with the concept of a singularity function, such as is exemplified by the unit step $u_{-1}(t)$ and unit ramp $u_{-2}(t)$. Based on this notation,

$$f_1(x) = \frac{1}{a}u_{-1}(x) - \frac{1}{a}u_{-1}(x-a) \qquad \text{density of } X$$

$$f_2(y) = \frac{1}{a}u_{-1}(y) - \frac{1}{a}u_{-1}(y-a) \qquad \text{density of } Y$$

$$\psi_x(s) = \mathcal{L}\frac{1}{a}u_{-1}(x) - \mathcal{L}\frac{1}{a}u_{-1}(x-a)$$

$$\psi_x(s) = \frac{1}{as} - \frac{1}{as}e^{-as}$$

$$\psi_y(s) = \frac{1}{as} - \frac{1}{as}e^{-as}$$

Since the two variables are independent, Eq. (6-69) gives

$$\psi_z(s) = \psi_x(s)\cdot\psi_y(s) = \frac{1}{a^2s^2}(1 - e^{-as})^2$$

The desired density function is the inverse Laplace transform of $\psi_z(s)$; thus,

$$f(z) = \mathcal{L}^{-1}\frac{1}{a^2s^2}(1 - 2e^{-as} + e^{-2as}) = \frac{1}{a^2}[u_{-2}(z) - 2u_{-2}(z-a) + u_{-2}(z-2a)]$$

The graphical representation of $f(z)$ follows directly from the definition of the unit step and unit ramp. This is illustrated in Fig. E6-14.

(b)

$$\psi_z(s) = -\frac{1}{a^2s^2}(1 - e^{-as})(1 - e^{as})$$

$$= -\frac{1}{a^2s^2}(2 - e^{-as} - e^{as})$$

$$f(z) = -\frac{1}{a^2}[2u_{-2}(z) - u_{-2}(z-a) - u_{-2}(z+a)]$$

It should be noted that for the convenience of calculation we have interchanged the defining terms of the Laplace transform with its inverse. Also note that the density function is similar to that of the previous case but displaced by a distance $-a$ along the axis of the random variable.

The simplicity of computation here lies in the symbolism used for describing the rectangular-shape function in a closed form through the use of singularity functions. Such methods are commonly practiced in electrical engineering problems. The reader, however, is warned against applying such techniques to density functions that are defined for $-\infty$ to $+\infty$. The proper extension of the above technique to such a problem requires the use of two-sided Laplace (or Fourier) transforms.

PROBLEMS

6-1. X and Y are random variables uniformly distributed in the interval 0 to 1. Find

(*a*) $P\{X < Y < +1\}$
(*b*) $P\{X \leq Y < -1\}$
(*c*) $P\{|X - Y| < 1\}$
(*d*) $P\{|X - Y| < \frac{1}{2}\}$
(*e*) $P\{X + Y < \frac{1}{2}\}$
(*f*) $P\{XY < \frac{1}{2}\}$

6-2. The random variable (X,Y) assumes only the three sets of values (0,0), (0,1), and (1,1), with equal probabilities. Find

$$[E(X - \tfrac{1}{4})(Y - \tfrac{3}{4})]$$

6-3. Let x have a Poisson distribution, that is,

$$f(x) = \frac{a^x}{x!} e^{-a} \qquad x = 0, 1, 2, \ldots$$

Show that

$$E(x^2) = aE(x + 1)$$

6-4. Let X and Y be independent random variables each with uniform distribution between 0 and 1. Find

(*a*) $E(X + Y)$
(*b*) $E[(X + Y)^2]$

6-5. Let X have a probability density

$$f(x) = \tfrac{1}{2}e^{-|x|}$$

Find the expectation and variance of X.

6-6. Let X and Y be normally distributed independent random variables, each with zero mean and unity variance. Find the expectation and variance of $\sqrt{x^2 + y^2}$.

6-7. Show that the sum of two independent binomial variables with the parameters (n_1,p) and (n_2,p) is also a binomial variable.

6-8. Same question for the sum of two independent random variables having Poisson distributions with means λ_1 and λ_2.

6-9. Let S be a random electric voltage varying between 0 and 1 volt, with a uniform probability distribution. The signal S is perturbed by an additive independent random noise N having a uniform distribution between 0 and 2 volts.

(a) Determine $\bar{X}$.

(b) Determine the average power when the voltage X is applied to a resistor of 2 ohms.

6-10. A number X is chosen at random from the integers 1, 2, 3, . . . , n; find $\bar{X}$ and its standard deviation.

6-11. A die is loaded in such a way that the probability of getting x is proportional to x ($x = 1, 2, 3, 4, 5, 6$). Find the smallest number of throwings n for which

$$P\{|X - E(X)| > \tfrac{1}{9}\} \leq 0.001$$

where X stands for the average of the sum in n throwings.

6-12. Two points A and B are chosen at random on the circumference of a circle with center C and radius R. Let X be the area of the triangle ABC. Find $\bar{X}$.

6-13. Let X and Y be standardized normally distributed random variables. Find the probability density of X/Y.

6-14. The independent variables X_k all have distributions of the Poisson type ($k = 1, 2, \ldots, n$).

$$X = X_1 + X_2 + \cdots + X_n$$

(a) Find the characteristic function of X_k.

(b) Find the characteristic function of X.

(c) Determine the distribution of X.

6-15. The probability density distribution

$$f(x) = \frac{1}{\Gamma(r/2)2^{r/2}} x^{r/2-1}e^{-x/2} \qquad 0 < x < \infty$$

$$f(x) = 0 \qquad \text{elsewhere}$$

is called a chi-square distribution. Γ stands for the familiar gamma function, that is,

$$\Gamma(K) = \int_0^\infty x^{K-1}e^{-x}\,dx$$

(a) Find the moment-generating function for the chi-square distribution.

(b) Find the first and the second moment.

6-16. Let

$$V_1 = \sin 2\pi F \qquad V_2 = \cos 2\pi F$$

where F is a random variable uniformly distributed between 0 and 1. Show that the two random voltages V_1 and V_2 are dependent but not correlated.

6-17. The joint moment-generating function of two random variables X and Y is given:

$$\psi_{x,y}(t,s) = [a(e^{t+s} + 1) + b(e^t + e^s)]^2 \qquad a > 0 \qquad b > 0 \qquad a + b = \tfrac{1}{2}$$

Determine

$E(X)$, $E(Y)$, var X, var Y, and correlation coefficient

6-18. From the moment-generating functions described in this chapter, derive the standard deviation of binomial, Poisson, uniform, and normal distributions in one and two dimensions.

6-19. The random variable X is normally distributed with $(0, \sigma)$, and the random variable Y is distributed uniformly in the interval $[-\pi, +\pi]$. Find the probability density of

$$Z = X \sin Y$$

A solution in closed form may be found in Pugachev (Chap. 5).

CHAPTER 7

NORMAL DISTRIBUTIONS AND LIMIT THEOREMS

This chapter is primarily concerned with the probability distribution of functions of a large number of independent random variables. The main results of the chapter are contained in two basic theorems with frequent applications. These are the law of large numbers and the central-limit theorem. Prior to the development of these theorems, we present in some detail the multidimensional gaussian distribution. The latter distribution is of particular significance in dealing with systems of several independent random variables.

7-1. Bivariate Normal Considered as an Extension of One-dimensional Normal Distribution. In this section we study the normal bivariate distribution as an extension of the one-dimensional normal distribution:

$$f(x) = \frac{1}{\sqrt{2\pi}\,\sigma_1} \exp\left[-\frac{(x-a_1)^2}{2\sigma_1^2}\right] \tag{7-1}$$

This density function is a symmetric function about the mean value a_1. The exponent is negative for all values of the real variable x except for $x = a_1$. This can be alternatively expressed by referring to the mathematical term *positive definite*. We say that $[(x - a_1)/\sqrt{2}\,\sigma_1]^2$ is a positive definite form in $x - a_1$. This implies that for all real values of the parameter $y = x_1 - a$ the above form is positive except for $y = 0$.

Some familiarity with linear algebra should suggest that as a generalization of the above concept, for a bivariate, the exponent will be of the form

$$f(x_1,x_2) = C \exp\left(-\tfrac{1}{2}Q\right) \tag{7-2}$$

where Q is a positive definite quadratic form in variables $(x_1 - a_1)$ and $(x_2 - a_2)$, that is,

$$Q = A_{11}(x_1 - a_1)^2 + 2A_{12}(x_1 - a_1)(x_2 - a_2) + A_{22}(x_2 - a_2)^2 \tag{7-3}$$

The real coefficients A_{ij} $(i, j = 1, 2)$ should be such that Q remains positive for all real values of x_1 and x_2, except for $x_1 - a_1 = x_2 - a_2 = 0$. The expression in Eq. (7-3), which is a quadratic form in variables $y_1 = x_1 - a_1$ and $y_2 = x_2 - a_2$, can be written as

$$Q(y_1,y_2) = \sum_{i,j=1}^{2} A_{ij}y_iy_j \qquad A_{ij} = A_{ji} \tag{7-4}$$

$$Q(y_1,y_2) = A_{11}y_1^2 + 2A_{12}y_1y_2 + A_{22}y_2^2 \tag{7-5}$$

The form shown in Eqs. (7-4) and (7-5) is the proper generalization of the one variable form $Q(y) = Ay^2$. The reader who wishes to acquire introductory information about quadratic forms may refer to texts on linear algebra and matrices.*

The coefficients A_{ij} can be conveniently arranged in a matrix form:

$$[A] = \begin{bmatrix} A_{11} & A_{12} \\ A_{12} & A_{22} \end{bmatrix} \tag{7-6}$$

The quadratic form associated with the real matrix can be written as

$$Q(y_1,y_2) = [y_1,y_2]\begin{bmatrix} A_{11} & A_{12} \\ A_{12} & A_{22} \end{bmatrix}\begin{bmatrix} y_1 \\ y_2 \end{bmatrix} = Y^tAY \tag{7-7}$$

where
$$Y = \begin{bmatrix} y_1 \\ y_2 \end{bmatrix}$$

and Y^t is the transposed Y matrix.

The standard texts on matrices point out that the necessary and sufficient conditions for a quadratic form to be positive definite are the positiveness of all leading principal minors of the associated matrix A, that is,

$$\begin{aligned} &A_{11} > 0 \\ &A_{22} > 0 \\ &|A| = A_{11}A_{22} - A_{12}^2 > 0 \end{aligned} \tag{7-8}$$

Next, one should examine Eq. (7-2) to justify the requirement for a two-dimensional density distribution. A comparison with Eq. (5-58) yields

$$\frac{1}{2}A_{11} = \frac{1}{a^2} \qquad \frac{1}{2}A_{12} = -\frac{1}{k} \qquad \frac{1}{2}A_{22} = \frac{1}{b^2} \tag{7-9}$$

$$C = \frac{1}{\pi abk}\sqrt{k^2 - a^2b^2} = \frac{\sqrt{|A|}}{2\pi}$$

$$f(x_1,x_2) = \frac{\sqrt{A}}{2\pi}\exp\left\{-\frac{1}{2}[A_{11}(x_1-a_1)^2 + 2A_{12}(x_1-a_1)(x_2-a_2) + A_{22}(x_2-a_2)^2]\right\} \tag{7-10}$$

In the above form one identifies $1/A_{11}(1-\rho^2)$ as the variance of X_1, $1/A_{22}(1-\rho^2)$ as the variance of X_2, and $-\rho^2/A_{12}(1-\rho^2)$ as the covari-

* See also F. M. Reza and Samuel Seely, "Modern Network Analysis," chap. 3, McGraw-Hill Book Company, Inc., New York, 1959.

ance between X_1 and X_2. The parameters a_1 and a_2 can be identified with the means of X_1 and X_2, respectively. The final result of this normalization procedure is

$$f(x_1,x_2) = \frac{1}{2\pi\sigma_1\sigma_2\sqrt{1-\rho^2}} \exp\left\{\frac{-1}{2(1-\rho^2)}\left[\frac{(x_1-a_1)^2}{\sigma_1^2} - 2\rho\frac{(x_1-a_1)(x_2-a_2)}{\sigma_1\sigma_2} + \frac{(x_2-a_2)^2}{\sigma_2^2}\right]\right\} \tag{7-11}$$

$$\sigma_1^2 = \frac{A_{22}}{|A|} \qquad \sigma_2^2 = \frac{A_{11}}{|A|} \qquad \rho = \frac{-A_{12}}{\sqrt{A_{11}A_{22}}} \tag{7-12}$$

This is the density function for the normal bivariate random variable in a normalized symmetric form exhibiting all its pertinent statistical parameters. The coefficient $\rho\sigma_1\sigma_2$ is the covariance of the two one-dimensional random variables X_1 and X_2, and ρ their correlation coefficient. In a particular case when the two random variables X_1 and X_2 are statistically independent (for instance, when two independent effects of one experiment are considered), $\rho = 0$ and

$$f(x_1,x_2) = \frac{1}{\sqrt{2\pi}\,\sigma_1}\exp\left[-\frac{(x_1-a_1)^2}{2\sigma_1^2}\right]\frac{1}{\sqrt{2\pi}\,\sigma_2}\exp\left[-\frac{(x_2-a_2)^2}{2\sigma_2^2}\right] \tag{7-13}$$

This result is of some interest in our subsequent work. It states that, when the two sampling variables of a normal bivariate are mutually independent, the two-dimensional normal distribution reduces to the product of two distributions of single variables.

7-2. Multinormal Distribution. The procedure developed in the previous section can be directly generalized to the case of n-dimensional normal distributions. The n-dimensional normal density function is of the form

$$f(x_1,x_2, \ldots ,x_n) = C_n \exp\left[-\tfrac{1}{2}Q(x_1,x_2, \ldots ,x_n)\right] \tag{7-14}$$

C_n is an appropriate constant and Q a quadratic polynomial in

$$y_K = x_K - a_K$$

that is,

$$Q(y_1,y_2, \ldots ,y_n) = \sum_{i,j=1}^{n} A_{ij}(x_i - a_i)(x_j - a_j) = \sum_{i,j=1}^{n} A_{ij}y_iy_j \tag{7-15}$$

The matrix A being a real symmetric matrix,

$$[A] = \begin{bmatrix} A_{11} & A_{12} & A_{13} & \cdots & A_{1n} \\ A_{21} & A_{22} & A_{23} & \cdots & A_{2n} \\ \cdots & \cdots & \cdots & \cdots & \cdots \\ A_{n1} & A_{n2} & A_{n3} & \cdots & A_{nn} \end{bmatrix} \tag{7-16}$$

the quadratic form can be written as

$$Q(y_1,y_2, \ldots ,y_n) = [y_1,y_2, \ldots ,y_n] \begin{bmatrix} A_{11} & A_{12} & \cdots & A_{1n} \\ A_{21} & A_{22} & \cdots & A_{2n} \\ \cdots & \cdots & \cdots & \cdots \\ A_{n1} & A_{n2} & \cdots & A_{nn} \end{bmatrix} \begin{bmatrix} y_1 \\ y_2 \\ \cdots \\ y_n \end{bmatrix} \tag{7-17}$$

$$Q = Y^tAY \tag{7-18}$$

where

$$Y = \begin{bmatrix} y_1 \\ y_2 \\ \cdots \\ y_n \end{bmatrix} \tag{7-19}$$

and Y^t is its transpose.

As a generalization of the concept of Eqs. (7-1) and (7-10), Q must remain positive for all nonzero real values of $y_1, y_2, \ldots, y_n$. This requirement is satisfied if, and only if, A is a positive definite matrix, that is, all its principal minors are nonnegative.

$$A_{11} > 0 \qquad \begin{vmatrix} A_{11} & A_{12} \\ A_{21} & A_{22} \end{vmatrix} > 0$$

$$\begin{vmatrix} A_{11} & A_{12} & A_{13} \\ A_{21} & A_{22} & A_{23} \\ A_{31} & A_{32} & A_{33} \end{vmatrix} > 0 \tag{7-20}$$

$$\cdots\cdots\cdots\cdots$$

$$\text{Determinant of } [A] > 0$$

In order that Eq. (7-14) represent a normal distribution, we must have

$$\int_{-\infty}^{\infty} \int_{-\infty}^{\infty} \cdots \int_{-\infty}^{\infty} f(x_1,x_2, \ldots ,x_n)\, dx_1\, dx_2 \cdots dx_n = 1 \tag{7-21}$$

$$C_n \int_{-\infty}^{\infty} \int_{-\infty}^{\infty} \cdots \int_{-\infty}^{\infty} \exp\left[-\tfrac{1}{2}Q(y_1,y_2, \ldots ,y_n)\right] dy_1\, dy_2 \cdots dy_n = 1 \tag{7-22}$$

The value of C_n can be determined from Eq. (7-22) for any given matrix $[A]$ and set of real numbers $a_1, a_2, \ldots, a_n$. The detailed calculation of C_n requires space which is not presently available. The interested reader is referred to Cramèr, Laning and Battin, or Wilks. It can be shown that the constant C_n has the value

$$C_n = \frac{\sqrt{|A|}}{(2\pi)^{n/2}} \tag{7-23}$$

$$\frac{\sqrt{|A|}}{(2\pi)^{n/2}} \int_{-\infty}^{\infty} \int_{-\infty}^{\infty} \cdots \int_{-\infty}^{\infty} \exp\left[-\tfrac{1}{2} \sum_{i,j} A_{ij} (x_i - a_i)(x_j - a_j)\right] dx_1 \cdots dx_n = 1 \tag{7-24}$$

The average value of the variable X_k is a_k. The variance of X_k and the covariance of X_i and X_j are found to be (Wilks)

$$\sigma_k^2 = E[(X_k - a_k)^2] = \frac{\text{cofactor of } A_{kk}}{|A|} \tag{7-25}$$

$$\text{Covariance of } X_i \text{ and } X_j = E[(X_i - a_i)(X_j - a_j)] = \frac{\text{cofactor of } A_{i,j}}{|A|} \qquad i \neq j \tag{7-26}$$

In the particular case, when all pertinent random variables X_1, X_2, . . . , X_n are mutually independent, we have

$$\text{Covariance of } X_i \text{ and } X_j = 0 \qquad A_{ij} = 0 \quad i \neq j \tag{7-27}$$

that is, A is a diagonal matrix. The variances become

$$\sigma_k^2 = \frac{1}{A_{kk}} \qquad k = 1, 2, \ldots, n \tag{7-28}$$

The normal multivariate density of n mutually independent random variables is given by

$$\begin{aligned} f(x_1,x_2, \ldots ,x_n) &= \frac{\sqrt{|A|}}{(2\pi)^{n/2}} \exp\left[-\frac{1}{2}\sum_{k=1}^{n} A_{kk}(x_k - a_k)^2\right] \\ &= \frac{1}{(2\pi)^{n/2}} \frac{1}{\sigma_1\sigma_2 \cdots \sigma_n} \exp\left[-\frac{1}{2}\sum_{k=1}^{n}\left(\frac{x_k - a_k}{\sigma_k}\right)^2\right] \\ &= \prod_{k=1}^{n} \frac{1}{\sqrt{2\pi}\,\sigma_k} \exp\left[-\frac{1}{2}\left(\frac{x_k - a_k}{\sigma_k}\right)^2\right] \end{aligned} \tag{7-29}$$

That is, in the case of mutually independent normal variables the joint density distribution is the product of n one-dimensional normal distributions.

7-3. Linear Combination of Normally Distributed Independent Random Variables. The object of this section is to exhibit a most useful property of gaussian distributions. It will be shown that any random variable consisting of the linear combination of several normally distributed independent random variables has itself a normal distribution.

Let X_1 and X_2 be independent random variables with normally distributed density functions with zero means and standard deviations σ_1 and σ_2:

$$f_1(x_1) = \frac{1}{\sqrt{2\pi}\,\sigma_1} e^{-x_1^2/2\sigma_1^2} \tag{7-30}$$

$$f_2(x_2) = \frac{1}{\sqrt{2\pi}\,\sigma_2} e^{-x_2^2/2\sigma_2^2} \tag{7-31}$$

Our problem is to find the density function for the random variable

$$Y = a_1X_1 + a_2X_2 \tag{7-32}$$

where a_1 and a_2 are real numbers. According to the rules of transformation of variables (Sec. 5-11), the density functions for random variables $Y_1 = a_1X_1$ and $Y_2 = a_2X_2$ are, respectively,

$$\frac{1}{\sqrt{2\pi}\,\alpha_1} \exp\left(-\frac{1}{2}\frac{y_1^2}{\alpha_1^2}\right) \qquad \alpha_1 = a_1\sigma_1 \tag{7-33}$$

and

$$\frac{1}{\sqrt{2\pi}\,\alpha_2} \exp\left(-\frac{1}{2}\frac{y_2^2}{\alpha_2^2}\right) \qquad \alpha_2 = a_2\sigma_2 \tag{7-34}$$

The density function for the random variable

$$Y = Y_1 + Y_2 \tag{7-35}$$

can be obtained by convolution, as described in Chap. 6, since Y_1 and Y_2 are independent variables. Thus,

$$\rho(y) = \int_{-\infty}^{+\infty} f_1(y - y_2) \cdot f_2(y_2)\,dy_2$$

where $f_1(y_1)$ and $f_2(y_2)$ are the densities of Y_1 and Y_2, respectively.

$$\begin{aligned}
\rho(y) &= \int_{-\infty}^{+\infty} \frac{1}{2\pi\alpha_1\alpha_2} \exp\left[-\frac{(y-y_2)^2}{2\alpha_1^2}\right] \exp\left(-\frac{y_2^2}{2\alpha_2^2}\right) dy_2 \\
&= \int_{-\infty}^{+\infty} \frac{1}{2\pi\alpha_1\alpha_2} \exp\left[-\frac{(\alpha_1^2+\alpha_2^2)y_2^2 - 2\alpha_2^2y_2y + \alpha_2^2y^2}{2\alpha_1^2\alpha_2^2}\right] dy_2 \\
&= \int_{-\infty}^{+\infty} \frac{1}{2\pi\alpha_1\alpha_2} \exp\left\{-\frac{\alpha_1^2+\alpha_2^2}{2\alpha_1^2\alpha_2^2}\left(y_2 - \frac{\alpha_2^2y}{\alpha_1^2+\alpha_2^2}\right)^2\right. \\
&\qquad\qquad \left. - y^2\left[\frac{1}{2\alpha_1^2} - \frac{\alpha_2^2}{2\alpha_1^2(\alpha_1^2+\alpha_2^2)}\right]\right\} dy_2 \\
&= \int_{-\infty}^{+\infty} \left\{\frac{\sqrt{\alpha_1^2+\alpha_2^2}}{\sqrt{2\pi}\,\alpha_1\alpha_2}\right. \\
&\qquad \left. \exp\left[-\frac{\alpha_1^2+\alpha_2^2}{2\alpha_1^2\alpha_2^2}\left(y_2 - \frac{\alpha_2^2y}{\alpha_1^2+\alpha_2^2}\right)^2\right]\right\} \\
&\qquad \cdot \left\{\frac{1}{\sqrt{2\pi(\alpha_1^2+\alpha_2^2)}} \exp\left[-\frac{1}{2(\alpha_1^2+\alpha_2^2)}\,y^2\right] dy_2\right\}
\end{aligned} \tag{7-36}$$

But

$$\int_{-\infty}^{+\infty} \frac{1}{\sqrt{2\pi}\,\sigma} \exp\left(-\frac{\mu^2}{2\sigma^2}\right) d\mu = 1$$

Thus the desired density function $\rho(y)$ is

$$\rho(y) = \frac{1}{\sqrt{2\pi(\alpha_1^2 + \alpha_2^2)}} e^{-y^2/2(\alpha_1^2+\alpha_2^2)} \tag{7-37}$$

That is, Y also has a normal distribution with zero mean and standard deviation

$$\sigma = \sqrt{a_1^2\sigma_1^2 + a_2^2\sigma_2^2}$$

The above procedure can be applied repeatedly in order to obtain the density function of the linear combinations of several normally distributed independent random variables. Thus the following theorem can be established by induction.

Theorem. Let X_k be normally distributed independent random variables with zero means and variance σ_k^2, that is,

$$f(x_k) = \frac{1}{\sqrt{2\pi}\,\sigma_k} e^{-x_k^2/2\sigma_k^2} \qquad k = 1, 2, \ldots, n \tag{7-38}$$

The random variable Y obtained by a linear combination of X_k,

$$Y = a_1X_1 + a_2X_2 + \cdots + a_nY_n \tag{7-39}$$

is a normally distributed random variable with zero mean and standard deviation σ, that is,

$$\text{Variance of } Y = \sigma^2 = a_1^2\sigma_1^2 + a_2^2\sigma_2^2 + \cdots + a_n^2\sigma_n^2 \tag{7-40}$$

$$\rho(y) = \frac{1}{\sqrt{2\pi(\alpha_1^2 + \alpha_2^2 + \cdots + \alpha_n^2)}} \exp\left[-\frac{y^2}{2(\alpha_1^2 + \alpha_2^2 + \cdots + \alpha_n^2)}\right] = \frac{1}{\sqrt{2\pi}\,\sigma} e^{-y^2/2\sigma^2} \tag{7-41}$$

where $\alpha_k = a_k\sigma_k \qquad k = 1, 2, \ldots, n$ (7-42)

7-4. Central-limit Theorem. In the previous sections we have discussed how the linear combination of n independent random variables with normal distribution leads to a normal distribution in n-dimensional space. The engineering implication of this theorem is of great significance. In many applications we may be able to assume linearity or *superposition* of the effects of several *independent* causes. In such problems, if each cause has a normal distribution, the interpretation of the above theorem is justified. We may be concerned with the addition of signals in an adder, or the series or parallel combination of a number of components (Fig. 7-1*a* and *b*). In each case some kind of law of addition may hold, and this law may be exploited to obtain the probability density of the over-all sum of the effects. In the stated theorem of Sec. 7-3 we have assumed a normal distribution for each variable. In this section we remove this restriction and show that under reasonably general circumstances the density distribution of the sum of n random variables approaches *normal* distribution as n is greatly increased. To be more exact, the following very significant theorem, called the *central-limit theorem*, holds.

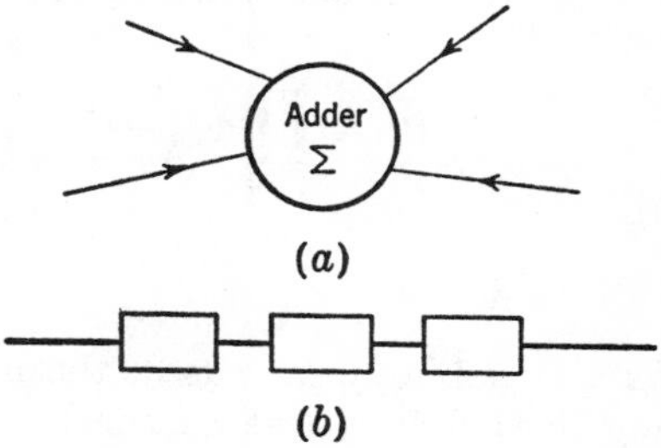

FIG. 7-1. (*a*) A number of random quantities summed up in an adder; (*b*) a series combination of a number of physical elements.

Theorem. Let X_k ($k = 1, 2, \ldots, n$) be mutually independent random variables with identical density distribution functions with a given finite average m_1 and standard deviation σ_1. Then as n is increased the density function of the variable

$$X = X_1 + X_2 + \cdots + X_n$$

will asymptotically approach a normal distribution with

$$m = nm_1 \qquad \sigma = \sqrt{n}\,\sigma_1 \tag{7-43}$$

That is, for any real pair of numbers ($a < b$)

$$\lim_{n\to\infty} P\left\{a < \frac{X - nm_1}{\sqrt{n}\,\sigma_1} < b\right\} = \frac{1}{\sqrt{2\pi}} \int_a^b e^{-x^2/2}\,dx \tag{7-44}$$

Proof. The proof of this theorem is somewhat lengthy. In the following we only sketch a proof, but for details the reader is referred to Cramèr (p. 214).

Consider the standardized random variable

$$X_0 = \frac{X - m}{\sigma} = \frac{1}{\sqrt{n}} \sum_{k=1}^{n} \frac{X_k - m_1}{\sigma_1} \tag{7-45}$$

According to Sec. 6-10, the characteristic function of X_0 can be determined by multiplying the characteristic functions of the variables $(X_k - m_1)/\sigma_1$ for $k = 1, 2, \ldots, n$. Thus the results of Example 6-10 suggest:

$$\text{Characteristic function of } X_k - m_1 = 1 + 0 - \sigma_1{}^2 \frac{t^2}{2} + t^2 R(t) \tag{7-46}$$

where $|R(t)| \to 0$ as $t \to 0$. (See Hardy, "Pure Mathematics," p. 289, sec. 151.)

Then the characteristic function of the random variable X_0 is

$$\Phi_{X_0}(t) = \prod_{i=1}^{n} \Phi_k\left(\frac{t}{\sigma_1\sqrt{n}}\right) = \left[1 - \frac{t^2}{2n} + \frac{t^2}{n}R\left(\frac{t}{\sqrt{n}}\right)\right]^n$$
$$= \left(1 - \frac{t^2}{2n} + \cdots\right)^n \tag{7-47}$$

As n is infinitely increased, the neglected terms become very small for any fixed finite t (see Cramèr). Now the limit of $(1 - t^2/n)^n$ when n approaches infinity is e^{-t^2}; therefore

$$\Phi_{X_0}(t)_{n\to\infty} = \lim_{n\to\infty}\left[1 - \frac{t^2}{2n} + \frac{t^2}{n}R\left(\frac{t}{\sqrt{n}}\right)\right]^n = e^{-t^2/2} \tag{7-48}$$

The characteristic function of X_0 will asymptotically approach the value of $e^{-t^2/2}$.

The above fundamental theorem was known to Laplace at the beginning of the nineteenth century, but its formal proof was given a hundred years later by Liapounoff. Today there is a large class of associated theorems known as central-limit theorems. It can be shown that it is not necessary for the random variables X_k to have the same type of distribution. The theorem holds under certain very general conditions which are beyond our present scope of interest.

In conclusion, the reader should have acquired the feeling that in engineering problems dealing with a large number of statistically defined components one may be able to study the over-all behavior of the system, subject to a number of plausible assumptions and constraints.

Example 7-1. In Fig. E7-1, a number of independent noise voltages V_i $(i = 1, 2, \ldots, n)$ are received in the adder; that is,

$$V = \sum_{i=1}^{n} V_i$$

Each noise voltage has a uniform distribution in the interval $[-A, +A]$:

$$v_i = \frac{1}{2A} \qquad \text{for } |v_i| \le A \quad i = 1, 2, \ldots, n$$
$$v_i = 0 \qquad \text{elsewhere} \quad i = 1, 2, \ldots, n$$

(a) Determine and plot the distribution function of V for $n = 2$.
(b) Same question as (a) for $n = 3$.
(c) Same question as (a) for n much larger than 3.

Solution

(*a*) Let the probability density of V_1, V_2, and $V = V_1 + V_2$ be, respectively,

$$f_1(v_1) = \frac{1}{2A} \qquad -A \le v_1 \le A$$
$$f_2(v_2) = \frac{1}{2A}$$
$$f(v) = f_1(v_1) * f_2(v_2)$$

As our readers are generally familiar with unilateral Laplace transforms, we employ that technique.

$$\mathcal{L}f_1(v_1) = F_1(s) = \frac{1}{2A}\left(\frac{1}{s}e^{As} - \frac{1}{s}e^{-As}\right)$$
$$\mathcal{L}f_2(v_2) = F_2(s) = \frac{1}{2A}\left(\frac{1}{s}e^{As} - \frac{1}{s}e^{-As}\right)$$
$$\mathcal{L}f(v) = F_1(s) \cdot F_2(s) = \frac{1}{4A^2s^2}(e^{2As} - 2 + e^{-2As})$$

Let $u_{-2}(v - k)$ and $u_{-3}(v - k)$ stand for unit ramp and unit parabola applied at $v = k$, respectively; then

$$f(v) = \frac{1}{4A^2}[u_{-2}(v + 2A) - 2u_{-2}(v) + u_{-2}(v - 2A)]$$

The sum of these three unit ramps has the triangular shape illustrated in Fig. E7-1*a*.

(*b*) The extension of the method described in part (*a*) yields

$$f(v) = \mathcal{L}^{-1}\left\{\frac{1}{8A^3s^3}(e^{As} - e^{-As})^3\right\}$$
$$= \frac{1}{8A^3}[u_{-3}(v + 3A) - 3u_{-3}(v + A) + 3u_{-3}(v - A) - u_{-3}(v - 3A)]$$

The sum of these four parabolas leads to the probability density curve shown in Fig. E7-1*b*.

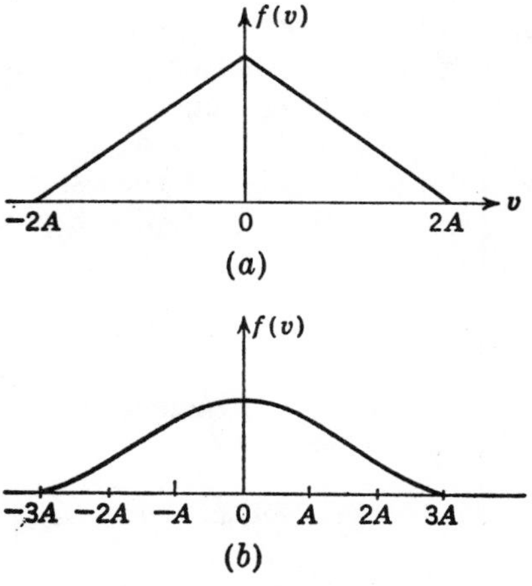

FIG. E7-1

(c) For very large n, the density of V is an asymptotically normal curve with the mean $m = nm_1$ and standard deviation $\sigma = \sqrt{n}\,\sigma_1$, where m_1 and σ_1 are the mean and the standard deviation of each random variable V_k.

$$m = nm_1 = 0$$

$$\sigma = \sqrt{n}\,\sigma_1 = \sqrt{n}\left(\frac{1}{2A}\int_{-A}^{A} x^2\,dx\right)^{1/2} = A\sqrt{\frac{n}{3}}$$

Example 7-2. An honest die is rolled 1,000 times. What is the probability that the total score is between 3,550 and 3,450?

Solution. Let X_k be the random variable associated with the kth throw of the die, and S_n the total score of n trials; then

$$E(X_k) = \frac{1+2+3+4+5+6}{6} = 3.5$$

$$\text{var}\,(X_k) = (1^2+2^2+3^2+4^2+5^2+6^2)\tfrac{1}{6} - 3.5 = {}^{35}\!/_{12}$$

$$S_n = \sum_{k=1}^{n} X_k$$

The standardized random variable X_0 is defined as

$$X_0 = \frac{S_n - 3.5n}{\sqrt{n \cdot {}^{35}\!/_{12}}}$$

When n is reasonably large, then, according to the central-limit theorem, the distributions of X_0 and S_n/n approach that of a normal variate. More specifically,

$$P\left\{3.450 < \frac{S_n}{n} \leq 3.550\right\} \cong 0.68$$

7-5. A Simple Random-walk Problem. As an application of the central-limit theorem, let us consider the problem of a random walk on a straight line. A moving object starts from the origin of the abscissa on this line and makes a sequence of random unit steps in either a positive or a negative direction. The probability of the object moving in the positive direction has a given value p for each step, independent of the previous step. If X_i denotes the ith step, the position of the object after n steps is given by the random variable:

$$X = X_1 + X_2 + \cdots + X_n \tag{7-49}$$

Each random variable X_i assumes the value of $+1$ or -1 with respective probabilities p and $1 - p$. Therefore,

$$\begin{aligned} \bar{X}_i &= 1 \cdot p - 1(1-p) = 2p - 1 & i &= 1, 2, \ldots, n \\ \overline{X_i^2} &= 1 \cdot p + 1(1-p) = 1 & i &= 1, 2, \ldots, n \\ \text{var}\,X_i &= 1 - (2p-1)^2 = 4p(1-p) & i &= 1, 2, \ldots, n \end{aligned} \tag{7-50}$$

The distribution of the variable

$$X_0 = \frac{X - n(2p - 1)}{2\sqrt{np(1 - p)}} \tag{7-51}$$

approaches a standard normal curve.

The probability that after n steps the object lies between two points with abscissa a and b ($b > a$) can be computed from a normal distribution table. For this, one first specifies the range (c,d) of X_0 corresponding to range (a,b) of X; the use of the normal table follows.

$$P\{a < X < b\} = P\{c < X_0 < d\} = \frac{1}{\sqrt{2\pi}} \int_c^d e^{-t^2/2}\, dt \tag{7-52}$$

For example, when $p = \frac{1}{2}$, the probability that, after n steps, the object lies in the range $[-2\sqrt{n}, 2\sqrt{n}]$ is

$$\frac{1}{\sqrt{2\pi}} \int_{-2}^{2} e^{-t^2/2}\, dt = 0.9546$$

Example 7-3. Find the probability that a moving object obeying the above random-walk law is found after n steps in each of the regions described below:

(*a*) $\left[0, \frac{\sqrt{n}}{2}\right]$ for $p = \frac{1}{2}$

(*b*) $\left[-\frac{\sqrt{n}}{2}, \frac{\sqrt{n}}{2}\right]$ for $p = \frac{1}{2}$

(*c*) $n = 400$ $p = \frac{3}{4}$
$[0,300]$

(*d*) $n = 400$ $p = \frac{3}{4}$
$[300,400]$

(*e*) $n = 400$ $p = \frac{3}{4}$
$[-\infty,0]$

Solution

(*a*)

$$X_0 = \frac{X - n(2P - 1)}{2\sqrt{nP(1 - P)}} = \frac{X}{\sqrt{n}}$$

$$P\left\{0 < X < \frac{\sqrt{n}}{2}\right\} = P\left\{0 < X_0 < \frac{1}{2}\right\} = 0.1915$$

(*b*) $$P\left\{-\frac{\sqrt{n}}{2} < X < \frac{\sqrt{n}}{2}\right\} = P\left\{-\frac{1}{2} < X_0 < \frac{1}{2}\right\} = 0.3830$$

(*c*) $P\{0 < X < 300\} = P\{-11.54 < X_0 < 5.77\} \approx 1$

(*d*) $P\{300 < X < 400\} = P\{5.77 < X_0 < 11.54\} \approx 0$

(*e*) $P\{-\infty < X < 0\} = P\{-\infty < X_0 < -11.54\} \approx 0$

7-6. Approximation of the Binomial Distribution by the Normal Distribution. You may have employed the binomial and the normal distribution for solving the same problem and have frequently noted that the two distributions lead to approximately the same numerical answers. In the present section we are in a position to investigate the circumstances under which these two distributions may lead to practically identical results.

Consider an experiment which has only two possible outcomes E and E'. Let

$$P\{E\} = p \qquad P\{E'\} = 1 - p = q \tag{7-53}$$

If the experiment is repeated n times, the probability of obtaining an outcome

$$E_r = (\overbrace{EE \cdots E}^{r})(\overbrace{E'E' \cdots E'}^{n-r})$$

irrespective of the order of the sequence is

$$\binom{n}{r} p^r q^{n-r} \tag{7-54}$$

The probability distribution is of the binomial type. This fact can be alternatively expressed in the following way.

Consider a random variable X_i associated with the ith experiment. If the experiment gives rise to E, we assign a value of 1 to X_i, otherwise zero. The probability distribution associated with X_i assumes either of the two values p and q. The average and the standard deviation of X_i are given by

$$\begin{aligned} E(X_i) &= 1 \cdot P\{X_i = 1\} + 0 \cdot P\{X_i = 0\} \\ &= 1 \cdot p + 0 \cdot q = p \end{aligned} \tag{7-55}$$

$$E(X_i^2) = 1^2 \cdot p + 0^2 \cdot q = p$$

$$\text{Variance of } X_i = p - p^2 = pq \tag{7-56}$$

Consider next the random variable X associated with the original experiment and its probability distribution. The random variable X has the binomial distribution given in Eq. (7-54). However, X can now be considered as the sum of a number of independent random variables, i.e.,

$$X = \sum_{i=1}^{n} X_i$$

According to Sec. 6-8 we have

$$E(X) = E\left(\sum_{i=1}^{n} X_i\right) = \sum_{i=1}^{n} E(X_i) = np \tag{7-57}$$

$$\sigma_X{}^2 = \sigma_{X_1}{}^2 + \sigma_{X_2}{}^2 + \cdots + \sigma_{X_n}{}^2 = npq \tag{7-58}$$

The central-limit theorem asserts that as n is increased the distribution of the standardized random variable X_0 asymptotically approaches a normal distribution. That is, the random variable

$$X_0 = \frac{X - np}{\sqrt{npq}}$$

is normally distributed with $\bar{X}_0 = 0$ and $\sigma_{X_0} = 1$. The binomial distribution can be approximated by

$$P\{a < \underset{\text{binomial}}{X} < b\} \cong \frac{1}{\sqrt{2\pi}} \int_{(a-np)/\sqrt{npq}}^{(b-np)/\sqrt{npq}} e^{-t^2/2}\, dt \tag{7-59}$$

The approximation is rather satisfactory even for small values of n if np is, say, larger than 10 or so. The result of this section can be expressed in the form

$$\sum_{\substack{a<r<b \\ \text{binomial}}} \binom{n}{r} p^r q^{n-r} \cong \frac{1}{\sqrt{2\pi}} \int_{(a-np)/\sqrt{npq}}^{(b-np)/\sqrt{npq}} e^{-t^2/2}\, dt \tag{7-60}$$

The right side of this equation presents the area under the normal curve of the standardized variable between the above limits. The left side can

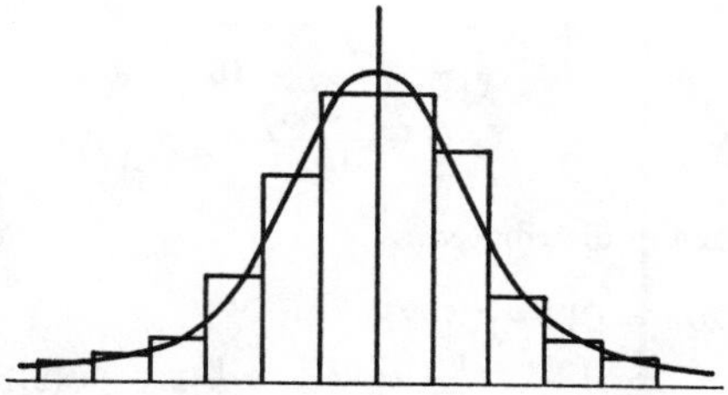

FIG. 7-2. A normal approximation to binomial distribution.

also be represented by an area. In fact, consider the elementary rectangles whose bases are equal to the difference between two successive r's and whose heights are $\sqrt{npq}\binom{n}{r} p^r q^{n-r}$ (Fig. 7-2), where the area of each elementary rectangle still equals $\binom{n}{r} p^r q^{n-r}$. The approximation is rather good when p is not close to either 0 or 1. Note that for $p = \frac{1}{2}$ the graph of the binomial is symmetric about its mean. This is not true for $p \neq \frac{1}{2}$. The normal curve, however, is of course symmetric about its mean value.

Example 7-4. Compare the binomial distribution

$$f(r)_{\text{binomial}} = \binom{10}{r} p^r q^{10-r}$$

with the normal distribution for each of the following cases:

$$p = \tfrac{1}{4}$$
$$p = \tfrac{1}{2}$$

For the binomial distribution assume an approximation of the following type:

$$f(a)_{\text{binomial}} = \int_{a-\frac{1}{2}}^{a+\frac{1}{2}} g(x)_{\text{normal}}\, dx$$

Solution. About each point with abscissa r ($r = 1, 2, \ldots, 10$) construct a rectangle with an area of $\binom{10}{r} p^r q^{10-r}$. The base of the rectangle is taken to be equal to $1/\sqrt{10pq}$. Equation (7-60) can be directly applied for computing the probabilities. The evaluation of the areas of the rectangles and that under the normal curve will show that the approximation is rather good.

Example 7-5. An ordinary coin is tossed 900 times. What is the probability that the number of heads will be less than 420?

Solution. The problem requires the evaluation of

$$P(X < 420)$$

where X is the *number* of heads. To evaluate this probability, we use a normal approximation to the binomial distribution.

$$n = 900 \qquad p = q = \tfrac{1}{2}$$
$$\sigma = \sqrt{npq} = 15$$

Then

$$X_0 = \frac{X - 450}{15}$$

is asymptotically normally distributed.

$$P\{X < 420\} = P\{X_0 < -2\}$$
$$\cong \int_{-\infty}^{-2} \frac{1}{\sqrt{2\pi}} e^{-t^2/2}\, dt = \frac{1}{2}[1 - 2\phi(2)] = 0.0227$$

where $\phi(a)$ denotes the area under the standardized normal curve between 0 and a.

7-7. Approximation of Poisson Distribution by a Normal Distribution.

As the second application of the central-limit theorem we approximate the Poisson distribution by a normal distribution. Let the random variable X have a Poisson distribution

$$f(x) = \frac{\lambda^x}{x!} e^{-\lambda} \tag{7-61}$$

In Chap. 6, it was shown that the first and the second moment of X are, respectively,

$$E(X) = \lambda$$
$$E(X^2) = \lambda + \lambda^2 \tag{7-62}$$

and

$$\sigma_X{}^2 = \lambda \tag{7-63}$$

The standardized random variable

$$X_0 = \frac{X - \lambda}{\sqrt{\lambda}}$$

has the following moment-generating function:

$$\psi_x(t) = \exp \lambda(e^t - 1) \tag{7-64}$$

$$\psi_{x-\lambda}(t) = \exp (\lambda e^t - \lambda - \lambda t) \tag{7-65}$$

$$\begin{aligned} \psi_{x_0}(t) &= \exp (\lambda e^{t/\sqrt{\lambda}} - \lambda - \sqrt{\lambda}\, t) \\ &= \exp \left(\lambda + \sqrt{\lambda}\, t + \frac{t^2}{2} + \frac{t^3}{3!\sqrt{\lambda}} + \cdots - \lambda - \sqrt{\lambda}\, t\right) \\ &= \exp \left(\frac{t^2}{2} + \frac{t^3}{3!\sqrt{\lambda}} + \cdots\right) \end{aligned} \tag{7-66}$$

As $\lambda \to \infty$, the moment-generating function of X_0 approaches that of a standardized normal random variable, that is,

$$\lim_{\lambda \to \infty} P\left\{a < \frac{X - \lambda}{\sqrt{\lambda}} \le b\right\} = \frac{1}{\sqrt{2\pi}} \int_a^b e^{-x^2/2}\, dx \tag{7-67}$$

7-8. The Laws of Large Numbers

Weak Law of Large Numbers. The interpretation of this law implies that, if a random experiment is repeated a large number of times, the average of the results will differ only slightly from the expected value of each experiment. For instance, if an honest coin is tossed n times, as $n \to \infty$ the average of the results of our experiment, say the frequency of the recorded number of heads, will tend to ½, which is the expected value of the variable. More exactly:

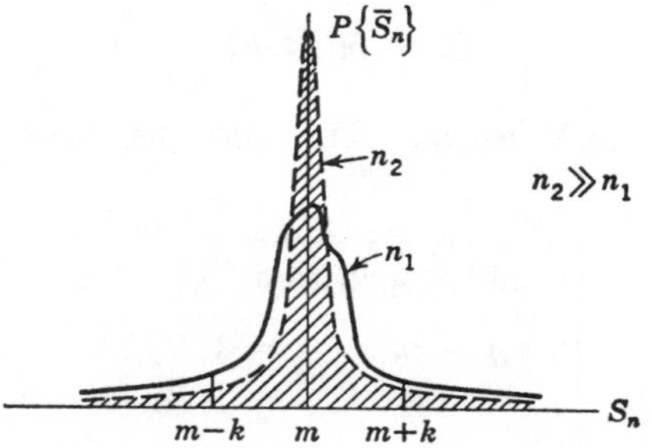

FIG. 7-3. An illustration of the law of large numbers.

Theorem. Let $X_1, X_2, \ldots, X_n$ be independent random variables such that

$$E(X_i) = m \qquad i = 1, 2, \ldots, n \tag{7-68}$$

$$\sigma_{X_i} = \sigma$$

Then for any positive k, the random variable

$$S_n = \frac{1}{n} \sum_{i=1}^{n} X_i \tag{7-69}$$

satisfies the inequality

$$P\{m - k < S_n < m + k\} \underset{n \to \infty}{\longrightarrow} 1 \tag{7-70}$$

That is, $(X_1 + X_2 + \cdots + X_n)/n$ will approach m with probability 1

Proof. The proof follows directly from Chebyshev's inequality. In fact, the statement of the theorem implies that

$$E(S_n) = m$$

$$\sigma_S = \frac{\sigma}{\sqrt{n}}$$

Thus, the application of Eq. (6-25) yields

$$P\{|S_n - m| > k\} \leq \frac{\sigma^2}{nk^2} \tag{7-71}$$

But, for a given pair of σ and k, the ratio of σ^2/nk^2 tends to zero as $n \to \infty$. Therefore

$$P\{|S_n - m| > k\} \underset{n \to \infty}{\longrightarrow} 0 \tag{7-72}$$

This inequality can also be written in the alternative form

$$P\{|S_n - m| \leq k\} \underset{n \to \infty}{\longrightarrow} 1 \tag{7-73}$$

Strong Law of Large Numbers. The following theorem is given without a proof.

Theorem. Let $X_1, X_2, \ldots, X_n, \ldots$ be an infinite sequence of independent random variables such that

$$E(X_i) = m \qquad i = 1, 2, \ldots$$

$$\sigma_{X_i} = \sigma \qquad i = 1, 2, \ldots$$

Let ω be the set of points of the sample space for which

$$\lim_{n \to \infty} \frac{X_1(\omega) + X_2(\omega) + \cdots + X_n(\omega)}{n} = m \tag{7-74}$$

Then ω has probability 1 (is said to be *almost certain*). The strong law implies that the limit of the average approaches the common expected value of the afore-mentioned independent variables. For proof see Loève and Fortet. It should be kept in mind that the laws of large numbers and the central-limit theorem can be stated under more general circumstances than those assumed in this section.

PROBLEMS

7-1. An honest coin is tossed 1,000 times.

(*a*) Find the probability of a head occurring in less than 500 times.

(*b*) Find the probability of a head occurring less than 500 but more than 450 times. Use the normal approximation.

(*c*) Same question as (*b*) but use Chebyshev's inequality.

7-2. X is a random variable with binomial distribution ($n = 300$, $p = \frac{1}{3}$). Approximate $P\{x \leq 100\}$.

7-3. For a random voltage V assume the following values with their respective probabilities:

$$\begin{bmatrix} -2, & -1, & 0, & 1, & 2 \\ 0.05, & 0.25, & 0.15, & 0.45, & 0.10 \end{bmatrix}$$

(*a*) Find $\bar{V}$.

(*b*) Find the standard deviation of V.

(*c*) If the voltage V is applied across a 20-ohm resistor, determine the average power W dissipated in the resistor in unit time.

(*d*) What is the probability of the power W being less than 0.10, 0.20, 0.50 watt, respectively?

(*e*) The voltage V is applied to the same resistor but only for a period of $\frac{1}{2}$ microsecond. Now suppose that this experiment is repeated 10^6 times. Find the probability that the total dissipated power will remain between w_1 and w_2 watts.

(*f*) Calculate part (*e*) for

$$\begin{aligned} w_1 &= 0.01 & w_2 &= 0.02 \\ w_1 &= 0.05 & w_2 &= 0.10 \\ w_1 &= 0.10 & w_2 &= 0.15 \end{aligned}$$

7-4. The independent noise voltages V_1 and V_2 are added to a d-c signal of 10 volts after going through amplifiers of parameters k_1 and k_2, respectively. Find the density of the output

$$V = k_1V_1 + k_2V_2 + 10$$

(*a*) V_1 and V_2 are normal with parameters (3,4) and (−2,3).

$$k_1 = 2 \qquad k_2 = 3$$

(*b*) V_1 and V_2 are uniformly distributed between 0 and 1 volt.

$$k_1 = 2 \qquad k_2 = 3$$

7-5. Let U and V be the output of two adders. The input to each adder is obtained through a number of linear networks not necessarily independent, that is,

$$U = \sum_{k=1}^{n} a_k x_k \qquad V = \sum_{n=1}^{n} b_k x_k$$

(*a*) Determine the standard deviation of U and V.

(*b*) Show that the covariance of the output of the adders is

$$\sum_{i=1}^{n} \sum_{j=1}^{n} a_i b_j \mu_{ij}$$

where μ_{ij} is the covariance of X_i and X_j.

7-6. Let S_n be the average of n identically distributed random variables with binomial distribution (p,q). Prove that

$$P\left\{|S_n - p| > \frac{2}{\sqrt{n}}\right\} \leq \frac{1}{16}$$

7-7. Let X be a random signal with (m,σ). Using Chebyshev's inequality, show

(a) $P\{|X - m| \leq \alpha\} \geq 95$ per cent for $\alpha \geq 4.5\sigma$

(b) $P\{|X - m| \leq \alpha\} \geq 99$ per cent for $\alpha \geq 10\sigma$

(c) $P\left\{\text{fluctuation} = \left|\frac{X - m}{m}\right| \leq \alpha\right\} \geq 95$ per cent for $\alpha \geq 4.5\frac{\sigma}{|m|}$

(d) $P\left\{\left|\frac{X - m}{m}\right| \leq \alpha\right\} \geq 99$ per cent for $\alpha \geq 10\frac{\sigma}{|m|}$

(e) When the signal is approximately normally distributed, the numbers 4.5 and 10 in the above inequalities could be respectively reduced to 1.96 and 2.58 [see Parzen (Chap. 8)].

7-8. Show that the characteristic function of an n-dimensional normal distribution is

$$\phi(t_1,t_2, \ldots ,t_n) = \exp\left(-\frac{1}{2}\sum_{i=1}^{n}\sum_{j=1}^{n} \alpha_{ij}t_it_j\right)$$

where α_{ij} is the second joint moment of X_i and X_j, and

$$\bar{X}_k = 0 \qquad \text{for } k = 1, 2, \ldots , n$$

7-9. Consider a normally distributed variable with the probability density

$$f(x_1,x_2, \ldots ,x_n) = C \exp(-\tfrac{1}{2}Q)$$

where, as usual,
$$Q = \sum_{i=1}^{n}\sum_{j=1}^{n} \alpha_{ij}(X_i - \bar{X}_i)(X_j - \bar{X}_j)$$

Find the probability density of Q by using the concept of moment-generating function and transformation of variables.

CHAPTER 8

CONTINUOUS CHANNEL WITHOUT MEMORY

8-1. Definition of Different Entropies. In a preceding chapter we studied the transmission of information by discrete symbols. In many practical applications the information is transmitted by continuous signals, such as continuous electric waves. That is, the transmitted signal (Fig. 8-1) is a continuous function of time during a finite time interval. During that interval the amplitude of the signal assumes a continuum of values with a specified probability density function.

The main object of this chapter is to outline some results for continuous channels similar to those discussed for discrete systems, principally the entropy associated with a random variable assuming a continuum of values. The extension of mathematical results obtained for finite, discrete systems to infinite systems or systems with continuous parameters is quite frequent in problems of mathematics and physics. Such extensions require a certain amount of care if mathematical difficulties and inaccuracies are to be avoided. For example, matrix algebra is quite familiar to most scientists in so far as finite matrices are concerned. The same concept can be extended to cover infinite matrices and Hilbert spaces, the extension being subject to special mathematical disciplines that require time and preparation to master. Similarly, in network theory, one is familiar with the properties of the rational driving-point impedance functions associated with lumped linear networks; but when dealing with impedances associated with transmission lines, one is far less knowledgeable as to the class of pertinent transcendental functions describing the impedances. (In fact, as yet there is a very limited body of work available on the extension of existing methods of network synthesis from lumped-parameter to distributed-parameter systems.)

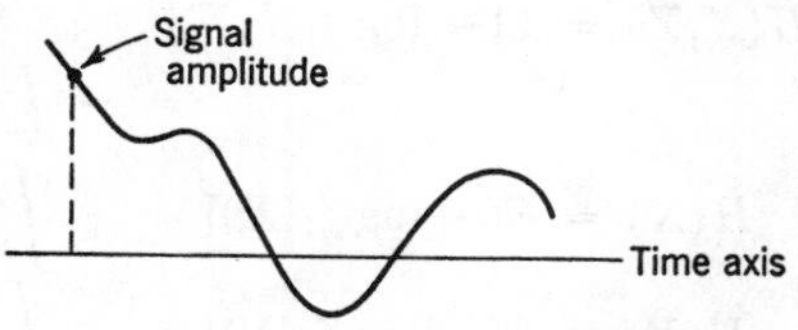

FIG. 8-1. A continuous signal.

This pattern of increased complexity of analysis, requiring special mathematical consideration for passage from the finite to the infinite and from the discrete to the continuous, also prevails in the field of information theory.

One method of presentation is to extend the definitions for entropies from discrete to continuous cases in a way similar to the presentation of the probability of discrete and continuous random variables. In fact, we have already used such a technique for defining the expectation of a continuous random variable. This analogous presentation has the apparent merit of simplicity and convenience, but at the expense of not always being well defined from a strictly mathematical point of view. Also, the engineering significance of the entropy of a continuous random variable becomes somewhat obscure, as is shown later. The mathematically inclined reader may find it more tenable to start this discussion with the definition of the mutual information between two random objects each assuming a continuum of values. The procedure has been outlined in Shannon's original paper as well as in a fundamental paper of A. N. Kolmogorov (see also the reference cited on page 289).

The definitions of the different entropies in the discrete case were based on the concept of different expectations encountered in the case of two-dimensional discrete distributions. In a similar way, we may introduce different entropies in the case of one-dimensional or multidimensional random variables with continuous distributions. For a one-dimensional random variable,

$$H(X) = E[-\log f(X)] = -\int_{-\infty}^{+\infty} f(x) \log f(x)\, dx \tag{8-1}$$

The different entropies associated with a two-dimensional random variable possessing a joint density $f(x,y)$ and marginal densities $f_1(x)$ and $f_2(y)$ are

$$H(X,Y) = E[-\log f(X,Y)] = -\int_{-\infty}^{+\infty}\int_{-\infty}^{+\infty} f(x,y) \log f(x,y)\, dx\, dy \tag{8-2}$$

$$H(X) = E[-\log f_1(X)] = -\int_{-\infty}^{+\infty} f_1(x) \log f_1(x)\, dx \tag{8-3}$$

$$H(Y) = E[-\log f_2(Y)] = -\int_{-\infty}^{+\infty} f_2(y) \log f_2(y)\, dy \tag{8-4}$$

$$H(X|Y) = E[-\log f_y(X|Y)] = -\int_{-\infty}^{+\infty}\int_{-\infty}^{+\infty} f(x,y) \log \frac{f(x,y)}{f_2(y)}\, dx\, dy \tag{8-5}$$

$$H(Y|X) = E[-\log f_x(Y|X)] = -\int_{-\infty}^{+\infty}\int_{-\infty}^{+\infty} f(x,y) \log \frac{f(x,y)}{f_1(x)}\, dx\, dy \tag{8-6}$$

Finally, as will be pointed out in Chap. 9, for an n-dimensional random variable possessing a probability density function $f(x_1,x_2, \ldots ,x_n)$, the entropy is defined as

$$H(X_1, \ldots, X_n) = E[-\log f(X_1, \ldots, X_n)]$$
$$= -\int_{-\infty}^{+\infty} \cdots \int_{-\infty}^{+\infty} f(x_1, \ldots, x_n) \log f(x_1, \ldots, x_n)\, dx_1 \cdots dx_n \quad (8\text{-}7)$$

All definitions here are contingent upon the existence of the corresponding integrals.

8-2. The Nature of Mathematical Difficulties Involved. In this section we describe the mathematical difficulties encountered in extending the concept of self-information from discrete to continuous models. There are at least three basic points to be discussed:

1. The entropy of a random variable with continuous distribution may be negative.
2. The entropy of a random variable with continuous distribution may become infinitely large. Furthermore, if the probability scheme under consideration is "approximated" by a discrete scheme, it can be shown that the entropy of the discrete scheme will always tend to infinity as the quantization is made finer and finer.
3. In contrast to the discrete case, the entropy of a continuous system does not remain necessarily invariant under the transformation of the coordinate systems.

Of these three difficulties, perhaps the first one is the most apparent. The second and third require more explanation. For this reason we treat item 1 in this section but defer discussion of topics 2 and 3 to a later section.

Negative Entropies. In the discrete case all the entropies involved are positive quantities because the probability of the occurrence of an event in the discrete case is a positive number less than or equal to 1. In the continuous case,

$$\int_{-\infty}^{\infty} f(x)\, dx = 1$$
$$\int_{-\infty}^{\infty} \int_{-\infty}^{\infty} f(x,y)\, dx\, dy = 1 \quad (8\text{-}8)$$

Evidently, the density functions need not be less than 1 for all values of the random variable; this fact may lead to a negative entropy. A situation leading to a negative entropy is illustrated in Example 8-1, where the entropy associated with the density function depends on the value of a parameter. *This is a reason why the concept of self-information no longer can be associated with $H(X)$ as in the discrete case. We call $H(X)$ the entropy function, but $H(X)$ no longer indicates the average self-information of the source.*

Similar remarks are valid for conditional entropies. Thus it follows that the individual entropies may assume negative values. However, it will be shown that the mutual information is not subject to this objection.

Example 8-1. A random variable has the density function shown in Fig. E8-1. Find the corresponding entropy.

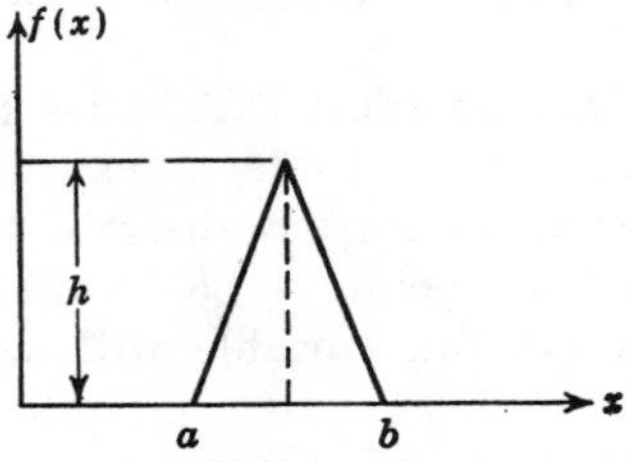

Fig. E8-1

Solution

$$f(x) = \frac{2h}{b-a}(x-a) \qquad \text{for } a \le x \le \frac{a+b}{2}$$

$$f(x) = \frac{2h}{b-a}(b-x) \qquad \text{for } \frac{a+b}{2} \le x \le b$$

$$H(X) = -\int_a^{(a+b)/2} \frac{2h}{b-a}(x-a) \ln \frac{2h}{b-a}(x-a)\,dx - \int_{(a+b)/2}^{b} \frac{2h}{b-a}(b-x) \ln \frac{2h}{b-a}(b-x)\,dx$$

The above integrals can be evaluated by parts. In so doing, note that

$$\int x \ln \lambda x\,dx = \frac{x^2}{2} \ln \lambda x - \frac{x^2}{4}$$

Thus, $$H(X) = \frac{-h}{b-a}\left[(x-a)^2 \ln \frac{2h}{b-a}(x-a) - \frac{(x-a)^2}{2}\right]_a^{(a+b)/2} + \frac{h}{b-a}\left[(b-x)^2 \ln \frac{2h}{b-a}(b-x) - \frac{(b-x)^2}{2}\right]_{(a+b)/2}^{b}$$

$$H(X) = \frac{-h}{b-a}\left\{\left[\frac{(b-a)^2}{4}\ln h - \frac{(b-a)^2}{8}\right] + \left[\frac{(b-a)^2}{4}\ln h - \frac{(b-a)^2}{8}\right]\right\}$$

$$= \frac{h(b-a)}{2}\left(-\ln h + \frac{1}{2}\right)$$

and since $$\frac{h(b-a)}{2} = \int_{-\infty}^{\infty} f(x)\,dx = 1$$

thus $$H(X) = -\ln h + \tfrac{1}{2}$$

The entropy depends on the parameter h, but a translation of the probability curve along the x axis does not change its value. Note also that

$$\begin{aligned} H(X) &> 0 \qquad \text{for } h < \sqrt{e} \\ H(X) &= 0 \qquad \text{for } h = \sqrt{e} \\ H(X) &< 0 \qquad \text{for } h > \sqrt{e} \end{aligned}$$

8-3. Infiniteness of Continuous Entropy. Let X be a one-dimensional random variable with a well-defined range $[a,b]$ and a probability density function $f(x)$, that is,

$$P\{c < X \leq d\} = \int_c^d f(x)\,dx = F(d) - F(c) \tag{8-9}$$

We propose to examine the entropy associated with this random variable, following a familiar mathematical routine. That is, we divide the interval of interest between a and b into nonoverlapping subintervals (see Fig. 8-2):

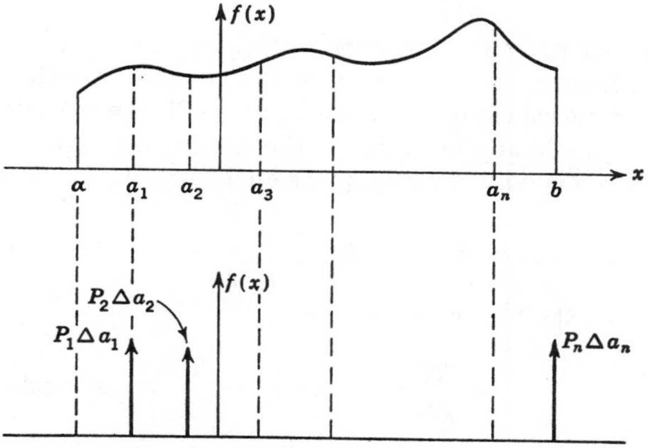

FIG. 8-2. A quantization of a continuous signal for computing entropy.

$$(a,a_1];\ (a_1,a_2];\ \ldots\ ;\ (a_n,b] \tag{8-10}$$

$$a < a_1 < a_2 < \cdots < a_n < a_{n+1} = b \tag{8-11}$$

$$a_1 - a = \Delta a_1, \ \ldots\ , a_k - a_{k-1} = \Delta a_k, \ \ldots\ , b - a_n = \Delta a_{n+1}$$

$$k = 1, 2, \ldots, n+1 \tag{8-12}$$

$$P(a < X \leq a_1) = \int_a^{a_1} f(x)\,dx = F(a_1) - F(a) = p_1\,\Delta a_1 \tag{8-13}$$

$$P(a_1 < X \leq a_2) = \int_{a_1}^{a_2} f(x)\,dx = F(a_2) - F(a_1) = p_2\,\Delta a_2 \tag{8-14}$$

. . . .

$$P(a_n < X \leq b) = \int_{a_n}^{b} f(x)\,dx = F(b) - F(a_n) = p_{n+1}\,\Delta a_{n+1} \tag{8-15}$$

Now we may define another random variable X_d, assuming only the discrete set of values

$$[a_1,a_2, \ldots ,a_n,b] \tag{8-16}$$

with respective probabilities

$$[p_1\,\Delta a_1, p_2\,\Delta a_2, \ \ldots\ , p_n\,\Delta a_n, p_{n+1}\,\Delta a_{n+1}] \tag{8-17}$$

According to Eq. (8-9) the events under consideration form a finite, complete probability scheme, as

$$\sum_{k=1}^{n+1} p_k \, \Delta a_k = F(b) - F(a) = 1 \tag{8-18}$$

Thus

$$H(X_d) = -\sum_{k=1}^{n+1} p_k \, \Delta a_k \log p_k \, \Delta a_k \tag{8-19}$$

Now, let the length of each interval in Eq. (8-19) become infinitely small by infinitely increasing n. It is reasonable to anticipate that in the limit, when every interval becomes vanishingly small, the entropy of this discrete scheme should approach that of the continuous model. The process can be made more evident by adopting an arbitrary level of quantization, say

$$\Delta a_1 = \Delta a_k = \Delta a_{n+1} = \Delta x \qquad k = 1, 2, \ldots, n+1 \tag{8-20}$$

and evaluating the above entropy,

$$H(X_d) = -\sum_{k=1}^{n+1} p_k \, \Delta x \log p_k - \sum_{k=1}^{n+1} p_k \, \Delta x \log \Delta x \tag{8-21}$$

But

$$H(X) = \lim_{\Delta x \to 0} H(X_d) \tag{8-22}$$

Therefore when Δx is made smaller and smaller, while $p_k \, \Delta a_k$, the area under the curve between a_{k-1} and a_k, tends to zero, the ratio of the area to Δa_k remains finite for a continuous distribution. In the limit

$$\lim_{\Delta x \to 0} H(X_d) = -\int_a^b f(x) \log f(x) \, dx - \lim_{\Delta x \to 0} \sum_{k=1}^{n+1} p_k \, \Delta x \log \Delta x \tag{8-23}$$

assuming that the first integral exists.

As the subintervals are made smaller by making n larger, the $p_k \, \Delta x$ become smaller but the entropy $H(X_d)$ increases. Thus, in the limit when an infinite number of infinitesimal subintervals are considered, the entropy becomes infinitely large. The interpretation is that the continuous distribution can potentially convey infinitely large amounts of information. We have used the word "potentially" since the information must be received by a receiver or an observer. The observer can receive information with a bounded accuracy. Thus $H(X)$ should preferably be written as $H_\epsilon(X)$, indicating the bounded level of accuracy of the observer. If the observer had an infinitely great level of accuracy, he could detect an infinitely large amount of information from a random signal assuming a continuum of values.

In a manner similar to the definition of entropy in the discrete case, we may define the entropy of a complete continuous scheme defined in $[a,b]$ as

$$H(X) = -\int_a^b f(x) \log f(x)\, dx$$

It is important to note that the integral of Eq. (8-1) defining the entropy of a continuous random variable is not necessarily infinite. However, the above limiting process, which introduces the concept of a discrete analog model with an infinitely large number of states, always leads to an infinite entropy.

8-4. Variability of the Entropy in the Continuous Case with Coordinate Systems. Consider a one-dimensional continuous random variable X with a density function $f(x)$. Let the variable X be transformed into a new variable Y by a continuous one-to-one transformation. The density function $\rho(y)$ is

$$\rho(y) = f(x) \left| \frac{dx}{dy} \right| \tag{8-24}$$

If it is assumed that the transformation $Y = g(X)$ is monotone and single-valued, the entropy associated with Y is

$$H(Y) = -\int_{-\infty}^{+\infty} \rho(y) \log \rho(y)\, dy \tag{8-25}$$

$$H(Y) = -\int_{-\infty}^{+\infty} \left[f(x) \left| \frac{dx}{dy} \right| \right] \log \left[f(x) \left| \frac{dx}{dy} \right| \right] dy \tag{8-26}$$

$$H(Y) = -\int_{-\infty}^{\infty} f(x) \log f(x)\, dx - \int_{-\infty}^{\infty} f(x) \log \left| \frac{dx}{dy} \right| dx \tag{8-27}$$

$$H(Y) = H(X) + \int_{-\infty}^{+\infty} f(x) \log \left| \frac{dy}{dx} \right| dx \tag{8-28}$$

The entropy of the new system depends on the associated function $\log |dx/dy|$. This is in contrast with the discrete case. In the discrete case, the values associated with the random variable do not enter into the computation. For instance, the entropy associated with the throwing of an ordinary die is

$$\begin{aligned} [X] &= [1,2,3,4,5,6] \\ [P] &= [\tfrac{1}{6},\tfrac{1}{6},\tfrac{1}{6},\tfrac{1}{6},\tfrac{1}{6},\tfrac{1}{6}] \\ H(X) &= 6(-\tfrac{1}{6} \log \tfrac{1}{6}) = \log 6 \end{aligned}$$

A change of variable, say $Y = X^2$, does not produce any change in the entropy since $[P]$ remains unchanged:

$$\begin{aligned} [X^2] &= [1,4,9,16,25,36] \\ [P] &= [\tfrac{1}{6},\tfrac{1}{6},\tfrac{1}{6},\tfrac{1}{6},\tfrac{1}{6},\tfrac{1}{6}] \\ H(X^2) &= \log 6 = H(X) \end{aligned}$$

When the transformation of the continuous random variable is linear, that is,

$$Y = AX + B \tag{8-29}$$

Eq. (8-28) yields

$$H(Y) = H(X) + \int_{-\infty}^{\infty} f(x) \log |A| \, dx \tag{8-30}$$

$$H(Y) = H(X) + \log |A| \qquad \text{in bits} \tag{8-31}$$

Equation (8-31) suggests that the entropy of a continuous random variable subjected to a linear transformation of the axis remains invariant within a constant $\log |A|$.

Example 8-2. Find the entropy of a continuous random variable with the density function as illustrated in Fig. E8-2.

$$\begin{aligned} f(x) &= bx^2 \qquad && 0 \le x \le a \\ &= 0 && \text{elsewhere} \end{aligned}$$

Determine the entropy $H(X_1)$ when $x_1 = x + d$, $d > 0$. Answer same question for the transformation $x_2 = 2x$.

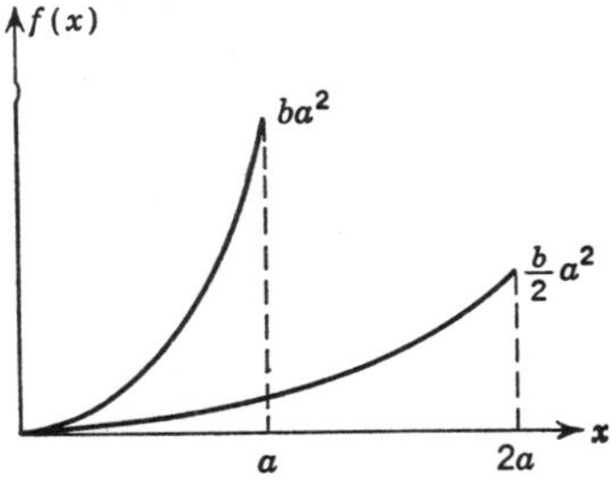

FIG. E8-2

Solution. The value of b which makes the above $f(x)$ a permissible probability density function is given by

$$\int_0^a f(x)\, dx = b \frac{x^3}{3} \Big]_0^a = \frac{ba^3}{3} = 1$$

$$H(X) = -\int_0^a bx^2 \ln bx^2 \, dx$$

Note that
$$\int x^2 \ln \lambda x \, dx = \frac{x^3}{3} \ln \lambda x - \frac{x^3}{9}$$

Thus
$$H(X) = -2b \left[\frac{x^3}{3} \ln \sqrt{b}\, x - \frac{x^3}{9} \right]_0^a = -2b \frac{a^3}{3} (\ln \sqrt{b}\, a - 1/3)$$

$$H(X) = -2 \left(\ln \sqrt{\frac{3}{a}} - 1/3 \right) = \frac{2}{3} + \ln \frac{a}{3}$$

The entropy may be positive, negative, or zero, depending on the parameter a.

$$\begin{aligned} H(X) &> 0 \qquad && a > 3e^{-2/3} \\ H(X) &= 0 && a = 3e^{-2/3} \\ H(X) &< 0 && a < 3e^{-2/3} \end{aligned}$$

Now consider the simple translation of the vertical axis,

$$x_1 = x + d \qquad d > 0$$

The probability density curve will be simply d units shifted to the left. The entropy becomes

$$H(X_1) = -\int_d^{d+a} b(x_1 - d)^2 \ln [b(x_1 - d)^2]\, dx_1$$

$$H(X_1) = -2b\left[\frac{(x_1 - d)^3}{3} \ln \sqrt{b}\,(x_1 - d) - \frac{(x_1 - d)^3}{9}\right]_d^{d+a}$$

$$H(X_1) = -2b\frac{a^3}{3}\left(\ln \sqrt{b}\, a - \frac{1}{3}\right) = \frac{2}{3} + \ln \frac{a}{3} = H(X)$$

This result could have been predicted from Eq. (8-31), as

$$\log A = \log 1 = 0$$

For the transformation $x_2 = 2x$, one finds

$$\rho(x_2) = \frac{bx^2}{2} = \frac{b}{8} x_2^2$$

$$H(X_2) = -\int_0^{2a} \frac{b}{8} x_2^2 \log \frac{b}{8} x_2^2\, dx_2$$

$$H(X_2) = -\frac{b}{4}\left[\frac{x_2^3}{3}\left(\ln \frac{\sqrt{b}}{2\sqrt{2}} x_2 - \frac{1}{3}\right)\right]_0^{2a} = \ln \frac{2}{3} a + \frac{2}{3}$$

This result is in agreement with Eq. (8-31), that is,

$$H(X_2) = H(X) + \int_0^a bx^2 \ln 2\, dx = H(X) + \ln 2$$

8-5. A Measure of Information in the Continuous Case. The material of the preceding two sections might lead one to think that the concept of entropy loses its usefulness for continuous systems. On the contrary, the concept of entropy is as important in the continuous case as in the discrete case. To put this concept into focus, one has slightly to reorient one's thoughts by putting the emphasis on the transinformation rather than on the individual entropies. The different entropies associated with a continuous channel, that is, $H(X)$, $H(Y)$, $H(X|Y)$, $H(Y|X)$, and $H(X,Y)$, have no direct interpretation as far as the information processed in the channel is concerned. However, it will be shown that the transinformation $I(X;Y)$ retains its information-theory significance. Owing to this, we use the concept of the transinformation $I(X;Y)$ of the random variables X and Y as the starting point in defining the entropy of continuous systems.

$$I(X;Y) = \int_{-\infty}^{\infty} \int_{-\infty}^{\infty} f(x,y) \log \frac{f(x,y)}{f_1(x)f_2(y)}\, dx\, dy \qquad (8\text{-}32)$$

Note that, as before, we have

$$I(X;Y) = H(X) - H(X|Y) = H(Y) - H(Y|X) = H(X) + H(Y) - H(X,Y) \quad (8\text{-}33)$$

Now we should be able to demonstrate how this measure of transinformation does not face the three mentioned difficulties encountered in dealing with individual entropies.

Transinformation Is Nonnegative. A proof of the validity of this property can be obtained by using the basic inequality for convexity of a logarithmic function.

$$\begin{aligned} I(X;Y) &= \int_{-\infty}^{\infty}\int_{-\infty}^{\infty} f(x,y) \log \frac{f(x,y)}{f_1(x)f_2(y)}\,dx\,dy \\ &= -\int_{-\infty}^{\infty}\int_{-\infty}^{\infty} f(x,y) \log \frac{f_1(x)}{f(x|y)}\,dx\,dy \\ &\geq -\int_{-\infty}^{\infty}\int_{-\infty}^{\infty} f(x,y) \left[\frac{f_1(x)}{f(x|y)} - 1\right] \log e\,dx\,dy \\ &= -\int_{-\infty}^{\infty}\int_{-\infty}^{\infty} f_2(y)f_1(x) \log e\,dx\,dy \\ &\qquad + \int_{-\infty}^{\infty}\int_{-\infty}^{\infty} f(x,y) \log e\,dx\,dy \\ &= 1 \cdot 1 \cdot \log e - \log e = 0 \end{aligned} \quad (8\text{-}34)$$

Hence

$$I(X;Y) \geq 0$$

Transinformation Is Generally Finite. Consider the expression

$$\log \frac{P\{x_i < X < x_i + \Delta x,\ y_j < Y < y_j + \Delta y\}}{P\{x_i < X < x_i + \Delta x\}P\{y_j < Y < y_j + \Delta y\}} \quad (8\text{-}35)$$

which is a direct extension of the definition of mutual information in the discrete case. As Δx and Δy are made smaller, each of these probability terms tends to zero. This was the reason for the individual continuous entropies to tend to infinity in our passage from the discrete to the continuous models. While each of the individual terms tends to zero, the above ratio remains finite for all cases of interest. In fact, in the limit, the expression becomes

$$\log \frac{f(x_i,y_j)}{f_1(x_i)f_2(y_j)} \quad (8\text{-}36)$$

It is certainly reasonable to exclude the degenerate cases corresponding to densities which are not absolutely continuous. (See references given in the footnotes on pages 277 and 295.)

To sum up, in passing from the discrete to the continuous model, each one of the entropies $H(Y)$, $H(X)$, and $H(X,Y)$ leads to the calculation of the logarithm of some infinitesimal probabilities, thus leading to infinite entropies. However, the expression of Eq. (8-32) generally remains finite and will lead to a finite measure of entropy for the mutual entropy $I(X;Y)$.*

Invariance of Transinformation under Linear Transformation. Finally, we should like to show that, in contrast with self-informations, our measure of mutual information, Eq. (8-32), remains invariant under all linear-scale transformations at the input and the output of the channel. The proof will follow by applying the general equations of transformation of variables; that is, let

$$\begin{aligned} X_1 &= aX + b \\ Y_1 &= cY + d \end{aligned} \tag{8-37}$$

Then

$$\begin{aligned} H(X_1) &= -\int_{-\infty}^{\infty} \rho_1(x_1) \log \rho_1(x_1)\, dx_1 \\ H(Y_1) &= -\int_{-\infty}^{\infty} \rho_2(y_1) \log \rho_2(y_1)\, dy_1 \\ H(X_1,Y_1) &= -\int_{-\infty}^{\infty}\int_{-\infty}^{\infty} \rho(x_1,y_1) \log \rho(x_1,y_1)\, dx_1\, dy_1 \end{aligned} \tag{8-38}$$

where $\rho_1(x_1)$ and $\rho_2(y_1)$ are the probability density functions associated with the variables X_1 and Y_1, respectively, and $\rho(x_1,y_1)$ is their joint density function. But according to Sec. 5-12,

$$\begin{aligned} \rho_1(x_1) &= \frac{f_1(x)}{|a|} \\ \rho_2(y_1) &= \frac{f_2(y)}{|c|} \\ \rho(x_1,y_1) &= \frac{f(x,y)}{|J(x_1,y_1/x,y)|} = \frac{f(x,y)}{\begin{vmatrix} a & 0 \\ 0 & c \end{vmatrix}} = \frac{f(x,y)}{|ac|} \end{aligned} \tag{8-39}$$

$$\begin{aligned} H(X_1) &= H(X) + \log |a| \\ H(Y_1) &= H(Y) + \log |c| \\ H(X_1,Y_1) &= H(X,Y) + \log |ac| \end{aligned} \tag{8-40}$$

Thus

$$\begin{aligned} I(X_1;Y_1) &= H(X_1) + H(Y_1) - H(X_1,Y_1) \\ &= H(X) + H(Y) - H(X,Y) + \log |a| \\ &\qquad + \log |c| - \log |ac| \\ &= I(X;Y) \end{aligned} \tag{8-41}$$

While the individual entropies may change under linear transformations, the transinformation remains intact.

* $I(X;Y)$ is finite when all densities are absolutely continuous—the transinformation may become infinite in some extrinsic circumstances (see, for instance, *IRE Trans. on Inform. Theory*, December, 1956, pp. 102–108, or theorem 1.1 of the Gel'fand and Iaglom reference cited on page 295.

Thus we have removed all three objections by selecting the concept of transinformation for the basis of our discussions.*

The following elementary properties of transinformation $I(X;Y)$ are self-evident:

$$I(X;Y) = I(Y;X) \tag{8-42}$$
$$I(X;Y) \geq 0 \tag{8-43}$$
$$I(X;Y) = 0 \quad \text{if } X \text{ and } Y \text{ are independent variables} \tag{8-44}$$
$$H(X) \geq H(X|Y) \tag{8-45}$$
$$H(Y) \geq H(Y|X) \tag{8-46}$$
$$H(X) + H(Y) \geq H(X,Y) \tag{8-47}$$

8-6. Maximization of the Entropy of a Continuous Random Variable. The maximum entropy of a complete discrete scheme occurs when all the events are equiprobable. This statement is not meaningful in the case of a random variable assuming a continuum of values. In this case, it is quite possible to have entropies which may not be finite. This might be interpreted as a pitfall for the definition of the channel capacity. However, the situation can be improved by assuming some plausible constraint on the nature of the density distributions. For instance, if the random variable has a finite range, then one may ask what type of density distribution leads to the greatest value of entropy. Such questions can be answered by using mathematical maximization techniques from the calculus of variations, such as the method of Lagrange multipliers. We shall employ this method in the subsequent sections for the following three basic constraints.

Case 1. What type of probability density distribution gives maximum entropy when the random variable is bounded by a finite interval, say $a \leq X \leq b$?

Case 2. Let X assume only nonnegative values, and let the first moment of X be a prespecified number a $(a > 0)$. What probability density distribution leads to the maximum entropy?

Case 3. Given a random variable with a specified second central moment (or a specified standard deviation σ), determine the probability density distribution that has the maximum entropy.

* This is indeed in accordance with our fundamental frame of reference. To measure something, one must have a basis of comparison. The "arithmetical ratio" of this comparison gives an indication of the relative measure of the thing that is being measured with respect to some adopted unit.

Similarly, in information theory, it is the difference of some a priori and a posteriori expectation of the system that provides us with a measure of the average gain or loss of knowledge or uncertainty about a system. In problems of information theory, the above "arithmetical ratio" in turn is translated into the difference of two entropies. This is of course due to the use of the logarithmic scale.

By using such techniques, the following answers will be obtained.

Case 1. The maximum entropy is associated with a random variable with a uniform probability density distribution between a and b.

Case 2. The maximum entropy corresponds to an exponential probability density distribution of the form

$$\frac{1}{a} e^{-x/a}$$

Case 3. Among the specified class of probability density functions the gaussian distribution

$$\frac{1}{\sqrt{2\pi}\,\sigma} e^{-(x^2/2\sigma^2)}$$

has the largest entropy.

8-7. Entropy Maximization Problems. Now we shall employ the variational technique for solving cases 1, 2, and 3 of the previous section.

1. One has to maximize

$$H(X) = -\int_a^b f(x) \ln f(x)\, dx \qquad \text{nats/sym} \tag{8-48)*$$

subject to the constraint

$$\int_a^b f(x)\, dx = 1$$

Let λ be a constant multiplier; the unknown solution $f(x)$ must satisfy

$$-\frac{\partial}{\partial f}(f \ln f) + \lambda \frac{\partial}{\partial f}(f) = 0 \tag{8-49}$$

that is,

$$-1 - \ln f(x) + \lambda = 0$$

$$f(x) = e^{\lambda - 1} \tag{8-50}$$

As λ is a constant, this equation shows that the required distribution must be uniformly constant in the interval (a,b). The value of this constant can be found directly.

$$f(x) = \frac{1}{b-a} \qquad a \le x \le b$$

The associated maximal entropy is

$$H(X) = -\int_a^b \frac{1}{b-a} \ln \frac{1}{b-a}\, dx$$

$$= \ln (b-a) \tag{8-51}$$

For a continuous random variable bounded to a finite interval, the uniform probability density provides the maximum entropy.

* The use of the natural logarithm here is for convenience in algebra.

2. When the expected value, that is, the first moment of the continuous random variable $X \geq 0$, is specified as $E(X) = a$, the unknown function $f(x)$ is subject to the following constraints:

$$H(X) = -\int_0^\infty f(x) \ln f(x)\, dx \tag{8-52}$$

$$\int_0^\infty f(x)\, dx = 1$$

$$\int_0^\infty x f(x)\, dx = a \qquad a > 0$$

Using the method of Lagrangian multipliers, we find

$$\frac{\partial}{\partial f}(-f \ln f) + \mu \frac{\partial}{\partial f}(f) + \lambda \frac{\partial}{\partial f}(fx) = -(1 + \ln f) + \mu + \lambda x = 0 \tag{8-53}$$

$$f(x) = e^{\mu - 1 + \lambda x} \tag{8-54}$$

The desired density distribution is of an exponential type. The values of λ and μ can be determined by direct substitution of $f(x)$ in the constraint relations:

$$e^{\mu-1} \int_0^\infty e^{\lambda x}\, dx = 1$$

$$e^{\mu-1} \int_0^\infty x e^{\lambda x}\, dx = a \tag{8-55}$$

Note that λ must not be positive; otherwise the probability constraint cannot be satisfied. Based on this remark, the above equations yield

$$e^{\mu-1} \left[\frac{1}{\lambda} e^{\lambda x}\right]_0^\infty = -\frac{e^{\mu-1}}{\lambda} = 1$$

$$e^{\mu-1} \left[\frac{1}{\lambda}\left(x - \frac{1}{\lambda}\right) e^{\lambda x}\right]_0^\infty = \frac{e^{\mu-1}}{\lambda^2} = a$$

$$e^{\mu-1} = -\lambda \qquad \lambda = -\frac{1}{a}$$

or

$$e^{\mu-1} = a\lambda^2 \qquad e^{\mu-1} = \frac{1}{a} \tag{8-56}$$

Finally

$$f(x) = -\lambda e^{\lambda x} = \frac{1}{a} e^{-x/a} \tag{8-57}$$

The extremal entropy has a value of

$$\begin{aligned} H(X) &= -\int_0^\infty \frac{1}{a} e^{-x/a} \ln \frac{1}{a} e^{-x/a}\, dx \\ &= \ln a \int_0^\infty \frac{1}{a} e^{-x/a}\, dx + \frac{1}{a} \int_0^\infty \frac{x}{a} e^{-x/a}\, dx \\ &= \ln a + 1 \end{aligned} \tag{8-58}$$

Thus the maximum possible entropy for all continuous random variables with prespecified first moment in $[0, \infty]$ is

$$H(X) = \ln ae$$

The logarithm is here computed to the natural base.

3. Let $f(x)$ be a one-dimensional probability density function, and let the random variable X have a preassigned standard deviation σ and zero mean. Which function $f(x)$ gives the maximum of the entropy $H(X)$?

Following the outlined procedure of the calculus of variations, one would maximize a linear combination of the constraints through evaluation of the constant multipliers of these constraints. To be specific,

$$\begin{aligned} \int_{-\infty}^{\infty} f(x)\,dx &= 1 \\ \int_{-\infty}^{\infty} x^2 f(x)\,dx &= \sigma^2 \\ H(X) &= -\int_{-\infty}^{\infty} f(x) \ln f(x)\,dx \end{aligned} \tag{8-59}$$

According to the previously mentioned technique,

$$-\frac{\partial}{\partial f}(f \ln f) + \frac{\partial}{\partial f}(\mu f) + \frac{\partial}{\partial f}(\lambda x^2 f) = 0 \tag{8-60}$$

$$-(1 + \ln f) + \mu + \lambda x^2 = 0 \tag{8-61}$$

$$f(x) = e^{\mu - 1} e^{\lambda x^2} \tag{8-62}$$

But

$$\int_{-\infty}^{\infty} e^{\mu - 1} e^{\lambda x^2}\,dx = 1 \tag{8-63}$$

$$\int_{-\infty}^{\infty} x^2 e^{\mu - 1} e^{\lambda x^2}\,dx = \sigma^2 \tag{8-64}$$

The latter equations yield

$$e^{\mu - 1} \sqrt{-\frac{\pi}{\lambda}} = 1 \tag{8-65}$$

$$\lambda = \frac{-1}{2\sigma^2}$$

$$e^{\mu - 1} = \sqrt{\frac{1}{2\pi}}\,\frac{1}{\sigma}$$

Finally,

$$f(x) = \frac{1}{\sqrt{2\pi}\,\sigma} e^{-(x^2/2\sigma^2)} \tag{8-66}$$

Among all one-dimensional density distributions with prespecified second-order moment (average power), the gaussian (normal) distribution provides the largest entropy. The maximum value of the entropy can be found directly.

$$\ln f(x) = -\ln \sqrt{2\pi}\,\sigma - \frac{x^2}{2\sigma^2} \tag{8-67}$$

$$H(X) = +\int_{-\infty}^{\infty} \frac{1}{\sqrt{2\pi}\,\sigma} e^{-(x^2/2\sigma^2)} \ln \sqrt{2\pi}\,\sigma\, dx + \int_{-\infty}^{\infty} \frac{1}{\sqrt{2\pi}\,\sigma} \frac{x^2}{2\sigma^2} e^{-(x^2/2\sigma^2)}\, dx \tag{8-68}$$

$$H(X) = \ln \sqrt{2\pi}\,\sigma + \frac{\overline{X^2}}{2\sigma^2} = \ln \sqrt{2\pi}\,\sigma + \frac{1}{2} \tag{8-69}$$

The maximum entropy in natural logarithmic units is

$$H(X) = \ln \sqrt{2\pi e}\,\sigma \tag{8-70}$$

The above three maximization problems can be generalized in a direct fashion to the case of multidimensional random variables under similar constraints. For example, it was shown by Shannon that in n-dimensional distributions, when all second-order moments are preassigned and the different variables are mutually independent, the maximum entropy will correspond to an n-dimensional gaussian distribution.

8-8. Gaussian Noisy Channels. As an example of the application of the preceding material, consider a continuous channel where the transmitted and the received signals have a joint gaussian density distribution.

$$f(x,y) = \frac{1}{2\pi\sigma_x\sigma_y\sqrt{1-\rho^2}} \exp\left[-\frac{1}{2(1-\rho^2)}\left(\frac{x^2}{\sigma_x^2} - 2\rho\frac{xy}{\sigma_x\sigma_y} + \frac{y^2}{\sigma_y^2}\right)\right] \qquad |\rho| \neq 1 \tag{8-71}$$

The marginal densities can be obtained directly as

$$f_1(x) = \frac{1}{\sqrt{2\pi}\,\sigma_x} \exp\left(-\frac{x^2}{2\sigma_x^2}\right) \tag{8-72}$$
$$f_2(y) = \frac{1}{\sqrt{2\pi}\,\sigma_y} \exp\left(-\frac{y^2}{2\sigma_y^2}\right)$$

The application of the defining equations for entropies yields

$$H(X) = -\int_{-\infty}^{\infty} f_1(x) \ln f_1(x)\, dx = \ln \sqrt{2\pi e}\,\sigma_x \tag{8-73}$$
$$H(Y) = -\int_{-\infty}^{\infty} f_2(y) \ln f_2(y)\, dy = \ln \sqrt{2\pi e}\,\sigma_y$$

$$\begin{aligned} H(X,Y) &= -\int_{-\infty}^{\infty}\int_{-\infty}^{\infty} f(x,y) \ln f(x,y)\, dx\, dy \\ &= \int_{-\infty}^{\infty}\int_{-\infty}^{\infty} f(x,y) \ln 2\pi\sigma_x\sigma_y\sqrt{1-\rho^2}\, dx\, dy \\ &\quad + \int_{-\infty}^{\infty}\int_{-\infty}^{\infty} f(x,y) \frac{1}{2(1-\rho^2)}\left(\frac{x^2}{\sigma_x^2} - 2\rho\frac{xy}{\sigma_x\sigma_y} + \frac{y^2}{\sigma_y^2}\right) dx\, dy \end{aligned} \tag{8-74}$$

The double integral of the last equation is the sum of three second-order moments, that is,

$$\frac{1}{2(1-\rho^2)}\left[\frac{1}{\sigma_x^2}E(X^2) - \frac{2\rho}{\sigma_x\sigma_y}E(XY) + \frac{1}{\sigma_y^2}E(Y^2)\right] = \frac{1}{2(1-\rho^2)}\left(1 - 2\rho\frac{\mu_{11}}{\sigma_x\sigma_y} + 1\right) \quad (8\text{-}75)$$

where $\mu_{11} = E(XY)$.

Now, recall that

$$\rho = \text{correlation coefficient} = \frac{\mu_{11}}{\sigma_x\sigma_y}$$

Thus

$$\begin{aligned} H(X,Y) &= \ln 2\pi\sigma_x\sigma_y\sqrt{1-\rho^2} + \frac{1}{2(1-\rho^2)}(2 - 2\rho^2) \\ &= \ln 2\pi\sigma_x\sigma_y e\sqrt{1-\rho^2} \end{aligned} \quad (8\text{-}76)$$

The mutual information in this channel is

$$\begin{aligned} I(X;Y) &= H(X) + H(Y) - H(X,Y) \\ &= \ln\sqrt{2\pi e}\,\sigma_x + \ln\sqrt{2\pi e}\,\sigma_y - \ln 2\pi\sigma_x\sigma_y e\sqrt{1-\rho^2} \end{aligned}$$

or

$$I(X;Y) = -\tfrac{1}{2}\ln(1-\rho^2) \qquad |\rho| \neq 1 \quad (8\text{-}77)$$

This equation indicates a measure of transinformation for the gaussian channel. The transinformation depends solely on the correlation coefficient between the transmitted and the received signals. When the noise is such that the received signal is independent of the transmitted signal we have

$$\rho = 0 \qquad \text{and} \qquad I(X;Y) = 0$$

When the correlation coefficient is increased, the mutual information will increase.

8-9. Transmission of Information in Presence of Additive Noise. The rate of transmission of information in a channel may be defined as the mean or the expected value of the function:

$$I = \log\frac{f(X,Y)}{f_1(X)f_2(Y)} \quad (8\text{-}78)$$

That is,

$$E(I) = I(X;Y) = \int_{-\infty}^{\infty}\int_{-\infty}^{\infty} f(x,y)\log\frac{f(x,y)}{f_1(x)f_2(y)}\,dx\,dy \quad (8\text{-}79)$$

The rate of transmission in bits provides a measure for the average information processed in the channel in the sense described previously. The maximum of $I(X;Y)$ with respect to all possible input probability densities, but under some additional constraints, leads to the concept of channel capacity in continuous channels. In other words, in contrast with the discrete case, the channel capacity in the continuous case is not an absolute quantity but depends on the constraint. The evaluation of the channel capacity is generally a difficult problem, and no general method can be given to cover all circumstances. However, in certain special cases we are able to study the rate of transmission and evaluate its maximum. One such case which is also of much practical significance is the channel with additive noise.

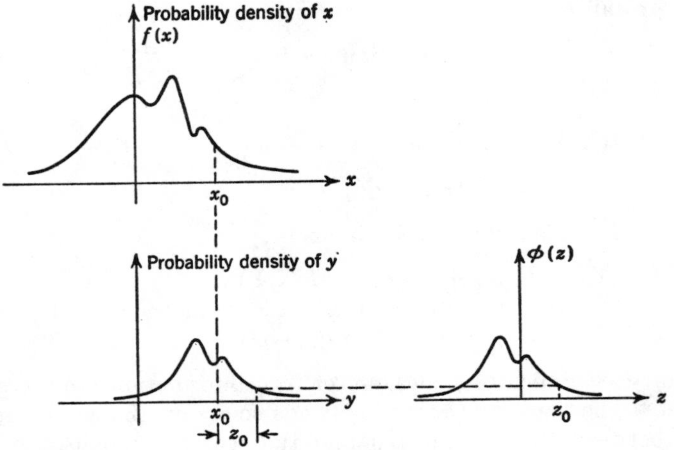

Fig. 8-3. An illustration of the performance of a continuous channel in the presence of additive noise.

Let X be the random variable describing the transmitted signal and Y the received signal. We assume that the noise in the channel is additive and statistically independent of X:

$$Y = X + Z$$
$$f_x(z + x|x) = \phi(z) \tag{8-80}$$

where Z is a random variable, with a probability density function $\phi(z)$. Equation (8-80) suggests some simplification in the relations among different density functions associated with X and Y. In fact, reference to Fig. 8-3 suggests that

$$P\{z_0 < Z < z_0 + dz_0\} = \phi(z_0)\, dz_0 \tag{8-81}$$
$$P\{y_0 < Y < y_0 + dy_0|X = x_0\} = P\{z_0 < Z < z_0 + dz_0|X = x_0\} \tag{8-82}$$

This conditional probability function is independent of x_0 and depends only on the noise structure. Therefore,

$$P\{y_0 < Y < y_0 + dy_0|X = x_0\} = P\{z_0 < Z < z_0 + dz_0\} = \phi(z_0)\, dz_0 \tag{8-83}$$

The conditional probability density $f_x(y|x)$ and the noise density function $\phi(z)$ are of identical structures. That is, in our familiar notation,

$$f_x(y|x) = f_x(x + z|x) = \phi(z) \tag{8-84}$$

The identical nature of the two probability functions here suggests identical entropies:

$$H(Y|X) = H(X + Z|X) = H(Z) \tag{8-85}$$

Thus the transinformation becomes

$$\begin{aligned} I(X;Y) &= H(Y) - H(Y|X) = H(Y) - H(Z) \\ &= H(Y) + \int_{-\infty}^{\infty} \phi(z) \log \phi(z)\, dz \end{aligned} \tag{8-86}$$

(For a schematic presentation, see Fig. 8-4.)

The channel capacity can be determined by finding the maximum of $I(X;Y)$ in Eq. (8-86) with respect to all possible probability density functions $f_1(x)$ subject to certain required constraints. The most common types of constraint are those limiting the peak or the average value of signal power at the transmitter (see Sec. 8-6). Such problems are generally tedious and require lengthy treatment. However, with appropriate further assumptions the problem may be simplified. An example of such a simplification is given in the next section.

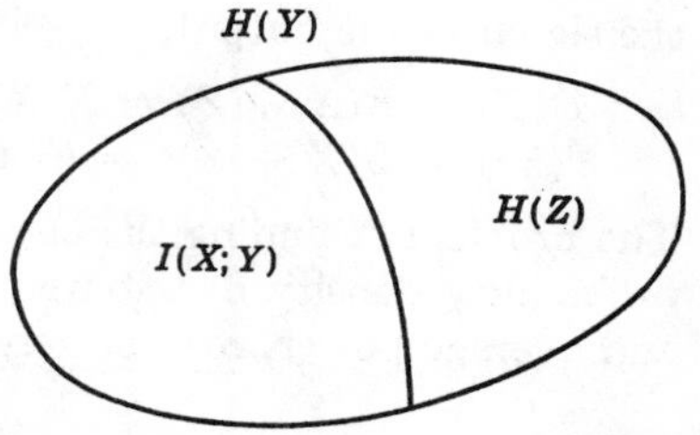

FIG. 8-4. A set-theory interpretation of the diagram suggests

$$I(X;Y) = H(Y) - H(Z)$$

(All entropies referred to are assumed to be finite and positive.)

8-10. Channel Capacity in Presence of Gaussian Additive Noise and Specified Transmitter and Noise Average Power. The following constraints are required:

$$\int_{-\infty}^{\infty} \frac{1}{\sqrt{2\pi}\,\sigma_z} \exp\left(-\frac{z^2}{2\sigma_z^2}\right) dz = 1 \tag{8-87}$$

$$\int_{-\infty}^{\infty} f_1(x)\, dx = 1 \tag{8-88}$$

$$\int_{-\infty}^{\infty} x f_1(x)\, dx = 0$$

$$\int_{-\infty}^{\infty} x^2 f_1(x)\, dx = \sigma_x^2$$

Equation (8-87) corresponds to the assumption that the noise is normally distributed with zero mean and average power σ_z^2. Equation (8-88) requires the transmitted signal to have zero mean and specified power σ_x^2 but is otherwise unrestricted. (The constraints on the mean of X and Z are included for convenience but they are not essential in the subsequent developments.)

According to Eq. (8-86), we have

$$I(X;Y) = H(Y) - H(Z) = H(Y) - \tfrac{1}{2} \ln 2\pi e\sigma_z^2 \tag{8-89}$$

The additive structure of noise in the channel provides us with a simpler means of computing the channel capacity, namely,

$$\max I(X;Y) = \max [H(Y) - H(Z)] \tag{8-90}$$

Observe that $I(X;Y)$ in the above form is only indirectly dependent on $f_1(x)$, the probability density at the input. The channel capacity may be computed by maximizing $H(Y)$ under the previously mentioned constraints. Let us first compute the mean and the standard deviation of the signal at the output.

$$E(Y) = E(X + Z) = E(X) + E(Z) = 0 \tag{8-91}$$
$$E(Y^2) = E(X + Z)^2 = E(X^2) + E(Z^2) = \sigma_x^2 + \sigma_z^2 = \text{const} \tag{8-92}$$

The problem of finding the channel capacity is now reduced to finding a probability density distribution having the following prespecified mean and average power, respectively:

$$0, \sigma_x^2 + \sigma_z^2$$

It was proved in Sec. 8-7 that, when the standard deviation is preassigned, the density distribution with the largest entropy is a normal distribution with zero mean [Eq. (8-66)]. The maximum obtainable entropy for $H(Y)$ in the considered class of input distributions is

$$H(Y) = \ln \sqrt{2\pi e(\sigma_x^2 + \sigma_z^2)} \qquad \text{nats/sym} \tag{8-93}$$

Thus the channel capacity is

$$C = \max I(X;Y) = \max [H(Y) - \tfrac{1}{2} \ln 2\pi e\sigma_z^2] \tag{8-94}$$
$$C = \tfrac{1}{2} \ln \frac{\sigma_x^2 + \sigma_z^2}{\sigma_z^2}$$

Finally, letting*

$$S = \sigma_x^2 \qquad N = \sigma_z^2$$

yields

$$C = \frac{1}{2} \ln \left(1 + \frac{S}{N}\right) \qquad \text{nats/sym} \tag{8-95}$$

* If X is a random voltage applied to a unit resistor and

$$\bar{X} = 0 \qquad \overline{X^2} = S$$

then the expected value of the instantaneous power delivered to the resistor is

$$E(X^2) = \overline{X^2} = S$$

This remarkably simple result (due to Shannon) gives the channel capacity as one-half of the logarithm of the ratio of the average signal power at the output to noise power. Note that Y is the sum of two independent random variables X and Z. Since Z and Y are normally distributed, one can show that X will also be normally distributed. To sum up, under additive gaussian noise and average power limitation on noise and on the output signal, the input and the output signals both must have normal distributions in order to give the highest rate of transmission of information. These considerations can be generalized to the case where the input and noise are multidimensional random variables. This will be discussed later.

8-11. Relation between the Entropies of Two Related Random Variables. In many physical problems we deal with systems whose input-output relationships are specified. In problems of probabilistic origin, it is often of interest to investigate the entropy of the output in relation to the entropy of the input. Consider the case where the output is a monotonic function of the input:

FIG. 8-5. An example of the transformation of random variables.

$$Y = g(X)$$

According to Sec. 5-11, the probability density $\rho(y)$ is given by

$$\rho(y) = f(x)\left|\frac{dx}{dy}\right| = f(\psi(y))|\psi'(y)|$$

where the unique function $\psi(y)$ stands for the inverse relationship between input and output, that is,

$$X = \psi(Y)$$

The entropy of the random variable Y describing the output of a transducer is found to be

$$\begin{aligned} H(Y) &= -\int_{-\infty}^{\infty} \rho(y) \log \rho(y)\, dy \\ &= -\int_{-\infty}^{\infty} f(\psi(y))|\psi'(y)| \log f(\psi(y))\, dy \\ &\quad - E\left[\log\left|J\left(\frac{X}{Y}\right)\right|\right] \end{aligned} \tag{8-96}$$

$J(x/y)$ stands for the so-called Jacobian of x with respect to y. The capital letters, as usual, denote random variables. Finally one finds

$$H(Y) = H(X) - E\left[\log\left|J\left(\frac{X}{Y}\right)\right|\right] \tag{8-97}$$

Thus the entropy at the output is equal to the entropy at the input less the average value of the logarithm of the absolute value of the Jacobian of the input with respect to the output. The above simple procedure may be applied in a direct fashion to show that the transinformation remains invariant under some quite general transformations. In fact, let X be subjected to a monotonic transformation $Z = g(X)$; according to the material of Sec. 5-12, we have

$$H(Z) = H(X) + \int f_1(x)\,dx \log\left|\frac{dz}{dx}\right|$$
$$H(Z|y_0) = H(X|y_0) + \int f_x(x|y_0)\,dx \log\left|\frac{dz}{dx}\right|$$

Note that

$$\begin{aligned} H(Z|Y) &= H(X|Y) + \int f_2(y_0)\,dy_0 \int f_x(x|y_0)\,dx \log\left|\frac{dz}{dx}\right| \\ &= H(X|Y) + \iint f(x,y_0)\,dx\,dy_0 \log\left|\frac{dz}{dx}\right| \end{aligned}$$

After the necessary calculations, one finds that

$$H(Z) - H(Z|Y) = H(X) - H(X|Y)$$

The procedure may be extended to the case of a transformation of two or more random variables.

An analogous situation holds if the transducer is a multiport, that is, X and Y are finite random vectors (multidimensional random variables). Under the conditions described in Sec. 5-11, we are able to compute the probability density function of the output. Using the Jacobian symbol, the result of Eq. (5-76) can be rewritten as

$$\rho(y_1,y_2, \ldots ,y_n) = \left| J\left(\frac{x_1,x_2, \ldots ,x_n}{y_1,y_2, \ldots ,y_n}\right)\right| f(x_1,x_2, \ldots ,x_n) \qquad (8\text{-}98)$$

The output entropy for this multiport is*

$$\begin{aligned} H(Y_1,Y_2, \ldots ,Y_n) &= -E[\log \rho(Y_1,Y_2, \ldots ,Y_n)] \\ &= H(X_1,X_2, \ldots ,X_n) - E\left[\log\left|J\left(\frac{X_1,X_2, \ldots ,X_n}{Y_1,Y_2, \ldots ,Y_n}\right)\right|\right] \end{aligned} \qquad (8\text{-}99)$$

In the particular case when the transducer achieves a simple linear algebraic operation, that is,

$$[Y] = [A][X]$$

* The entropy of a multidimensional random variable will be discussed shortly in more detail. We trust that the reader will not be inconvenienced by the injection of this paragraph slightly ahead of schedule.

the term $|J(Y/X)|$ is equal to the absolute value of the determinant Δ of the transformation, and the Jacobian $J(X/Y)$ is equal to $1/|\Delta|$ since

$$\left| J\left(\frac{X}{Y}\right) \right| = |\text{det of } A^{-1}| = \left| \frac{1}{\Delta} \right| \tag{8-100}$$

In this case the relation between the two entropies assumes the following simpler form:

$$H(Y_1,Y_2, \ldots ,Y_n) = H(X_1,X_2, \ldots ,X_n) + \log |\Delta| \tag{8-101}$$

An example of the application of this formula occurs when the entropy of a random vector is known and a linear transformation of coordinate axes takes place. For instance, for all distance-preserving transformations, such as the rotation of a coordinate system, the entropy of the input remains unchanged. The extension of the above to the case of a general linear system where input and output are related by a linear differential or integral equation requires much care. For such systems, the validity of many of the known results should be more closely examined.

8-12. Note on the Definition of Mutual Information. A more precise definition of the communication entropy of a source and channel has been subsequently used by the Russian mathematicians A. N. Kolmogorov, A. M. Iaglom, and I. M. Gel'fand.* While this definition has a more fundamental appeal than the one discussed earlier in this chapter, it relies on more complex mathematical tools, not generally available in engineering courses. For the sake of *reference* this definition is included here. However, its understanding is not mandatory in an introductory presentation.

Let ξ and ζ be random objects (vectors, functions, or even generalized functions) defined over X and Y with appropriate probability distributions (not necessarily continuous densities).

$$\begin{aligned} P_\xi\{A\} &= P\{\xi \in A\} \\ P_\eta\{B\} &= P\{\eta \in B\} \\ P_{\xi\eta}\{C\} &= P\{(\xi,\eta) \in C\} \end{aligned} \tag{8-102}$$

The quantity of information in the random object ξ relative to η is defined as

$$I(\xi;\eta) = \int_x \int_y P_{\xi\eta}\, dx\, dy \log \frac{P_{\xi\eta}\{dx\, dy\}}{P_\xi\{dx\}P_\eta\{dy\}} \tag{8-103}$$

* The above presentation is based on A. N. Kolmogorov, On the Shannon Theory of Information Transmission in the Case of Continuous Signals, *IRE Trans. on Inform. Theory,* vol. IT-2, pp. 102–108, December, 1956. See also A. N. Kolmogorov, A. M. Iaglom, and I. M. Gel'fand, "Quantity of Information and Entropy for Continuous Distributions," report given at Third All-Union Mathematics Conference, 1956.

The mathematical superiority of this formula to the one originally employed by Shannon lies in the fact that it is of a much more general nature. The following properties for the mutual information function can be established when ξ, η, and ζ are random objects of a fairly general nature.

$$I(\xi;\eta) = I(\eta;\xi)$$
$$I(\xi;\eta) \geq 0 \tag{8-104}$$

Equality holds if, and only if, ξ and η are independent objects. If (ξ_1,η_1) and (ξ_2,η_2) are two independent pairs, then

$$I((\xi_1,\xi_2);\, (\eta_1,\eta_2)) = I(\xi_1;\eta_1) + I(\xi_2;\eta_2)$$
$$I((\xi,\eta);\, \zeta) \geq I(\xi;\zeta)$$

$I((\xi,\eta);\zeta) = I(\eta;\zeta)$ if, and only if, the conditional distribution of ζ depends only on η for a fixed ξ and η.

PROBLEMS

8-1. A source transmits pulses of constant duration but different heights. The height of a pulse X varies between a_1 and a_2 volts. The source is connected to a channel; at the receiving end the height of the pulse can be considered as a random variable Y varying between b_1 and b_2 volts. The joint probability density of X and Y is

$$p(x,y) = \frac{1}{(a_2 - a_1)(b_2 - b_1)}$$

(*a*) Determine the source entropy $H(X)$.
(*b*) Determine the entropy $H(Y)$.
(*c*) Determine the entropy $H(X,Y)$.
(*d*) Determine the transinformation $I(X;Y)$ and discuss the results.

8-2. X is a random variable uniformly distributed between $+1$ and -1.
(*a*) Find the means, the standard deviations, and the correlation coefficients between the following variables: (1) X and X^2 and (2) X and X^3.
(*b*) Calculate the entropy associated with each individual random variable.

8-3. Consider a continuous communication system where the joint probability density is described as

$$f(x,y) = \frac{1}{\pi ab} \qquad \text{for } \frac{x^2}{a^2} + \frac{y^2}{b^2} < 1 \quad a > b > 0$$
$$= 0 \qquad \text{elsewhere}$$

(*a*) Determine the source entropy.
(*b*) Determine the equivocation entropy.
(*c*) Determine the entropy at the receiver.
(*d*) Determine the transinformation.

8-4. Let X and Y, the cartesian coordinates of a point M, be independent random variables uniformly distributed in [0,1].
(*a*) Study the marginal and the joint distribution of

$$R = (X^2 + Y^2)^{1/2} \qquad \text{and} \qquad \phi = \tan^{-1}\frac{Y}{X}$$

(*b*) Determine different entropies for the communication model (R,ϕ).
(*c*) Evaluate the transinformation.

8-5. A continuous channel has the following characteristics:

$$f(y|x) = \left(\frac{1}{\alpha}\sqrt{3\pi}\right) e^{-(y-\frac{1}{2}x)^2/3\alpha^2}$$

The input to the channel is a random voltage with density

$$f_1(x) = \frac{1}{2\alpha\sqrt{\pi}} e^{-x^2/4\alpha^2}$$

X and Y may assume any values from $-\infty$ to $+\infty$. Determine
(*a*) The entropy of the source
(*b*) The conditional entropy of the channel
(*c*) The transinformation

8-6. Answer the same questions as in Prob. 8-5 for the sources and the channels described below:

(*a*)
$$f_1(x) = x(4 - 3x) \qquad 0 \le x \le 1$$
$$f(y|x) = \frac{6y(2 - x - y)}{4 - 3x} \qquad 0 \le y \le 1$$

(*b*)
$$f_1(x) = e^{-x} \qquad 0 \le x \le \infty$$
$$f(y|x) = xe^{-xy} \qquad 0 \le y \le \infty$$

(*c*)
$$f(x,y) = \frac{1}{2x^2y} \qquad 1 \le x < \infty \qquad \frac{1}{x} \le y \le x$$

CHAPTER 9

TRANSMISSION OF BAND-LIMITED SIGNALS

9-1. Introduction. A major aim of this chapter is to derive and discuss the Shannon-Hartley fundamental channel-capacity formula for band-limited time functions. This formula states that under certain plausible conditions the maximum rate of transmission of information for band-limited signals perturbed by independent gaussian noise, when the signal and the noise *power* are limited to S and N, respectively, is

$$C_t = W \log \left(1 + \frac{S}{N}\right) \qquad \text{bits per second} \tag{9-1}$$

where $(-2\pi W, +2\pi W)$ specifies the frequency range of the class of band-limited signals under consideration. Since the class of band-limited signals constitutes the most important class of signals applied to any communication apparatus, the above equation forms a central theme for the study of optimum performance of communication devices transmitting continuous messages. While this well-known formula is astonishingly simple and intuitively could be accepted as a direct extension of Eq. (8-95), its derivation is rather complicated. Actual derivation of Eq. (9-1) is based on a number of assumptions which must be carefully examined and some mathematical developments which require detailed attention. While the information-theory content of this chapter will be primarily devoted to the derivation and physical interpretation of this basic formula, we shall digress and present some basic relevant mathematical techniques. Adequate acquaintance with these techniques will enable the reader to broaden his view and be prepared for undertaking similar problems.

Before proceeding, it is worthwhile to organize our thoughts by making a sketch of the development to come.

1. In Sec. 9-2 we continue our study of continuous channels without memory, when the input is a multidimensional random variable.

2. Section 9-3 presents the maximum rate of transmission of information for a class of multidimensional random variables perturbed by independent gaussian noise under certain power constraints.

3. Sections 9-4 to 9-7 are devoted to building a bridge for transition from a class of continuous signals to a class of multidimensional random

variables. The ensemble of continuous signals forms what is called a stochastic process. There are generally an infinite number of random variables involved in such signals. We therefore have to devise certain mathematical techniques enabling us to reduce a problem of such complexity to a problem dealing with finite multidimensional random variables. These sections present the plausible assumptions under which Eq. (9-1) holds. Thus we have transformed the problem of communication of band-limited continuous signals over noisy channels into the study of entropies associated with a multidimensional random variable.

4. Section 9-8 is somewhat of a digression. There we present the elements of an important mathematical tool, namely, the theory of normed vector spaces. In the long run, the reader will realize the impact of this important concept in many communication problems. As an immediate application of the idea of vector spaces, a presentation of the familiar Fourier series and sampling theorem will be given. The patient reader will be rewarded later with a deeper understanding of Shannon's geometric model of the encoding of continuous messages.

5. In Sec. 9-13 some possible encoding procedures will be discussed for a geometric model of communication of continuous messages. It will be shown that under favorable circumstances one may transmit at a rate arbitrarily close to C_t, as described in Eq. (9-1).

9-2. Entropies of Continuous Multivariate Distributions. The concept of the entropy of a continuous single variate was presented in some detail in Chap. 8. This concept can be directly generalized for defining the entropy associated with a multidimensional continuous random variable. For instance, let X stand for an n-dimensional multivariate

$$X = [X_1, X_2, \ldots, X_n]$$

with a probability density function

$$f_1(x_1, x_2, \ldots, x_n)$$

The associated entropy is defined as

$$\begin{aligned} H(X_1, X_2, \ldots, X_n) &= E[-\log f_1(X_1, X_2, \ldots, X_n)] \\ &= -\underbrace{\int_{-\infty}^{\infty} \int_{-\infty}^{\infty} \cdots \int_{-\infty}^{\infty}}_{n} f_1(x_1, x_2, \ldots, x_n) \\ &\qquad \log f_1(x_1, x_2, \ldots, x_n)\, dx_1\, dx_2 \cdots dx_n \end{aligned} \tag{9-2}$$

In order to save space, we may write this equation symbolically as

$$H(\tilde{X}) = E[-\log f_1(\tilde{X})] \tag{9-2a}$$

The meaning and the properties of $H(X)$ are similar to those described in Chap. 8. In a similar fashion we may describe the transinformation in a multidimensional continuous channel without memory. For instance, assume the output of the channel to be an m-dimensional random variable

$$Y = [Y_1, Y_2, \ldots, Y_m]$$

with corresponding density function

$$f_2(y_1, y_2, \ldots, y_m)$$

Then the entropy at the output can be obtained as

$$H(Y_1, Y_2, \ldots, Y_m) = E[-\log f_2(Y_1, Y_2, \ldots, Y_m)] \tag{9-3}$$

or, symbolically,

$$H(\tilde{Y}) = E[-\log f_2(\tilde{Y})]$$

The noise characteristic of the channel and the joint probability density function can be similarly defined as an extension of the same concepts in the two-dimensional case:

$$P\{y_1 < Y_1 < y_1 + dy_1, \ldots, y_m < Y_m < y_m + dy_m | X_1 = x_1, \ldots, X_n = x_n\} = f(\tilde{y}|\tilde{x})\, d\tilde{y}$$

$$P\{y_1 < Y_1 < y_1 + dy_1, \ldots, y_m < Y_m < y_m + dy_m \cap x_1 < X_1 < x_1 + dx_1, \ldots, x_n < X_n < x_n + dx_n\} = f(\tilde{x},\tilde{y})\, d\tilde{x}\, d\tilde{y}$$

The physical interpretation of the situation is quite simple. Suppose that, say, the heights and the ages of a group of people are being communicated over a noisy channel. At the input we know the two-dimensional probability density of the heights and the ages of the group. Then the input-output joint density is a function of four variables $f(x_1,x_2; y_1,y_2)$. The following relations are self-explanatory:

$$f_1(x_1,x_2) = \int_{y_1=-\infty}^{\infty} \int_{y_2=-\infty}^{\infty} f(x_1,x_2;\, y_1,y_2)\, dy_1\, dy_2$$

$$f_2(y_1,y_2) = \int_{x_1=-\infty}^{\infty} \int_{x_2=-\infty}^{\infty} f(x_1,x_2;\, y_1,y_2)\, dx_1\, dx_2$$

Each set of transmitted data $[x_1,x_2]$ is independent of the previously transmitted set (no memory involved); however, x_1 and x_2 of each particular set may be interdependent. In general, if the noise effect on a particular sequence, say the kth $[x_1,x_2, \ldots, x_n]$, is independent of the noise effect on any other transmitted sequence, say the jth $[x_1,x_2, \ldots, x_n]$, then we say that the transmitter has no memory. In the opposite case the transmitter exhibits a memory. The mathematical study of channels with memory is rather involved. In Chap. 11 we shall discuss briefly discrete channels with memory, but presently we shall confine ourselves to sources and channels without memory. In the light of the previous discussion, the transinformation in channels without memory can be symbolically written as

$$I(\tilde{X};\tilde{Y}) = E\left[\log \frac{f(\tilde{X},\tilde{Y})}{f_1(\tilde{X})f_2(\tilde{Y})}\right] \tag{9-4}$$

It is possible to derive formulas for channel capacity under certain constraints similar to those discussed in Chap. 8. A particular case of such constraints is discussed in the following section.

9-3. Mutual Information of Two Gaussian Random Vectors.* In many problems of the physical world we are interested in investigating mutual information of two complex random phenomena. Each of these phenomena may be expressed by a random vector, that is, an n variate. In this section, we investigate the simplest and perhaps most frequent case when two random vectors are normally distributed. The following derivation of the mutual information conveyed by a multidimensional gaussian random variable about another such variable is due to Gel'fand and Iaglom.† Let $X = [X_1,X_2, \ldots ,X_n]$ and $Y = [Y_1,Y_2, \ldots ,Y_m]$ be n- and m-dimensional normal random variables (random vectors), respectively. Let Z be the random vector describing their joint behaviors.

$$Z = [X;Y] = [X_1,X_2, \ldots ,X_n; Y_1,Y_2, \ldots ,Y_m]$$

Without loss of generality, we assume

$$\overline{X_k} = 0 \qquad k = 1, 2, \ldots, n$$
$$\overline{Y_k} = 0 \qquad k = 1, 2, \ldots, m$$

According to Sec. 7-2, the n-dimensional normal probability of X can be written as

$$f_1(x) = \frac{1}{(2\pi)^{n/2}(\det A)^{1/2}} \exp\left[-\tfrac{1}{2}(A^{-1}x,x)\right] \tag{9-5a}$$

where $[A]$ is the moment matrix; that is, its elements a_{ij} are defined as

$$a_{ij} = \text{moment } X_iX_j$$

This may be written symbolically as

$$a_{ij} = \int x_ix_jf(z)\,dz = \int x_ix_jf_1(x)\,dx$$

$f_1(x)$ symbolically refers to the n-dimensional probability function of x.

* See S. Kullback, "Information Theory and Statistics," chap. 9, John Wiley & Sons, Inc., New York, 1959. The proof of this section may be skipped in a first reading. The proof of the statement, although basically simple, assumes familiarity with quadratic functions, the partitioning of a multinormal variate into two sets, and the determination of their partitioned covariance matrix.

† I. M. Gel'fand and A. M. Iaglom, Calculation of the Amount of Information about a Random Function Contained in Another Such Function, *Uspekhi Mat. Nauk S.S.S.R.*, new series, vol. 12, 1957. English translation in *Trans. Am. Math. Soc.*, ser. 2, vol. 12, pp. 199–246, 1959. See also A. N. Kolmogorov, On the Shannon Theory of Information Transmission in the Case of Continuous Signals, *IRE Trans. on Inform. Theory,* vol. IT-2, pp. 102–108, December, 1956.

Similarly, for the distributions of Y and Z we have, respectively,

$$f_2(y) = \frac{1}{(2\pi)^{m/2}(\det B)^{1/2}} \exp\left[-\tfrac{1}{2}(B^{-1}y,y)\right] \tag{9-5b}$$

$$f_3(z) = \frac{1}{(2\pi)^{(n+m)/2}(\det C)^{1/2}} \exp\left[-\tfrac{1}{2}(C^{-1}z,z)\right] \tag{9-5c}$$

where the elements of $[B]$ and $[C]$ are defined, respectively, as:

$$b_{ij} = \int y_i y_j f_2(y)\,dy$$
$$c_{ij} = \int z_i z_j f(z)\,dz$$

The moment matrix $[C]$ for the joint gaussian distribution is found to be

$$[C] = \begin{bmatrix} A & D \\ D^t & B \end{bmatrix}$$

where t stands for matrix transposition and the elements of $[D]$ are defined as

$$d_{ij} = \int x_i y_j f(z)\,dz$$

(A, B, and C are assumed to be nonsingular matrices.)

In order to compute the mutual information between the two random vectors one has to employ Eq. (9-4). To this effect, we write symbolically

$$I(X;Y) = \int \left\{ \frac{1}{2} \log \frac{\det A \det B}{\det C} - \frac{1}{2}\left[(C^{-1}z,z) - (A^{-1}x,x) - (B^{-1}y,y)\right] \right\} f(z)\,dz \tag{9-6}$$

But note that

$$\begin{aligned} \int (A^{-1}x,x) f(z)\,dz &= \int (A^{-1}x,x) f_1(x)\,dx = n \\ \int (B^{-1}y,y) f(z)\,dz &= m \\ \int (C^{-1}z,z) f(z)\,dz &= n + m \end{aligned} \tag{9-7}$$

Finally the mutual information becomes

$$I(X;Y) = \frac{1}{2} \log \frac{\det A \det B}{\det C} \tag{9-8}$$

This compact formula could be of considerable use in application problems.

The mutual information of two gaussian random vectors can alternatively be expressed in terms of their correlation coefficients.

$$I(X;Y) = -\tfrac{1}{2} \log (1 - \rho_1^2)(1 - \rho_2^2) \cdots (1 - \rho_l^2) \tag{9-8a}$$

where ρ_j is the correlation coefficient between X'_j and Y'_j and $l = \min(n,m)$.

A formal proof of this statement can be established subject to an appropriate transformation of (X,Y) to a mutually independent set (X',Y'). The reader who wishes to forgo such an exercise may satisfy himself by considering this equation as an extension of Eq. (8-77).

9-4. A Channel-capacity Theorem for Additive Gaussian Noise. In order to obtain a simple formulation for the maximum transmission of information, we make the following plausible assumptions. These assumptions are of a practical nature, and they lead to a simplified mathematical formulation.

1. Let the input and the output be n-dimensional variates $\tilde{X}$ and $\tilde{Y}$, respectively, with

$$\overline{X_k} = 0 \qquad \overline{X_k^2} = \sigma_{x_k}^2 \qquad k = 1, 2, \ldots, n$$

2. Let the noise $\tilde{Z}$ also be an n-dimensional normal variate with

$$\overline{Z_k} = 0 \qquad \overline{Z_k^2} = \sigma_{z_k}^2 \qquad k = 1, 2, \ldots, n$$

Covariance of (Z_k and Z_j) $= 0$ for $k \neq j$ (independent components).

3. The noise is assumed to be additive, that is,

$$\tilde{Y} = \tilde{X} + \tilde{Z}$$

With these assumptions, we compute the transinformation and obtain the associated channel capacity. First, one can compute the entropy $H(\tilde{Y}|\tilde{X})$ similar to Eq. (8-84). In fact,

$$f_x(\tilde{y}|\tilde{x}) = f_x(\tilde{x} + \tilde{z}|\tilde{x}) = \varphi(\tilde{z}) \tag{9-9}$$

where $\varphi(\tilde{z})$ is the density of noise.

$$\varphi(\tilde{z}) = \prod_{k=1}^{n} \left[\frac{1}{(2\pi)^{1/2}} \frac{1}{\sigma_{z_k}} e^{-z_k^2/2\sigma_{z_k}^2} \right] \tag{9-10}$$

The entropy associated with $\tilde{Z}$ is the sum of the entropies of each of its components since the components are statistically independent; thus

$$H(\tilde{Y}|\tilde{X}) = -\int_{-\infty}^{\infty} \varphi(\tilde{z}) \log \varphi(\tilde{z})\, d\tilde{z} = H(\tilde{Z}) = \sum_{k=1}^{n} \log \sqrt{2\pi e}\, \sigma_{z_k} \tag{9-11}$$

The transinformation is

$$I(\tilde{X};\tilde{Y}) = H(\tilde{Y}) - H(\tilde{Y}|\tilde{X}) = H(\tilde{Y}) - H(\tilde{Z})$$
$$I(\tilde{X};\tilde{Y}) = H(\tilde{Y}) - \sum_{k=1}^{n} \log \sqrt{2\pi e}\, \sigma_{z_k} \tag{9-12}$$

In order to find the channel capacity, we need to maximize $I(\tilde{X};\tilde{Y})$ under assumed constraints. This coincides with the maximization of the entropy $H(\tilde{Y})$. Therefore, we examine more closely the random vector $\tilde{Y}$. Each component of $\tilde{X}$ is affected by an independent gaussian perturbation, that is,

$$Y_k = X_k + Z_k$$

Furthermore, we have specified that X_k has a distribution density with zero mean and given standard deviation. Therefore, according to Eqs. (6-10) and (6-24), Y_k has a distribution with zero mean and standard deviation σ_{y_k} such that

$$\sigma_{y_k}^2 = \sigma_{x_k}^2 + \sigma_{z_k}^2 \tag{9-13}$$

In order to make $H(\tilde{Y})$ a maximum we note that the entropy associated with a multidimensional scheme is greatest when:

1. All the dimensions are independent random variables.
2. Each dimension has the greatest entropy under the specified constraint (Sec. 8-7).

Condition 1 implies independence of sampling points Y_k, and condition 2 requires that each sample have a gaussian distribution (see Sec. 8-7). Thus we are led to the case where the received signal Y has statistically independent components each normally distributed (with zero mean and the specified standard deviation). The maximum possible transinformation under these conditions is

$$\begin{aligned} I_{\max}(\tilde{X};\tilde{Y}) &= \sum_{k=1}^{n} \log \sqrt{2\pi e}\,\sigma_{y_k} - \sum_{k=1}^{n} \log \sqrt{2\pi e}\,\sigma_{z_k} \\ &= \sum_{k=1}^{n} \log \frac{\sigma_{y_k}}{\sigma_{z_k}} \\ &= \sum_{k=1}^{n} \frac{1}{2} \log \left(1 + \frac{\sigma_{x_k}^2}{\sigma_{z_k}^2}\right) \end{aligned} \tag{9-14}$$

This formula obviously gives the channel capacity under the specified constraints. The source $\tilde{X}$ achieving such a rate of transinformation for the channel under discussion is also an independent n-dimensional gaussian source as its density can be directly obtained by the convolution of two such densities ($x_k = y_k - z_k$).

A final simplification is required before reaching the compact formulation of Eq. (9-1). We may assume that the variances (power) for all the X_k are identical, also that the variances of the noise samples are equal.

$$\begin{aligned} \sigma_{x_k} &= \sigma_x \\ \sigma_{z_k} &= \sigma_z \end{aligned} \qquad k = 1, 2, \ldots, n$$

In this case the maximum of the transinformation is

$$I_{\max}(\tilde{X};\tilde{Y}) = \sum_{k=1}^{n} \frac{1}{2} \log \left(1 + \frac{\sigma_x^2}{\sigma_z^2}\right) \tag{9-15}$$

In Sec. 6-3, it was pointed out that, when $\bar{X} = 0$, the variance σ_x^2 represents the average power dissipated in a unit resistor under the application of a random voltage X. Thus σ_x^2 and σ_z^2 can be replaced by S and N, respectively, the signal and the noise power.

$$I_{\max}(\tilde{X};\tilde{Y}) = \frac{n}{2} \log \left(1 + \frac{S}{N}\right)$$

The engineering significance of this equation lies in the fact that it gives an upper limit for transinformation in a communication channel under "reasonable" assumptions. Furthermore, the value of that upper bound is described in terms of signal and noise power which can both be measured in the laboratory.

9-5. Digression. While a continuation of the more basic approach to the mathematical study of multivariate channels with or without memory

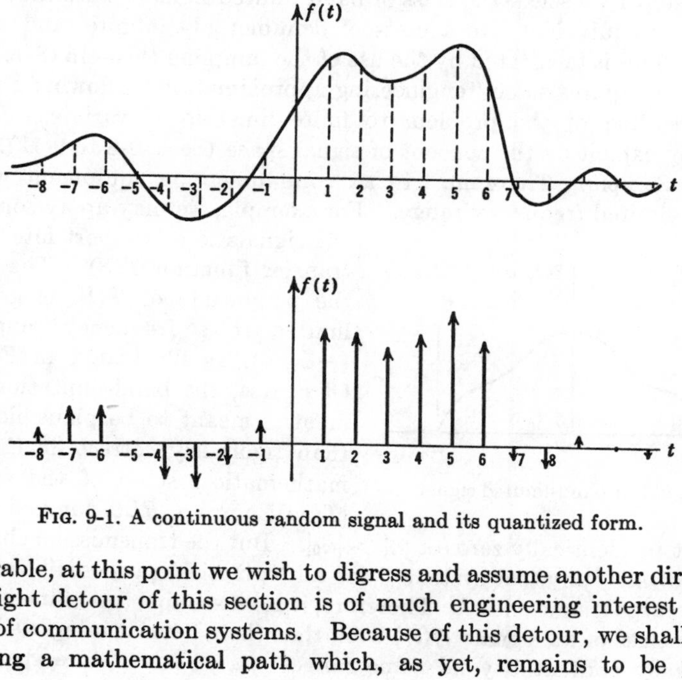

FIG. 9-1. A continuous random signal and its quantized form.

is desirable, at this point we wish to digress and assume another direction. The slight detour of this section is of much engineering interest in the study of communication systems. Because of this detour, we shall avoid traveling a mathematical path which, as yet, remains to be paved.

Meanwhile we shall have an opportunity to see the machinery which has already been supplied for the future building of such a road.

In order to appreciate the engineering problem which has forced us to this detour, consider what comes to the mind of an engineer thinking of a continuous signal. A random continuous signal $f(t)$ is represented in Fig. 9-1; its value at any instant of time is unpredictable. That is, at each instant t_k, $f(t_k)$ is a random variable. The signal is a member of a class of signals that are referred to as *stochastic*. The study of such sources requires the statistical knowledge of generally nondenumerably infinite numbers of random variables such as $f(t_k)$. This is a tremendous task and considerably beyond our present scope. However, on our detour two important steps are emphasized that will help the reader to obtain some useful results, namely, a simplified version of this complex problem. These steps are as follows:

1. Describe a class of continuous signals, the study of which can be "practically" reduced to the study of a discrete problem.
2. Once the problem is reduced to a discrete case, it may be further reduced by plausible approximation to sources of the type of finite multidimensional random variables. Thereafter, the methods of Secs. 9-2 and 9-3 may be employed.

For step 1 we select the class of band-limited signals, since they are, in a sense, "equivalent" to a class of denumerably infinite random variables. This is facilitated by the use of the sampling theorem (Sec. 9-6).

Step 2 requires some "engineering approximation," allowing a further simplification of the problem to finite-dimensional variates. This is done by exploiting the concept of signal space (Secs. 9-8 to 9-12).

9-6. Sampling Theorem. In all communication equipment we deal with a limited frequency range. For example, we may apply some electric signals to a two-port filter with a transfer function $T(S)$. The plot of the magnitude of $T(S)$ is generally limited to a frequency range, say $(-\omega_0,\omega_0)$, as illustrated in Fig. 9-2. Of course, the band-limitation statement is meant to be plausible rather than rigorously correct, in the strict mathematical sense of the word, as $|T(j\omega)|$ for an *RLC* lumped system cannot be identically zero for $|\omega| > |\omega_0|$. But the transmission characteristics of all "physical systems" for all "practical purposes" vanish for "very large frequencies." Therefore, it is reasonable to confine ourselves to the class of all signals $\{f(t)\}$ such that their Fourier integrals have no frequency content beyond some range $(-\omega_0,+\omega_0)$. More specifically,

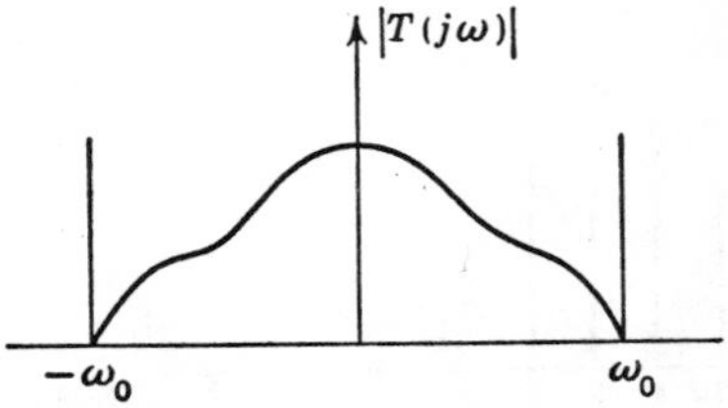

FIG. 9-2. A band-limited signal.

a signal $f(t)$ is said to be *band-limited* when

$$F(j\omega) = \mathfrak{F}[f(t)] = \int_{-\infty}^{\infty} f(t)e^{-j\omega t}\,dt = 0 \qquad \text{for } |\omega| > |\omega_0| \neq 0 \quad (9\text{-}16)$$

The frequencies of the human voice are generally between a few cycles and 4,000 cycles per second. The frequency range of the human eye and ear is also limited. The bandwidths of telephone, telegraph, and television are other examples of band-limited communication equipment.

While it takes a continuum of values to identify an arbitrary continuous signal in the real interval $[-\infty, +\infty]$, we shall show that the restriction of Eq. (9-16) will reduce the identification problem to that of a real function specified by a denumerable set of values. The sampling theorem below states that, if a signal is band-limited, it can be completely specified by its values at a sequence of discrete points. This theorem serves as a basis for the transition from a problem of continuum to a problem of discrete domain.

Theorem. Let $f(t)$ be a function of a real variable, possessing a band-limited Fourier integral transform $F(j\omega)$ such that*

$$F(j\omega) = 0 \qquad \text{for } |\omega| > |\omega_0|$$

Then $f(t)$ is completely determined by knowledge of its value at a sequence of points with abscissas equal to $\pi n/\omega_0$,

$$[n] = [\ldots, -2, -1, 0, 1, 2, \ldots]$$

Furthermore, $f(t)$ can be expressed in the following form:

$$f(t) = \sum_{n=-\infty}^{\infty} f\left(\frac{\pi n}{\omega_0}\right) \frac{\sin \omega_0(t - \pi n/\omega_0)}{\omega_0(t - \pi n/\omega_0)} \qquad (9\text{-}17)$$

Proof. Consider the pair of Fourier integrals:

$$F(j\omega) = \int_{-\infty}^{\infty} f(t)e^{-j\omega t}\,dt \qquad (9\text{-}18a)$$

$$f(t) = \frac{1}{2\pi} \int_{-\omega_0}^{\omega_0} F(j\omega)e^{j\omega t}\,d\omega \qquad (9\text{-}18b)$$

Obtaining the Fourier series expansion of the function $F(j\omega)$ in its funda-

* For mathematical convenience we confine ourselves to cases where $F(j\omega)$ does not contain delta functions. Such restrictions could be removed by special mathematical considerations. $F(j\omega)$ may contain delta functions but not at points $\omega = \pm\omega_0$. The theorem is also valid when the interval is not necessarily centered at the origin.

mental period of $2\omega_0$ yields*

$$F(j\omega) = \sum_{n=-\infty}^{\infty} c_n e^{(jn\pi/\omega_0)\omega} \qquad \text{for } |\omega| < |\omega_0| \tag{9-19}$$

where c_n are Fourier coefficients, that is,

$$c_n = \frac{1}{2\omega_0} \int_{-\omega_0}^{\omega_0} F(j\omega) e^{-j(n\pi/\omega_0)\omega}\, d\omega \tag{9-20}$$

Equation (9-18*b*) suggests the following values for the Fourier series coefficients in Eq. (9-20):

$$f\left(-\frac{n\pi}{\omega_0}\right) = \frac{1}{2\pi} \int_{-\omega_0}^{\omega_0} F(j\omega) e^{-j(n\pi/\omega_0)\omega}\, d\omega$$

Thus

$$c_n = \frac{\pi}{\omega_0} f\left(-\frac{\pi n}{\omega_0}\right) \tag{9-21}$$

These equations show that the Fourier coefficients are completely determined by a knowledge of the values of the original function $f(t)$, sampled at intervals of time π/ω_0 apart. Thus $F(j\omega)$ is uniquely determined by a knowledge of the values of the sampled ordinates. This, in turn, guarantees the unique determination of $f(t)$ through Eq. (9-18*b*), as Fourier integral pairs uniquely determine each other.†

In order to prove the identity of Eq. (9-17), we note that the right-hand member of the equation is a time function which assumes the value of $f(\pi n/\omega_0)$ at time $t = \pi n/\omega_0$. Indeed, all the terms of the summation of Eq. (9-17) vanish for $t = \pm(\pi k/\omega_0)$, $k = 1, 2, \ldots$, except for $k = n$, for which

$$f\left(\frac{\pi n}{\omega_0}\right) \frac{\sin \omega_0(\pi n/\omega_0 - \pi n/\omega_0)}{\omega_0(\pi n/\omega_0 - \pi n/\omega_0)} = f\left(\frac{\pi n}{\omega_0}\right) \tag{9-22}$$

Thus the right-hand member of Eq. (9-22) coincides with $f(t)$ at the sampling points. But according to the first part of the theorem, proved earlier, the function $f(t)$ is completely determined through its values at these sampled points, whence the identity of the two sides of Eq. (9-17) is proved. The symbolic equation (9-23) serves as a reminder of the

* The reader is assumed to be familiar with Fourier series and integrals. Fourier series expansion of a real-valued function is quite common. By the Fourier series expansion of $F(j\omega) = A(\omega) + jB(\omega)$, we mean the sum of the Fourier expansions of A and B in the same interval. Note that c_n and c_{-n} are not necessarily conjugate for all n unless $F(j\omega)$ is a real-valued function.

† As a more direct proof, substitute for c_n in Eq. (9-19) its values taken from Eq. (9-21); then substitute this result in Eq. (9-18*b*), in order to establish the identity of the two sides of Eq. (9-17).

fact that, when the sampling theorem holds, the coefficients of the Fourier series expansion of $F(j\omega)$ lead to the sampling values of $f(t)$.

$$\left[\frac{\omega_0}{\pi}\right][\ldots,c_{-2},c_{-1},c_0,c_1,c_2,\ldots] = \left[\ldots,f\left(\frac{-2\pi}{\omega_0}\right),f\left(\frac{-\pi}{\omega_0}\right),f(0),f\left(\frac{\pi}{\omega_0}\right),f\left(\frac{2\pi}{\omega_0}\right),\ldots\right] \quad (9\text{-}23)$$

An equivalent way of expressing the sampling theorem is the following:

Theorem. Let $F(j\omega)$ be a function of the real variable ω possessing a time-limited Fourier inverse integral transform $f(t)$, that is,

$$|f(t)| = 0 \qquad \text{for } |t| > |t_0|$$

Then $F(j\omega)$ is completely determined by its ordinates at a sequence of points with abscissas equal to $n\pi/t_0$, where n assumes the following values:

$$[\ldots,-2,-1,0,1,2,\ldots]$$

Proof. The proof of this theorem is analogous to the proof of the previous one. The theorem may be derived from the previous one in the following way:

1. In the hypothesis of the sampling theorem in the frequency domain, interchange t with ω, t_0 with ω_0, and f with F.

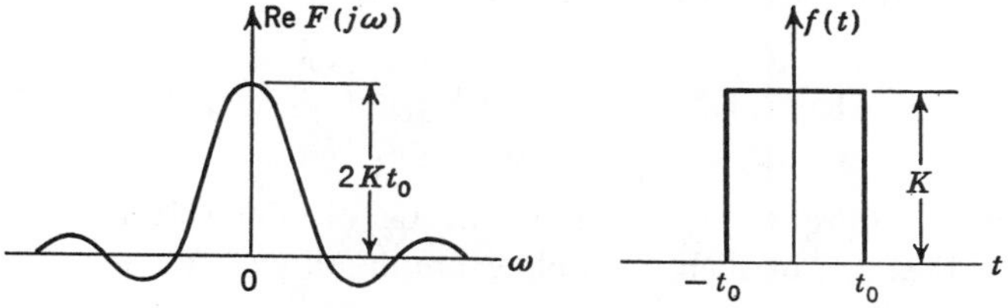

FIG. 9-3. A time-limited signal and its Fourier transform.

2. Then the conclusion of the theorem will coincide with the statement of the present one.

$$F(j\omega) = \sum_{n=-\infty}^{\infty} F\left(j\frac{\pi n}{t_0}\right) \frac{\sin t_0(\omega - \pi n/t_0)}{t_0(\omega - \pi n/t_0)} \qquad (9\text{-}24)$$

As an example of the application of the sampling theorem, consider a time-limited function $f(t)$ constrained by

$$f(t) = 0 \qquad \text{for } |t| > |t_0|$$

According to the sampling theorem in the frequency domain, $F(j\omega)$, the Fourier transform of $f(t)$, will be completely determined by its values at the doubly infinite sequence

$$\left[\ldots, -\frac{2\pi}{t_0}, -\frac{\pi}{t_0}, 0, \frac{\pi}{t_0}, \frac{2\pi}{t_0}, \ldots\right]$$

For example, if we assume that at these points the function $F(j\omega)$ takes on the sequence of values

$$[\ldots, 0, 0, 2Kt_0, 0, 0, \ldots]$$

then, because of the band-limited character of $f(t)$, $F(j\omega)$ will be given by

$$F(j\omega) = \sum_{-\infty}^{\infty} F\left(j\frac{n\pi}{t_0}\right) \frac{\sin t_0(\omega - n\pi/t_0)}{t_0(\omega - n\pi/t_0)} = 2Kt_0 \frac{\sin t_0\omega}{t_0\omega} \tag{9-25}$$

A check of the admissibility of this answer is provided by a consultation of Fourier integral tables (or direct derivation). In fact, from Fourier integral tables one obtains

$$f(t) = k \qquad \text{for } |t| < |t_0|$$
$$f(t) = 0 \qquad \text{elsewhere}$$

Thus $f(t)$ is indeed a band-limited function, as originally assumed.

Similarly, we may consider an ideal low-pass filter as an example for checking the validity of the sampling theorem in the time domain. Let

$$F(j\omega) = 0 \qquad \text{for } |\omega| \geq |\omega_0|$$
$$\text{Re } F(j\omega) = k \qquad \text{for } |\omega| < |\omega_0| \quad k > 0$$
$$\text{Im } F(j\omega) = 0 \qquad \text{for } |\omega| < |\omega_0|$$

The corresponding time function can be obtained from the Fourier integral tables (or, in such a simple example, directly):

$$f(t) = \omega_0 k \frac{\sin \omega_0 t}{\pi\omega_0 t} \tag{9-26}$$

Note that, because of the band-limited character of $F(j\omega)$, $f(t)$ should be completely determined by the following domain and range.

$$\left[\ldots, -\frac{2\pi}{\omega_0}, -\frac{\pi}{\omega_0}, 0, \frac{\pi}{\omega_0}, \frac{2\pi}{\omega_0}, \ldots\right]$$
$$\left[\ldots, 0, 0, \frac{\omega_0 k}{\pi}, 0, 0, \ldots\right]$$

This result is indeed in agreement with the expression for $f(t)$ suggested by the sampling theorem.

$$f(t) = \sum_{n=-\infty}^{\infty} f\left(\frac{\pi n}{\omega_0}\right) \frac{\sin \omega_0(t - \pi n/\omega_0)}{\omega_0(t - \pi n/\omega_0)} = \omega_0 k \frac{\sin \omega_0 t}{\pi\omega_0 t} \tag{9-27}$$

9-7. A Physical Interpretation of the Sampling Theorem. A physical interpretation of the sampling theorem can readily be proposed. Suppose that a continuous band-limited voltage signal $v(t)$ is given. We could quantize $v(t)$ at times

$$\left[\ldots, -\frac{2\pi}{\omega_0}, -\frac{\pi}{\omega_0}, 0, \frac{\pi}{\omega_0}, \frac{2\pi}{\omega_0}, \ldots\right]$$

The quantized voltages are successively applied to an ideal low-pass filter as impulses of appropriate magnitude at the specified times. The response of an ideal low-pass filter with cutoff at ω_0 to a unit impulse $u_0(t)$ is found to be $k(\sin \omega_0 t)/\pi t$, where k is the constant of the filter. Thus

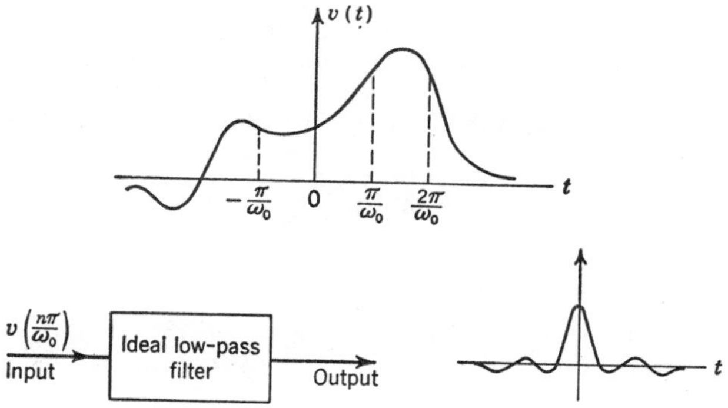

FIG. 9-4. Physical interpretation of the sampling theorem.

the total output of the filter, $v_0(t)$, will represent the original time function $v(t)$ within some scale factor.

$$v_0(t) = k\frac{\omega_0}{\pi} \sum_{n=-\infty}^{\infty} v\left(\frac{\pi n}{\omega_0}\right) \frac{\sin \omega_0(t - \pi n/\omega_0)}{\omega_0(t - \pi n/\omega_0)} \tag{9-28}$$

If the constant of the filter is π/ω_0, then $v_0(t)$ and $v(t)$ become identical.

The concept of the sampling theorem has frequently been employed in communication problems such as the extensive work done at the Bell Telephone Laboratories on speech transmission and also by other investigators prior to Shannon, such as Nyquist, Küpfmüler, and Gabor. The mathematical statement of the sampling theorem has been made by E. T. Whittaker, J. M. Whittaker, and several other mathematicians. Subsequent to Shannon's use of the sampling theorem, a number of interesting articles on this subject have appeared in the literature of engineering as well as in mathematical journals (see N-2 of the Appendix).

Example 9-1. We illustrate the sampling theorem with the following example. Consider a band-limited signal with a triangular frequency distribution, as shown in Fig. E9-1.

From a table of Fourier transforms (or by performing the integration) we find that the time-domain function describing such a signal is given by

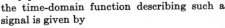

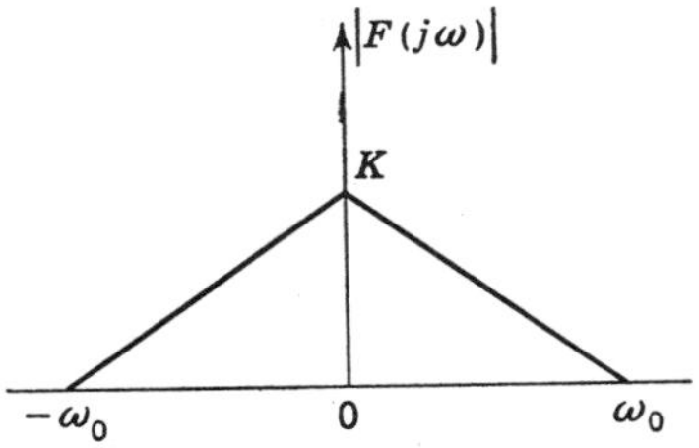

FIG. E9-1

$$f(t) = \frac{K\omega_0}{2\pi}\left[\frac{\sin(\omega_0 t/2)}{\omega_0 t/2}\right]^2 \tag{a}$$

Now, the sampling theorem states that

$$f(t) = \sum_{n=-\infty}^{\infty} f\left(\frac{\pi n}{\omega_0}\right)\frac{\sin \omega_0(t - \pi n/\omega_0)}{\omega_0(t - \pi n/\omega_0)} \tag{b}$$

where $f(\pi n/\omega_0)$ is the value of $f(t)$ at the respective sampling points.

We illustrate the theorem by showing that the summation of Eq. (*b*) does, in fact, yield the function in Eq. (*a*) when the appropriate values are inserted in Eq. (*b*).

First, evaluate $f(\pi n/\omega_0)$ at the sampling points by setting t in Eq. (*a*) equal to $\pi n/\omega_0$.

$$f(0) = \frac{K\omega_0}{2\pi} \tag{c}$$

$$f\left(\frac{n\pi}{\omega_0}\right) = \frac{K\omega_0}{2\pi}\left(\frac{\sin \pi m}{\pi m}\right)^2 = 0 \tag{d}$$

where $m = n/2$, $n = \pm 2, \pm 4, \pm 6, \ldots$.

$$f\left(\frac{n\pi}{\omega_0}\right) = \frac{K\omega_0}{2\pi}\left[\frac{\sin(2n/4)\pi}{(2n/4)\pi}\right]^2 = \frac{K\omega_0}{2\pi}\left(\frac{4}{2\pi n}\right)^2 = \frac{K\omega_0}{2\pi}\frac{4}{n^2\pi^2} \tag{e}$$

where n is odd.

Next, evaluate $\dfrac{\sin \omega_0(t - n\pi/\omega_0)}{\omega_0(t - n\pi/\omega_0)}$ for each sampling point by letting n take on integral values from $-\infty$ to ∞.

By a trigonometric identity, $\sin \omega_0(t - n\pi/\omega_0) = \sin \omega_0 t \cos n\pi - \cos \omega_0 t \sin n\pi$. Therefore

$$\frac{\sin \omega_0(t - n\pi/\omega_0)}{\omega_0(t - n\pi/\omega_0)} = \sin \omega_0 t \frac{(-1)^n}{\omega_0 t - n\pi}$$

Equation (*b*) can now be written

$$f(t) = \sin \omega_0 t \sum_{n=-\infty}^{\infty} f\left(\frac{n\pi}{\omega_0}\right)\frac{(-1)^n}{\omega_0 t - \pi n} \tag{f}$$

Since $f(n\pi/\omega_0) = f(-n\pi/\omega_0)$ [from Eq. (*e*)], we can add pairs of terms having the same $|n|$ and take the summation from $n = 1$ to $n = \infty$. The pairwise sums can be written

$$f\left(\frac{\pi n}{\omega_0}\right)\left[\frac{(-1)^n}{\omega_0 t - \pi n} + \frac{(-1)^n}{\omega_0 t + \pi n}\right] = f\left(\frac{\pi n}{\omega_0}\right)\left[\frac{(-1)^n \cdot 2\omega_0 t}{(\omega_0 t)^2 - \pi^2 n^2}\right] \qquad \text{for } n > 0 \tag{g}$$

Then, substituting Eq. (*g*) into Eq. (*f*),

$$f(t) = \sin \omega_0 t \left[\frac{1}{\omega_0 t} f(0) + 2\omega_0 t \sum_{n=1}^{\infty} f\left(\frac{\pi n}{\omega_0}\right)\frac{(-1)^n}{(\omega_0 t)^2 - (\pi n)^2}\right] \tag{h}$$

But from Eqs. (c) to (e) we know the values of $f(n\pi/\omega_0)$ to substitute into (h). Furthermore, we know that $f(n\pi/\omega_0) = 0$ (n even) from Eq. (d), and so we can substitute a new index, m, into the summation, where

$$n = 2m - 1 \tag{i}$$

thus summing only over the odd values of n.

Performing these substitutions in (h) we obtain

$$f(t) = \frac{K\omega_0}{2\pi} \sin \omega_0 t \left[\frac{1}{\omega_0 t} + 2\omega_0 t \sum_{m=1}^{\infty} \frac{-4}{\pi^2(2m-1)^2} \frac{1}{(\omega_0 t)^2 - \pi^2(2m-1)^2} \right] \tag{j}$$

We substitute the following trigonometric identity for $\sin \omega_0 t$:

$$\sin \omega_0 t = 2 \sin \frac{\omega_0 t}{2} \cos \frac{\omega_0 t}{2} \tag{k}$$

$$f(t) = \frac{K\omega_0}{2\pi} 2 \sin \frac{\omega_0 t}{2} \cos \frac{\omega_0 t}{2} \left(\frac{1}{\omega_0 t} - 2\omega_0 t \sum_{m=1}^{\infty} \frac{1}{\pi^2(2m-1)^2\{(\omega_0 t/2)^2 - [(2m-1)/2]^2 \pi^2\}} \right) \tag{l}$$

Performing some algebraic manipulations, we get

$$f(t) = \frac{K\omega_0}{2\pi} \frac{\sin (\omega_0 t/2) \cos (\omega_0 t/2)}{\omega_0 t/2} \left[1 + \frac{2(\omega_0 t)^2}{\pi^2} \sum_{m=1}^{\infty} \frac{1}{(2m-1)^2} \frac{1}{(m-\frac{1}{2})^2\pi^2 - (\omega_0 t/2)^2} \right] \tag{m}$$

Then by bringing $(\omega_0 t)^2/\pi^2$ inside the summation and dividing both numerator and denominator by Eq. (d),

$$f(t) = \frac{K\omega_0}{2\pi} \frac{\sin (\omega_0 t/2) \cos (\omega_0 t/2)}{\omega_0 t/2} \left[1 + 2 \sum_{m=1}^{\infty} \frac{1}{\pi^2(m-\frac{1}{2})^2} \frac{(\omega_0 t/2)^2}{\pi^2(m-\frac{1}{2})^2 - (\omega_0 t/2)^2} \right] \tag{n}$$

Rewriting the second term in the summation, using the following identity,

$$\frac{x^2}{a^2 - x^2} = \frac{a^2}{a^2 - x^2} - 1$$

$$f(t) = \frac{K\omega_0}{2\pi} \frac{\sin (\omega_0 t/2) \cos (\omega_0 t/2)}{\omega_0 t/2} \left\{ 1 + 2 \sum_{m=1}^{\infty} \frac{1}{\pi^2(m-\frac{1}{2})^2} \left[\frac{\pi^2(m-\frac{1}{2})^2}{\pi^2(m-\frac{1}{2})^2 - (\omega_0 t/2)^2} - 1 \right] \right\} \tag{o}$$

Separating the parts of the summation,

$$f(t) = \frac{K\omega_0}{2\pi} \frac{\sin (\omega_0 t/2) \cos (\omega_0 t/2)}{\omega_0 t/2} \left[1 + 2 \sum_{m=1}^{\infty} \frac{1}{\pi^2(m-\frac{1}{2})^2 - (\omega_0 t/2)^2} - \frac{8}{\pi^2} \sum_{m=1}^{\infty} \frac{1}{(2m-1)^2} \right] \tag{p}$$

But

$$\tan z = 2z \sum_{n=0}^{\infty} \frac{1}{\pi^2(n+\frac{1}{2})^2 - z^2} \tag{q}$$

(see, for instance, E. C. Titchmarsh, "Theory of Functions," p. 113, Oxford University Press, New York, 1950), which can be rewritten with the substitution $m = n + 1$.

$$\frac{\tan z}{z} = 2 \sum_{m=1}^{\infty} \frac{1}{\pi^2(m - \frac{1}{2})^2 - z^2} \tag{r}$$

Widder* shows

$$\sum_{m=1}^{\infty} \frac{1}{(2m-1)^2} = 1 + \frac{1}{3^2} + \frac{1}{5^2} + \cdots = \frac{\pi^2}{8} \tag{s}$$

Substituting $z = \omega_0 t/2$ in Eq. (r) and substituting Eqs. (r) and (s) in Eq. (p), we get

$$f(t) = \frac{K\omega_0}{2\pi} \frac{\sin(\omega_0 t/2)\cos(\omega_0 t/2)}{\omega_0 t/2} \left[1 + \frac{\tan(\omega_0 t/2)}{\omega_0 t/2} - \frac{8}{\pi^2}\frac{\pi^2}{8}\right] \tag{t}$$

and finally

$$f(t) = \frac{K\omega_0}{2\pi} \left[\frac{\sin(\omega_0 t/2)}{\omega_0 t/2}\right]^2 \tag{u}$$

This solution was derived by several students, in particular, S. Rubin, during a course on information theory.

9-8. The Concept of a Vector Space. The use of n-dimensional real or complex space is quite common in engineering problems. For instance, in the study of signals in communication theories one frequently employs the concept of vector space, although this may not appear in its strict mathematical frame of reference. The concept of *power content* of signals in engineering texts is an example of tacitly exploiting a basic product of vector-space theory. It is safe to assume that not all the readers are familiar with these concepts. For the benefit of such readers, we include this section as a digression from the main stream of thought. *The section is designed to provide a brief glimpse into vector-space theory.* Meanwhile the use of basic axioms required by a vector space and the properties specified for a norm are presented in a way parallel to the treatment of some of the material presented in Chap. 2.

Spaces. A set or collection of elements, generally called *points*, is said to form a *space.* This space is not what is generally understood by geometrical points. This may be a collection of points, or vectors, or functions, etc. For example, the set of all functions

$$y = x \cos \alpha$$

where x is a given real number and α a real variable, forms a space S. For each value of the parameter x we have a point in this space. The

* "Advanced Calculus," p. 340, Example *B*, Prentice-Hall, Inc., Englewood Cliffs, N.J.

points $y_1 = 2\cos\alpha$, $y_2 = -3\cos\alpha$ are elements of the set under consideration. This is usually written in the form

$$y_1 \in S \qquad y_2 \in S$$

Vector Space. A vector space is defined in the following way: Let F be the set of real or complex numbers. V is said to be a *vector space* if, for every pair of points x and y, $x \in V$, $y \in V$, and a number $\alpha \in F$, the operations of addition and multiplication by a number are defined so that

$$\begin{aligned} x + y &\in V \\ \alpha x &\in V \end{aligned} \tag{9-29}$$

Furthermore, the following properties of the space are required:

1. There exists an element "zero" in V such that, for each $x \in V$, $x + 0 = x$.
2. For every $x \in V$ there corresponds a unique point $-x \in V$ such that $-x + x = 0$.
3. Addition is associative, for example,

$$(x + y) + z = x + (y + z) = x + y + z$$

4. $x + y = y + x$ (commutative law) for all $x, y \in V$.
5. $\alpha(x + y) = \alpha x + \alpha y$.
6. $(\alpha + \beta)x = \alpha x + \beta x$.
7. $(\alpha\beta)x = \alpha(\beta x)$.
8. $1 \cdot x = x$.
9. $0 \cdot x = 0$.

It is worthwhile to note here that the above properties are those of ordinary vectors as encountered in undergraduate courses in applied sciences, and for this reason any set of elements that has these properties is called a *vector space*. In mathematical terminology, V is said to form an *abstract space* over F.

EXAMPLE. The most familiar example of vector space is given by an ordinary two-dimensional rectangular coordinate, that is, the set of all ordered pairs (a_1,a_2) of real numbers, addition being defined as follows: Given

$$x = (a_1,a_2) \qquad y = (b_1,b_2)$$

then $x + y = (a_1 + b_1,\ a_2 + b_2)$ and $\alpha x = (\alpha a_1,\alpha a_2)$.

$$\text{Vector } 0 = (0,0) \qquad x = (a_1,a_2) \qquad -x = (-a_1,-a_2)$$

All the properties listed above being satisfied, the set forms a vector space. One has to be rather careful here in interpreting the pair (a_1,a_2). It is familiar from analytical geometry that such a pair represents a "point" in the plane, and so addition of pairs as defined above will be meaningless in geometrical language. But there is no harm in letting a pair (a_1,a_2), say, represent a vector issuing from the origin in the plane to the point (a_1,a_2); then any misunderstanding can be obviated.

Linear and Linear Normed Spaces. A space S is said to be linear if, for every pair of numbers $(\alpha \in F, \beta \in F)$ and any pair x, y of S, $\alpha x + \beta y \in S$. In the application of the theory of linear vector spaces to physical problems, a most significant type of vector space is encountered. This is called *normed linear space.* Normed linear spaces exhibit a natural generalization of the familiar euclidean space. We first define what is meant by *norm.*

The norm of an element x of V, denoted by $||x||$, is a nonnegative real number satisfying the following properties.

$$\begin{aligned}
&\text{(I)} && ||x|| = 0 \qquad \text{if } x = 0 \\
&\text{(II)} && ||x|| \neq 0 \qquad \text{if } x \neq 0 \\
&\text{(III)} && ||x + y|| \leq ||x|| + ||y|| \\
&\text{(IV)} && ||\alpha x|| = |\alpha|\ ||x||
\end{aligned} \tag{9-30}$$

A vector space with a norm is called *normed linear space.*

Inner Product. As is known from ordinary vector analysis, the "inner product" of two vectors $x = (a_1,a_2)$ and $y = (b_1,b_2)$ in the real two-dimensional rectangular system is defined as $\langle x,y\rangle = a_1b_1 + a_2b_2$.

If we adopt the $\sqrt{\langle x,x\rangle}$ of a vector as its norm, we have

$$\langle x,x\rangle = a_1^2 + a_2^2 = ||x||^2$$

If the two-dimensional normed linear space is a complex space, then a natural extension of the above definition for the inner product leads to the following. Let

$$x = (\xi_1,\eta_1)$$
$$y = (\xi_2,\eta_2)$$

where ξ_1, η_1, ξ_2, η_2 are complex numbers. Then the inner product of x and y is

$$\langle x,y\rangle = \xi_1\bar{\xi}_2 + \eta_1\bar{\eta}_2$$

where the bar stands for a complex conjugate. The norm of x is defined as

$$\langle x,x\rangle = ||x||^2 = \xi_1\bar{\xi}_1 + \eta_1\bar{\eta}_1 = |\xi_1|^2 + |\eta_1|^2$$

Note that when the space is real this more general definition of the inner product is obviously valid. A vector space for which the inner product is defined is sometimes called an *inner-product space.* The following properties for the inner product are given:

$$\begin{aligned}
&\text{(I)} && \langle x,y\rangle = \overline{\langle y,x\rangle} \\
&\text{(II)} && \langle \alpha_1x_1 + \alpha_2x_2, y\rangle = \alpha_1\langle x_1,y\rangle + \alpha_2\langle x_2,y\rangle \\
&\text{(III)} && \langle x,x\rangle \geq 0 \qquad \langle x,x\rangle = 0 \qquad \text{if, and only if, } x = 0
\end{aligned} \tag{9-31}$$

N-dimensional Real or Complex Inner-product Spaces. We have already defined the linear spaces with an inner product. Here we should like to reiterate and establish the notation that is commonly used for real and complex spaces.

An n-dimensional real or complex space will be denoted by R^n and C^n, respectively.

An example of R^1 is given by the space of points on a straight line. A point x has a norm equal to the absolute value of its abscissa. The inner product of two points x and y is xy. As an example of the C^1, one may consider the set of all complex numbers.

The space R^n is the n-dimensional ordinary euclidean space. The sum of two elements $x \in R^n$ and $y \in R^n$,

$$x = (a_1,a_2, \ldots ,a_n) \qquad y = (b_1,b_2, \ldots ,b_n)$$

is defined as

$$x + y = (a_1 + b_1,\ a_2 + b_2,\ \ldots,\ a_n + b_n)$$

Then one sees that the square of a natural norm for an element here is $||x||^2 = \sum_{k=1}^{n} a_k{}^2$ and the inner product of x and y is

$$\langle x,y \rangle = \sum_{i=1}^{n} a_ib_i \tag{9-32}$$

For a complex normed space C^n, each element $x \in C^n$ is defined by its complex coordinates.

$$x = (x_1,x_2, \ldots ,x_n) \qquad \text{where each } x_i \text{ is complex}$$

Expressions for norm and inner products can also be readily obtained.

Hilbert Space. A direct generalization of R^n will require defining R^∞, where R^∞ denotes the set of all vectors of the form $(a_1,a_2,a_3, \ldots)$ with a denumerably infinite number of components. Here, addition and scalar multiplication will be defined as usual and the norm as

$$||x|| = \sqrt{\sum_{i=1}^{\infty} a_i{}^2}$$

Evidently, the norm will have a meaning if $\sum_{i=1}^{\infty} a_i{}^2$ is convergent, i.e., $\sum_{i=1}^{\infty} a_i{}^2 < \infty$. Hence only those vectors with an infinite number of components and possessing finite norms are allowed. The inner product of elements x and y is defined as usual:

$$\langle x,y \rangle = \sum_{i=1}^{\infty} a_i b_i$$

This definition has a meaning because

$$|a_i b_i| \leq (a_i^2 + b_i^2)$$

Thus

$$\sum_{i=1}^{\infty} |a_i b_i| \leq \left(\sum_1^{\infty} a_i^2 + \sum_1^{\infty} b_i^2\right) < \infty$$

The totality of all such vectors which have a denumerably infinite number of real components and whose norm has meaning is called a real *Hilbert space.**

It is to be noted that the Hilbert space need not be a real space. A space C^∞ with an inner product also forms a Hilbert space. In this case we have a complex Hilbert space. The inner product of two elements x and y is defined by

$$\langle x,y \rangle = \sum_{k=1}^{\infty} x_k \bar{y}_k = \overline{\langle y,x \rangle} \tag{9-33}$$

At the close of this short and incomplete digression into the field of vector space, it seems appropriate to describe in passing one or two more related terms. A set of elements

$$[A] = [A_1, A_2, \ldots, A_n]$$

of a vector space S is said to form a *linearly independent* set if there exist no nonzero $k_i \in F$ numbers

$$[k] = [k_1, k_2, \ldots, k_n]$$

such that

$$[A][k]^t = k_1 A_1 + \cdots + k_n A_n = 0$$

A *basis* in a vector space is a set of linearly independent elements

$$B = \{x_1, x_2, \ldots, x_m\}$$

such that every element x of the space can be generated in a unique manner by a linear combination of the elements of B. (B spans S.)

$$x = k_1 x_1 + k_2 x_2 + \cdots + k_m x_m \qquad \text{all } k\text{'s taken from } F$$

If m is finite, the vector space is a finite-dimensional space. It can be shown that every linear space has a basis.

Two elements x and y of S are said to be *orthogonal* if their inner product is zero:

$$\langle x,y \rangle = \sum_{k=1}^{\infty} x_k \bar{y}_k = 0 \tag{9-34}$$

* For complete definition of a Hilbert space, see textbooks on linear spaces.

Furthermore, if

$$||x|| = ||y|| = 1 \tag{9-35}$$

then the two elements are said to be *orthonormal*. For example, in the ordinary space R^2 the elements $x = (1,2)$ and $y = (3,-\frac{3}{2})$ are orthogonal, while $x = (\frac{1}{2},\sqrt{3}/2)$, $y = (-\sqrt{3}/2,\frac{1}{2})$ are orthonormal elements. The latter points form a basis for R^2.

As a direct extension of the above, subject to some mathematical care (according to the concept of convergence), one can consider a class of well-defined functions (say real signals with finite power in a finite time interval) as points in a Hilbert space. Equations (9-34) and (9-35) suggest the following definitions for orthogonality and orthonormality of two elements $f(x)$ and $g(x)$:

$$\int_a^b [f(x)]^2\,dx = ||f||^2 \qquad \text{norm}$$

$$\int_a^b f(x)\,g(x)\,dx = \langle f,g \rangle = 0 \qquad \text{orthogonality}$$

$$\int_a^b [f(x)]^2\,dx = \int_a^b [g(x)]^2\,dx = 1 \qquad \text{orthonormality}$$

9-9. Fourier-series Signal Space. As an immediate engineering application of the vector-space concept, we shall show that the function space associated with the class of ordinary communication signals possessing a Fourier series expansion is a Hilbert space. On the basis of this consideration, the reader may appreciate the use of the powerful mathematical tools of vector spaces for subsequent research in the field of electrical communications.

Obviously, at present, we shall not be concerned with the fact that the reader may not be immediately rewarded by the use of this modern tool. To be able to apply the theory requires the pedagogic development of many examples of applications. This is beyond our present objective. However, the two cases of Fourier series and the sampling theorem will be discussed briefly.

Fourier Series and the Hilbert Space. Consider a function $f(x)$ defined in the interval $[-\pi,+\pi]$ and expressible in that interval in the form of a convergent Fourier series.

$$\begin{aligned} f(x) &= \frac{a_0}{2} + \sum_{k=1}^{\infty} a_k \cos kx + \sum_{k=1}^{\infty} b_k \sin kx \\ &= \sum_{k=-\infty}^{+\infty} c_k e^{jkx} \end{aligned} \tag{9-36}$$

where
$$a_0 = \frac{1}{\pi}\int_{-\pi}^{\pi} f(x)\,dx$$
$$a_k = \frac{1}{\pi}\int_{-\pi}^{\pi} f(x)\cos kx\,dx \qquad k = 1, 2, \ldots \tag{9-37}$$
$$b_k = \frac{1}{\pi}\int_{-\pi}^{\pi} f(x)\sin kx\,dx$$

and
$$c_k = \frac{1}{2\pi}\int_{-\pi}^{\pi} f(x)e^{-jkx}\,dx \tag{9-38}$$

[For positive k, $c_k = (a_k - jb_k)/2$, and for negative k, $c_k = (a_k + jb_k)/2$.]

Without undue concern about the conditions for the existence and convergence of such series, we merely assume that the function $f(x)$ has a Fourier series expansion and belongs to the class of square summable functions, that is,

$$\int_{-\pi}^{\pi} [f(x)]^2\,dx < \infty \tag{9-39}$$

For notational convenience, let us denote this class of functions by L^2, the superscript 2 being a reminder of the fact that the integral of the second power of $f(x)$ in the interval of definition is a finite number. Now a function $f \in L^2$ can be represented by a point in a Hilbert space. In fact, the doubly infinite sequence of numbers c_k in Eq. (9-36) uniquely defines the function $f(x)$.

$$[\ldots,c_{-2},c_{-1},c_0,c_1,c_2,\ldots]$$

The addition of elements and the zero element are defined in a straightforward manner. Thus the reader can check for himself that all requirements of a vector space are satisfied. Furthermore, the inner product and the norm also can be defined. That is, let $f \in L^2$ be a point in the vector space with the coordinate $[c_k]$. The norm of f is

$$||f|| = \left(\sum_{k=-\infty}^{\infty} c_k \cdot \bar{c}_k\right)^{1/2} = \left(\sum_{k=-\infty}^{\infty} |c_k|^2\right)^{1/2} \tag{9-40}$$

By a direct expansion of $f(x)$ in a Fourier series, it can be seen that

$$||f|| = \left\{\frac{1}{2\pi}\int_{-\pi}^{\pi} [f(x)]^2\,dx\right\}^{1/2} \tag{9-41}$$

For instance, the coordinates of the point A_1 representing the function $(\sin x)/\sqrt{\pi}$ are

$$\left[\ldots,0,0,\frac{j}{2\sqrt{\pi}},0,\frac{-j}{2\sqrt{\pi}},0,0,\ldots\right]$$

The coordinates of the point B_1 representing the function $(\cos x)/\sqrt{\pi}$ are

$$\left[\ldots,0,0,\frac{1}{2\sqrt{\pi}},0,\frac{1}{2\sqrt{\pi}},0,0,\ldots\right]$$

Note that

$$||A_1|| = ||B_1|| = \frac{1}{\sqrt{2\pi}}$$
$$<A_1,B_1> = \frac{j}{2\sqrt{\pi}}\frac{1}{2\sqrt{\pi}} - \frac{j}{2\sqrt{\pi}}\frac{1}{2\sqrt{\pi}} = 0 \tag{9-42}$$

Elements A_1 and B_1 are orthogonal, and if the scale of the distance is normalized with a factor of $\sqrt{2\pi}$, then A_1 and B_1 will also be an orthonormal pair. The same argument applies to other functions listed in Eq. (9-44). That is, for any two points A_k and B_k representing $(\cos kx)/\sqrt{\pi}$ and $(\sin kx)/\sqrt{\pi}$, respectively, we have

$$||A_k|| = \left[\frac{1}{2\pi}\int_{-\pi}^{\pi}\left(\frac{\cos kx}{\sqrt{\pi}}\right)^2 dx\right]^{1/2} = \frac{1}{\sqrt{2\pi}} = ||B_k|| \qquad k = 1, 2, \ldots$$
$$<A_k,B_h> = 0 \qquad k \neq h \tag{9-43}$$

Thus, every point of the space can be generated in a unique way by a linear combination of the elements of its basis: $1, \sqrt{2}\cos x, \sqrt{2}\sin x, \sqrt{2}\cos 2x, \sqrt{2}\sin 2x, \ldots$.

At times, it is more convenient to employ directly the functional space concept, that is, to consider $f(t)$ as a point f in a Hilbert space without regard to its Fourier expansion. For instance, for all signals in L^2 defined in $[-\pi,\pi]$, we may accept the square of the norm as $\int_{-\pi}^{\pi} [f(t)]^2\,dt$. A basis for this space is

$$\frac{1}{\sqrt{2\pi}}, \frac{\cos x}{\sqrt{\pi}}, \frac{\sin x}{\sqrt{\pi}}, \frac{\cos 2x}{\sqrt{\pi}}, \frac{\sin 2x}{\sqrt{\pi}}, \ldots \tag{9-44}$$

An interpretation of the foregoing material in terms of electrical signals is in order. If f_1 and f_2 are two electric time signals expressible in terms of a Fourier series, $f_1, f_2 \in L^2$ in the time interval $[-T/2,T/2]$, and K is a real constant, then the following analogy between the class of L^2 signals and the corresponding signal space is instructive. (By signal space we mean the function space pertinent to the class of signals under consideration.) Table 9-1 will bring into focus some of this analogy.

9-10. Band-limited Signal Space. In Sec. 9-6 we derived the sampling theorem. In the light of that material, it is instructive to look at the class of band-limited functions $|\omega_0|$ which are square-integrable. We shall denote this class of function by BL^2 and consider them in signal space R^∞. The norm is defined as before. It can be shown that all the requirements of a normed space are fulfilled.

TABLE 9-1

Time signals in L^2 possessing convergent Fourier series expansions	Vector space of the signal
$f(t)$	Point f
Average power dissipated by signal $f(t)$ in the unit resistor in time $[-T/2, T/2]$	T^{-1} times the square of the length of f
$f_2(t) = kf(t)$. Power is multiplied by k^2	Length of $f_2 = k$ times the length of f_1
$\int_{-T/2}^{T/2} f_1(t) \cdot f_2(t)\, dt = 0$ Power $(f_1 + f_2) =$ Power $f_1 +$ Power f_2	f_1 and f_2 are orthogonal elements
All signals with average power P in time $\left[-\frac{T}{2}, +\frac{T}{2}\right]$	All points on the sphere with the radius $\sqrt{TP}$ and center at the origin
rms power associated with $f_1(t) + f_2(t)$ cannot be more than the sum of the individual rms powers of $f_1(t)$ and $f_2(t)$	$\|f_1 + f_2\| \leq \|f_1\| + \|f_2\|$

Let us consider the point representing the function

$$f_0(t) = \frac{\sin t}{t} \qquad ||f_0||^2 = \int_{-\infty}^{\infty} \left(\frac{\sin t}{t}\right)^2 dt = \pi \tag{9-45}$$

The functions

$$f_1(t) = \frac{\sin(\omega_0 t - \pi n)}{\omega_0 t - \pi n} \qquad f_2(t) = \frac{\sin(\omega_0 t - \pi m)}{\omega_0 t - \pi m}$$

are orthogonal elements; in fact,

$$\begin{aligned} \langle f_1, f_2 \rangle &= \int_{-\infty}^{\infty} \frac{\sin(\omega_0 t - \pi m)}{\omega_0 t - \pi m} \frac{\sin(\omega_0 t - \pi n)}{\omega_0 t - \pi n}\, dt = 0 \qquad \text{if } m \neq n \\ &= \frac{\pi}{\omega_0} \qquad \text{for } m = n \end{aligned} \tag{9-46}$$

Next consider a function $f(t) \in BL^2$.

$$f(t) = \sum_{n=-\infty}^{\infty} x_n \frac{\sin(\omega_0 t - n\pi)}{\omega_0 t - n\pi}$$

where
$$x_n = f\left(\frac{\pi n}{\omega_0}\right)$$

Each signal $f(t) \in BL^2$ has a point representation in our function space. The total energy of the signal, that is, the energy dissipated in infinite time $[-\infty, +\infty]$ in a unit resistor under the effect of a voltage $f(t)$, is

$$\int_{-\infty}^{\infty} [f(t)]^2 \, dt = ||f||^2 = \int_{-\infty}^{\infty} \left[\sum_{n=-\infty}^{\infty} x_n \frac{\sin (\omega_0 t - n\pi)}{\omega_0 t - n\pi} \right]^2 dt \qquad (9\text{-}47)$$

Application of Eq. (9-46) shows that

$$||f||^2 = \frac{\pi}{\omega_0} \sum_{n=-\infty}^{\infty} x_n^2 < \infty \qquad (9\text{-}48)$$

Thus, the electric energy of the signal $f(t)$ is numerically equal to the square of the distance of the corresponding element f. The class of signals with specified energy content P and in BL^2 corresponds to the points on the surface of the sphere of radius $\sqrt{P}$ centered at the origin.

The ordinary algebraic operations on members of BL^2 are self-evident. For instance,

$$\begin{aligned} f_1(t) &\to f_1 = [\ldots, x_{-1}, x_0, x_1, \ldots] \\ f_2(t) &\to f_2 = [\ldots, y_{-1}, y_0, y_1, \ldots] \\ [f_1(t) + f_2(t)] &\to f_1 + f_2 \\ &= [\ldots, x_{-1} + y_{-1}, x_0 + y_0, x_1 + y_1, \ldots] \qquad (9\text{-}49) \\ \int_{-\infty}^{\infty} f_1(t) \cdot f_2(t) \, dt &\to \langle f_1, f_2 \rangle = \frac{\pi}{\omega_0} \sum_{-\infty}^{\infty} x_k y_k \end{aligned}$$

The concept of Table 9-1, of course, holds for the space of band-limited signals. To sum up, through the use of the sampling theorem, we have been able to transfer the study of the class of band-limited signals to the study of similar problems in an infinite-dimensional space. Furthermore, if we also inject the qualification of randomness, we shall be able to consider a multidimensional random variable X in lieu of a class of continuous signals.

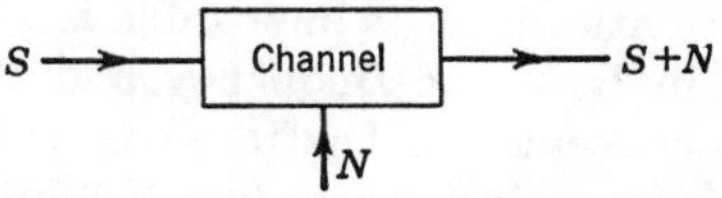

FIG. 9-5. A channel with an additive noise.

9-11. Band-limited Ensembles. The change of framework from randomly defined continuous signals to infinite-dimensional variates has somewhat simplified the problem and provided a suitable physical interpretation. However, the simplification is not yet adequate. In fact, we are still faced with an infinite-dimensional random variable, or what is generally referred to as a stochastic process. The study of such processes will be taken up in Chaps. 10 and 11. Thus, if we wish to rely only on our acquired background of finite-dimensional random variables, some further simplification will be required. To this end, Shannon considers only those signals in BL^2 such that their energy content is "principally" contained in a finite time interval [say, $-(T/2)$ to $+(T/2)$].

We consider the class of time functions $f(t)$ which are band-limited, in the range $-W$ to $+W$ cycles per second. According to the sampling theorem, each member of this class is fully determined by its values at the sampling points $1/2W$ apart.

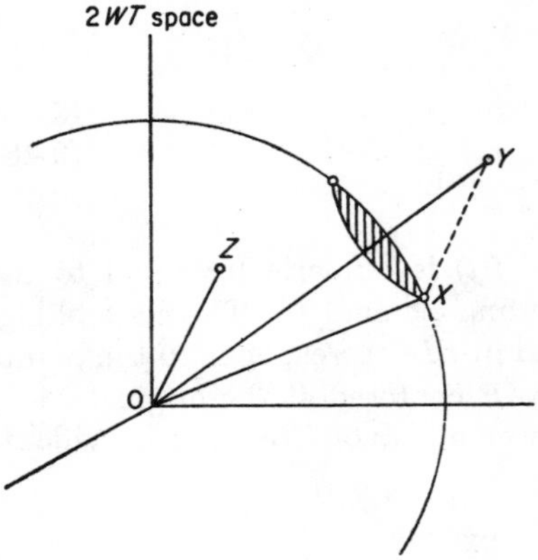

FIG. 9-6. An illustration of band-limited signals with specified power contents in the signal space. X is a transmitted signal, Z is a noise, and Y is the received signal.

$$\|X\| = \sqrt{2TWS} \qquad \|Z\| = \sqrt{2TWN}$$

$$\|Y\| = \sqrt{2TW(S+N)}$$

$$f(t) = \sum_{-\infty}^{\infty} x_n \frac{\sin \pi(2Wt - n)}{\pi(2Wt - n)}$$

Furthermore, we shall make the "practical" assumption that $f(t)$ is negligible outside a time interval $(-T/2, T/2)$,

$$f(t) \cong \sum_{-T/2}^{+T/2} x_n \frac{\sin \pi(2Wt - n)}{\pi(2Wt - n)} \tag{9-50}$$

T being a large integer. We have qualified our assumption as a practical rather than a theoretical one. Indeed, it is reasonable in the evaluation of $f(t)$ to stop at a place where the summation of Eq. (9-50) has negligible terms. However, mathematically, it is impossible to reconcile the idea of a function being limited in both the frequency and the time domain. The difficulties in this concept can be traced to the principle of uncertainty in the work of Gabor and others (see N-3 in the Appendix).

An interesting observation can be made with respect to frequency band-limited signals that are limited to the time interval $(-T/2, +T/2)$. Such signals are represented by points in a $2WT$-dimensional space. The average power S associated with a *typical signal* (that is, the power dissipated in time T by a voltage-stimulated signal applied to a unit resistor) is

$$S = \text{average power in time } T$$

$$= \frac{1}{T} \int_{-T/2}^{T/2} [f(t)]^2 \, dt \cong \frac{1}{2WT} \sum_{-WT}^{WT} x_n^2 \tag{9-51}$$

Then we have for the length or the norm of the associated vector

$$d = ||f|| = \sqrt{2WTS} \tag{9-52}$$

All signals whose power content is less than S are represented by points in the $2WT$-dimensional space within a sphere of radius d.

To sum up, we have established a means of studying a reasonably general class of signals which have a preassigned power content by studying their representative points of $2WT$-dimensional space. Note that our present confinement to band- and time-limited signals provides only the convenience of dealing with finite-dimensional spaces. This restriction can be removed if one tolerates the use of spaces with an infinite number of dimensions. As long as the integral of Eq. (9-51) converges, the concept of distance holds and our model can be used.

Now assume that a member of this class of band-limited signals with specified power S is applied to a noisy channel. Let the noise also be a member of this type of time functions but with a power content N. Then the output signal has a power content that satisfies the triangle inequality of Eq. (9-53),

$$||x + z|| \leq ||x|| + ||z|| \tag{9-53}$$

(equality holds for independent signal and noise). That is, the representative point for the output signal remains on the sphere of radius r centered at the origin:

$$r = \sqrt{2WT(S + N)} \tag{9-54}$$

Let Y be a received signal point; any point on a noise sphere centered at Y could be considered as a possible original signal. However, if there is only one possible signal near Y and listed in the transmission vocabulary, there will be no error in the decoding. Thus, care should be applied in the selection of the transmission signals. If the latter signals have some reasonable mutual distance, the effect of noise perturbation in decoding will not be too serious. A heuristic estimate of the size of the transmission alphabet can be obtained by calculating the largest number of points on the sphere with a mutual distance of $\sqrt{2WTN}$. When the number of dimensions becomes very large, the volume of the sphere lies very close to its surface and the ratio of the volume of the two spheres gives an estimate of the largest possible number of distinguishable signals. That is,

$$M = \frac{\text{vol sphere with rad } \sqrt{2WT(S + N)}}{\text{vol sphere with rad } \sqrt{2WTN}} \tag{9-55}$$

The volume of an n-dimensional sphere with radius R is of the form

$$\frac{\pi^{n/2}}{\Gamma(n/2+1)}R^n$$

where
$$\Gamma(x) = \int_0^\infty t^{x-1}e^{-t}\,dt$$

(See D. M. Y. Sommerville, "An Introduction to the Geometry of N-dimensions," p. 135, E. P. Dutton & Co., Inc., New York, 1929; and S. Goldman, "Information Theory," Chap. 6, Prentice-Hall, Inc., Englewood Cliffs, N.J., 1953.)

Therefore, we find

$$\begin{aligned} M &\approx \left(\frac{S+N}{N}\right)^{TW} \\ C_t &= \lim_{T\to\infty}\frac{1}{T}\log M \approx W\log\left(1+\frac{S}{N}\right) \end{aligned} \tag{9-56}$$

In the next section we employ this geometric interpretation for the determination of the entropies of this particular class of signals in the signal space.

From the probability point of view, when the time limitation is removed, we are actually dealing with a class of signals that form a stochastic process. The entropy of such processes will be discussed in Chap. 11. For the time being, we deal with time-limited ensembles. That is, based on our simplified model, the signal and the noise are multidimensional random variables.

$$\begin{aligned} \{s(t)\} &= [X_1,X_2, \ldots ,X_n] \\ \{n(t)\} &= [Z_1,Z_2, \ldots ,Z_n] \end{aligned} \tag{9-57}$$

The random variables X_k and N_k are defined at the same time, each with specified probability distributions.

9-12. Entropies of Band-limited Ensemble in Signal Space. Pursuant to the material of the preceding section we consider a point X in the signal space:

$$X\colon X_1,\, X_2,\, \ldots\,,\, X_n$$

We use the notation X to indicate a point in the n-dimensional space. Similarly, the notation Y represents a point in another n-dimensional space pertinent to output signals.

$$Y\colon Y_1,\, Y_2,\, \ldots\,,\, Y_n$$

For convenience we may consider a $2n$-dimensional space describing the behavior of the multidimensional random variable (X,Y). This is, in a way, similar to our ordinary two-dimensional random variables. In a $2n$-dimensional product space we have five main probability density functions of interest and, consequently, five main entropies. These densities and entropies can be symbolically represented by

$$\begin{array}{ll} f(\tilde{x},\tilde{y}) & H(\tilde{X},\tilde{Y}) \\ f_1(\tilde{x}) & H(\tilde{X}) \\ f_2(\tilde{y}) & H(\tilde{Y}) \\ f_x(\tilde{y}|\tilde{x}) & H(\tilde{Y}|\tilde{X}) \\ f_y(\tilde{x}|\tilde{y}) & H(\tilde{X}|\tilde{Y}) \end{array} \tag{9-58}$$

The rate of information transmission in the signal space is given by

$$I(\tilde{X};\tilde{Y}) = \iint f(\tilde{x},\tilde{y}) \log \frac{f(\tilde{x},\tilde{y})}{f_1(\tilde{x}) \cdot f_2(\tilde{y})} \, d\tilde{x} \, d\tilde{y} \tag{9-59}$$

In the above symbolic vector presentation we are simply extending our definitions from a one- or a two-dimensional space to an n- or a $2n$-dimensional signal space. The meaning of this symbolic notation has already been described in this section. The details are left to the reader for full justification and comprehension. Now once more, as in Sec. 9-4, the problem is reduced to developing a formula for the maximum rate of information transmission.

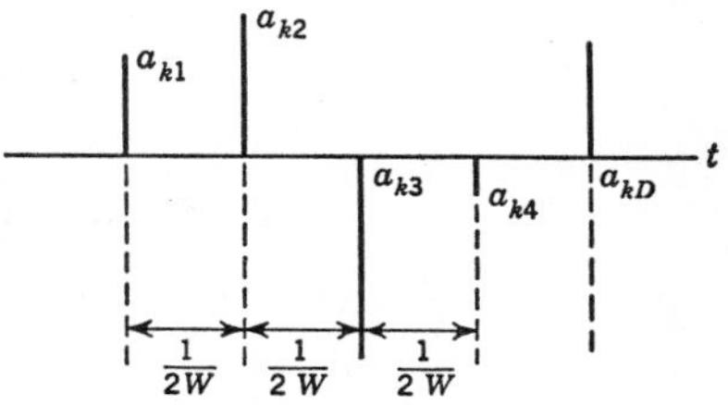

FIG. 9-7. The sequence of random variables $[a_{k1},a_{k2}, \ldots ,a_{kD}]$ represents a point in the signal space for the original continuous random signal.

Such a development can be achieved in a direct way. For example, if we make the further assumptions that

$$\begin{aligned} \overline{X_k} &= 0 \qquad k = 1, 2, \ldots , n \\ \overline{Z_k} &= 0 \\ \sigma_{x_k} &= \sigma_x \\ \sigma_{z_k} &= \sigma_z \end{aligned} \tag{9-60}$$

Noise has independent n-dimensional normal distribution.

$$\text{Output} = \text{input} + \text{noise} \qquad \tilde{Y} = \tilde{X} + \tilde{Z}$$

Then the direct application of the method of Sec. 9-4 will lead to Shannon's celebrated formula

$$\text{max transinformation} = W \log \left(1 + \frac{S}{N}\right)$$

It is only under the listed sequence of assumptions that the formula holds. If we do not wish to confine ourselves to any particular type of noise, we still can use the gaussian noise as a basis of comparison. Such a comparison is discussed in Shannon (I and II).

9-13. A Mathematical Model for Communication of Continuous Signals. In the previous section we ascertained that for band-limited continuous channels without memory, under certain plausible conditions, the maximum rate of transinformation is given by Eq. (9-1). The proof, so far, is an existence proof rather than a constructive one, since it does not present a method for transmitting information at the ideal rate. The following geometric model suggested by Shannon (III) is aimed at providing a more general proof for the possibility of encoding and decoding continuous signals for transmission over noisy memoryless channels at a rate as close to the ideal rate as desired. Shannon, under some general conditions, derives some bounds for the average probability of error for

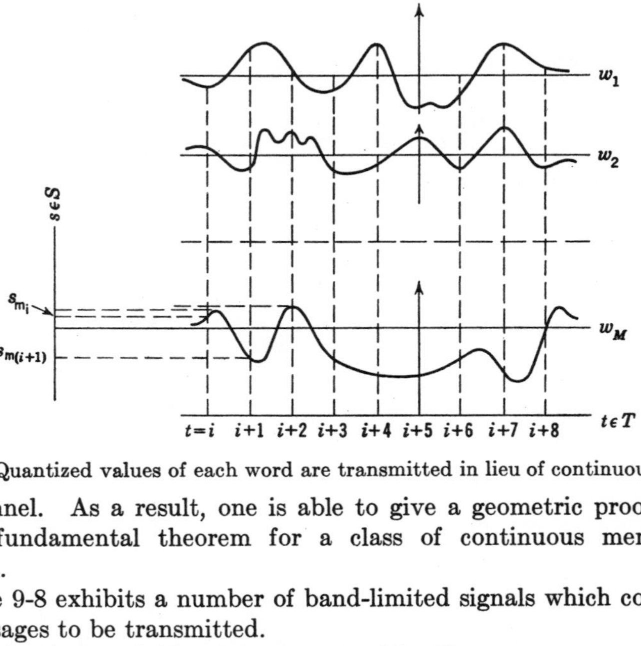

FIG. 9-8. Quantized values of each word are transmitted in lieu of continuous signals.

the channel. As a result, one is able to give a geometric proof of the second fundamental theorem for a class of continuous memoryless channels.

Figure 9-8 exhibits a number of band-limited signals which constitute the messages to be transmitted.

The following model has been suggested by Shannon.

Source. Let T be the set of integers and S the set of real numbers. At every instant $t \in T$, the source selects a signal $s \in S$ with some prespecified probability. Actually the source transmits words of a code book as discussed below.

Block Code. A block code consists of M band-limited words w_1, w_2, . . . , w_M. Each word consists of n letters from S, that is,

$$w_k = [s_{k1}, s_{k2}, \ . \ . \ . \ , s_{kn}]$$

where the ordinates s_{ki} are chosen at the proper sampling intervals (Fig.

9-8). For simplicity, suppose that the sampling terms beyond the above n terms are negligible. In order further to simplify the model, we make the tentative assumption that all words of the code book are equally probable, i.e.,

$$P\{w_k\} = \frac{1}{M} \qquad k = 1, 2, \ldots, M \tag{9-61}$$

Channel. The channel is assumed to be of a continuous type with additive noise. A transmitted letter s_{ki} will be received as

$$s_{ki} + X_{ki} \tag{9-62}$$

The noise X_{ki} is a random variable with specified probability distribution. Here we assume that the noise has a gaussian distribution centered about the value of the transmitted letter. From a physical point of view, this assumption is quite reasonable (see Fig. 8-3).

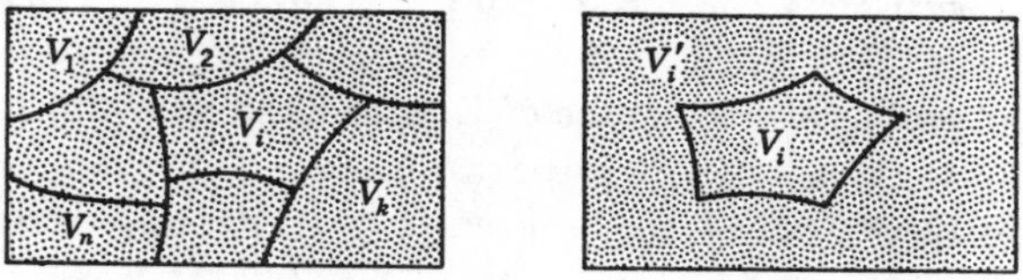

FIG. 9-9. A decoding scheme: All words received in the region V_k will be decoded as W_k.

Decoding Scheme. If the transmitter has M words, we may partition the receiving signal space into M disjoint regions V_k such that each w_k corresponds to a well-defined region V_k. If the decoding is to be an intelligent one, it must be devised with an eye to reducing the "probability of error" in a certain sense. In other words, the partitioning of the receiving universe should not be done at random. Let P_{ei} be the error in the decoding of the word w_i, that is, the probability of transmitting w_i and receiving v_i not in V_i.

$$P_{ei} = P\{v_i \in V_i' | w_i\} \tag{9-63}$$

The average error probability for the code block is

$$E = \overline{P_{ei}} = \sum_i^M P\{w_i\}P_{ei} = \frac{1}{M}\sum_i^M P_{ei} \tag{9-64}$$

9-14. Optimal Decoding. An optimal decoding procedure is the decision scheme for the partitioning of the receiving universe leading to a minimum of E under some assumed constraints. The constraints assumed here are those suggested in the previous section (signals with specified power content, additive gaussian noise, etc.). Reference is made to the geometric presentation of each word in an n-dimensional vector space. Each transmitted word w_k has a norm (distance from the origin) r_k. An optimal decision scheme consists of the following: If a word has been received with a point representation, say U_k, in the signal space, we compare its distance to points representing any permissible transmitted word, and assume that it corresponds to the closest such point. This procedure requires the partitioning of the n-dimensional space into M disjoint regions in such a manner that to each region V_k are assigned all those signal points that are closer to w_k than to any other point w_j. If a signal point is at an equal distance from two or more points of the w set, we may assign it arbitrarily to any one of the associated regions. It remains to show that this decision scheme corresponds to an optimal decoding procedure. In fact, for independent gaussian distribution in n-dimensional space, we have

$$(2\pi\sigma^2)^{-n/2} \iint \cdots \int \exp\left(\frac{-r^2}{2\sigma^2}\right) dx_1\, dx_2 \cdots dx_n \qquad \text{(9-65)}$$

[This can be seen as a special case of Eqs. (5-84) and (7-29).]

$$P_{ei} = (2\pi\sigma^2)^{-n/2} \underset{\text{over } V_i'}{\iint \cdots \int} \exp\left(\frac{-r^2}{2\sigma^2}\right) dx_1\, dx_2 \cdots dx_n \qquad \text{(9-66)}$$

Now suppose that we compute two terms, say $P_{ei} + P_{ej}$, first for this decision scheme and then for any other decision scheme. For the latter scheme, consider the following decoding procedure: Assume that a signal point A which is closest to $w_j \in V_j$ should be assigned to the region V_i instead of V_j. Since $r_{Aj} < r_{Ai}$ and the gaussian probability distribution is a monotonically decreasing function of r, according to Eq. (9-67) one can see that the above decision scheme leads to a lower E.

$$\frac{1}{M}(P'_{ej} + P'_{ei}) > \frac{1}{M}(P_{ej} + P_{ei}) \qquad \text{(9-67)}$$

This reasoning can be further extended in order to show that the above decoding scheme (also called maximum-likelihood detector) is an optimal decoding procedure [see Gilbert (I)]. An example of this partitioning is suggested in Fig. 9-10. If we had only three signal points U_1, U_2, and U_3 in the signal space, the three would determine the regions V_1, V_2, and V_3. Note that, if one makes the further assumption that all the sig-

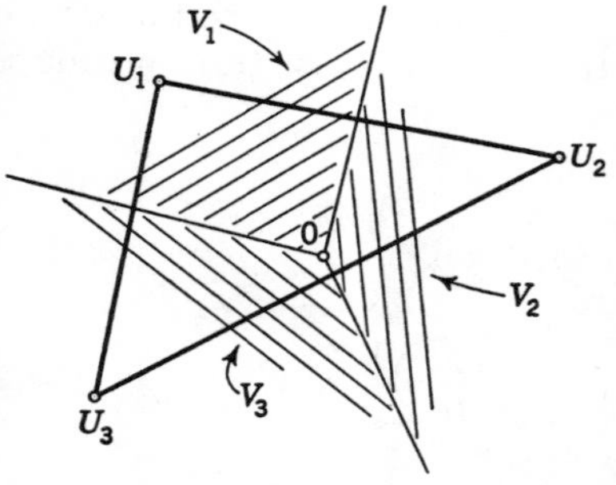

FIG. 9-10. Minimum-distance decoder: All received words closest to U_k are decoded as U_k.

nal points (in n-dimensional space) have equal power, then the maximum-likelihood decoding regions will consist of n-dimensional polyhedra with apexes at the origin of the coordinates. Each polyhedron is bounded by μ hyperplanes [$(n - 1)$-dimensional, $\mu \leq M - 1$] (Fig. 9-11 for three-dimensional space).

An Encoding Problem. We have now established a working model of a continuous channel. For instance, if it is desired to transmit waveforms selected from a finite set of band-limited continuous signals, each word w_k may be chosen as the vector representing the totality of the sampled values of the kth signal. The central problem for this communication model is to devise suitable codes, that is, a decision scheme that minimizes E subject to certain plausible constraints. If no additional constraints were imposed, then the best solution would consist of some sort of equidistant placement of M points in an n-dimensional space.

Most of the practically useful and theoretically interesting constraints stem from the fact that the original signals must have a limited power content. The following set of three hypotheses was considered by Shannon.

1. All signals (words) have the same power content P.
2. The power content of signals (words) is smaller than or equal to a specified power P.
3. The average power content of all signals (words) is smaller than or equal to a specified P.

Thus the corresponding encoding problem is to minimize E by proper placement of M points in the n-dimensional space with the following constraints:

1. M points on the sphere centered at the origin with radius $\sqrt{nP}$, respectively
2. M points within or on the sphere centered at the origin with radius $\sqrt{nP}$, respectively
3. M points such that their average squared distance to the origin is nP

9-15. A Lower Bound for the Probability of Error. We focus our attention, along with Shannon, on the communication of continuous signals in the presence of additive noise, when the code words lie on a sphere of radius $\sqrt{nP}$ (case 1) centered at the origin. Let N be the variance of noise at each sampling point, n the number of sampling points, and M the number of code words. Familiarity with the material of Chap. 8 makes it clear that, for an optimum encoding, the average error probability will be a function of M, n, and P/N. Because of the additivity of noise, only the ratio of the variances of P and N will enter the picture. Shannon's procedure for obtaining some bounds of E can be described in terms of a geometric model. The details of his derivations require much more space than that available here. We shall quote the method of attack and some of the results obtained. For details of the proof the reader is referred to the original reference [Shannon (III)]. The suggested method for obtaining a lower bound for the error probability is rather interesting. It is based on the following two direct steps:

1. Consider the error associated with this minimum-distance decoding scheme, that is, when the signal space is divided into appropriate polyhedra. Then evaluate the corresponding probability of error.

2. The exact evaluation of the error probability for the above geometric model seems to be a complex problem. Therefore, one may approximate the probability of error for this scheme by comparing it with the error in any similar model which may be subjected to simpler computation.

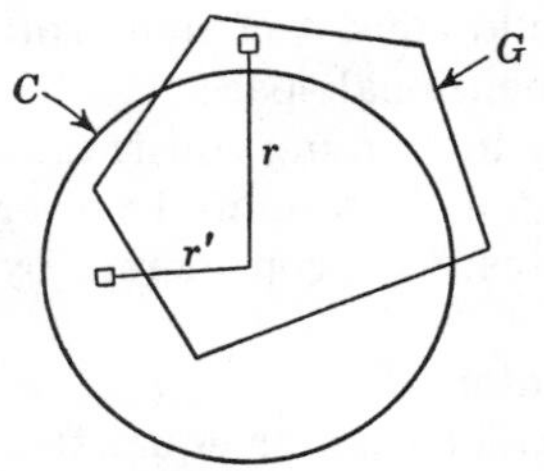

Fig. 9-11. Comparison of probability within a circle and a polygon of equal area for a monotonically decreasing probability density.

The probability of error associated with a decoding scheme, for a specific word w_i, is equal to the product of $P\{w_i\}$ and $P\{\text{received word not in the } i\text{th polyhedron } w_i\}$. In our adopted model, $P\{w_i\} = 1/M$, and the other probability is rather difficult to compute. The following basis for comparison may be established. Consider a two-dimensional monotonically decreasing probability density function $f(r)$ with $f(\infty) = 0$. Compare the probability of the random variable R assuming values in a circle C centered at the origin with the probability of the random variable assuming values in a polygon G of the same area (see Fig. 9-11). Let J stand for the common area between C and G; then

$$P\{R \in C\} = P\{R \in J\} + P\{R \in C \cap R \bar{\in} G\} \tag{9-68}$$

Now, if we compare the probability associated with any two elements of equal area, one in the polygon but not in the circle and the other in the circle but not in the polygon, we shall conclude that a smaller probability is associated with the former element. Thus

$$P\{(R \in G) \cap (R \bar{\in} C)\} \leq P\{(R \in C) \cap (R \bar{\in} G)\} \tag{9-69}$$

This reasoning can be extended to an analogous situation in the n-dimensional space. We may compare the probability of a signal point being in the n-dimensional polyhedron (or pyramid) with the similar probability for a right-angle n-dimensional cone, both with solid angle Ω_i. (Ω_i is the area cut by the pyramid or the cone on the unit n-dimensional sphere.

Both the pyramid and the cone have their apexes at the origin.) The signal point w_i lies on the axis of the cone at a distance $\sqrt{nP}$. Thus, we arrive at a practical method for obtaining a lower bound for the probability of error, as the computation of the latter probability can be directly accomplished.

$$E \geq \frac{1}{M} \sum_{i=1}^{M} P\{\text{signal being moved outside } i\text{th cone}\} = \frac{1}{M} \sum_{i=1}^{M} F(\Omega_i)$$

The probability function $F(\Omega_i)$ is a monotonically decreasing function of distance and also convex in the n-dimensional space. Thus

$$\frac{1}{M} \sum_{i=1}^{M} F(\Omega_i) \geq F\left(\frac{\Omega_0}{M}\right) \tag{9-70}$$

where

$$\Omega_0 = \sum_{i=1}^{M} \Omega_i$$

Shannon finds it easier to compute the probability of a signal point being displaced to outside a cone of half angle θ than to designate

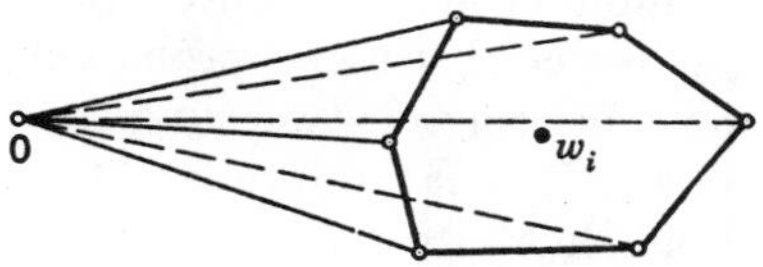

FIG. 9-12. A schematic diagram of the minimum-distance decoding pyramid in n-space.

the cone by its solid angle. He refers to this probability function by $Q(\theta)$. Having this new variable θ in mind, we find

$$E \geq F\left(\frac{\Omega_0}{M}\right) = Q(\theta_1) \tag{9-71}$$

where θ_1 corresponds to the solid angle $(1/M)\Omega(\pi)$, that is,

$$\Omega(\theta_1) = \frac{1}{M} \Omega(\pi)$$

The actual computation of $Q(\theta)$ turns out to be quite long and more complex than can be presented in a few pages. In Sec. 9-17, we shall quote Shannon's results without taking the space for their derivations.

9-16. An Upper Bound for the Probability of Error. We have pointed out more than once that the proper selection of the words in the code book is a most important factor in reducing the probability of error. In this section it will be shown that, even if the code-points are selected at random, the average probability of error of the code cannot surpass some limiting value. The evaluation of this bound is the subject of the present section.

Consider a circular cone with half angle θ about a received word v_i corresponding to a transmitted word w_i. If the cone does not surround any signal point, the word will be unambiguously detected as w_i. However, if there are other signal points in this cone, they may be incorrectly decoded as the original message. The probability of a code-point being in this cone is the ratio of the solid angle of the cone to that of the total space surrounding v_i, that is,

$$P\{\text{one code-point inside cone}\} = \frac{\Omega(\theta)}{\Omega(\pi)} \tag{9-72}$$

Assuming the code-points are independently (at random) distributed on the sphere of radius $\sqrt{nP}$, we find

$$P\{M-1 \text{ code-points in cone}\} = \left[\frac{\Omega(\theta)}{\Omega(\pi)}\right]^{M-1}$$

$$P\{\text{none of } M-1 \text{ code-points in cone}\} = \left[1 - \frac{\Omega(\theta)}{\Omega(\pi)}\right]^{M-1} \tag{9-73}$$

Actually, in order to compute the average error probability for the decoding scheme, we must first compute the probability of the transmitted signal being displaced in the region between a cone of half angle θ and one with half angle $\theta + d\theta$. This latter probability of the original signal being displaced by noise has already been designated by the notation $-dQ(\theta)$. Thus, the average probability of error for a random code is given exactly by

$$E = \int_{\theta=0}^{\pi} - \left\{1 - \left[1 - \frac{\Omega(\theta)}{\Omega(\pi)}\right]^{M-1}\right\} dQ(\theta) \tag{9-74}$$

Some simplification will occur if this exact formula is somewhat weakened by using the inequality $(1-x)^2 \geq 1 - nx$, which suggests

$$\left\{1 - \left[1 - \frac{\Omega(\theta)}{\Omega(\pi)}\right]^{M-1}\right\} \leq (M-1)\left[\frac{\Omega(\theta)}{\Omega(\pi)}\right] \leq M\left[\frac{\Omega(\theta)}{\Omega(\pi)}\right] \tag{9-75}$$

By dividing the range of integration into two parts $[0,\theta']$ and $[\theta',\pi]$, one finds

$$E \leq -\int_0^{\theta'} M\left[\frac{\Omega(\theta)}{\Omega(\pi)}\right] dQ(\theta) - \int_{\theta'}^{\pi} dQ(\theta) \tag{9-76}$$

Finally, we obtain an upper bound for the average probability of error by assigning to the arbitrary angle θ' the value of the cone half angle that corresponds to finding only one signal point in the cone, on an average. That is, $\theta' = \theta_1$, where θ_1 is defined by

$$\Omega(\theta_1) = \frac{1}{M}\,\Omega(\pi)$$

Thus, the average probability of error satisfies the inequalities

$$Q(\theta_1) \leq E \leq Q(\theta_1) - \int_0^{\theta_1} \frac{\Omega(\theta)}{\Omega(\theta_1)}\,dQ(\theta) \tag{9-77}$$

In the following section, we discuss the asymptotic behavior of this error probability when n is increased indefinitely.

9-17. Fundamental Theorem of Continuous Memoryless Channels in Presence of Additive Noise. Using the inequalities of Eq. (9-77), we wish to study the asymptotic behavior of the error probability when the rate of transmission of information approaches the channel capacity. Without going into Shannon's elaborate derivation of the probability function $Q(\theta_1)$, we merely quote his result when n asymptotically approaches infinity. Shannon shows that if $n \to \infty$ and if the signaling rate approaches the channel capacity (per degree of freedom), that is,

$$R = \frac{1}{n}\log M \to \frac{1}{2}\log\left(1 + \frac{P}{N}\right) = C$$

then the upper and the lower bound of Eq. (9-77) will coincide. In this case he has shown that the probability of error approaches the CDF of a standard normal distribution:

$$\Phi\left[\sqrt{n}\,\sqrt{\frac{2P(P+N)}{N(P+2N)}}\,(R - C)\right] \tag{9-78}$$

Expression (9-78) exhibits Shannon's fundamental theorem for continuous channels in the presence of additive gaussian noise. For any arbitrarily small but positive $\epsilon = C - R$, the probability of error for the codes approaches 0 as $n \to \infty$. From expression (9-78) it can also be noted that, with a fixed positive ϵ, it is not possible to encode messages such that one may transmit at a rate greater than the channel capacity with arbitrarily small error probability.

Note. An early conception of this idea appears in Shannon's article in the 1949 *Proceedings of the IRE.* The treatment in the recent (1959) *Bell System Technical Journal* article is quite elaborate. In the preceding pages we have tried to present Shannon's basic method of proof without recourse to the more complex techniques initiated in that paper. The omission was felt necessary in order to remain within the bounds of an elementary presentation. The reader who wishes to appreciate fully Shannon's techniques should consult the original paper.

The following supplementary articles for this chapter are suggested: Gilbert [I], Rice, and Thomasian. The first article makes use of the concept of distance and power for the transmission of a class of band-limited signals in the signal space. The second article, which historically is the closest one to Shannon [II], contains several interesting results. For instance, it is shown that the "reliability" of the transmission of band-limited signals in the presence of independent noise when the transmission rate is close to the channel capacity is approximately

$$\tfrac{1}{2}\left(1 + \frac{N}{P}\right)(C - R)^2$$

Shannon's equivalent result suggests a sharper estimate [Shannon (III, p. 642)],

$$\frac{(P + N)^2}{P(P + 2N)}(C - R)^2$$

The third afore-mentioned paper derives a relatively simple bound for the probability of error. So far, the sharpest results (but not the simplest) have been obtained in Shannon [III].

9-18. Thomasian's Estimate. A. J. Thomasian has recently made a study of continuous channels in the presence of additive noise when power contents of the signal and noise are limited (Sec. 9-15, case 2). A summary of his results is given below.

Let

$$u_k = [x_{k1}, x_{k2}, \ldots, x_{kn}]$$

be a possible input word. We assume

$$\frac{1}{n}\sum_{j=1}^{n} x_{kj}^2 \le P$$

As before, we designate the corresponding output word by

$$v_k = [y_{k1}, y_{k2}, \ldots, y_{kn}]$$

If the average noise power per coordinate, $N > 0$, is specified for additive independent gaussian noise with mean zero and variance N, the output of the channel is constrained by

$$P\{v|u\} = \prod_{j=1}^{n} \frac{1}{\sqrt{2\pi N}} \exp\left[-\frac{1}{2N}(y_j - x_j)^2\right] \qquad (9\text{-}79)$$

The following version of the fundamental theorem for the above communication system (additive independent gaussian noise) is that of Thomasian.

Theorem. Subject to the above assumptions, there exist M distinct input words $[u_1,u_2, \ldots ,u_M]$

$$\frac{1}{n}\sum_{j=1}^{n} x_{kj}^2 \leq P \qquad u_k = [x_{k1},x_{k2}, \ldots ,x_{kn}] \tag{9-80}$$

$$M \leq 2^{Rn} \qquad k = 1, \ldots , M$$

such that

$$\frac{1}{M}\sum_{k=1}^{M} P\{D'_k|u_k\} \leq 3 \exp\left\{-\frac{n}{4}\left[\sqrt{1+\frac{0.3(C-R)^2(P+N)}{P}}-1\right]\right\}$$

where C = channel capacity in bits per signal coordinate
R = suitable rate of signaling; i.e., $0 < R < C = \log (1 + P/N)^{1/2}$
D_k = output word decoded when input word is u_k and optimal decoding procedure is employed; D'_k is set complement to D_k

This theorem gives an upper bound for the average error probability in the described communication model. The decoding criterion is based on minimum distance. For the sake of simplicity we have disregarded the case where several words can be associated with a single transmitted word. Although Thomasian's formulation of the problem is similar to Shannon's approach, his derivation and proofs are quite different. Thomasian's proof is based on some basic lemmas and inequalities similar to those employed by Feinstein and Wolfowitz (see Chap. 12). A full derivation of this interesting theorem requires more time and space than is available at present. The interested reader is referred to the original article.*

PROBLEMS

9-1. A microphone does not let through sounds above 4,000 cycles per second. Determine the sampling interval, allowing reconstruction of the output waveform from sampled values.

9-2. Prove the sampling theorem for a time function $f(t)$ which is identically zero outside the interval $a < t < b$.

9-3. (*a*) Show that the volume of an n-dimensional hypersphere of radius R is

$$v = R^n f(R)$$

where

$$f(R) = \frac{(2\pi)^{n/2}}{n\int_0^\infty R^{n-1}\exp\left[-(R^2/2)\right] dR}$$

(*b*) For large n, plot the integrand in the denominator as a function of R.

* A. J. Thomasian, Error Bounds for Continuous Channels, *Proc. Fourth International Symposium on Information Theory,* to be published in 1961. See also Blackwell, Breiman, and Thomasian.

9-4. Compare the ratio of volumes of two hyperspheres of the same radius but dimensions n and $n - 2$, respectively.

9-5. Let $X = [X_1, X_2, \ldots, X_n]$ be a random vector with gaussian distribution:

$$\begin{aligned}&\text{Mean of } X_k = 0\\ &\text{Moment } X_k X_j = 0 \qquad\qquad k = 1, 2, \ldots, n \quad j \neq k\\ &\text{Standard deviation of } X_k = \sigma\end{aligned}$$

X may be represented by a point X in an n-dimensional space.

(*a*) Given a point X of this space with a distance R from the origin, discuss the relation between R and the power content of the signal.

(*b*) What is the probability of having points representing signals of this ensemble within a sphere of radius R about the origin?

(*c*) Same question as in part (*b*) for the point being between two spheres of respective radius R and $R + dR$.

9-6. Verify if a bandpass function restricted to the frequency interval λw, $(\lambda + 1)w$ can be expanded as

$$f(t) = \sum_n f\left(\frac{n}{2w}\right) \frac{\sin 2\pi w(\lambda + 1)(t - n/2w) - \sin 2\pi w\lambda(t - n/2w)}{2\pi w(t - n/2w)}$$

9-7. Verify the validity of the following identity due to C. R. Cahn. If $f(t)$ is a periodic function of period T and if all the Fourier coefficients vanish above the nth harmonic, then

$$f(t) = \sum_{k=0}^{2N} f\left(\frac{kT}{2N + 1}\right) \frac{\sin \{(2N + 1)(\pi/T)[t - kT/(2N + 1)]\}}{(2N + 1) \sin \{(\pi/T)[t - kT/(2N + 1)]\}}$$

(S. Goldman, "Information Theory," p. 83, Prentice-Hall, Inc., Englewood Cliffs, N.J., 1953.)

9-8. Show that the three vectors

$$[\tfrac{2}{3}, \tfrac{2}{3}, -\tfrac{1}{3}] \qquad [\tfrac{1}{3}, -\tfrac{2}{3}, -\tfrac{2}{3}] \qquad [\tfrac{2}{3}, -\tfrac{1}{3}, \tfrac{2}{3}]$$

form an orthonormal basis for R^3.

9-9. Show the validity of

$$||X + Y|| \leq ||X|| + ||Y||$$

in R^3 by a simple study of an associated triangle.

9-10. Prove that, if

$$||X + Y|| = ||X|| + ||Y||$$

then X and Y are linearly independent.

9-11. Let $f(t)$ be a function of period 2π such that

$$\begin{aligned}f(t) &= 1 \qquad && 0 < t < \pi\\ f(t) &= -1 && -\pi < t < 0\\ f(t) &= 0 && t = 0 \quad \text{or} \quad t = \pi\end{aligned}$$

Show that

$$f(t) = \frac{4}{\pi} \sum_{n=1}^{\infty} \frac{\sin (2n - 1)t}{2n - 1}$$

9-12. Show that the functions

$$\frac{1}{\sqrt{\pi}} \qquad \sqrt{\frac{2}{\pi}} \cos t \qquad \sqrt{\frac{2}{\pi}} \cos 2t \qquad \cdots$$

form an orthonormal set in L^2 with respect to the interval $[0,\pi]$. Verify the distance inequality.

9-13. Prove that if A, B, and X are voltages in a two- or three-dimensional signal space (using, for example, the space corresponding to the sampling theorem), and

$$|A - X| = |B - X|$$

then the signals $B - A$ and $X - \dfrac{A+B}{2}$ are orthogonal.

9-14. Show that the surface of an n-dimensional sphere of radius R is

$$\frac{n\pi^{n/2}R^{n-1}}{\Gamma(n/2 + 1)}$$

9-15. Using the formula of the preceding problem, with the notation of Sec. 9-15, show that

(*a*)
$$\frac{\Omega(\theta_1)}{\Omega(\Omega)} \leq \frac{\Gamma(n/2+1)(\sin\theta_1)^{n-1}}{n\Gamma[(n+1)/2]\pi^{1/2}\cos\theta_1}$$

where θ_1 is the cone angle such that the solid angle $\Omega(\theta_1)$ of the n-dimensional cone is $(1/M)\Omega(\pi)$.

(*b*)
$$\frac{\Omega(\theta_1)}{\Omega(\theta)} \leq \frac{(\sin\theta)^{n-1}}{(\sin\theta_1)^{n-1}}$$

9-16. Let $[Y_{-N} \cdots Y_{-1}Y_0Y_1 \cdots Y_N]$ be a random variable corresponding to the sampling intervals of a band-limited signal. Assume all Y_k mutually independent variables with normal distributions and equal standard deviation $\sigma = 1$. Show that the probability density of the square of the distance, that is,

$$X = \sum_{k=-N}^{k=N} Y_k^2$$

depends only on the sum of the square of the averages of Y_k's, but not explicitly on any individual variables.

$$X_0 = \sum_{k=-N}^{k=N} \bar{Y}_k^2$$

Can you derive the probability density function $P\{X\}$? (See Rice, particularly Sec. 4 and Appendix I.)

9-17. Let $X_1, X_2, \ldots, X_n$ be mutually independent random variables with a common-probability density function $f(x)$, $g(x)$ a real-valued function, and S the set of all points in the n-space such that

$$\sum_{k=1}^{n} g(x_k) \geq nd \qquad d \text{ is a fixed number}$$

Then prove that for any $t \geq 0$

$$\int \cdots \int_S f(x_1) \cdots f(x_n)\, dx_1 \cdots dx_n \leq \left[e^{-td} \int_{-\infty}^{\infty} e^{tg(x_1)} f(x_1)\, dx_1 \right]^n$$

This inequality is used by Thomasian.* Its proof follows directly. (See also Loève, p. 158.)

9-18. In the preceding problem, let all variables have standard normal distributions and $d > 1$. Consider $g(x) = x^2$ and show that

$$P\left\{ \sum_{k=1}^{n} X_k^2 \geq nd \right\} \leq \frac{e^{-2td}}{1 - 2t}$$

* *Ibid.*

PART 3

SCHEMES WITH MEMORY

The bringing together of theory and practice leads to the most favorable results; not only does practice benefit, but sciences themselves develop under the influence of practice, which reveals new subjects for investigation and new aspects of familiar subjects.

P. L. Chebyshev

Quoted by Khinchin in *Uspekhi Mat. Nauk*, vol. 8, no. 3, p. 3, 1953.

CHAPTER 10

STOCHASTIC PROCESSES

The mathematical theory of probability has grown tremendously during the past three decades. Today probability encompasses several professional fields of mathematical endeavor, such as game theory, decision theory, time series, and Markov chains.

Since 1940 several mathematicians have made fundamental contributions to the establishment of the new science of statistical theory of communications. Perhaps two of the most significant landmarks of communication theory which have immediate bearing on this subject are the Wiener-Kolmogorov theory of filtering and prediction and Shannon's information theory. In both cases, stochastic processes occupy an important place. In communication theories, messages that are transmitted in time intervals are generally dealt with; that is, the raw material consists of time series. This immediately gives rise to questions on the statistical nature of these time series and their accurate description, both at the entry and at the exit of physical systems.

The theory of filtering and prediction has been primarily concerned with problems of determining optimum linear systems in the sense of the least-square criterion, for extraction of signals from particular mixtures of signals and noise. This is a major problem in communication theory with frequent practical application. However, it seems that this specific topic has been somewhat overemphasized. At present a broader outlook, the general study of linear systems under stochastic regime, appears desirable. Such a study seems to be the most natural extension of the ordinary network theory which is concerned with the study of linear systems under deterministic time functions. Today this well-developed area of deterministic linear systems occupies an important position in the technological development of our applied sciences. Therefore, further knowledge of stochastic processes is essential to physical scientists interested in extending the study of deterministic networks to probabilistic systems.

Information theory deals with messages and their ensemble transformations. There again a fundamental study of the problems involved requires a knowledge of stochastic processes.

The present chapter constitutes an introductory survey of the theory of stochastic processes. The application of such processes to systems is the general theme of the statistical theory of communications.

This chapter is aimed at giving the communications engineer a short systematic treatment of the subject without too great a sacrifice of mathematical rigor. Those professionally interested in this field will find a large number of recent books and articles available for a more specialized coverage of the subject (see Middleton and Davenport and Root).

10-1. Stochastic Theory. In a first attempt to acquire some knowledge of probability theory, the subject must be confined to what may be called the *static* part of probability theory. This chapter will present an exposé of the *dynamic*, or, more appropriately, *stochastic*, part of the theory. It will study probabilistic phenomena which depend on time or any other real parameter. In the mathematical literature this part is referred to as the study of time series, or, more technically, a study of stochastic processes. The immediate objective is to acquaint the reader with a method of analysis of time series. This will require the introduction of some new terms and a reappraisal of the more elementary concepts of probability theory. Subsequent to defining time series, there will be a study of averages and expectations. Finally, we relate the study of the averages to the well-established theory of Fourier integrals, in much the same line of thought as relating the concept of moments to characteristic functions through Fourier integrals.

The following intuitive definition of a stochastic process can be given.* A time-dependent stochastic process is a random process whose outcomes are infinite sequences or functions, in contrast to a simple random variable whose outcomes are numbers or vectors. In other words, a stochastic process $\{X(t)\}$ is a probability process whose outcomes are functions of t. The values of the process at times $t_1, t_2, \ldots, t_n$, that is, $X(t_1), X(t_2), \ldots, X(t_n)$, form a sequence of random variables.

At each instant of time t_i there is a random variable $X(t_i)$ with a specified probability distribution. The stochastic process can have a discrete or continuous time parameter. A discrete process consists of a finite or an infinite sequence of random variables each defined at a discrete time. Without loss of generality, assume the time sequence to be integers $\ldots, -2, -1, 0, 1, 2, 3, \ldots$. Then the process will be designated by

$$X(t) = \{\ldots, X_{-2}, X_{-1}, X_0, X_1, X_2, X_3, \ldots\}$$

For a continuous stochastic process, the outcome at any desired instant

* See J. L. Doob, *Am. Math. Monthly*, vol. 49, no. 10, pp. 649–653, 1942.

$t_i \in t$, X_{ti}, is a random variable. Examples of a stochastic process in discrete and continuous time are sketched in Fig. 10-1.

In order completely to specify a random process, the joint probability distribution of any number of random variables of the process must be known. To be specific, for any integer n and any set of real numbers $t_1, t_2, t_3, \ldots, t_n$ belonging to the time interval of the process there must be a set of random variables $X(t_1), X(t_2), X(t_3), \ldots, X(t_n)$ with a known joint probability distribution*

$$P\{X_{t_1} < x_1, X_{t_2} < x_2, \ldots, X_{t_n} \leq x_n\}$$

As an example of a discrete process, consider the repeated throws of a biased coin, which are head or tail, with respective probabilities p and

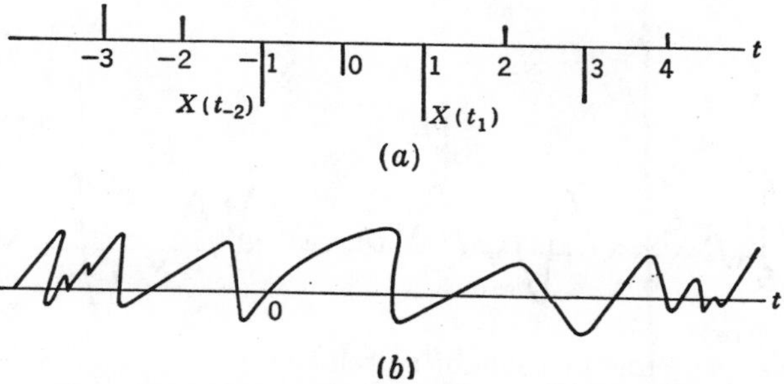

FIG. 10-1. (a) Example of a discrete-time stochastic process; (b) example of a continuous-time stochastic process.

$1 - p$. Call the result of the kth throw a random variable X_k, which assumes one of the two numerical values, say 1 corresponding to a head and 0 corresponding to a tail. Here the family of the random variables X consists of

$$X(t) = \ldots, X_{-2}, X_{-1}, X_0, X_1, X_2, X_3, \ldots$$

The continuation of the throws in both numerical directions is a matter of mathematical convenience rather than anything else. In this example, note that any two members of the family are independent of each other. In general, these random variables need not be independent. Each random variable of the family has a well-defined probability distribution, and all the joint probabilities for two or more members of the family are completely defined by the binomial law.

* It is to be noted that, while for the continuous processes the selection of the time sequence $\ldots, t_{-1}, t_0, t_1, t_2, t_3, \ldots$ is arbitrary, for a discrete process the time sequence must be selected appropriately from the doubly infinite sequence of the discrete times of the process. In the latter case, the number of variables is, at most, denumerable.

For a simple illustration of a continuous process, consider a large number of radio receivers registering a stochastic noise voltage, as sketched in Fig. 10-2. The voltage registered by each receiver is an *outcome* of the stochastic process, or a member of the *ensemble* function.

Select an infinite number of time-sampling points on the t axis about some time reference $t = 0$, such as $t_{-2}, t_{-1}, t_0, t_1, t_2, \ldots$ At a sampling

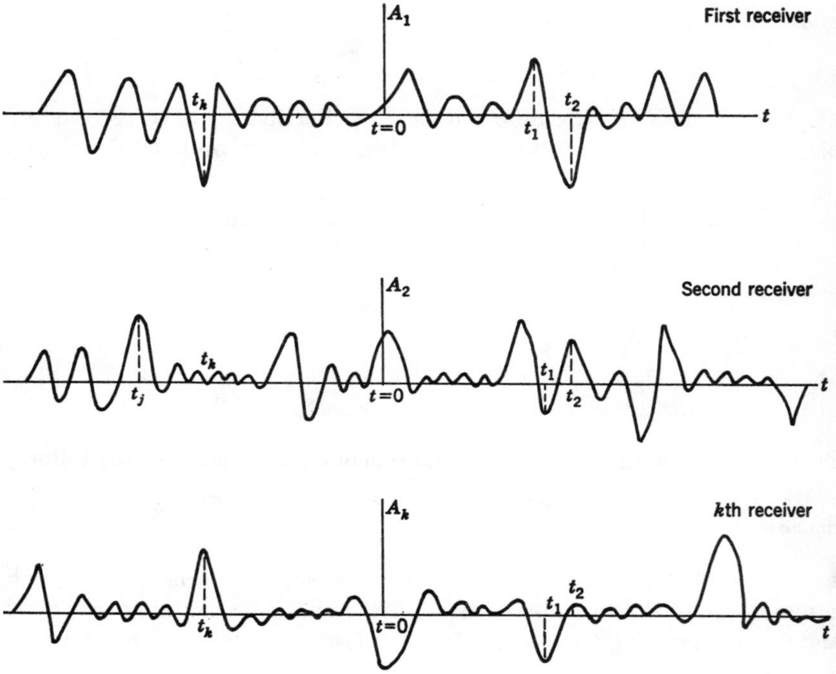

FIG. 10-2. An illustration of different outcomes of a process.

time t_i the values of the registered noise for different receivers will be designated by

$$A_1(t_i),\ A_2(t_i),\ \ldots,\ A_k(t_i),\ \ldots \qquad i = -2, -1, 0, 1, 2, \ldots$$

Assuming that a great many receivers are available, it should be possible, at least theoretically, to estimate the probability distribution of the noise amplitude $A(t_i) = X(t_i)$:

$$P\{x < X(t_i) < x + dx\} = f(x;t_i)\,dx \tag{10-1}$$

the probability that at time t_i the registered noise voltage lies between x and $x + dx$ volts. This is called the *first probability density distribution* of the process. To be more exact, the first probability density distribution is

$$f(x;t)\,dx$$

defined for all sampling times. Similarly, the second probability density distribution is defined as

$$f(x_1,x_2;\,t_1,t_2)\,dx_1\,dx_2$$

for any values of t_1 and t_2 [the joint probability of the noise voltage remaining between a given interval $(x_1,\, x_1 + dx_1)$ at sampling time t_1, and within the interval $(x_2,\, x_2 + dx_2)$ at another specified sampling time (t_2)]. This function should be defined for any finite pair of time points (t_k,t_j). Finally, the nth-order probability density function of the process can be defined as

$$f(x_a, \ldots ,x_{-2},x_{-1},x_0,x_1,x_2,x_3, \ldots ,x_b,t_a, \ldots , t_{-2},t_{-1},t_0,t_1,t_2,t_3, \ldots ,t_b) \qquad (10\text{-}2)$$

For example,

$$f(x_a, \ldots ,x_{-2},x_{-1},x_0,x_1, \ldots ,x_b,t_a, \ldots , t_{-2},t_{-1},t_0,t_1,t_2, \ldots ,t_b)\,dx_a \cdots dx_b$$

means the probability of finding simultaneously the values of the following random variables with the specified times in the range specified below:

$$\begin{array}{l} \cdots\cdots\cdots\cdots\cdots\cdots \\ x_{-2} < X(t_{-2}) < x_{-2} + dx_{-2} \\ x_{-1} < X(t_{-1}) < x_{-1} + dx_{-1} \\ x_0 < X(t_0) < x_0 + dx_0 \\ x_1 < X(t_1) < x_1 + dx_1 \\ x_2 < X(t_2) < x_2 + dx_2 \\ \cdots\cdots\cdots\cdots\cdots\cdots \end{array} \qquad (10\text{-}3)$$

The order n implies the consideration of the joint distribution of n random variables at specified sampling points. When the joint probability distributions are known for any selected finite k points of the interval of the process and for $k = 1, 2, \ldots , n$, then we consider that the process is known up to order n. Evidently the given data must be collectively consistent. The following section presents examples of stochastic processes.

10-2. Examples of a Stochastic Process. Consider the ensemble of the time functions

$$X = \sin\,(t + \theta) \qquad (10\text{-}4)$$

where t is the time and θ a random variable with given distribution density $\rho(\theta)$. For each value of t, say $t_{-1}, t_0, t_1, t_2, \ldots$, the function X assumes values associated with different random variables:

$$\begin{array}{lll} \cdots & \cdots & \cdots \\ t_{-1} & X(t_{-1}) & \sin(t_{-1} + \theta) \\ t_0 & X(t_0) & \sin(t_0 + \theta) \\ t_1 & X(t_1) & \sin(t_1 + \theta) \\ t_2 & X(t_2) & \sin(t_2 + \theta) \\ \cdots & \cdots & \cdots \end{array} \tag{10-5}$$

The probability density function, say at time $t = t_2$, can be computed from the knowledge of $\rho(\theta)$ and according to the rules of transformation of a random variable (see Sec. 5-11). Similarly, to compute the joint probability density,

$$f(x_1,x_{-2};t_1,t_{-2})\,dx_1\,dx_{-2} = P\begin{Bmatrix} x_1 < \sin(t_1 + \theta) < x_1 + dx_1 \\ x_{-2} < \sin(t_{-2} + \theta) < x_{-2} + dx_{-2} \end{Bmatrix}$$

Note that, in this example, $X_1, X_2, \ldots$ are actually functionally dependent, which is stronger than statistical dependence. For processes whose sample values at any t_k and t_j are independent, the desired joint probability density distribution is the product of the two individual first-order densities. As in the above, these sampled variables need not be independent in general, but the nature of their interdependence must be specified.

As a second example, consider the process

$$X = A + Bt \tag{10-6}$$

where A and B are independent random variables with normal distributions (zero mean and σ standard deviation). Then,

$$\begin{array}{lll} \cdots & \cdots & \cdots \\ t_{-1} & X(t_{-1}) & A + Bt_{-1} \\ t_0 & X(t_0) & A + Bt_0 \\ t_1 & X(t_1) & A + Bt_1 \\ \cdots & \cdots & \cdots \end{array} \tag{10-7}$$

As t_k is a constant number for each sampled variable, X is the linear combination of two independent variables, each with a given normal distribution. Thus, the first-order density for the ensemble is well-defined.

For example, at $t = 2$,

$$X_2 = X(t_2) = A + 2B$$

The random variable X_2, consisting of the linear combination of two normally distributed random variables, will itself have a normal distribution:

$$f(x_2) = \frac{1}{\sigma_0 \sqrt{2\pi}} \exp\left(-\frac{x_2^2}{2\sigma_0^2}\right)$$

with
$$\sigma_0 = \sqrt{(1+4)\sigma^2} = \sqrt{5}\,\sigma$$

Similarly, the joint density for any two or more sampled variables can be computed without any basic difficulties.

10-3. Moments and Expectations. This section develops some averaging considerations for stochastic processes. This is similar to the concept of mean values for ordinary functions, or the concept of moments and expectation of random variables.

There are two kinds of averaging involved in stochastic processes. The first kind deals with moments of the nth-order density distribution of the stochastic process; in the second kind of averaging the different averages associated with one or more time functions which are members of the process are taken.

Ensemble Averages. The reader is already familiar with the moments of different orders associated with a random variable. In the case of a stochastic process, the same idea applies. For example, the kth-order moment of the first-order density distribution of the process is defined as

$$m_k(t) = E\{[X(t)]^k\} = \int_{-\infty}^{\infty} x^k f(x;t)\, dx \tag{10-8}$$

The first- and the second-order moments for the random variable $X(t_i)$ are

$$\bar{X}(t_i) = m_1(t_i) = \int_{-\infty}^{\infty} x f(x;t_i)\, dx \tag{10-9}$$

$$\overline{[X(t_i)]^2} = m_2(t_i) = \int_{-\infty}^{\infty} x^2 f(x;t_i)\, dx \tag{10-10}$$

The familiar concept of the central moments and standard deviation, as discussed in Chap. 6, may also be employed:

$$\begin{aligned} \mu_2 &= E\{[X(t_i) - m_1(t_i)]^2\} \\ \mu_2 &= \int_{-\infty}^{\infty} [x - m_1(t_i)]^2 f(x;t_i)\, dx \end{aligned} \tag{10-11}$$

Next, it is natural to study the different moments associated with the second-order joint distribution of the process; for instance,

$$\overline{X(t_k)\cdot X(t_j)} = \int_{-\infty}^{\infty}\int_{-\infty}^{\infty} x_k x_j \cdot f(x_k,x_j;t_k,t_j)\, dx_k\, dx_j \tag{10-12}$$

The idea of ensemble averaging can be extended in an obvious manner to higher moments and higher-order joint probability densities of the process.

Time Averages. The concept of time averaging deals with the different averages computed from one of the time function members of the ensemble if such averages exist. Let $\{X(t)\}$ be a process depending on several random variables $A_1, A_2, \ldots, A_k$ and time t. For a given set of values of $A_1, A_2, \ldots, A_k$, we have what is referred to as an *outcome*. For example, to concentrate on one particular outcome of the process, $\{X(t)\}$, an average value for this outcome can be defined as

$$\langle X \rangle = \lim_{T \to \infty} \frac{1}{2T} \int_{-T}^{T} X(t)\, dt \qquad (10\text{-}13)$$

The definition is contingent on the existence of a finite value for the limit. For some outcomes of $X(t)$, $\langle X \rangle$ may not exist.

Similarly, introduce a second-order averaging for each member of the ensemble as

$$\langle X(t) \cdot X(t + \tau) \rangle = \lim_{T \to \infty} \frac{1}{2T} \int_{-T}^{T} X(t) \cdot X(t + \tau)\, dt \qquad (10\text{-}14)$$

This average gives a measure for the interaction or coherence between the values of the time function under consideration at a time t and time $t + \tau$, where τ is any real time interval. This type of averaging is very common in engineering problems. It may be added that time averaging is associated with a known member of the ensemble. The procedure does not exploit the knowledge of the probability distributions of the totality of the ensemble. It simply indicates the averages pertinent to a particular member. These averages may or may not be identical for all members of the ensemble.

10-4. Stationary Processes. In nontechnical language, a stochastic process is said to be stationary when the process is temporally homogeneous; i.e., its statistical properties remain invariant under *every* translation of the time scale. In technical language it is implied by a stationary process that

$$\begin{aligned} P\{X_1 \le a_1, X_2 \le a_2, X_3 \le a_3, \ldots, X_n \le a_n\} \\ &= F(a_1,a_2,a_3, \ldots ,a_n, t_1,t_2,t_3, \ldots ,t_n) \\ &= F(a_1,a_2,a_3, \ldots ,a_n, t_{1+T},t_{2+T}, \ldots ,t_{n+T}) \qquad (10\text{-}15) \end{aligned}$$

T being any real number, the identity must hold for every finite T and all appropriate choices of $t_1, t_2, \ldots, t_n$ and $X_1, X_2, \ldots, X_n$. (If this relationship holds for every finite integer n for $n = 1, 2, \ldots$ the process is *strictly stationary;* otherwise the process is stationary of order k, k being the highest integer for which the above relationship holds.) For many practical problems one is often confined to the cases where stationarity holds in the first- and second-order distributions. An equivalent interpretation of the stationarity is the fact that all joint probability densities (X_k,X_j) depend on the time difference $(k-j)$ but not on the absolute value of the time. Any outcome of the strictly stationary process, that is, any member of such ensemble, if shifted in time, will lead to another outcome of the same ensemble. The stationarity of the first order implies

$$E[X(t_k)] = E[X(t_{k+m})] \qquad \text{for every } k \text{ and every } m \tag{10-16}$$

The biased coin of Sec. 10-1 gives an example of a stationary stochastic process as successive trials are assumed to be independent. In the example of Sec. 10-2, $X = \sin(t + \theta)$, the density distribution for $X(t_k)$ and $X(t_j)$, is identical for the particular case when θ is uniformly distributed in an interval of length 2π.

The second-order stationarity implies that for any set of (t_k,t_j)

$$\begin{aligned} E\{[X(t_k) - \overline{X(t_k)}][X(t_k+\tau) - \overline{X(t_k+\tau)}]\} \\ = E\{[X(t_j) - \overline{X(t_j)}][X(t_j+\tau) - \overline{X(t_j+\tau)}]\} = \rho(\tau) \end{aligned} \tag{10-17}$$

This is an immediate consequence of the invariance of the second-order joint probability distribution under any shift of the origin of time. However, the converse of this statement is not necessarily true. A process may obey the latter equation without being stationary of the second order. Because of the importance of the concept of stationarity of the second order, one must supplement the above definition, which is based on the invariance of the joint probability distribution, with a less restrictive definition, that is, the invariance of the second-order expectation. The following definition is generally accepted in technical literature: A stochastic process is said to be stationary of the second order (or stationary *in the wide sense*) when the following two conditions are satisfied:

$$\begin{aligned} &\text{(I)}\ \ E[Y(t)]^2 < \infty \qquad \text{for all } t \in T \\ &\text{(II)}\ \ E[Y(t_i)Y(t_i+\tau)] = \rho(\tau) = \text{function of } \tau \text{ only, for all } t \end{aligned} \tag{10-18}$$

The advantage of this less restrictive definition is the fact that it generally provides a simpler means of determining the statistical character of a process. The second-order moment described in condition II is basically the same as the second-order central moment described earlier. In fact, consider the transformation of the variable

$$Y(t) = \frac{X(t) - E[X(t)]}{\sqrt{E\{X(t) - E[X(t)]\}^2}} \tag{10-19}$$

and note that

$$\overline{Y(t)} = 0$$

Thus the condition previously described by the second-order central moment of the process $\{X(t)\}$ will lead to condition II when the first-order moments are selected to be zero for all values of time.

Example 10-1. Consider the stochastic process

$$\{X(t)\} = At + Bt^2$$

where A and B are independent random variables with zero means and equal standard deviations σ. Compute the autocovariance function of the process.

Solution. The random variable $X(t_i)$ is a linear combination of two independent random variables:

$$X(t_i) = At_i + Bt_i^2$$
$$E[X(t_i)] = t_iE(A) + t_i^2E(B) = 0$$

The second-order expectation is

$$\begin{aligned} E[X(t_i) \cdot X(t_k)] &= E[(At_i + Bt_i^2)(At_k + Bt_k^2)] \\ &= E[A^2t_it_k + ABt_i^2t_k + BAt_it_k^2 + B^2t_i^2t_k^2] \\ &= t_it_kE(A^2) + (t_i + t_k)t_it_kE(AB) + t_i^2t_k^2E(B^2) \\ &= t_it_k(1 + t_it_k)\sigma^2 \end{aligned}$$

The second-order central moments are

$$E[X(t_i) - \overline{X(t_i)}]^2 = E[X^2(t_i)] = (t_i^2 + t_i^4)\sigma^2$$
$$E[X(t_k) - \overline{X(t_k)}]^2 = E[X^2(t_k)] = (t_k^2 + t_k^4)\sigma^2$$

The autocorrelation becomes

$$\rho_{xx}(t_i,t_k) = \frac{E\{[X(t_i) - \overline{X(t_i)}][X(t_k) - \overline{X(t_k)}]\}}{\sqrt{E\{[X(t_i) - \overline{X(t_i)}]^2\} \cdot E\{[X(t_k) - \overline{X(t_k)}]^2\}}}$$

$$\rho_{xx}(t_i,t_k) = \frac{t_it_k(1 + t_it_k)\sigma^2}{t_it_k\sqrt{(1 + t_i^2)(1 + t_k^2)\sigma^4}} = \frac{1 + t_it_k}{\sqrt{(1 + t_i^2)(1 + t_k^2)}}$$

The process is evidently not stationary, as the autocovariance function, that is, the numerator, depends on t_i and t_k, not merely the difference $t_i - t_k$.

Example 10-2. A stochastic process is described by

$$\{X(t)\} = A \sin t + B \cos t$$

where A and B are independent random variables with zero means and equal standard deviations. Show that the process is stationary of the second order.

Solution. The first and the second moments can be directly computed:

$$\begin{aligned} E[X(t)] &= E(A \sin t + B \cos t) \\ &= \sin tE(A) + \cos tE(B) = 0 \\ E[X(t_1)X(t_2)] &= E[(A \sin t_1 + B \cos t_1)(A \sin t_2 + B \cos t_2)] \\ &= E[A^2 \sin t_1 \sin t_2 + AB \sin (t_1 + t_2) + B^2 \cos t_1 \cos t_2] \\ &= \sin t_1 \sin t_2E(A^2) + \sin (t_1 + t_2)E(AB) + \cos t_1 \cos t_2E(B^2) \\ &= \sigma^2 \cos (t_1 - t_2) \end{aligned}$$

The second-order central moment depends only on $t_1 - t_2$; therefore the process is stationary. (In this example it is easy to show that the second-order probability distributions also depend only on the time difference $t_1 - t_2$. Thus, even according to the more restricted definition of the term, the process is stationary of the second order.)

10-5. Ergodic Processes.* In order to define an ergodic stochastic process $X(t)$, a new associated random variable will first be defined:

$$X_e = \lim_{r \to \infty} \frac{X(t_k) + X(t_{k+1}) + X(t_{k+2}) + \cdots + X(t_{k+r})}{r+1} \qquad \text{for a discrete process} \tag{10-20}$$

$$\text{and} \quad X_e = \lim_{T \to \infty} \frac{1}{2T} \int_{-T}^{T} X(t)\, dt \qquad \text{for a continuous process}$$

This random variable is somewhat indicative of the average values of all the occurring random variables of the family. For stationary processes it is generally true that this limit exists. Furthermore, assume that, for a large value of r, the above random variable does not further indicate any randomness but approaches a constant number. For an ergodic process these requirements must be met. For every outcome of the ergodic process the time average X_e should exist and should equal the expected value of any specific sampled random variable of the sequence:

$$E[X(t_k)] = \langle X_e \rangle \qquad \text{for every } k$$

According to the symbols adopted in the previous section, it may be stated that the first-order ergodic property of a stationary process implies

$$\overline{X(t_k)} = \langle X(t) \rangle \qquad \text{for every } k \tag{10-21}$$

Similarly, the ergodicity of the second order implies

$$\overline{X(t_k) \cdot X(t_k + \tau)} = \langle X(t) \cdot X(t + \tau) \rangle \tag{10-22}$$

From the mathematical point of view, the above "definition" for ergodicity remains somewhat incomplete. To be more exact, one should qualify the equality of the ensemble average and time average for the stationary process with the reservation "almost everywhere." The latter terminology has a specific mathematical meaning which will not be discussed here. The reader interested in the practical application of the theory may rest assured that he is not faced with such unusual circumstances in the study of ordinary physical systems.

Another mathematical point to be brought out here is the fact that, while the most important implication of ergodicity has been stated, there is a tacit omission of the delicate mathematical definition for such a process. The readers interested in a more precise definition may find the following presentation more satisfactory. Let

$$\{X\} = (\ldots, X_{-1}, X_0, X_1, X_2, \ldots)$$

* The material of this section was communicated in essence by Dr. L. Cote. The section may be omitted in a first reading. All equations referring to ergodicity imply "almost everywhere."

be a random sequence of a discrete stationary process, and $g(y_1,y_2, \ldots ,y_r)$ be any real single-valued function of r variables. Now define a function of r sampled random variables of the sequence

$$G(N,X) = \frac{1}{N+1} \sum_{k=0}^{N} g(X_{i_1+k},X_{i_2+k}, \ldots ,X_{i_r+k})$$

Let

$$\lim_{N\to\infty} G(N,X) = G(X)$$

The function $G(X)$ in the limit may or may not exist. If it exists and has a constant value G independent of X, and moreover if this constant value of G is equal to $E[g(X_1,X_2, \ldots ,X_r)]$ for all choices of sequence X_k, $X_{k+1}, \ldots , X_{k+r}$, then the sequence is called ergodic.

It is easy to see that the above-mentioned two cases are encompassed by this more general definition. In fact, letting

$$g(y_1,y_2, \ldots ,y_n) = g(y) = y \qquad n = 1$$

there results

$$G(N,X) = \frac{1}{N+1} \sum_{n=0}^{N} X_{i+n}$$

$$\lim_{N\to\infty} G(N,X) = E(X)$$

The process lim used here differs from the usual limit since G is a function of a random variable. (For a discussion of the concept, see Sec. 10-12.) Similarly, the second-order ergodicity can be illustrated by letting

$$g(y_1,y_2, \ldots ,y_n) = g(y_1,y_2) = [y_1 - E(X)][y_2 - E(X)]$$

$$G(N,X) = \frac{1}{N+1} \sum_{n=0}^{N} [X_i - E(X)][X_{i+K} - E(X)]$$

Since the process is stationary to begin with, it follows that

$$\lim_{N\to\infty} G(N,X) = \rho(K)$$

$\rho(K)$ does not depend on X_i but depends only on the time interval K. The foregoing defining procedure employs the concept of a discrete ergodic process, for convenience. The same conceptual pattern applies for defining continuous ergodic processes.

During the past three decades a vast amount of mathematical work has been produced on the subject of ergodicity. The famous Birkhoff's ergodic theorem is a classical landmark in this specialized field. The interested reader is referred to specialized articles on the subject.

The diagram of Fig. 10-3 is a reminder that the ergodic processes as defined here are a subclass of the stationary processes which are in turn included in the more general class of stochastic processes. This introductory presentation is not concerned with nonstationary processes.

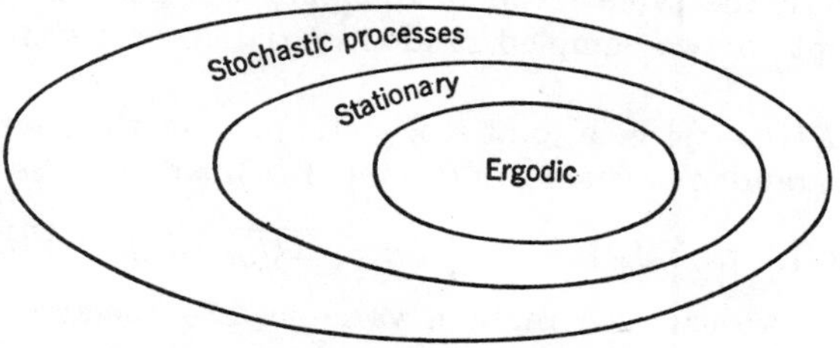

FIG. 10-3. A classification of different processes as discussed in the text.

In fact, the subsequent study of linear systems will be confined principally to ergodic processes.

Example 10-3. Determine whether or not the stochastic process

$$\{X(t)\} = A \sin t + B \cos t$$

is ergodic. A and B are normally distributed independent random variables with zero means and equal standard deviation.

Solution. The second-order central moment of the process was computed in Example 10-2:

$$E[X(t_1)X(t_2)] = \sigma^2 \cos (t_1 - t_2)$$

In order to obtain the time average, consider two specific members of the ensemble:

$$A_1 \sin t + B_1 \cos t$$

and

$$A_2 \sin t + B_2 \cos t$$

where A_1, A_2 and B_1, B_2 are some specific permissible values that the variables A and B may assume. Next compute the second-order time average.

$$\text{Time average} = \langle (A_1 \sin t + B_1 \cos t)[A_2 \sin (t + \tau) + B_2 \cos (t + \tau)]\rangle$$

It can be shown that this time average depends on the selection of the member in the ensemble. In fact, let $\tau = 0$, and inspection will show that

$$\text{Time average} = \langle A_1A_2 \sin^2 t\rangle + \langle B_1B_2 \cos^2 t\rangle = \tfrac{1}{2}(A_1A_2 + B_1B_2)$$

It has been found that there is at least *one* second-order time average that depends on the selected members of the ensemble, rather than being a constant. Thus the process is not ergodic although it is stationary.

10-6. Correlation Coefficients and Correlation Functions. This section begins with the introduction of the correlation coefficient ρ which is commonly used in the study of the interdependence of two random variables X and Y.

$$\text{Correlation coefficient} = \rho = \frac{E[(X - \bar{X}) \cdot (Y - \bar{Y})]}{\sqrt{E[(X - \bar{X})^2] \cdot E[(Y - \bar{Y})^2]}} \tag{10-23}$$

The numerator of this expression is called the *covariance* of the two variables. If the two variables are independent, their covariance is zero. If the correlation coefficient is zero, the variables are uncorrelated, but not *necessarily* independent. (This of course does not imply statistical independence.) In the same trend of thought, it is natural to extend this useful concept to two sampled random variables of a simple or a joint process.

Letting $\{X(t),Y(t)\}$ be a joint stochastic process, the covariance of the two sampled random variables $X(t_i)$ and $Y(t_k)$ is defined as

$$\text{Covariance } \{X(t_i),Y(t_k)\} = E\{[X(t_i) - \overline{X(t_i)}][Y(t_k) - \overline{Y(t_k)}]\} \qquad (10\text{-}24)$$

When the two sampled random variables are selected from a single process, the covariance coefficient is more specifically called the *autocovariance*. The autocovariance indicates a measure of interdependence or coherence between the two sampled random variables of the process $X(t_i)$ and $X(t_k)$, that is,

$$\begin{aligned}&\text{Autocovariance } \{X(t_i),X(t_k)\}\\ &\qquad = E\{[X(t_i) - \overline{X(t_i)}][X(t_k) - \overline{X(t_k)}]\}\end{aligned} \qquad (10\text{-}25)$$

An obvious simplification occurs if a convenient choice is made:

$$\begin{aligned}\overline{X(t_i)} &= 0 \qquad & \overline{X(t_k)} &= 0\\ \overline{Y(t_i)} &= 0 \qquad & \overline{Y(t_k)} &= 0\end{aligned} \qquad (10\text{-}26)$$

This "simplification" can be considered as an effect of a linear change of the variables. Such an operation has no significant effect on our studies except the simplification of results. In such a case,

$$\begin{aligned}\text{Covariance of } \{X(t_i),Y(t_k)\} &= E[X(t_i) \cdot Y(t_k)]\\ \text{Autocovariance of } \{X(t_i),X(t_k)\} &= E[X(t_i) \cdot X(t_k)]\end{aligned} \qquad (10\text{-}27)$$

For a stationary process of second order with a density function $f(x_i,x_k;t_i,t_k)$, the autocovariance depends only on the time lag $t_i - t_k$ and gives a measure of the effect of the past of the process on its present and future states:

$$E\{[X(t_i) - \overline{X(t_i)}][X(t_k) - \overline{X(t_k)}]\} = R_{xx}(|t_i - t_k|) \qquad (10\text{-}28)$$

The autocovariance of a stationary process is sometimes called the autocorrelation function in engineering literature. This terminology is not generally in agreement with the mathematical definition of correlation coefficient:

$$\rho_{xx}(\tau) = \frac{E\{[X(t_i) - \overline{X(t_i)}][X(t_i + \tau) - \overline{X(t_i + \tau)}]\}}{\sqrt{E[X(t_i) - \overline{X(t_i)}]^2 E[X(t_i + \tau) - \overline{X(t_i + \tau)}]^2}} \qquad (10\text{-}29)$$

However, if the assumption is made of zero first-order averages for $X(t_i)$ and $X(t_k)$, there occurs

$$\rho_{xx}(\tau) = \frac{E[X(t_i) \cdot X(t_i + \tau)]}{\sqrt{E[X(t_i)]^2 \cdot E[X(t_i + \tau)]^2}} \tag{10-30}$$

$$\rho_{xx}(\tau) = \frac{E[X(t_i) \cdot X(t_i + \tau)]}{R_{xx}(0)} = \frac{R_{xx}(|t_i - t_k|)}{R_{xx}(0)} \tag{10-31}$$

Under the above assumption the autocorrelation function will be identical with the autocovariance *within a constant.*

In order to avoid confusion, the normalized autocovariance function of a stationary process, that is, $R_{xx}(\tau)/R_{xx}(0)$, will be called the autocorrelation function of that process. This will be in conformity with the engineering literature. It will be tacitly assumed that the first-order averages are zero and a normalization is done to make

$$R_{xx}(0) = 1 \tag{10-32}$$

With this reservation in mind, the autocorrelation of a stationary process function is

$$\rho_{xx}(\tau) = E[X(t_i)X(t_i + \tau)] = \int_{-\infty}^{\infty} \int_{-\infty}^{\infty} x_1 x_2 f(x_1,x_2)\, dx_1\, dx_2 \tag{10-33}$$

Similarly, for a joint stationary process $\{X(t),Y(t)\}$ a cross-correlation function may be defined as

$$\rho_{xy}(\tau) = E[X(t_i) \cdot Y(t_i + \tau)] \tag{10-34}$$

To sum up, the autocorrelation and cross-correlation functions are special covariances for stationary processes of the second order or higher, provided that the means of the sampled variables are taken to be zero and $R_{xx}(0) = 1$, $R_{xy}(0) = 1$.

Finally, since later studies are confined to ergodic ensembles in linear systems, there is one more simplification ahead, the one for ergodic ensembles with zero first-order averages:

$$\begin{aligned} \rho_{xx}(\tau) &= \overline{X(t) \cdot X(t + \tau)} = \langle X(t) \cdot X(t + \tau)\rangle \\ \rho_{xy}(\tau) &= \overline{X(t) \cdot Y(t + \tau)} = \langle X(t) \cdot Y(t + \tau)\rangle \\ \rho_{yx}(\tau) &= \overline{Y(t) \cdot X(t + \tau)} = \langle Y(t) \cdot X(t + \tau)\rangle \\ \rho_{yy}(\tau) &= \overline{Y(t) \cdot Y(t + \tau)} = \langle Y(t) \cdot Y(t + \tau)\rangle \end{aligned} \tag{10-35}$$

In the study of ergodic ensembles, the correlation functions may be computed from either probabilistic considerations (joint density of samples) or from the deterministic point of view, that is, the time averaging of the two specific time functions under consideration. The equivalence of the two results is assumed by *ergodic hypothesis.* When dealing with two or more sampled random variables of a simple or a joint stationary stochastic process (τ units of time apart) a good picture is obtained of the interdependence of the different samples by computing the correlation functions and tabulating the results in a useful matrix form. For example, for a joint stationary process $\{X(t), Y(t)\}$, we have

$$\begin{bmatrix} \rho_{xx}(\tau) & \rho_{xy}(\tau) \\ \rho_{yx}(\tau) & \rho_{yy}(\tau) \end{bmatrix} \tag{10-36}$$

When dealing with ergodic ensembles, compute the correlation functions based on the concept of time averaging; that is, form the product of one member of the ensemble by another member shifted by a time interval τ. The product should be averaged as indicated previously. Care should be exercised, however, to determine that the process is ergodic to begin with, and this is not generally a simple task.

10-7. Example of a Normal Stochastic Process. Consider the stochastic process

$$\{X(t)\} = \{A \cos \omega t + B \sin \omega t\} \tag{10-37}$$

A and B being normally distributed random variables with zero means, unit variances, and zero correlation coefficient; then

$$\begin{aligned} X(t_i) &= X_i = A \cos \omega t_i + B \sin \omega t_i = \alpha_i A + \beta_i B \\ X(t_k) &= X_k = A \cos \omega t_k + B \sin \omega t_k = \alpha_k A + \beta_k B \end{aligned} \tag{10-38}$$

The density distribution for the random variable X_i is normal, as the variable is the linear sum of two normal variables. The same is true for the distribution of the random variable X_k.

X_i normal with zero mean $\sigma_i = \sqrt{\alpha_i^2 + \beta_i^2} = 1$ standard deviation
X_k normal with zero mean $\sigma_k = \sqrt{\alpha_k^2 + \beta_k^2} = 1$ standard deviation

The joint density distribution of X_i and X_k is a two-dimensional normal distribution also with (0,0) mean and a covariance matrix which can be determined in the following way:

$$[\rho] = \begin{bmatrix} \rho_{11} & \rho_{12} \\ \rho_{21} & \rho_{22} \end{bmatrix}$$

ρ_{12} is the covariance coefficient between X_i and X_k:

$$\begin{aligned} \rho_{12} &= E[(X_i - \bar{X}_i)(X_k - \bar{X}_k)] = E(X_i X_k) \\ &= E[A^2 \cos \omega t_i \cos \omega t_k + AB \cos \omega t_i \sin \omega t_k \\ &\qquad + BA \sin \omega t_i \cos \omega t_k + B^2 \sin \omega t_i \sin \omega t_k] \end{aligned} \tag{10-39}$$

Since A and B are uncorrelated random variables,

$$E(AB) = E(BA) = E(A)E(B) = 0$$

Thus

$$\rho_{12} = \cos \omega t_i \cos \omega t_k \, E(A^2) + \sin \omega t_i \sin \omega t_k \, E(B^2) \qquad (10\text{-}40)$$
$$\rho_{12} = \cos \omega(t_i - t_k)$$
$$[\rho] = \begin{bmatrix} 1 & \cos \omega(t_i - t_k) \\ \cos \omega(t_i - t_k) & 1 \end{bmatrix}$$

Now the joint distribution of X_i and X_k can be found by a direct substitution of pertinent quantities in the quadratic form associated with a two-dimensional normal variable. This distribution function is normal with (0,0) mean and covariances which depend solely on the time difference $t_i - t_k$; that is, choice of the origin of time is irrelevant. Thus the process is stationary of the second order. Meanwhile it has been shown in the above that the normalized correlation function ρ_{12} depends only on the time difference $t_i - t_k$, as it should be in a second-order stationary process. A generalization of this normal process can also serve as an example of a strictly stationary process; that is, the density distribution f_n for any number n of sampled variables remains intact under any translation of the time scale. The proof is left for the reader.

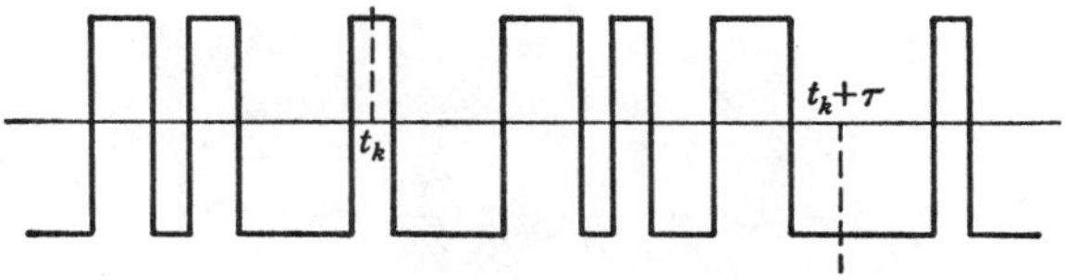

FIG. 10-4. A Poisson-type pulse process, described by Eqs. (10-42).

10-8. Examples of Computation of Correlation Functions. Consider the following stochastic process, which has many engineering applications. A pulse generator transmits pulses with random durations but with heights of $+E$ or $-E$. The occurrence of a pulse is supposed to follow the Poisson law, with $k/2$ the average number of pulses in the unit time.

In order to compute the autocorrelation function of this process, the familiar probabilistic definition of the ensemble averages is employed:

$$\rho_{11}(\tau) = E[X(t_1) \cdot X(t_1 + \tau)] = \Sigma\Sigma x_1 x_2 f(x_1,x_2) \qquad (10\text{-}41)$$

The variables are of a discrete nature, each one assuming binary values of plus or minus E. It is necessary to have the joint probability function for the two variables. For this, the probabilistic nature of the problem should first be clarified.

The mean number of zero crossings in a time interval T is kT. The probability of getting a specified number n of crossings in a time interval T assumes the Poisson distribution with an average of $\lambda = kT$. The probability of n crossings in time T is $[(kT)^n/n!]e^{-kT}$. Thus the probability of having $+E$ at time $t + \tau$ (provided that there was $+E$ at time t) is equal to the probability of having an even number of zero crossings between the two instants of time.

$$
\begin{aligned}
P\{+E \text{ at } t \text{ and } +E \text{ at } t+\tau\} &= P\{E \text{ at } t\} \times P\{E \text{ at } t+\tau | E \text{ at } t\} \\
&= \tfrac{1}{2}\left[\sum_{n=0,2,4,\ldots} \frac{(k\tau)^n}{n!} e^{-k\tau}\right] \\
P\{+E \text{ at } t \text{ and } -E \text{ at } t+\tau\} &= \tfrac{1}{2}\left[\sum_{n=1,3,5,\ldots} \frac{(k\tau)^n}{n!} e^{-k\tau}\right]
\end{aligned}
\tag{10-42}
$$

Note that the joint probability function depends only on the time lag τ. The factor $\frac{1}{2}$ appears as a consequence of the fact that, on an average, the probability of having $+E$ or $-E$ is assumed to be equal. Thus

$$
\rho_{11}(\tau) = \overline{[X(t_1) \cdot X(t_1+\tau)]} = E^2 \left[\sum_{n \text{ even}} \frac{(k\tau)^n}{n!} e^{-k\tau} - \sum_{n \text{ odd}} \frac{(k\tau)^n}{n!} e^{-k\tau}\right] \tag{10-43}
$$

$$
\rho_{11}(\tau) = E^2 e^{-k\tau}\left[1 - k\tau + \frac{(k\tau)^2}{2!} - \frac{(k\tau)^3}{3!} + \cdots\right] = E^2 e^{-2k\tau} \tag{10-44}
$$

In this equation it is tacitly assumed that τ is a positive number. The

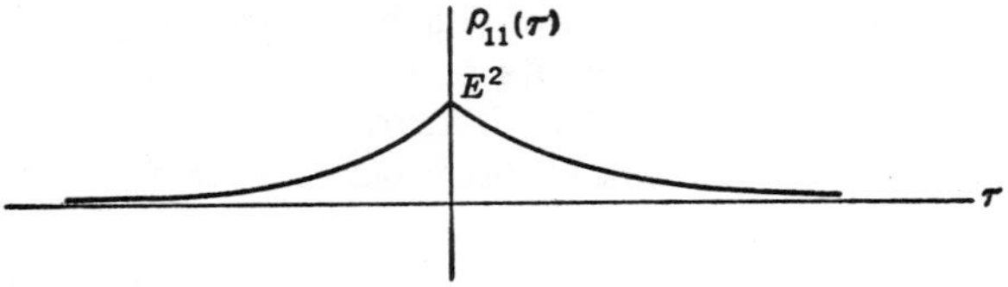

FIG. 10-5. Autocorrelation function of the Poisson process of Fig. 10-4.

same procedure of course applies when τ is negative. To encompass the two cases, write

$$
\rho_{11}(\tau) = E^2 e^{-2k|\tau|}
$$

The process is indeed a stationary process of the second order, and the normalized autocorrelation function $E[X(t_1) \cdot X(t_1 + \tau)]$ remains invariant under any translation of the time axis. A sketch of the autocorrelation function is given in Fig. 10-5.

As a second example, consider a particular noise process which has four distinct equally probable outcomes, as illustrated in Fig. 10-6. It is desired to compute the following data:

1. First-order probability distribution $F[X(t)]$ for $t = 2$
2. First-order probability distribution $F[X(t)]$ for $t = 4$
3. Second-order probability distribution $F[X(t_1),X(t_2)]$ for $t_1 = 2$, $t_2 = 4$
4. $E[X_2]$, $E[X_4]$, $E[X_2 \cdot X_4]$
5. Autocorrelation coefficient $\rho_{xx}[X_2,X_4]$

The probability distributions required in parts (1) and (2) are shown in Fig. 10-7*a* and *b*, respectively.

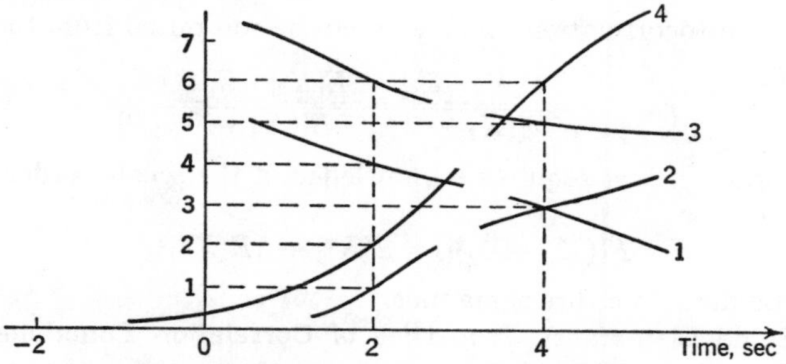

FIG. 10-6. A simplified example of a process.

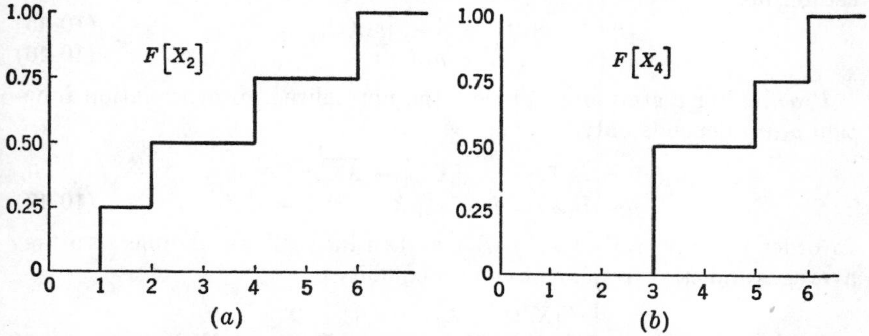

FIG. 10-7. (*a*) The CDF of the random variable X_2 of Fig. 10-6; (*b*) the CDF of the random variable X_4 of Fig. 10-6.

The following probability "density" table can easily be obtained:

X_4 \ X_2	1	2	4	6
3	$1/8$	$1/8$	$1/8$	$1/8$
5	$1/16$	$1/16$	$1/16$	$1/16$
6	$1/16$	$1/16$	$1/16$	$1/16$

The probability distribution function can easily be determined from the above; for example,

$$P\{X_2 \le 2, X_4 \le 3\} = \tfrac{1}{4}$$

The expected values are derived below:

$$\begin{aligned}
E[X_2] &= \tfrac{1}{4}(1 + 2 + 4 + 6) = \tfrac{13}{4} \\
E[X_4] &= \tfrac{1}{2}\cdot 3 + \tfrac{1}{4}\cdot 5 + \tfrac{1}{4}\cdot 6 = \tfrac{17}{4} \\
E[X_2 \cdot X_4] &= \tfrac{1}{8}\cdot 3 + \tfrac{1}{8}\cdot 6 + \tfrac{1}{8}\cdot 12 + \tfrac{1}{8}\cdot 18 + \tfrac{1}{16}\cdot 5 \\
&\quad + \tfrac{1}{16}\cdot 10 + \tfrac{1}{16}\cdot 20 + \tfrac{1}{16}\cdot 30 + \tfrac{1}{16}\cdot 6 \\
&\quad + \tfrac{1}{16}\cdot 12 + \tfrac{1}{16}\cdot 24 + \tfrac{1}{16}\cdot 36 = \tfrac{221}{16}
\end{aligned}$$

The desired autocorrelation coefficient can be computed from the relation

$$\rho_{xx} = \frac{E[X_2 \cdot X_4] - E[X_2] \cdot E[X_4]}{\sqrt{E[(X_2 - \tfrac{13}{4})^2]E[(X_4 - \tfrac{17}{4})^2]}}$$

The computation first requires a knowledge of the second-order expectations of the type

$$E[(X - \bar{X})^2] = E[X^2] - [E(X)]^2$$

This can be done in a direct manner.

10-9. Some Elementary Properties of Correlation Functions of Stationary Processes. The following simple properties can be directly established:

$$\rho_{11}(0) = 1 \ge |\rho_{11}(\tau)| \tag{10-45}$$
$$\rho_{11}(\tau) = \rho_{11}(-\tau) \tag{10-46}$$

Proof. For a stationary process the normalized autocorrelation function $\rho_{11}(\tau)$ depends only on τ; hence

$$\begin{aligned}
\rho_{11}(\tau) &= E[X_i - \bar{X}_i][X_{i+k} - \overline{X_{i+k}}] \\
&= E[X_{i-k} - \bar{X}_{i-k}][X_i - \bar{X}_i] = \rho_{11}(-\tau)
\end{aligned} \tag{10-47}$$

In order to show Eq. (10-45) (Schwartz's inequality), assume zero first averages and start from an obvious inequality:

$$\begin{gathered}
E[X(t) \pm X(t + \tau)]^2 \ge 0 \\
E[X(t)]^2 \pm 2E[X(t) \cdot X(t + \tau)] + E[X(t + \tau)]^2 \ge 0 \\
\rho_{11}(0) \pm 2\rho_{11}(\tau) + \rho_{11}(0) \ge 0 \\
\rho_{11}(0) \ge |\rho_{11}(\tau)|
\end{gathered}$$

The normalized autocorrelation of the sum of a number of stationary processes follows the symbolic rule for the ordinary product. For example, let

$$\begin{aligned}
Z(t) &= X(t) + Y(t) \\
\rho_{zz}(t_1,t_2) &= E[X(t_1) + Y(t_1)][X(t_2) + Y(t_2)] \qquad (10\text{-}48) \\
&= E[X(t_1) \cdot X(t_2)] + E[X(t_1)Y(t_2)] + E[X(t_2) \cdot Y(t_1)] \\
&\qquad + E[Y(t_2) \cdot Y(t_1)]
\end{aligned}$$

The first averages are generally assumed to be zero:

$$EX(t_1) = EX(t_2) = EY(t_1) = EY(t_2) = 0$$

Then

$$\rho_{zz}(t_1,t_2) = \rho_{xx}(t_1,t_2) + \rho_{xy}(t_1,t_2) + \rho_{yx}(t_1,t_2) + \rho_{yy}(t_1,t_2)$$

Since this chapter is confined to stationary processes, write

$$\rho_{zz}(\tau) = \rho_{xx}(\tau) + \rho_{xy}(\tau) + \rho_{yx}(\tau) + \rho_{yy}(\tau) \tag{10-49}$$

Therefore, under the above assumptions, the rule for deriving the normalized autocorrelation function of the sum of a number of stationary processes will symbolically follow the rule for the ordinary product of the sum of a number of terms; for example,

$$\begin{aligned}(X + Y)(X + Y) &= XX + XY + YX + YY \\ (X + Y + U)(X + Y + U) &= XX + XY + XU + YX \\ &\quad + YY + YU + UX + UY + UU\end{aligned} \tag{10-50}$$

When two processes $\{X\}$ and $\{Y\}$ are independent, then $\rho_{xy}(\tau) = 0$.

10-10. Power Spectra and Correlation Functions.* The readers are undoubtedly familiar with the significance of linear integral transformations (particularly Fourier and Laplace transforms) in engineering problems. The dual relationship of the frequency and the time domain is a most significant concept in the study of linear systems. Whenever a linear problem becomes involved in one of these domains it is possible alternatively to try to solve the problem in the other domain. In this respect the reader may recall the familiar development of network theory such as the concept of impedance functions and network synthesis procedures. These developments have taken a primary place in the light of the theory of Laplace transformation.

It is natural to develop and explore relationships between specific time averages of interest and their Fourier integrals. Such a development mathematically is very fruitful; meanwhile, from an engineering point of view, the idea of Fourier integrals, amplitude, and power spectra has certain physical significance.

For a real second-order stationary process (wide sense), the power spectrum is defined as the Fourier transform of the normalized autocorrelation function of the process:

$$\phi_{xx}(\omega) = \int_{-\infty}^{\infty} \rho_{xx}(\tau)e^{-j\omega\tau}\, d\tau \tag{10-51}$$

$$\rho_{xx}(\tau) = \frac{1}{2\pi}\int_{-\infty}^{\infty} \phi_{xx}(\omega)e^{j\omega\tau}\, d\omega \tag{10-52}$$

provided that the integrals exist. Since the autocorrelation function is

* The author wishes to make acknowledgement to Drs. R. A. Johnson and S. Jutila for valuable insights gained in discussions with them on the stochastic behavior of linear systems.

an even function, it may alternatively be written

$$\phi_{xx}(\omega) = 2 \int_0^\infty \rho_{xx}(\tau) \cos \omega\tau \, d\tau \tag{10-53}$$

$$\rho_{xx}(\tau) = \frac{1}{\pi} \int_0^\infty \phi_{xx}(\omega) \cos \omega\tau \, d\omega \tag{10-54}$$

These equations are sometimes called Wiener-Khinchin relations.

As an application of the Wiener-Khinchin relation, we compute the power spectrum of the Poisson process of Sec. 10-8.

$$\begin{aligned} \phi_{xx}(\omega) &= 2 \int_0^\infty E^2 e^{-2K|\tau|} \cos \omega\tau \, d\tau \\ &= \frac{4KE^2}{4K^2 + \omega^2} \end{aligned}$$

When dealing with two such processes, the cross power spectrum, which has a similar relationship to the cross-correlation function, may also be defined.*

Next, consider an ergodic process $\{X(t)\}$ and define an average power associated with any member of this ensemble. Let $X(t)$ be a specific member of the process and assume $X(t)$ to be a real-valued function of time. The average power associated with the truncated X_T during the interval $(-T,+T)$ is

$$\text{Average power in time } (-T,+T) = \frac{1}{2T} \int_{-T}^{T} [X_T(t)]^2 \, dt \tag{10-55}$$

If $X_T(t)$ were a current flowing into a 1-ohm resistor, the above expression would indicate the average power dissipated in that resistor. The average power associated with an ergodic process can be defined as

$$\langle [X(t)]^2 \rangle = \lim_{T \to \infty} \frac{1}{2T} \int_{-T}^{T} [X_T(t)]^2 \, dt \tag{10-56}$$

This average will be the same for almost all members of the ergodic ensemble.

Now let $F_T(j\omega)$ be the Fourier transform of the truncated time function $X_T(t)$ for a specific member of the ensemble:

$$\begin{aligned} X_T(t) &= 0 \qquad t > +T \\ X_T(t) &= 0 \qquad t < -T \end{aligned}$$

$$F_T(j\omega) = \int_{-T}^{T} X_T(t) e^{-j\omega t} \, dt \tag{10-57}$$

The immediate purpose is to derive a relationship between the average power and the function $\phi_{xx}(\omega)$, where

$$\phi_{xx}(\omega) = \lim_{T \to \infty} \frac{1}{2T} |F_T(j\omega)|^2 \tag{10-58}$$

* W. R. Bennett, Methods of Solving Noise Problems, *Proc. IRE*, May, 1956, pp. 609–638; J. H. Laning, Jr., and R. H. Battin, "Random Processes in Automatic Control," McGraw-Hill Book Company, Inc., New York, 1956.

assuming such a unique limit exists for all members of the process. To this end, the power spectrum of any general process can be alternatively defined as

$$\phi_{xx}(\omega) = E \lim_{T\to\infty} \frac{1}{2T} [|F_T(j\omega)|^2]$$

Subsequent to some algebraic manipulations, one can show that, for an ergodic process, this definition is consistent with the one given earlier.

By Plancherel's relation, we have

$$\frac{1}{2\pi} \lim_{T\to\infty} \frac{1}{2T} \int_{-T}^{\infty} |F_T(j\omega)|^2 \, d\omega = \lim_{T\to\infty} \frac{1}{2T} \int_{-T}^{T} [X_T(t)]^2 \, dt = \langle [X(t)]^2 \rangle \tag{10-59}$$

Using the defining equation of power spectrum and the relation

$$\rho_{xx}(0) = \langle [X(t)]^2 \rangle = \frac{1}{2\pi} \int_{-\infty}^{\infty} \phi_{xx}(\omega) \, d\omega < \infty \tag{10-60}$$

we find that*

$$\phi_{xx}(\omega) = \lim_{T\to\infty} \frac{1}{2T} E|F_T(j\omega)|^2 = \lim_{T\to\infty} \frac{1}{2T} |F_T(j\omega)|^2 \tag{10-61}$$

For an ergodic process the power spectrum is the limit of $1/2T$ times the square of the magnitude of the Fourier transform of the truncated time function $X_T(t)$ when T is increased indefinitely. This is a unique deterministic function for all outcomes of the process.

Finally it is to be noted that the power spectrum of the sum of several independent processes is the sum of their individual spectra. In this sense "linearity" holds, as in the case of the autocorrelation function of the sum of a number of independent processes.

Thus it has been shown that there is a close relationship between power spectrum and correlation functions. When dealing with problems of an ergodic nature, it is convenient to use either correlation function or power spectrum, depending on circumstances.

10-11. Response of Linear Lumped Systems to Ergodic Excitation.† Electrical engineers are quite familiar with the study of ordinary linear lumped bilateral networks under a periodic deterministic regime. In such systems, when initially relaxed, if a unit impulse excitation $U_0(t)$ applied at

* The assumptions leading to Eq. (10-61) are involved with mathematical complexities not given here (see Middleton, Chap. 3; and Davenport and Root, Chap. 6).

† The statistical design of linear systems has been developed during the past two decades. Among those who have made significant contributions to this development are Wiener, Kolmogorov, Shannon, Rice, Bode, Middleton, Zadeh, Lee, and many others. Today the subject of filtering and prediction occupies the nucleus of a course in the graduate curriculum of many electrical engineering departments around the world. An adequate coverage of this topic, which should include the work of a great many scientists, is completely outside the scope of the present book. Those interested in a full treatment of the subject are referred to **Laning and Battin, Wiener, Blanc-Lapierre, Davenport and Root, and Middleton.**

time $t = 0$ produces an output of $h(t)$, then the response of the same system to an excitation $x(t)$ is given by the familiar superposition integral:

$$\text{Output} = y(t) = \int_{-\infty}^{t} x(\tau) \cdot h(t - \tau)\, d\tau \qquad \text{with } h(t) = 0 \quad \text{for } t < 0 \tag{10-62}$$

This is due to the fact that a unit impulse $U_0(t - \tau)$ applied to the system at time $t = \tau$ will give rise to an output of $h(t - \tau)$.

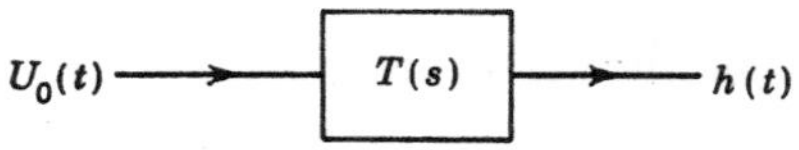

FIG. 10-8. A passive linear system specified by its response to unit impulse.

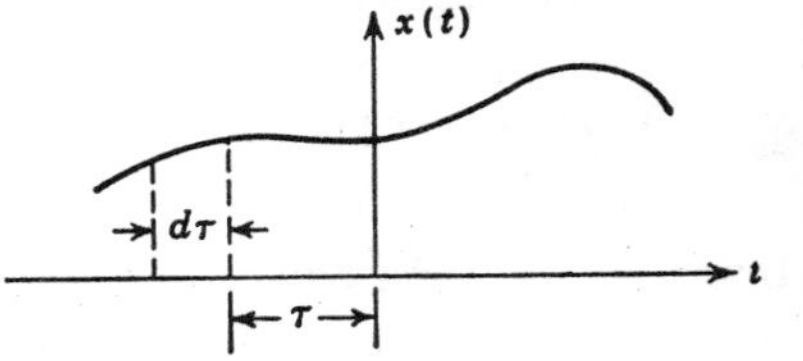

FIG. 10-9. A method for obtaining the response of linear systems.

The use of Fourier or Laplace transforms is most natural. In fact, the convolution of two functions in the time domain corresponds to the product of their Laplace transforms:

$$\begin{aligned} \mathfrak{L}\{h(t)\} &= T(s) \\ \mathfrak{L}\{x(t)\} &= A(s) \\ \mathfrak{L}\{y(t)\} &= B(s) \end{aligned} \tag{10-63}$$

Then for an initially relaxed system,

$$\boxed{B(s) = A(s) \cdot T(s)} \tag{10-64}$$

From this latter relation the inverse Laplace transform may be computed; thus, the problem is at least theoretically solved. The above fundamental relation is perhaps the most significant equation of the linear system theory.

The principal aim in this section is to explore the possibility of deriving a fundamental relation for the performance of a linear lumped bilateral system under ergodic regimes. Fortunately, an equally simple and elegant relationship exists. This simplicity is the main reason for the existence of the vast literature on the subject.

Let $\{X(t)\}$ be an ergodic input to a linear system initially at rest with a unit impulse response of $h(t)$. For a moment, concentrate on one specific member of the input ensemble, $x(t)$. The corresponding output will be a specific time function $y(t)$ such that

$$y(t) = \int_{-\infty}^{t} x(\tau)h(t - \tau)\, d\tau = \int_{0}^{t} x(t - \alpha)h(\alpha)\, d\alpha \tag{10-65}^*$$

* By letting $t - \tau = \alpha$, $y(t) = \int_0^\infty x(t - \alpha)h(\alpha)\, d\alpha$ is obtained.

If the specific input is shifted in time by τ, the output will undergo an equal time translation. This will enable a computation of the autocorrelation function of the output ensembles. The autocorrelation function can be written in the form of a double integral instead of the product of two integrals:*

$$\begin{aligned}\rho_{yy}(\tau) &= \langle Y(t) \cdot Y(t+\tau)\rangle \\ &= \int_0^\infty \int_0^\infty \langle X(t-\alpha)X(t+\tau-\beta)\rangle h(\alpha) \cdot h(\beta)\, d\alpha\, d\beta \qquad (10\text{-}66)\end{aligned}$$

Note that

$$\begin{aligned}\langle X(t-\alpha) \cdot X(t+\tau-\beta)\rangle &= \langle X(t-\alpha) \cdot X(t-\alpha+\tau-\beta+\alpha)\rangle \\ &= \rho_{xx}(\tau-\beta+\alpha) \qquad (10\text{-}67)\end{aligned}$$

But
$$\rho_{xx}(\tau-\beta+\alpha) = \frac{1}{2\pi}\int_{-\infty}^{\infty} \phi_{xx}(\omega)e^{j\omega(\tau-\beta+\alpha)}\, d\omega$$

Consequently,

$$\begin{aligned}\rho_{yy}(\tau) &= \frac{1}{2\pi}\int_0^\infty \int_0^\infty \int_{-\infty}^{\infty} \phi_{xx}(\omega)e^{j\omega(\tau-\beta+\alpha)}h(\alpha) \cdot h(\beta)\, d\alpha\, d\beta\, d\omega \\ \rho_{yy}(\tau) &= \frac{1}{2\pi}\int_{-\infty}^{\infty} \phi_{xx}(\omega)e^{j\omega\tau}\, d\omega \int_0^\infty h(\alpha) \cdot e^{j\alpha\omega}\, d\alpha \int_0^\infty h(\beta)e^{-j\beta\omega}\, d\beta\end{aligned} \qquad (10\text{-}68)$$

According to the fundamental equation of linear systems, the system function is the Laplace transform (Fourier transform if $s = j\omega$) of the unit impulse response $h(t)$:

$$\begin{aligned}\int_0^\infty h(\alpha)e^{j\alpha\omega}\, d\alpha &= T(-j\omega) \\ \int_0^\infty h(\beta)e^{-j\beta\omega}\, d\beta &= T(j\omega)\end{aligned} \qquad (10\text{-}69)$$

Therefore

$$\rho_{yy}(\tau) = \frac{1}{2\pi}\int_{-\infty}^{\infty} \phi_{xx}(\omega)e^{j\omega\tau}\, d\omega\, [T(j\omega) \cdot T(-j\omega)] \qquad (10\text{-}70)$$

Using the definition of the power spectrum for autocorrelation functions,

$$\begin{aligned}\int_{-\infty}^{\infty} \phi_{yy}(\omega)e^{j\omega\tau}\, d\omega &= \int_{-\infty}^{\infty} \phi_{xx}(\omega)e^{j\omega\tau}\, d\omega\, [T(j\omega) \cdot T(-j\omega)] \\ \phi_{yy}(\omega) &= \phi_{xx}(\omega) \cdot T(j\omega) \cdot T(-j\omega) \\ \boxed{\phi_{yy}(\omega) = \phi_{xx}(\omega)|T(j\omega)|^2} & \qquad (10\text{-}71)\dagger\end{aligned}$$

* The interchanging of the integral sign and the averaging sign can be justified. For additional proof see Davenport and Root, Secs. 9-2 and 9-3.

† The use of the Fourier transform in this section is in conformity with the mathematical literature on the subject, particularly the defining of equations of power spectra. This should not deter the reader from making a comparison with the fundamental deterministic equation of the linear system as given previously in the notation of the Laplace transform.

This is the fundamental relation for the performance of linear systems under ergodic regimes. The power spectrum of the output can be directly computed from the knowledge of the system function and the power spectrum of the input. Statistical information about the input thus will lead to a statistical identification of the output.

To sum up, the reader should feel that he has obtained some concrete results; namely, when faced with an ergodic stochastic input to a linear (or linearized) system, he may proceed with the following steps:

1. Determine the autocorrelation function of the input.
2. Compute the power spectrum of the input $\phi_{xx}(\omega)$.
3. Obtain the system function $T(s)$ between the input and the output ports of the system.
4. Compute the square of the magnitude of the system function for different values of ω, $|T(j\omega)|^2$.
5. The power spectrum of the output is $|T(j\omega)|^2 \cdot \phi_{xx}(\omega)$.
6. The autocorrelation function of the output process can be determined by taking the inverse Fourier transform of the expression obtained in step 5.

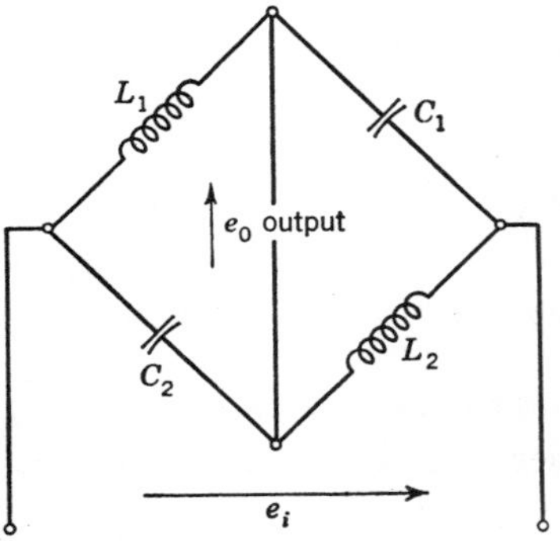

FIG. 10-10. Example of a linear system under stationary regime.

Example of White Noise. Consider the white noise applied to the input terminals of the network of Fig. 10-10. It is desired to give an indication of the statistical nature of the output.

The term *white noise* refers to a stochastic process with a constant power spectrum over all frequency ranges.

$$\phi_{xx}(\omega) = \alpha$$

(This concept implies infinite power, which is not realistic. At present, ignore this unrealistic implication.) The required system function is found to be

$$\frac{E_o}{E_i} = \frac{L_1S}{L_1S + 1/C_1S} - \frac{1/C_2S}{L_2S + 1/C_2S} = \frac{C_1L_1S^2 - 1}{(C_1L_1S^2 + 1)(C_2L_2S^2 + 1)}$$

The power spectrum of the output process is

$$\phi_{yy}(\omega) = \left[\frac{C_1L_1\omega^2 + 1}{(1 - C_1L_1\omega^2)(1 - C_2L_2\omega^2)}\right]^2 \alpha$$

An example of an application of Eq. (10-71) to communication theory is the effect of thermal resistor noise. In light of physical considerations, one concludes that the voltage fluctuations due to thermal noise may be considered as a gaussian process. These considerations lead to the fact that a linear resistor of resistance R can be replaced by a passive noiseless resistor in series with a stochastic voltage source having a flat power spectrum of $2KTR$. (T is the temperature of the system in degrees Kelvin, and K is referred to as Boltzmann's constant. In most applications the value of KT is taken as 4×10^{-21}.)

The problem of the study of noise in linear systems is treated in detail by many authors, for example, Lawson and Uhlenbeck, Rice, Davenport and Root, and Freeman. The general approach to these problems consists in replacing noisy resistors by noiseless ones fed by gaussian sources. Consequently, at any output port of the linear system, one can calculate the output power spectrum by proper application of Eq. (10-71).

Example 10-4. A stationary input $\{X(t)\}$ with an autocorrelation function

$$\rho_{xx}(\tau) = Ae^{-\alpha|\tau|}$$

is applied to a system such as that described by the differential equation

$$\frac{d^2y}{dt^2} + a\frac{dy}{dt} + by = \{X(t)\}$$

Determine the power spectrum and the variance of the output process.

Solution. The power spectrum of the input is

$$\phi_{xx}(\omega) = \frac{2A\alpha}{\alpha^2 + \omega^2}$$

The power spectrum of the output $\phi_{yy}(\omega)$ can be directly obtained by the application of Eq. (10-71). The variance of the output may be obtained in the following manner:

$$\rho_{yy}(\tau) = \frac{1}{2\pi}\int_{-\infty}^{\infty} \phi_{xx}(\omega)|T(j\omega)|^2 e^{j\omega\tau}\, d\omega$$

$$\text{Variance of the output} = \frac{1}{2\pi}\int_{-\infty}^{\infty} \phi_{xx}(\omega)|T(j\omega)|^2\, d\omega$$

The integration may be accomplished by consulting appropriate integral tables.

10-12. Stochastic Limits and Convergence. For the sake of mathematical completeness, we now supplement the introductory study of stochastic processes with a discussion of the concept of integration and differentiation of such processes. This can be accomplished after some limiting procedure for the process has been defined.

From a physical point of view, in engineering problems, a noise process is generally subject to integration and differentiation when it passes through a linear system. The ordinary integrators and differentiators which are so commonly used in servomechanisms give the simplest examples of such situations.

This section defines the stochastic limits and the stochastic convergence. The following modes of stochastic convergence are frequently used:

Convergence in Probability (Abbreviated Form i.p.). An infinite sequence of random variables,

$$X_1, X_2, X_3, \ldots, X_n, \ldots$$

converges to the random variable X in probability if, for any positive number ϵ,

$$\lim_{n \to \infty} P\{|X_n - X| > \epsilon\} = 0 \tag{10-72}$$

Convergence in the Mean-square Sense (Abbreviated Form m.s.). This convergence is defined as follows:

$$\underset{n \to \infty}{\text{l.i.m.}}\ X_n = X \tag{10-73}$$

if

$$\lim_{n \to \infty} E|X_n - X|^2 = 0$$

where l.i.m. stands for the limit in the mean.

Almost Certain Convergence (Abbreviated a.c.). For a.c. convergence, the set of realized sequences of $X_1, X_2, \ldots, X_n, \ldots$ converges to X with probability 1 when n approaches infinity. The a.c. convergence, sometimes called the *strong convergence*, implies i.p. convergence (but not conversely). There are also other modes of convergence defined in the literature (for necessary and sufficient conditions for each mode of convergence, see, for example, M. S. Bartlett, "An Introduction to Stochastic Processes," or J. E. Moyal, *J. Roy. Statist. Soc.*, ser. B, vol. 11, no. 2, 1949).

The above concept of convergence applied to a discrete sequence can be extended to the case of stochastic functions. For example, for a process $\{X(t)\}$, define $X(\tau)$ as the stochastic limit for $X(t)$, abbreviated as

$$\underset{t \to \tau}{\text{l.i.m.}}\ X(t) = X_0(\tau)$$

when it satisfies the condition

$$\lim_{t \to \tau} E[X(t) - X_0(\tau)]^2 = 0 \tag{10-74}$$

This is the definition for stochastic m.s. convergence of the process. (There are also a.c. convergence and convergence in probability for stochastic functions, but a discussion will not be undertaken here.)

Example 10-5. Consider the tossing of an honest coin. Let $X(n)$ be the number of heads in n throwings divided by n; that is, $X(n)$ is a random variable obeying the binomial distribution:

$$E[X(n)] = \tfrac{1}{2} \qquad \text{average}$$

$$\sigma^2[X(n)] = \frac{1}{4n} \qquad \text{standard deviation}$$

It can be shown that the sequence

$$X(1), X(2), \ldots, X(n), \ldots$$

converges, in the sense of 1, 2, and 3 below, to $\frac{1}{2}$.

1. By Chebyshev's inequality (Feller, p. 219, or Loève, p. 14):

$$P\{|X(n) - \tfrac{1}{2}| \geq \epsilon\} \leq \frac{1/4n}{\epsilon^2}$$

In the limit:

$$\underset{n\to\infty}{P}\{|X(n) - \tfrac{1}{2}| \geq \epsilon\} = 0$$

2.
$$E[X(n) - \tfrac{1}{2}]^2 = \sigma^2[X(n)] = \frac{1}{4n}$$

For $n \to \infty$:

$$\text{l.i.m. } X(n) = \tfrac{1}{2}$$

as

$$\lim_{n\to\infty} E[X(n) - \tfrac{1}{2}]^2 = 0$$

3. Using the strong law, one can also prove that $\underset{n\to\infty}{[X(n)]}$ converges to $\frac{1}{2}$ in the a.c. sense. The proof is somewhat complex and will be omitted here (see Loève, p. 19).

10-13. Stochastic Differentiation and Integration. The rigorous treatment of modes of convergence, continuity, differentiability, and integrability for stochastic processes is beyond the scope of this text. For the immediate purpose, it seems sufficient to give rudimentary definitions similar to the concepts acquired in ordinary courses in analysis.

A real process $X(t)$ is said to be continuous at a time t in m.s. if

$$\lim_{h\to 0} E|X(t + h) - X(t)|^2 = 0 \tag{10-75}$$

The m.s. derivative of a continuous process shown by $X(t)$ is defined as

$$\underset{h\to 0}{\text{l.i.m.}} \frac{X(t + h) - X(t)}{h} = X'(t) \tag{10-76}$$

As in the case of the integration of deterministic functions, different types of stochastic integrals can be defined. The Riemann stochastic integral as well as the Lebesgue and the Stieltjes stochastic integrals is frequently used in the technical literature. At present, this discussion is confined to giving the definition of stochastic integrals in the Riemann sense.

Divide the real interval of integration (a,b) into n arbitrary subintervals dt_i. The Riemann integral of a stochastic process $\{X(t)\}$ can be approached in the following way:

$$\int_a^b X(t)\, dt = \underset{n\to\infty}{\text{l.i.m.}} \sum_1^n X(t_i)\, dt_i = R(a,b) \tag{10-77}$$

This is contingent upon the convergence of the sum into the function $R(a,b)$ in the m.s. sense. It can be shown that the integral exists if the double integral of the autocovariance function of the process, $\rho_{xx}(t_1,t_2)$, exists over the square $[(a,b)$ by $(a,b)]$ in the (t_1,t_2) plane (see J. E. Moyal, *J. Roy. Statist. Soc.*, ser. B, vol. 11, no. 2, 1949, pp. 168–169).

It is not intended to delve further into this problem; however, it is of interest to note that, once the concept of differentiation and integration is introduced, the territory of the subject of discussion can be extended to cover the stochastic differential, integral, and difference equations. For example, the differential equation relating current and voltage in an *RLC* series network is

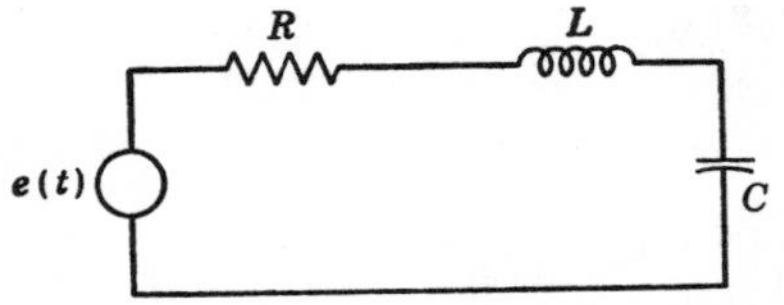

FIG. 10-11. Example of an *RLC* network under stochastic regime.

$$L\frac{d^2i}{dt^2} + R\frac{di}{dt} + \frac{1}{C}i = \frac{de}{dt} \tag{10-78}$$

If the driving voltage is a stochastic process $\{E(t)\}$, the current response will also be a stochastic process $\{I(t)\}$, the two processes being related by the stochastic differential equation

$$L\{\ddot{I}(t)\} + R\{\dot{I}(t)\} + \frac{1}{C}\{I(t)\} = \{\dot{E}(t)\} \tag{10-79}$$

The solutions to such equations are somewhat similar to the solution of their counterparts in the ordinary theory of differential equations.

The different moments and correlation functions of a stochastic process and its derivative are interrelated. Let $\{X(t)\}$ be a process and $\{X'(t)\}$ its first derivative. Then, with some mathematical care* it is possible to show that

$$X'(t) - \overline{X'(t)} = \frac{d}{dt}[X(t) - \overline{X(t)}]$$

$$E[X'(t_i) - \overline{X'(t_i)}][X'(t_j) - \overline{X'(t_j)}] = \frac{\partial^2}{\partial t_i\,\partial t_j}E[X(t_i) - \overline{X(t_i)}][X(t_j) - \overline{X(t_j)}]$$

$$R_{x'x'}(t_i,t_j) = \frac{\partial^2 R_{xx}(t_i,t_j)}{\partial t_i\,\partial t_j}$$

Thus, the correlation function of the derivative is equal to the second mixed derivative of the correlation of the original process. Furthermore, if $\{X(t)\}$ is stationary it follows that its derivative is also a stationary process.

$$R_{x'x'}(\tau) = \frac{\partial^2 R_{xx}(\tau)}{\partial t_i\,\partial t_j} = -\frac{\partial^2 R_{xx}(\tau)}{\partial \tau^2}$$

* The process $\{X(t)\}$ is assumed to have a derivative and differentiable averages and correlation function. Although an informal derivation of the above relations is straightforward, a rigorous derivation is somewhat involved with convergence considerations.

Similar observations can be made about an integral of a random process. Let $\{X(t)\}$ be a known process and $f(s,t)$ a suitable kernel. We are interested in studying the integral process $\{Y(s)\}$.

$$\{Y(s)\} = \int f(s,t)\{X(t)\}\, dt$$

Taking into consideration some mathematical concepts (existence of finite averages, correlation, basic definition of the process $\{Y(s)\}$, etc.) which are not covered here, one finds

$$Y(s_i) - \overline{Y(s_i)} = \int f(s_i,t_i)[X(t_i) - \overline{X(t_i)}]\, dt$$

$$E[Y(s_i) - \overline{Y(s_i)}][Y(s_j) - \overline{Y(s_j)}]$$
$$= \int\int f(s_i,t_i)f(s_j,t_j)E[X(t_i) - \overline{X(t_i)}][X(t_j) - \overline{X(t_j)}]\, dt_i\, dt_j$$
$$R_{yy}(s_i,s_j) = \int\int f(s_i,t_i)f(s_j,t_j)R_{xx}(t_i,t_j)\, dt_i\, dt_j$$

10-14. Gaussian-process Examples of a Stationary Process. In the study of the noise in communication systems, the engineer is frequently faced with periodic processes which have converging Fourier series expansions in a given interval, say $(-T,T)$:

$$X(t) = \sum_{k=1}^{\infty} a_k \cos \omega_k t + b_k \sin \omega_k t \qquad (10\text{-}80)^*$$

a_k and b_j are generally mutually independent random variables for all positive integers k and j. In many instances it is convenient to assume that random variables a_k and b_j have normal distributions, with zero means and specified variances:

(I) $$E(a_k) = 0 \qquad k = 1, 2, \ldots$$
$$E(b_k) = 0$$
(II) $$E(a_i b_k) = E(a_i) \cdot E(b_k) = 0$$
(III) $$E(a_k^2) = E(b_k^2) = \sigma_k^2 \qquad (10\text{-}81)$$
$$\sum_1^{\infty} \sigma_k^2 < \infty$$

The first objective is to show that such a process is stationary. For this, compute the second-order joint density function at times, say, t_1 and t_2. According to the central-limit theorem, the random variables $X(t_1)$ and $X(t_2)$ will approach normal distributions having zero means and equal standard deviations. Furthermore, a bivariate central-limit theorem will show that these two normal random variables also have a joint normal distribution. Under assumption III, the standard deviation of each term (a_k,b_k) of the sum $X(t)$ will be independent of t_1. Thus the joint density of $X(t_1)$ and $X(t_2)$ is independent of t_1 and t_2. It can also

* D-c component assumed to be zero.

be similarly concluded that the joint density for any number of sampling points depends only on the pertinent time differences and that the process is strictly stationary.

Now compute the autocorrelation function of the process:

$$\begin{aligned} E[X(t_1)X(t_2)] &= \sum_{m=1}^{\infty}\sum_{n=1}^{\infty} \{E[a_m a_n \cos m\omega t_1 \cos n\omega t_2] \\ &\qquad + E[b_m b_n \sin m\omega t_1 \sin n\omega t_2]\} \\ &= \sum_{1}^{\infty} \{E[a_n{}^2 \cos n\omega t_1 \cos n\omega t_2] \\ &\qquad + E[b_n{}^2 \sin n\omega t_1 \sin n\omega t_2]\} \end{aligned} \tag{10-82}$$

$$E[X(t_1) \cdot X(t_2)] = \sum_{1}^{\infty} E[a_n{}^2 \cos n\omega(t_1 - t_2)]$$

By letting $t_2 = t_1 + \tau$,

$$E[X(t_1) \cdot X(t_1 + \tau)] = \rho_{xx}(t_1, t_1 + \tau) = \rho_{xx}(\tau) = \sum_{1}^{\infty} \cos n\omega\tau\, E(a_n{}^2) \tag{10-83}$$

This function, of course, depends only on τ. The process is stationary of second order.

This process is an example of a larger class of processes which are generally called stationary gaussian in the literature.

10-15. The Over-all Mathematical Structure of Stochastic Processes. This section will give an over-all picture and a summary of the discussed stochastic processes. The major source of information for the content of this section is the talk by J. L. Doob which was given before the International Congress of Mathematicians at the Amsterdam meeting in 1954. According to this source, a satisfactory definition of a stochastic process is that it consists of a family of random variables. (In most cases, the variables are real-valued functions; this is referred to as the *standard* process.)

Standard Stochastic Processes. As discussed previously, these processes are defined for the T:

$$\{X(t), t \in T\}$$

where T is the real space. The joint distributions of the finite sets of the random variables of the process at different times are given by definition of the process:

$$X(t_1), X(t_2), \ldots, X(t_n)$$

The standard process is stationary if, for every finite parameter set $t_1, t_2, \ldots, t_n$, the joint distribution of

$$X(t_1 + h), X(t_2 + h), \ldots, X(t_n + h)$$

does not depend on the number h.

Standard Process with Mutually Independent Random Variables. Consider a process $\{X(t), t \in T\}$; if $t_1 < t_2 \cdots < t_n$ is a set of parameter points on T and if the variables

$$X(t_2) - X(t_1); X(t_3) - X(t_2); \ldots ; X(t_n) - X(t_{n-1})$$

are mutually independent, then the process is said to be a standard process with mutually independent random variables. The most important subclass of these processes is the Brownian-motion process. In a Brownian motion, the variable $X(t_k) - X(t_{k-1})$ is assumed to have normal distribution with zero mean and variance that is proportional to the parameter $t_k - t_{k-1}$ for all values of the integer k.

Standard Markov Process. The standard process $\{X(t), t \in T\}$ is said to be Markovian if the conditional probability of a future state depends only on the present state but not on the past history of the process. In mathematical language, for $t_1 < t_2 < \cdots t_n$,

$$P\{X(t_n) \in A|X(t_1), X(t_2), \ldots , X(t_{n-1})\} = P\{X(t_n) \in A|X(t_{n-1})\} \quad (10\text{-}84)$$

A Markov process is defined by the conditional probability distribution $P[X(t_k)|X(t_{k-1})]$ for $k = 1, 2, \ldots , n$ together with $P[X(t_1)]$, the initial probability of the process.

The so-called Markov chain which frequently appears in the literature of communication engineering is a particular case of the standard Markov process having a finite number of discrete states:

$$S_1, S_2, \ldots , S_n$$

The probability of the process going to the kth state depends solely on the immediate preceding state, S_{k-1}. This information is generally conveyed either by a transition probability matrix or by a state diagram.

A typical question on Markov chains is how to determine the probability of reaching state j from state k in m steps. The probability of the process being initially in a certain state along with the transition probability matrix provides a simple answer for this type of problem.

Standard Martingale and Semimartingale. About 1940, J. L. Doob studied the concept of martingales and semimartingales. A process $\{X(t), t \in T\}$ is a martingale if the conditional expectation of its future state, given the past and present, is equal to the value of its present state. More specifically, for $t_1 < t_2 \cdots < t_n$,

$$E\{X(t_n)|X(t_1), \ldots ,X(t_{n-1})\} = X(t_{n-1}) \quad (10\text{-}85)$$

with probability 1.

For defining a semimartingale, replace the equality by inequality ($>$). Not much application of these processes has yet appeared in the technical literature except in connection with information theory.

10-16. A Relation between Positive Definite Functions and Theory of Probability*

Definition of a Positive Definite Function.† A continuous function $f(x)$ real on the x axis is said to be positive definite if it satisfies the following requirement. Let $x_1, x_2, \ldots, x_n$ be n real numbers and $a_1, a_2, \ldots, a_n$ complex numbers. Then for all values of $n > 2$ it is required to have

$$\sum_{h=1}^{n} \sum_{k=1}^{n} f(x_h - x_k)a_h\bar{a}_k \geq 0 \tag{10-86}$$

This definition can be interchanged with the following when $a(x)$ is a continuous function in a given interval, $a \leq x \leq b$:

$$\int_a^b \int_a^b f(x - y)a(x) \cdot \overline{a(y)}\, dx\, dy \geq 0 \tag{10-87}$$

Bochner's Theorem. In 1932 S. Bochner introduced the following basic theorem: Any positive definite function $f(x)$ can be represented by the Stieltjes integral:

$$f(x) = \int_{-\infty}^{\infty} e^{jxy}\, dF(y) \qquad -\infty < x < +\infty \tag{10-88}$$

$F(y)$ being a real bounded nondecreasing function. Conversely, any function represented by such an integral is a positive definite function. On the basis of this theorem (proof is omitted), it can be seen that the characteristic function of a random variable is a positive definite function. In fact, if $F(y)$ is a CDF, then obviously $f(x)$ will be, by definition, the associated characteristic function. The converse is not true, since $F(y)$ needs to satisfy the additional conditions $F(-\infty) = 0$ and $F(+\infty) = 1$ in order to be a permissible CDF.

Khinchin's Theorem. The necessary and sufficient condition for a function $\rho(\tau)$ to be the autocorrelation function of a stationary stochastic process is that $\rho(\tau)$ could be represented as

$$\rho(\tau) = \int_{-\infty}^{\infty} \cos \tau x\, dF\,(x) \qquad -\infty < x < \infty \tag{10-89}$$

where $F(x)$ is a CDF.

An equivalent statement is the fact that the necessary and sufficient

* Positive definite functions are commonly used in physics and engineering. The material of this section may serve as a reminder of the existence of links between the theory of positive definite functions and probability. The interested reader is referred to S. Bochner, "Harmonic Analysis and the Theory of Probability," University of California Press, Berkeley, Calif., 1955. The section may be omitted in a first reading.

† Original definition introduced by M. Mathias Uber, Positive Fourier-Integral, *Math. Z.*, Bd. 16, pp. 103–125, 1923. See also K. Fan, Les Fonctions definies-positives et les fonctions complètement monotones, *Mém. sci. math.*, fascicule 114, Paris, 1950.

condition for a continuous real function $\rho(\tau)$ to be a permissible autocorrelation function for a stationary process is that $\rho(\tau)$ should be positive definite and $\rho(0) = 1$. The proof of sufficiency will not be presented here. However, the proof of the necessity which is comparatively simple is given:

The autocorrelation function $\rho(\tau)$ must be continuous and satisfy the three conditions of Sec. 10-9.

$$\sum_{h=1}^{n}\sum_{k=1}^{n} \rho(\tau_h - \tau_k)a_h\bar{a}_k = \sum_{h=1}^{n}\sum_{k=1}^{n} \int_{-\infty}^{\infty} e^{j(\tau_h-\tau_k)x} \cdot a_h\bar{a}_k \, dF(x)$$

$$= \int_{-\infty}^{\infty} \left(\sum_{h=1}^{n} e^{j\tau_h x}a_h\right)\left(\sum_{k=1}^{n} e^{-j\tau_k x}\bar{a}_k\right) dF(x)$$

$$= \int_{-\infty}^{\infty} \left|\sum_{h=1}^{n} e^{j\tau_h x}a_h\right|^2 dF(x) > 0 \qquad (10\text{-}90)$$

Thus there is the significant result that, if $\rho(\tau)$ is representable in the above form, $\rho(\tau)$ is positive definite.

PROBLEMS

10-1. Consider the stochastic process

$$\{X(t)\} = A \cos \omega t + B \sin \omega t$$

where

$$\bar{A} = \bar{B} = 0$$
$$\sigma_A = \sigma_B = \sigma$$
$$\overline{AB} = 0$$

(*a*) Is this process stationary?

(*b*) Study the autocorrelation function of the process when A and B are normally distributed.

10-2. Prove that the following process is stationary if A_k and B_k are uncorrelated random variables with zero means and standard deviation σ.

$$\{X(t)\} = \sum_{k=1}^{n} (A_k \sin \omega_k t + B_k \cos \omega_k t)$$

10-3. Consider a stochastic process consisting of rectangular pulses. The height of the pulse is a random variable varying between 0 and 1 volt with uniform probability distribution in that interval. The widths of the pulses are all equal, and the heights of successive pulses are independent.

(*a*) Find the autocorrelation function of the process.

(*b*) Is the process stationary?

(*c*) Find the power spectrum.

10-4. Is the process described in Prob. 10-1 ergodic?

10-5. Find the correlation function of a stationary random process whose spectral density is

$$\psi_{xx}(\omega) = k = \text{const}$$

10-6. A stationary stochastic voltage process with a correlation function $e^{-k|\tau|}$ is applied to an *RLC* series network. Find the power spectrum of the current flowing in this network.

10-7. Study the behavior of a general lossless one-port under the effect of a stationary-process driving force whose autocorrelation function is $e^{-k|\tau|}$. (Employ Foster's reactance theorem.)

10-8. Which one of the following functions is admissible as an autocovariance or autocorrelation function of a second-order stationary process?

(*a*) $+\dfrac{1}{\sqrt{|\tau|}}$.

(*b*) Graph of Fig. P10-8.

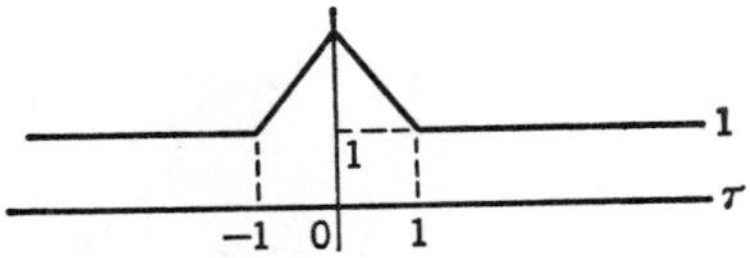

FIG. P10-8

10-9. The correlation function of a stationary process is given by

$$\rho_{xx}(\tau) = Ae^{-k|\tau|}$$

where A and k are appropriate positive constants. Find the spectral density of the process.

10-10. A stochastic process is described by

$$\{x(t)\} = A \cos (t + \alpha)$$

where A and α are statistically independent random variables. A is normally distributed with zero mean and standard deviation equal to 1. α is distributed between 0 and 2π with a uniform density of $1/2\pi$.

(*a*) Find the first- and the second-order ensemble averages.
(*b*) Find the first- and the second-order time averages.
(*c*) Find the autocovariance function.
(*d*) Find the autocorrelation function.
(*e*) Is this process stationary?
(*f*) Is this process ergodic?

10-11. A noise process $\{X(t)\}$ goes through a delay network; that is, the output of the network at time t is

$$\{X(t - \tau)\}$$

Consider a linear combination of the two processes:

$$\{Y(t)\} = K_1\{X(t)\} + K_2\{X(t - \tau)\}$$

where K_1 and K_2 are specified constants. Find the autocorrelation function and the power spectral density of $\{Y(t)\}$ in terms of those parameters of $\{X(t)\}$.

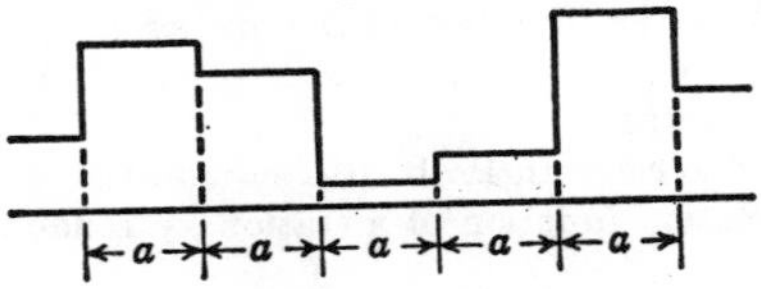

FIG. P10-3

10-12. Find the cross-correlation function between the two processes given below:

$$\{X(t)\} = A \sin \omega t + B \cos \omega t$$
$$\{Y(t)\} = -A \cos \omega t + B \sin \omega t$$

10-13. Study the band-limited process

$$\{X(t)\} = \sum_{K=-\infty}^{\infty} A_k \frac{\sin \pi(2wt - k)}{\pi(2wt - k)}$$

where A_k's are normally distributed independent random variables with zero means and equal variance.

10-14. Consider an AM signal carrier $\{y(t)\} = \alpha \cos (\omega_0 t + \phi)$, where $\alpha(t)$ and $\phi(t)$ are, respectively, the corresponding envelope and phase. Show that, if $\alpha(t)$ and $\phi(t)$ are ergodic, the same is true for $y(t)$ (see Middleton, Sec. 1.6).

Note: For additional problems consult Davenport and Root, Chaps. 6, 8, and 9; Middleton, Chaps. 1, 2, and 3; and Laning and Battin.

CHAPTER 11

COMMUNICATION UNDER STOCHASTIC REGIMES

11-1. Stochastic Nature of Communication. In the study of probability theory it was pointed out that, when the number of random variables is not finite, we deal with stochastic processes. It was also pointed out that dealing with stochastic processes and their related problems requires some special techniques. The preceding chapter was concerned with an introductory presentation of the theory of stochastic processes. This chapter provides a brief application of the subject to information theory.

In dealing with physical sources of communications, such as teletype or radio communication, we are generally confronted with time series. For example, when a simple binary source transmits 0's and 1's, theoretically we have a doubly infinite sequence of 0's and 1's which must be considered as a member of an ensemble of such doubly infinite sequences,

$$\begin{array}{l} \ldots, 1, 1, 0, 0, 0, 1, 0, 1, 1, 0, 1, 1, \ldots \\ \ldots, 1, 0, 1, 0, 0, 1, 0, 0, 0, 1, 0, 0, \ldots \\ \ldots\ldots\ldots\ldots\ldots\ldots\ldots\ldots\ldots \end{array}$$

If for convenience we assume that the letter X_k is transmitted at a specific instant t_k, then X_k is a random variable assuming either one of the values 0 or 1. A random message of this source is written as

$$\ldots, X_{-2}, X_{-1}, X_0, X_1, X_2, \ldots$$

Similarly for a teletype we have an analogous doubly infinite series where each X_k can assume any of a teletype's characters. The characterization of such information sources directly follows from the characterization of their time series. For example, as discussed in the preceding chapter, the joint probability distribution of, say, X_k and X_j should be obtained from either an analytical or experimental description of the process.

In Chaps. 3, 4, 8, and 9, we confined ourselves to independent sources where successive symbols were selected independently from a finite alphabet. In the present chapter we relax this constraint and allow sources with some sort of interdependence among the transmission probabilities of the symbols. Therefore, it is natural to seek to establish an information theory dealing with such time series. More specifically, it is our task to associate a communication entropy with a stochastic information source (with memory) and to study the transmission of information in channels as before. Furthermore, we wish to establish some theorems analogous to the fundamental theorems of transmission of information in discrete and continuous channels.

N. Wiener was one of the first scientists who clearly described the stochastic nature of communication problems. Wiener put in focus the fact that the communication of information is primarily of a statistical nature. That is, at a given time a message is drawn from a universe of possible messages according to some probability law. At the next moment another message from this universe will be transmitted, and so on. The joint probability distribution of the transmitted and the received messages contains all the mathematical information necessary for the study of a communication channel. This joint probability distribution is not generally known, although usually we know the input and the conditional probabilities (noise structures). The development of an information theory for stochastic sources and channels began with the classical work of Shannon, who considered sources of the Markov type. Considerable mathematical clarification was brought forth later in the work of B. McMillan, who further defined stationary sources, their entropy, and the fundamental theorems of information theory. Further elaborate proofs and treatment are due to A. Feinstein and A. I. Khinchin. These contributions have considerably clarified the subtle concepts of stationary sources, entropy, and the fundamental theorems initiated in Shannon's work.

A complete presentation of stochastic information theory is not advanced here. The reader with a professional interest will find the above references indispensable. The following is aimed at giving an introductory treatment of the subject at a level suitable for the present first course in the subject. For those who may not be able to afford the time to study this part of the theory, the following preview will be of interest. The main portion of this chapter deals with the justification by mathematical techniques of the validity of our intuitive concepts of information theory of discrete and continuous random variables when applied to sources and channels with memory. The contributors have also shown under what hypotheses such generalizations are valid.

In the subsequent discussion, we shall first define and briefly describe a simple class of stochastic processes of great interest in engineering problems, the so-called Markov chains. We give examples of stationary and ergodic Markov chains and will point out that in many communication problems the sources are of these types. The natural next step is to investigate the information-theory aspect of simple Markov chains. Given a stationary or ergodic Markov source, its communication entropy will be defined. Finally we shall present a brief description of the performance of a communication channel driven by a discrete stationary source.

11-2. Finite Markov Chains. Consider a Markovian stochastic process with a finite number of states:

$$[S] = [S_1, S_2, \ldots, S_n] \tag{11-1}$$

The chain consists of a sequence of states from $[S]$ such as

$$\ldots, S_2, S_2, S_1, S_6, S_5, \ldots$$

where the probability of moving from any state S_i to the immediately succeeding state S_j is prespecified by p_{ij} as an element of the so-called transition probability matrix $[P] = [p_{ij}]$.

$$[P] = \begin{bmatrix} p_{11} & p_{12} & \cdots & p_{1n} \\ p_{21} & p_{22} & \cdots & p_{2n} \\ \cdots & \cdots & \cdots & \cdots \\ p_{n1} & p_{n2} & \cdots & p_{nn} \end{bmatrix} \tag{11-2}$$

Note that all elements of $[P]$ are nonnegative and that the sum of the elements of each row is unity. If all the elements of $[P]$ do not depend on time, the associated Markov chain is said to be stationary. We deal only with stationary chains.

Let $[P^{(r)}]$ be the row probability matrix corresponding to the chain reaching different states from all possible given initial states in r steps; it was shown in Eq. (2-117) that

$$[P^{(r)}] = [P^{(0)}][P]^r \tag{11-3}$$

where $[P^{(0)}]$ is the row probability matrix of the initial states. When $[P]^r$ for some values of the positive integer r has only strictly positive elements (no zeros) the chain is referred to as a *regular chain*. Of course, if $[P]^r$ has no zero entry for $r = r_0$, then all powers of $[P]$ with $r \geq r_0$ will not have any zero entry. Then clearly, starting with a nonzero initial probability, from each state S_k one can reach any state S_j with a nonzero probability.

A state S_k is said to be an *absorbing state* if it would be impossible to leave that state:

$$P\{S_k|S_k\} = 1$$

A Markov chain containing at least one absorbing state is referred to as an *absorbing Markov chain*. Obviously, an absorbing Markov chain is not a regular chain. A Markov chain is said to be *ergodic* if, after a certain finite number of steps, it is possible to go from any state to any other state with a nonzero probability. Thus, a regular chain is ergodic but the converse is not necessarily true (see Prob. 11-4).

Example 11-1. Determine if the chain illustrated in Fig. E11-1 is ergodic.

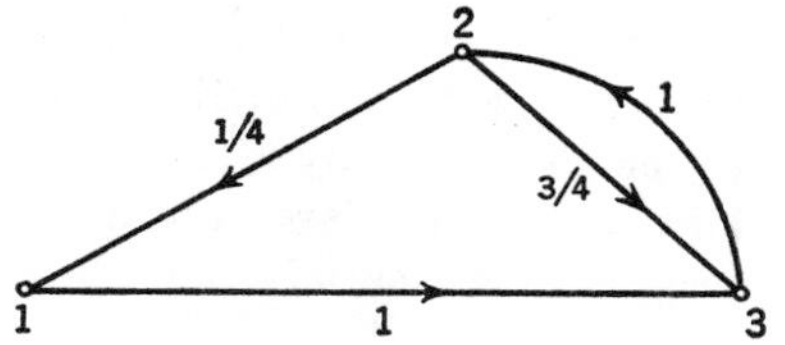

FIG. E11-1

Solution. The transition probability matrix is

$$\begin{bmatrix} 0 & 0 & 1 \\ \frac{1}{4} & 0 & \frac{3}{4} \\ 0 & 1 & 0 \end{bmatrix}$$

One has to determine if any power of this matrix has any zero entry. A simple way of checking this would be to replace all the nonzero entries by, say, the number 1 and call the new matrix $[X]$. Next derive $[X]^2$, $[X]^4$, $[X]^8$, etc. These multiplications are simple, and all nonzero entries in $[X]^{2^k}$ can also be replaced by 1, as the actual values of the entries are not of concern. In the present example it can be seen that $[X]^4$ has a zero entry, but $[X]^8$, $[X]^{16}$, . . . , have no zero entries. Therefore the chain is a regular Markov chain and is thus also ergodic.

Example 11-2. Same question as in Example 11-1 for the chain illustrated in Fig. E11-2.

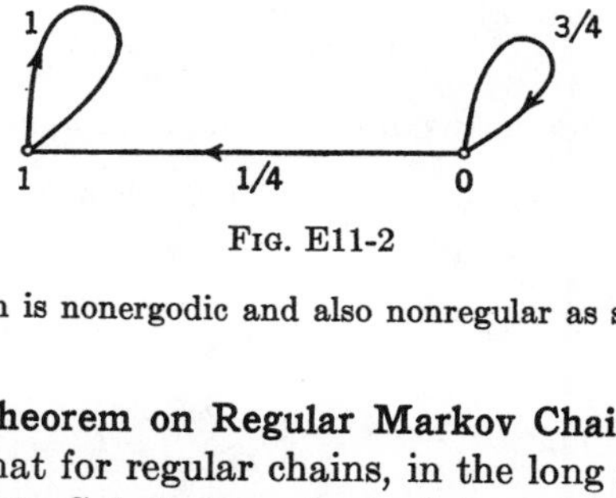

FIG. E11-2

Solution. The chain is nonergodic and also nonregular as state 1 is an absorbing state.

11-3. A Basic Theorem on Regular Markov Chains. In this section we wish to show that for regular chains, in the long run, the probability of reaching any state S_j is independent of the initial state and furthermore that for large values of r the probability of reaching a state S_j in r steps is independent of r. In other words, the statistical properties of the chain are somewhat homogeneous. More specifically:

Theorem. Let $[P]$ be the transition matrix of a regular Markov chain; then

$$\lim_{k\to\infty} \begin{bmatrix} p_{11} & p_{12} & \cdots & p_{1n} \\ p_{21} & p_{22} & \cdots & p_{2n} \\ \cdots & \cdots & \cdots & \cdots \\ p_{n1} & p_{n2} & \cdots & p_{nn} \end{bmatrix}^k = \begin{bmatrix} t_1 & t_2 & \cdots & t_n \\ t_1 & t_2 & \cdots & t_n \\ \cdots & \cdots & \cdots & \cdots \\ t_1 & t_2 & \cdots & t_n \end{bmatrix} = [t_i] = T \quad (11\text{-}4)$$

where $[t_i]$ is a permissible transition probability matrix. In order to prove this theorem, we first establish the following lemma:

Lemma. Let P be an $n \times n$ transition probability matrix with no zero entry and V a column matrix with n positive elements v_k. Denote the largest and the smallest element of V by v_M and v_m, respectively, and the smallest element of P by α_m; then the largest and the smallest elements of $PV = U$, u_M and u_m, respectively, satisfy the following inequality:

$$u_M - u_m \leq (1 - 2\alpha_m)(v_M - v_m) \tag{11-5}$$

Proof. The proof follows by first finding a lower bound for u_m and an upper bound for u_M by a direct inspection of the matrix product

$$\begin{bmatrix} p_{11} & p_{12} & \cdots & p_{1n} \\ p_{21} & p_{22} & \cdots & p_{2n} \\ \cdots & \cdots & \cdots & \cdots \\ p_{n1} & p_{n2} & \cdots & p_{nn} \end{bmatrix} \begin{bmatrix} v_1 \\ v_2 \\ \cdots \\ v_n \end{bmatrix} = \begin{bmatrix} u_1 \\ u_2 \\ \cdots \\ u_n \end{bmatrix}$$

It will be shown that*

$$\begin{aligned} u_m &\geq \alpha_m v_M + (1 - \alpha_m) v_m \\ u_M &\leq \alpha_m v_m + (1 - \alpha_m) v_M \end{aligned} \tag{11-6}$$

To show this, let $u_k = u_m$ be the smallest element of the U matrix.

$$\begin{aligned} u_k = u_m = p_{k1}v_1 + p_{k2}v_2 + \cdots + p_{km}v_m \\ + \cdots + p_{kM}v_M + \cdots + p_{kn}v_n \end{aligned} \tag{11-7}$$

On the right-hand side let us add and subtract $\alpha_m v_M + \alpha_m v_m$:

$$\begin{aligned} u_m = \alpha_m v_M - \alpha_m v_m + p_{k1}v_1 + \cdots + (p_{km} + \alpha_m)v_m + \cdots \\ + (p_{kM} - \alpha_m)v_M + \cdots + p_{kn}v_n \geq \alpha_m v_M - \alpha_m v_m \\ + \left(\sum_{i=1}^{n} p_{ki}\right) v_m \end{aligned} \tag{11-8}$$

but $\sum_{i=1}^{n} p_{ki} = 1$. Thus

$$u_m \geq \alpha_m v_M + (1 - \alpha_m) v_m$$

Similarly, if $u_k = u_M$ is the largest element of the U matrix, one may write

$$\begin{aligned} u_M &= p_{k1}v_1 + p_{k2}v_2 + \cdots + p_{km}v_m + \cdots + p_{kM}v_M + \cdots + p_{kn}v_n \\ &= \alpha_m v_m - \alpha_m v_M + p_{k1}v_1 + \cdots + (p_{km} - \alpha_m)v_m + \cdots \\ &\quad + (p_{kM} + \alpha_m)v_M + \cdots + p_{kn}v_n \leq \alpha_m v_m - \alpha_m v_M \\ &\quad + \left(\sum_{i=1}^{n} p_{k1}\right) v_M = \alpha_m v_m + (1 - \alpha_m) v_M \end{aligned} \tag{11-9}$$

* The proof given here was set forth by José Perini during a course on information theory. (See also Kemeny and Snell.)

From these inequalities one concludes that

$$u_M - u_m \le (1 - 2\alpha_m)(v_M - v_m)$$

Having established this relation, it is now possible to show that as k is increased the difference between the largest and the smallest element in, say, the first column of $[U]$ will be reduced. In fact, for a positive integer k the matrix $[P]^k$ is a transition probability matrix. Therefore the product of $[P]^k$ by the first column of $[P]$ will obey the above lemma.

$$u_M{}^{(k)} - u_m{}^{(k)} \le [1 - 2\alpha_m{}^{(k)}][v_M{}^{(k)} - v_m{}^{(k)}] \tag{11-10}$$

But note that $u_M{}^{(k)}$ and $u_m{}^{(k)}$ are, respectively, the largest and the smallest term in the first column of $[P]^{k+1}$. In other words,

$$v_M^{(k+1)} - v_m^{(k+1)} \le [1 - 2\alpha_m{}^{(k)}][v_M{}^{(k)} - v_m{}^{(k)}] \tag{11-11}$$

The iteration of this method suggests that

$$\begin{aligned} v_M^{(k+2)} - v_m^{(k+2)} &\le [1 - 2\alpha_m{}^{(k)}][v_M^{(k+1)} - v_m^{(k+1)}] \\ &\le [1 - 2\alpha_m{}^{(k)}]^2[v_M{}^{(k)} - v_m{}^{(k)}] \end{aligned} \tag{11-12}$$

But $\alpha_m{}^{(k)}$, being the smallest number in a transition probability matrix, cannot be larger than $\frac{1}{2}$. Therefore,

$$[1 - 2\alpha_m{}^{(k)}]^2 \le [1 - 2\alpha_m{}^{(k)}] \tag{11-13}$$

It becomes evident that the difference between the largest and the smallest element of each column in $[P]^k$ becomes smaller as k is increased. This shows that there is a limiting column matrix with all elements equal to t_1 such that for large k the product of $[P]^k$ and the first column of $[P]$ approaches the n element column matrix with elements equal to t_1. By a similar reasoning, we find that there exists an $n \times n$ matrix T as described by the above theorem. With some additional mathematical computation, it can be shown that a regular Markov chain has a unique probability matrix T. In fact, let $[t]$ be any row of T; then

$$\lim_{k \to \infty} [P]^k = T \tag{11-14}$$

Note that

$$\lim_{k \to \infty} \{[P]^{k-1} \cdot [P]\} = [T][P]$$

Hence

$$[T] = [T][P] \tag{11-15}$$

Now let us compute the probability row matrix for a regular chain reaching any state k after a large number of steps n, having started with an initial row probability matrix $[P^{(0)}]$.

$$[P^{(n)}] = [P^{(0)}][P]^n = [P^{(0)}][T] \tag{11-16}$$

$$[P_1{}^0 \quad P_2{}^0 \quad \cdots \quad P_n{}^0] \begin{bmatrix} t_1 & t_2 & \cdots & t_n \\ t_1 & t_2 & \cdots & t_n \\ \cdots & \cdots & \cdots & \cdots \\ t_1 & t_2 & \cdots & t_n \end{bmatrix} = [t_1 \quad t_2 \quad \cdots \quad t_n] \tag{11-17}$$

Under such circumstances the probability of reaching any particular state after a large number of steps will be the same; that is, it does not depend on initial probability.

For a regular Markov chain, the average number of times of being in any state approaches the corresponding probability entry in the T matrix. That is, the probability of occurrence of a state approaches the value specified by $[t]$ irrespective of the initial probability. This fact is a direct conclusion of the law of large numbers. Example 11-3 will exhibit the correctness of this statement. (For proof see Kemeny and Snell, "Finite Markov Chains," p. 73.)

In closing this section it is useful to point out that Eq. (11-15) suggests a simple method for obtaining the elements of the T matrix.

$$\begin{aligned} t_1p_{11} + t_2p_{21} + \cdots + t_np_{n1} &= t_1 \\ t_1p_{12} + t_2p_{22} + \cdots + t_np_{n2} &= t_2 \\ \cdots\cdots\cdots\cdots\cdots\cdots & \\ t_1p_{1n} + t_2p_{2n} + \cdots + t_np_{nn} &= t_n \\ t_1 + t_2 + \cdots + t_n &= 1 \end{aligned} \tag{11-18}$$

The values of $t_1, t_2, \ldots, t_n$ can be readily computed from Eq. (11-18).

11-4. Entropy of a Simple Markov Chain. Consider a simple stationary Markov chain with a finite number of states

$$[A_1, A_2, \ldots, A_n] \tag{11-19}$$

and the transition probability matrix

$$[P\{A_j|A_i\}] = [p_{ij}] = \begin{bmatrix} p_{11} & p_{12} & \cdots & p_{1n} \\ p_{21} & p_{22} & \cdots & p_{2n} \\ \cdots & \cdots & \cdots & \cdots \\ p_{n1} & p_{n2} & \cdots & p_{nn} \end{bmatrix} \tag{11-20}$$

If the system is initially in state A_i, the probabilities corresponding to a transition of one step to any other state form a set of complete and exhaustive probability schemes. (Thus, the sum of the elements in each row of the transition probability matrix is unity.)

$$\begin{gathered} [(A_1|A_i), (A_2|A_i), \ldots, (A_n|A_i)]^* \\ [p_{i1}, p_{i2}, \ldots, p_{in}] \end{gathered} \tag{11-21}$$

* The symbols $(A_k|A_i)$ and $(A_k^{(r)}|A_i)$, respectively, will be used in this section as a short-hand notation for the following chain of events:

$$(X_1 = A_k) \cap (X_0 = A_i) \qquad (X_{0+r} = A_k) \cap (X_0 = A_i)$$

For the purpose of computing entropies [Eq. (11-26)], all events with specified i, k, and r must be considered distinct. For example, $(A_3|A_1)^{(2)}$ may consist of $A_1A_1A_3$, $A_1A_2A_3$, $A_1A_3A_3$, $A_1A_4A_3$, etc. It is for this reason that the notation $p_{ik}^{(r)}$ will be used to denote the set of probabilities of every individual member of $(A_k|A_i)^{(r)}$.

This notation seems to be helpful in the initial discussion, in order to avoid more complex formulation. However, since it is mathematically awkward, it will be dropped after it has served its purpose.

With this scheme, we associate an entropy H_i indicating the average amount of uncertainty of the system for moving one step ahead when starting with state A_i.

$$H_i^{(1)} = - \sum_{j=1}^{n} p_{ij} \log p_{ij} \tag{11-22}$$

If the probability of the system being initially in state A_i is designated by p_i, it is natural to calculate the average uncertainty of the chain for moving one step ahead from any initial state, when the initial states have specified probabilities, that is,

$$H(X) = \overline{H_i^{(1)}} = \sum_{i=1}^{n} p_i H_i^{(1)} = - \sum_{i=1}^{n} \sum_{j=1}^{n} p_i p_{ij} \log p_{ij} \tag{11-23}$$

We may call $H(X)$, or more specifically $H^{(1)}$, the entropy of the chain with specified initial probabilities for moving one step.

More generally, the set of events of going from A_i to any other state in r steps,

$$[(A_1^{(r)}|A_i), (A_2^{(r)}|A_i), \ldots, (A_n^{(r)}|A_i)] \tag{11-24}$$

constitutes a finite complete probability scheme. The entropy of this finite scheme is

$$H_i^{(r)} = - \sum_{j=1}^{n} p_{ij}^{(r)} \log p_{ij}^{(r)} \tag{11-25}$$

where $p_{ij}^{(r)}$ stands for the probability of any one of a discrete chain moving from the ith to the jth state in r steps. Thus, the entropy of the chain for moving r steps ahead from the initial states when the initial probabilities are specified is

$$H^{(r)} = \overline{H_i^{(r)}} = \sum_{i=1}^{n} p_i H_i^{(r)} = - \sum_{i=1}^{n} \sum_{j=1}^{n} p_i p_{ij}^{(r)} \log p_{ij}^{(r)} \tag{11-26}$$

For example, in order to compute the entropy of the set of all meaningful three-letter English words (assuming simple Markovian structure) the following procedure would apply. Compute the entropy of all permissible three-letter words out of the set *aaa*, *aab*, *aac*, . . . , *aba*, *abb*, *abc*, ... Do the same for three-letter English words beginning with *b*, *c*, etc. The application of Eq. (11-26) will lead to the average entropy of three-letter English words. The entropy per letter is $\frac{1}{3}H^{(r)}$. If very long messages are considered, the entropy per letter will provide an estimate of the entropy of the English language.

In Sec. 11-5 it will be shown that $\lim_{r \to \infty} [H^{(r)}/r]$ exists for any ergodic Markov source. The mathematical impact of this fact stems from the statistically homogeneous structure of the source output. (This is in a way analogous to saying that $[P]^n$ approaches a limit for $n \to \infty$ for regular Markovian sources.)

A simple theorem to be given here is as follows:

Theorem. The different-order entropies of a regular Markov chain with initial probability [t] (as described above) are additive, that is,

$$H^{(\alpha+\beta)} = H^{(\alpha)} + H^{(\beta)} \qquad \alpha, \beta \text{ arbitrary positive integers} \tag{11-27}$$

We present a simple proof suggested by Khinchin.

Let us start from the state A_i and go to A_k in $r + 1$ steps. We do this by going first one step ahead and then r steps. The entropy will be

$$H_i^{(r+1)} = H_i^{(1)} + \sum_{k=1}^{n} p_{ik} H_k^{(r)} \tag{11-28}$$

The associated average entropy for the system to move $r + 1$ steps is

$$H^{(r+1)} = \sum_{i=1}^{n} p_i H_i^{(r+1)} \tag{11-29}$$

where p_i is the initial probability of the chain starting with the ith state. Substituting Eq. (11-28) in Eq. (11-29) we find

$$\begin{aligned} H^{(r+1)} &= \sum_{i=1}^{n} p_i H_i^{(1)} + \sum_{i=1}^{n} p_i \sum_{k=1}^{n} p_{ik} H_k^{(r)} \\ &= H^{(1)} + \sum_{i=1}^{n} p_i (p_{i1} H_1^{(r)} + p_{i2} H_2^{(r)} + \cdots + p_{in} H_n^{(r)}) \end{aligned} \tag{11-30}$$

Thus, due to Eq. (11-18)

$$H^{(r+1)} = H^{(1)} + H^{(r)} \tag{11-31}$$

As an immediate application of this theorem we have

$$H^{(r)} = H^{(1)} + H^{(r-1)} = 2H^{(1)} + H^{(r-2)} = rH^{(1)} = rH(X)$$

The entropy of the Markov chain for moving r steps ahead is equal to r times the entropy of the chain for moving one step ahead. Since $H^{(1)} = H(X)$ is the basic entropy associated with the Markov scheme, it is interesting to note that the entropy relation of such a chain for moving r steps ahead is similar to the entropy of an extension of a discrete independent source without memory, as discussed in Chap. 4. Based on the foregoing, it is possible to define the communication entropy of a regular Markovian source starting with any arbitrary initial probabilities as

$$H(X) = \lim_{n\to\infty} \frac{H^{(n)}}{n} \tag{11-32}$$

The same defining equation applies when the source is stationary but not necessarily Markovian. It is of mathematical interest to establish the existence of this source entropy for stationary sources with a finite alphabet and a finite memory (finite intersymbol effect). The existence proof is given in Feinstein (I, p. 85).

Example 11-3. Consider the Markov state diagram of Fig. E11-3. If the initial probability matrix is $P^{(0)} = [\frac{1}{4} \quad \frac{3}{4}]$, find

(*a*) $P_A^{(1)}$, the probability of reaching state A in one step.
(*b*) $P_B^{(1)}$, the probability of reaching state B in one step.
(*c*) $P_A^{(2)}$, the probability of reaching state A in two steps.
(*d*) $P_B^{(2)}$, the probability of reaching state B in two steps.
(*e*) $P_A^{(3)}$, the probability of reaching state A in three steps.
(*f*) $P_B^{(3)}$, the probability of reaching state B in three steps.
(*g*) The t matrix.
(*h*) Compute the entropy $H^{(1)}$.
(*i*) Compute the entropy $H^{(2)}$.
(*j*) Compute the entropy $H^{(3)}$.
(*k*) Compare $H^{(1)} + H^{(2)}$ with $H^{(3)}$.
(*l*) Same question as in part (*k*) but with the initial probability matrix $[\frac{3}{7} \quad \frac{4}{7}]$.

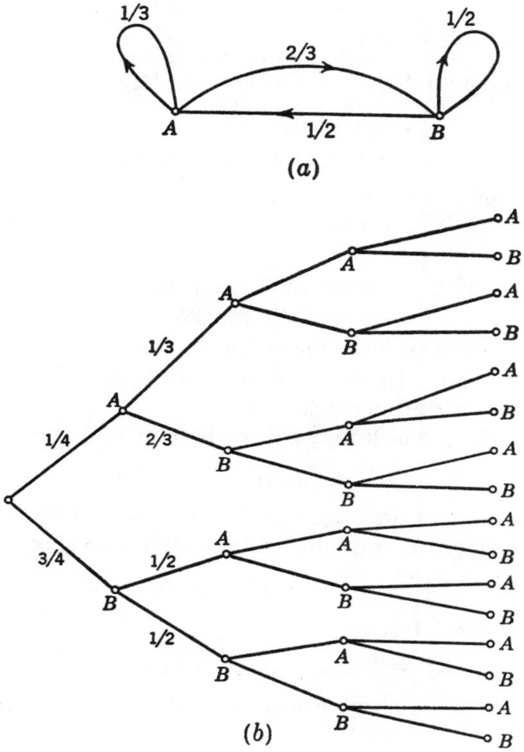

FIG. E11-3

Solution. The solutions for parts (*a*) to (*f*) can be obtained by the matrix method or by drawing a tree and computing probability measures, as discussed in Chap. 2.

(*a*) and (*b*)
$$[\tfrac{1}{4} \quad \tfrac{3}{4}]\begin{bmatrix} \tfrac{1}{3} & \tfrac{2}{3} \\ \tfrac{1}{2} & \tfrac{1}{2} \end{bmatrix} = [\tfrac{11}{24} \quad \tfrac{13}{24}]$$

(*c*) and (*d*)
$$[\tfrac{1}{4} \quad \tfrac{3}{4}]\begin{bmatrix} \tfrac{1}{3} & \tfrac{2}{3} \\ \tfrac{1}{2} & \tfrac{1}{2} \end{bmatrix}^2 = [\tfrac{61}{144} \quad \tfrac{83}{144}]$$

(*e*) and (*f*)
$$[\tfrac{1}{4} \quad \tfrac{3}{4}]\begin{bmatrix} \tfrac{1}{3} & \tfrac{2}{3} \\ \tfrac{1}{2} & \tfrac{1}{2} \end{bmatrix}^3 = [\tfrac{371}{864} \quad \tfrac{493}{864}]$$

(*g*)
$$\begin{aligned} \tfrac{1}{3}t_1 + \tfrac{1}{2}t_2 &= t_1 \\ \tfrac{2}{3}t_1 + \tfrac{1}{2}t_2 &= t_2 \\ t_1 + t_2 &= 1 \end{aligned}$$

The above equations yield

$$t = [\tfrac{3}{7} \quad \tfrac{4}{7}] \qquad T = \begin{bmatrix} \tfrac{3}{7} & \tfrac{4}{7} \\ \tfrac{3}{7} & \tfrac{4}{7} \end{bmatrix}$$

Note that the answers to the previous parts approach t rather rapidly.

(*h*)
$$\begin{aligned} H_A^{(1)} &= -\tfrac{2}{3} + \log 3 \\ H_B^{(1)} &= 1 \\ H^{(1)} &= \tfrac{7}{12} + \tfrac{1}{4}\log 3 \end{aligned}$$

(*i*)
$$\begin{aligned} H_A^{(2)} &= \tfrac{1}{9}\log 9 + \tfrac{2}{9}\log \tfrac{9}{2} + \tfrac{1}{3}\log 3 + \tfrac{1}{3}\log 3 = \tfrac{4}{3}\log 3 - \tfrac{2}{9} \\ H_B^{(2)} &= \tfrac{1}{6}\log 6 + \tfrac{1}{3}\log 3 + \tfrac{1}{4}\log 4 + \tfrac{1}{4}\log 4 = \tfrac{1}{2}\log 3 + \tfrac{7}{6} \\ H^{(2)} &= \tfrac{1}{4}H_A^{(2)} + \tfrac{3}{4}H_B^{(2)} = \tfrac{17}{24}\log 3 + \tfrac{118}{144} \end{aligned}$$

(*j*)
$$\begin{aligned} H_A^{(3)} &= \tfrac{16}{9}\log 3 + \tfrac{1}{27} \\ H_B^{(3)} &= \tfrac{11}{12}\log 3 + \tfrac{53}{36} \\ H^{(3)} &= \tfrac{163}{144}\log 3 + \tfrac{53}{48} + \tfrac{1}{108} \end{aligned}$$

(*k*) $H^{(1)} + H^{(2)}$ differs very little from $H^{(3)}$.

(*l*) The initial probability matrix is taken to be $[t]$; thus $H^{(1)} + H^{(2)} = H^{(3)}$ holds as an identity.

11-5. Entropy of a Discrete Stationary Source. Shannon's original work was confined to sources of the Markov type (Shannon [I], Secs. 2 and 4). The concept of more general sources and their characterization are due to McMillan. In this section we discuss McMillan's characterization of a discrete stationary source.

Let A be a finite set of letters called the alphabet of the source.

$$[A] = [a_1, a_2, \ . \ . \ . \ , a_N]$$

Consider a source transmitting one letter from $[A]$ at each instant t_k, where t_k is an element of a doubly infinite time sequence t:

$$[t] = [. \ . \ . \ , t_{-2}, t_{-1}, t_0, t_1, t_2, \ . \ . \ .]$$

A typical transmitted message is

$$\{X\} = \{. \ . \ . \ , x_{-2}, x_{-1}, x_0, x_1, x_2, \ . \ . \ .\}$$

where X_k is a random variable assuming any one of values $a_k \in A$ for $k \in K$.

$$[K] = [. \ . \ . \ , -2, -1, 0, 1, 2, \ . \ . \ .]$$

Our first task is to study the probabilistic nature of this source from its output. For this, we need to define what is referred to as a *cylinder set* of events. Among all members of the ensemble $\{X\}$, consider those sequences that have specified outputs at certain specified instants. More precisely, let, for example,

$$x_{-1} = a_1 \qquad x_0 = a_2 \qquad x_2 = a_5 \qquad x_k = a_n$$

Then all the doubly infinite sequences satisfying these specifications form a set E which is called a cylinder set. Each one of these sequences is an element of the cylinder.

$$E = [. \ . \ . \ , x_{-2}, a_1, a_2, x_1, a_5, \ . \ . \ . \ , x_k = a_n, \ . \ . \ .] \tag{11-33}$$

Now suppose that the output of the source is a stationary process; then it will be homogeneous in time, that is, any cylinder set E will be carried onto a cylinder set with identical probabilities after a shift of, say, one time unit.

$$TE = [. \ . \ . \ , x_{-2}, x_{-1}, a_1, a_2, x_2, a_5, \ . \ . \ . \ , x_k, x_{k+1} = a_n, \ . \ . \ .]$$

In other words, messages of this type will most likely be among different output sequences of the source. The shift of time axis obviously may be in either time direction (T,T^{-1}).

Furthermore, we assume that a probability measure is defined for the space of the message ensemble and that it is such that the probability measure of any cylinder S is equal to that of the shifted cylinder TS.

$$\begin{aligned} P\{TS\} &= P\{S\} \\ P\{TE\} = P\{T^{-1}E\} &= P\{E\} \end{aligned} \tag{11-34}$$

The next step is to define the source entropy, that is, the per-symbol rate at which the source emits information. Consider all the sequences of the type

$$E = [. \ . \ . \ ,x_k,x_{k+1},x_{k+2}, \ . \ . \ . \ ,x_{k+n-1}, \ . \ . \ .] \tag{11-35}$$

Now assign n specific letters from the alphabet $[A]$ to the n positions $x_k, x_{k+1}, \ldots, x_{k+n-1}$, that is, the transmitter transmits one specific letter at a specified instant. Each of the N^n distinct sequences of this type with a defined probability measure forms a cylinder S. Let $\mu\{S\}$ be the probability measure of a particular cylinder S; then the communication entropy of the set of N^n possible n-term sequences can be defined in the usual manner.

$$H_n = -\sum_S \mu(S) \log \mu(S) \tag{11-36}$$

The stationarity hypothesis asserts that H_n remains independent of the initial moment t_k but of course is dependent on n. The crucial point in

McMillan's definition of the entropy of a stationary source is the fact that the quantity

$$H(X) = \lim_{n\to\infty} \frac{H_n}{n} \tag{11-37}$$

exists and most naturally represents the entropy of the source much in the same way that Shannon defines the entropy of an independent and a Markovian source. The following proof of the existence of the entropy for stationary sources is Khinchin's simplified version of McMillan's more general results.

Consider the following cylinder sets of messages:

A_m, the cylinder set with specified letters in m specified positions

A_n, the cylinder set with specified letters in n specified positions

A_{m+n}, the cylinder set with specified letters in the previously specified positions

Evidently, the entropies of these cylinder-set families satisfy the relations

$$\begin{aligned} H(A_{m+n}) &= H(A_n) + H(A_m|A_n) \\ H(A_m|A_n) &\leq H(A_m) \end{aligned} \tag{11-38}$$

Thus, using the notation of Eq. (11-36) in the above, we find

$$H_n \leq H_{m+n} \leq H_m + H_n \tag{11-39}$$

For the special cases of cylinders with $m = 1$ and $m = n$ we have, respectively,

$$\begin{aligned} H_n &\leq H_{n+1} \\ H_{2n} &\leq 2H_n \end{aligned} \tag{11-40}$$

The latter inequality is easily generalized to

$$H_{nk} \leq nH_k \tag{11-41}$$

or, for $k = 1$, $H_n \leq nH_1$,

$$\frac{H_n}{n} \leq H_1 < +\infty \tag{11-42}$$

Equation (11-42) implies that an upper and a lower limit exist for H_n/n by virtue of the Boltzano-Weirstrauss theorem, since H_n/n is bounded above and below. Let

$$a = \liminf \frac{H_n}{n} < +\infty \qquad n \to \infty \tag{11-43}$$

Now to prove the convergence of the sequence H_n/n, it is sufficient to show that $\limsup\limits_{n\to\infty} (H_n/n) = a$. We may choose, for any $\epsilon > 0$, an integer q such that

$$\left| \frac{H_q}{q} - a \right| < \epsilon$$

In particular,

$$\frac{H_q}{q} < a + \epsilon \tag{11-44}$$

Then, for any n, select an integer $k > 1$ such that

$$(k - 1)q < n \leq kq$$

Now it follows that

$$\begin{aligned} H_n &\leq H_{kq} \leq kH_q \\ \frac{H_n}{n} &\leq \frac{kq}{n}\frac{H_q}{q} < \frac{k}{k-1}(a + \epsilon) \end{aligned} \tag{11-45}$$

Thus for sufficiently large n, $k \to \infty$,

$$a - \epsilon < \frac{H_n}{n} < \frac{k}{k-1}(a + \epsilon) < a + 2\epsilon$$

Therefore $$\limsup \frac{H_n}{n} = a \qquad n \to \infty \tag{11-46}$$

It follows that as n approaches infinity the entropy per symbol of the stationary source X tends to a or $H(X)$, which is the entropy per letter at the source. Further restriction on the output of a stationary source is required in order to define an ergodic source. Ergodic sources are discussed, for example, in Khinchin (pp. 49–54). More general definitions of entropy for sources which are not necessarily stationary have appeared in the literature [Rozenblatt-Rot (I, II)].

Note on AEP. The foregoing extension of the theory was in context based on the so-called *asymptotic equipartition property* (AEP). A mathematical description of this important property, which was given by McMillan, may be found in Khinchin (Chap. 2) and Feinstein (I, Chap. 6). The following heuristic and brief description is included here. For a given finite source alphabet $[A]$ of N symbols, consider the cylinder set C with n specified letters. Each sequence in C may be regarded as one of N^n elementary events of a finite probability scheme. Every sequence of C is a cylinder of the infinite space A^I of the source with a definite probability measure $\mu(C)$. Of course, for stationary sources, this probability depends on n but not on the time. Next, in line with the material of Sec. 7-8, one may define a random variable

$$Z_n = -\frac{1}{n}\log \mu(c)$$

For stationary sources, the expected value of Z_n is

$$\bar{Z}_n = -\frac{1}{n}\sum_c \mu(c)\log\mu(c)$$

The right-hand side of this equation can be identified with the per-symbol entropy H_n/n of the described n sequence. That is,

$$\bar{Z}_n = \frac{H_n}{n}$$

The AEP states that as $n \to \infty$ for stationary sources, $\bar{Z}_n$ approaches a definite limit (convergence in probability) called the *source entropy.* That is, for any arbitrarily small $\epsilon > 0$ and $\delta > 0$, we can find a sufficiently large n such that

$$p\{|Z_n - H| > \epsilon\} < \delta$$

Stronger statements are possible in the case of ergodic sources. For a formal proof, see references cited before.

11-6. Discrete Channels with Finite Memory. In the preceding section we discussed a source with a finite memory emitting discrete signals from a finite alphabet. For physical consideration, one desires to feed the output of a source into a transmission medium which is called a channel. The first step in this direction is to define a channel and its behavior under a stationary regime.

A channel is a two-port with an input and output. The input to the channel has an alphabet (a source alphabet if driven by the source) which we assume to be finite. Similarly, we assume for convenience that the output of the channel also uses sequences of letters drawn from a finite alphabet. Let the input and output alphabets be $\{A\}$ and $\{B\}$, respectively. In memoryless channels the noise structure is generally specified by a conditional probability matrix

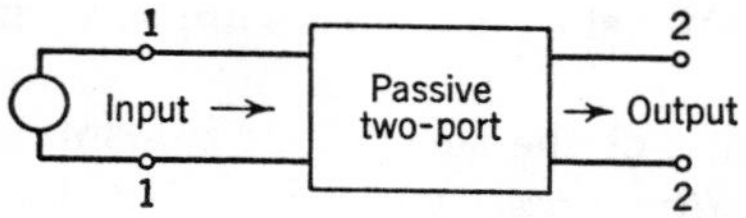

FIG. 11-1. A two-port analog of a communication system.

$$P\{b_j|a_k\} \qquad \text{for all } b_j \in B \quad a_k \in A$$

When the channel has no memory, the noise probability matrix is independent of the life history of the channel. When the channel has a finite memory, then the noise probability depends on the life history of the transmitted sequences up to the finite memory time prior to the emission of the signal. For example, for a Markovian channel the noise matrix is of the form

$$P\{Y_k = b|, \ldots, X_{-1}, X_0, X_1, \ldots, X_k\} = P\{Y_k = b|X_k\} \quad (11\text{-}47)$$

As our objective is to give a more general description of a channel, a more general method for describing the noise is essential. For this purpose, consider a member of an input ensemble x and its corresponding mate at the output y (that is, if x is transmitted, y is received).

$$\begin{array}{cc} \text{input} & \text{output} \\ [A] & [B] \\ \{X\} = \{\ldots, x_{-2}, x_{-1}, x_0, x_1, \ldots\} & \{Y\} = \{\ldots, y_{-2}, y_{-1}, y_0, y, \ldots\} \end{array}$$

Let X^I and Y^I be all possible source and received sequences, respectively. In X^I, let us focus attention on a cylinder $x^{4,1}$ which has a specific letter, say a_1, at a specific position, say x_4.

$$x^{4,1} = \ldots, x_{-1}, x_0, x_1, x_2, x_3, a_1, x_5, \ldots$$

Similarly, for a moment, concentrate on a particular cylinder at the output, say $y^{1,2}$, which has a specified letter b_2 at the position y_1.

$$y^{1,2} = \ldots, y_{-1}, y_0, b_2, y_2, \ldots$$

To know the noise characteristic we must know the conditional probability of cylinder $y^{1,2}$ being received when $x^{4,1}$ is transmitted:

$$P\{y^{1,2}|x^{4,1}\}$$

More specifically, for all possible cylinders $S_A \subset X^I$ at the input, we must have the conditional probability corresponding to any possible cylinder for messages at the output $S_B \subset Y^I$. To sum up, the following requirements are necessary in order to specify a general channel.

1. Input alphabet $[A]$
2. Output alphabet $[B]$
3. $$P\{S_B|S_A\} = \nu_x \qquad \text{for all } S_A \in X^I \text{ and } S_B \in Y^I \qquad (11\text{-}48)$$

Thus a discrete channel is specified by the set of triple data

$$[A, \nu_x, B] \qquad (11\text{-}49)$$

If a channel is such that its noise structure remains invariant with respect to a time shift, that is,

$$\nu_{Tx}(TS) = \nu_x(S) \qquad (11\text{-}50)$$

(T being the shift operator), then the channel is said to be stationary.

11-7. Connection of the Source and the Discrete Channel with Memory. In the terminology of information theory, a channel is driven by a source in much the same way as a passive electric circuit is driven by an electric source. The information source and the channel must have a common alphabet in order to provide a meaningful coupling. When the source transmits a letter $x_k \in A$, then at the output of the channel the letter will be received as a letter $y_k \in B$. If a sequence of letters

$$\ldots, x_{-1}, x_0, x_1, x_2, \ldots$$

is transmitted, a sequence of letters

$$\ldots, y_{-1}, y_0, y_1, y_2, \ldots$$

will be received. The probability distribution of Y_k in the latter sequence obviously depends on the statistical properties of the input sequence, or what could be referred to as the probability measure ν_x. If the probability distribution of Y_k depends only on the statistical properties of the sequence $\ldots, x_{k-1}, x_k$, we say that the channel is without *anticipation.* This implies that the statistical information about the present state at the receiver is specified by the past and the present states at the input. If, furthermore, the distribution of Y_k depends only on $x_{k-m}, \ldots, x_k$, then we say that the channel has a finite memory of m units.

The situation of connecting a source to a channel is quite similar to many familiar deterministic setups. For example, when a passive two-port network is driven by an ordinary electric source, one has the setup of Fig. 11-1. Here we have the following basic specifications:

1. An ideal source, ideal in the sense that its characteristic does not depend on the network to which it is to be connected
2. A passive network, that is, no output at 22 unless a source is connected to 11

The performance of the source and channel is specified by deterministic laws. Here, we know how to describe the output in terms of the source and the network parameters. Our present problem presents the information-theory analog of the afore-mentioned situation. The source transmits at random messages $x \in A^I$. The correspondence between the output and the input of the channel is a random one because of the effect of noise, but the statistical description of this random effect is governed by the distribution ν_x. As in the case of channels without memory, we consider the product space C^I of the pair, input signal x and output signal y.

$$\begin{aligned} x &\in A^I \\ y &\in B^I \\ (x,y) &\in C^I \end{aligned} \tag{11-51}$$

The specification of a probability distribution on C^I is, in fact, similar to specifying the joint probability matrix associated with a product space in the case of discrete random variables.

In a similar fashion one can describe a basic cylinder $E \in C^I$ as the product of a cylinder $E_1 \in A^I$ and $E_2 \in B^I$. It can be shown that we can associate a probability measure with each such cylinder of C^I. A more general event is split up into basic cylinders or limits of them in order to find its probability. The mathematical development for defining the probability measure of the product space is not given here. But one may visualize that in essence the treatment is very similar to the discrete case where a joint (product) probability is obtained as the product of the marginal and the conditional probabilities.

$$P\{x \cap y\} = P\{x\} \cdot P\{y|x\}$$

To sum up, the source and the channel may be described as a new source $[C,\omega]$, where C is the product of the two alphabets, $A \times B$, and ω the appropriate probability measure.* The product space C acts as a source in a product space similar to the space $\{X,Y\}$, and its probability measure ω is analogous to the joint probability matrix defined in Chap. 2.

11-8. Connection of a Stationary Source to a Stationary Channel. A most interesting result of the previous discussion is the study of the stationary regime of source and channel. The following presentation is due to Khinchin.

1. First of all, it can be shown that, if $[A,\mu]$ and $[A,\nu_x,B]$ are stationary, the product source $[C,\omega]$ is also stationary. [For proof, see Khinchin (pp. 80–82).]

2. Each stationary source has an entropy; therefore

$$[A,\mu],\ [B,\eta],\ [C,\omega]$$

each have definite entropies. $[B,\eta]$ is the equivalent output source of the channel.

3. Let these entropies first be defined for all n-term sequences $x_0, x_1, \ldots, x_{n-1}$ emitted by the source and transmitted through the channel; more specifically,

$$\begin{aligned} H_n(X) &\leftarrow \{x_0,x_1, \ldots ,x_{n-1}\} \\ H_n(Y) &\leftarrow \{y_0,y_1, \ldots ,y_{n-1}\} \\ H_n(X,Y) &\leftarrow \{(x_0,y_0), (x_0,y_1), \ldots , (x_{n-1},y_{n-1})\} \\ H_n(X|Y) &\leftarrow \{(x_0|Y), (x_1|Y), \ldots , (x_{n-1}|Y)\} \\ H_n(Y|X) &\leftarrow \{(X|y_0), (X|y_1), \ldots , (X|y_{n-1})\} \end{aligned} \tag{11-52}$$

Therefore

$$\begin{aligned} H_n(X,Y) &= H_n(X) + H_n(Y|X) \\ H_n(X,Y) &= H_n(Y) + H_n(X|Y) \end{aligned} \tag{11-53}$$

* The probability measure for the joint event $x \in S_1$, $y \in S_2$ is defined as

$$\omega(S) = \omega(S_1 \cap S_2) = \int_{S_1} \nu_x(S_2)\, d\mu(x)$$

Now consider the same relations for the entropies per symbol, that is,

$$\frac{1}{n} H_n(X,Y) = \frac{1}{n} H_n(X) + \frac{1}{n} H_n(Y|X)$$
$$\frac{1}{n} H_n(X,Y) = \frac{1}{n} H_n(Y) + \frac{1}{n} H_n(X|Y) \qquad (11\text{-}54)$$

When the sequences are made larger and larger, then,

$$\lim \frac{1}{n} H_n(X,Y) = H(X,Y) \qquad n \to \infty$$
$$\lim \frac{1}{n} H_n(X) = H(X) \qquad n \to \infty$$
$$\lim \frac{1}{n} H_n(Y) = H(Y) \qquad n \to \infty \qquad (11\text{-}55)$$

Note that according to Eqs. (11-54) the two channel entropies also exist,

$$\lim \frac{1}{n} H_n(Y|X) = H(Y|X) \qquad n \to \infty$$
$$\lim \frac{1}{n} H_n(X|Y) = H(X|Y) \qquad n \to \infty$$

That is, the basic relations among different entropies hold under a stationary regime.

$$H(X,Y) = H(X) + H(Y|X)$$
$$H(X,Y) = H(Y) + H(X|Y)$$

The important mathematical conclusion to be pointed out is that, for stationary regimes:

1. The existence of different entropies is a fact.
2. The significance of the entropies and their interrelationships is just the same as in the case of a regime under an independent source and memoryless channel.
3. Finally, the rate of transinformation exists and is the same as defined before.

$$I(X;Y) = H(X) + H(Y) - H(X,Y)$$

4. The stationary capacity of the channel C_s is the max $I(X;Y)$ over all stationary sources.

Without including more complex mathematical machinery, a few suggestions for further reading are offered. Shannon's original contributions were primarily concerned with Markovian sources. In the light of recent contributions it has been established that the first and second fundamental theorems hold for stationary (not necessarily Markovian or ergodic) sources and channels with or without memory. These subsequent results are primarily due to McMillan, Feinstein, and Khinchin. According to Khinchin, up to 1953 no adequate proofs of these theorems existed. Some of the more recent developments are due to Gel'fand, Kolmogorov, Blackwell, Feinstein, Rozenblatt-Rot, Perez, and Driml. The transactions of the first and the second Prague Conference (Czechoslovak Academy of Sciences, 1958) contain a considerable amount of research related to the entropy of the stochastic signals. These developments are of a mathematical nature, outside the scope of the present undertaking.

PROBLEMS

11-1. Determine which of the following transition probability matrices correspond to regular chains. Find the T matrix.

(a) $\begin{bmatrix} 0 & 1 \\ 1 & 0 \end{bmatrix}$ (b) $\begin{bmatrix} 1 & 0 \\ 0 & 1 \end{bmatrix}$ (c) $\begin{bmatrix} 1/4 & 3/4 \\ 3/4 & 1/4 \end{bmatrix}$ (d) $\begin{bmatrix} 0 & 1 \\ 2/3 & 1/3 \end{bmatrix}$

(e) $\begin{bmatrix} 1/3 & 2/3 \\ 0 & 1 \end{bmatrix}$ (f) $\begin{bmatrix} 1 & 0 \\ 1/4 & 3/4 \end{bmatrix}$ (g) $\begin{bmatrix} 1/3 & 0 & 2/3 \\ 0 & 1 & 0 \\ 0 & 3/4 & 1/4 \end{bmatrix}$ (h) $\begin{bmatrix} 0 & 1 & 0 \\ 0 & 0 & 1 \\ 1/4 & 3/4 & 0 \end{bmatrix}$

11-2. Which of the following Markov transition probability matrices are regular? x means any positive nonzero number. The rows are assumed to add to 1.

(a) $\begin{bmatrix} 0 & x & 0 \\ 0 & 0 & x \\ x & 0 & 0 \end{bmatrix}$ (b) $\begin{bmatrix} x & x & 0 \\ x & x & 0 \\ 0 & 0 & x \end{bmatrix}$ (c) $\begin{bmatrix} 0 & 0 & x & 0 \\ 0 & x & 0 & 0 \\ x & 0 & x & 0 \\ 0 & x & 0 & x \end{bmatrix}$

(d) $\begin{bmatrix} 0 & x & 0 & 0 & 0 \\ 0 & 0 & 0 & 0 & x \\ 0 & x & 0 & x & x \\ x & x & x & x & x \\ 0 & 0 & 0 & x & 0 \end{bmatrix}$

11-3. Study the Markov chain of Fig. P11-3. Determine transient and ergodic states. Is the chain ergodic?

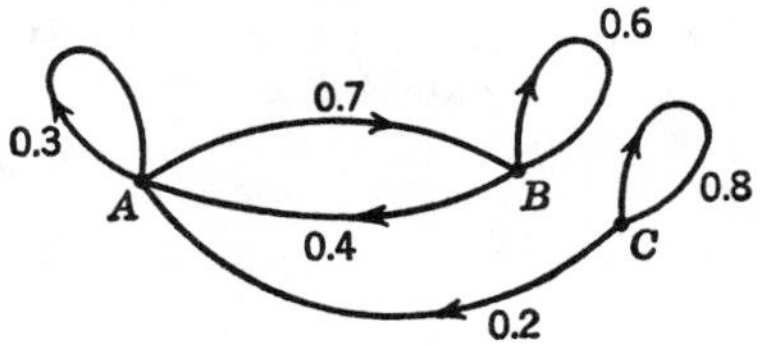

Fig. P11-3

11-4. Show that the chain illustrated in Fig. P11-4 is ergodic but not regular.

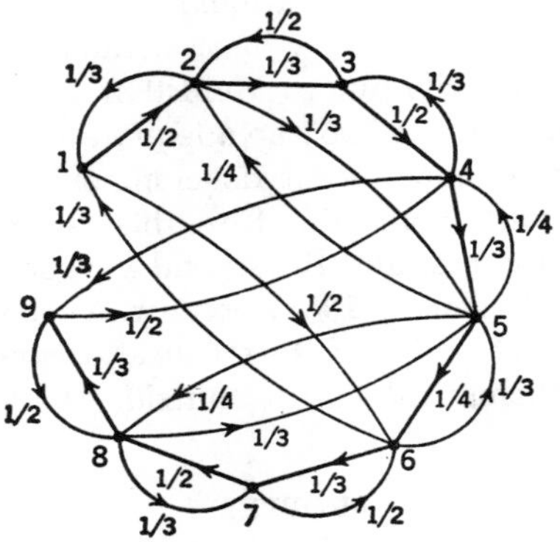

FIG. P11-4

11-5. Consider the chain represented by the transition probability matrix

$$\begin{bmatrix} 0 & 1 \\ 1 & 0 \end{bmatrix}$$

(*a*) Is the chain regular?
(*b*) Is the chain ergodic?
(*c*) Find the T matrix.

11-6. Draw a Markov diagram for a game of tennis. Assume that a player has a fixed probability P for winning each point. Determine:
(*a*) Probability of winning a game when $P = 0.60$.
(*b*) Probability of winning a set when $P = 0.60$.
(*c*) Parts (*a*) and (*b*) for $P = 0.51$.

J. L. Snell, Finite Markov Chains and Their Applications, *Am. Math. Monthly*, vol. 66, no. 2, pp. 99–104, February, 1959.

11-7. (*a*) Find the T matrix for the probability transition matrix below.

$$\begin{bmatrix} \frac{1}{4} & \frac{1}{4} & \frac{1}{2} \\ 0 & \frac{2}{3} & \frac{1}{3} \\ \frac{3}{4} & \frac{1}{4} & 0 \end{bmatrix}$$

(*b*) Assuming an initial probability matrix of $[\frac{1}{3}, \frac{1}{3}, \frac{1}{3}]$, compute the entropies $H^{(1)}$, $H^{(2)}$, and $H^{(3)}$ and compare $H^{(3)}$ with $H^{(1)} + H^{(2)}$.

(*c*) Discuss and verify the situation when $H^{(1)} + H^{(2)} = H^{(3)}$.

11-8. For Prob. 11-1 determine the entropy for each regular Markov chain.

11-9. Show that, for a regular Markov chain, there is a limiting probability α_j of being in the state S_j independent of the starting state; α_j is also the fraction of times that the process can be expected in state S_j (see Kemeny and Snell, Chap. 4).

PART 4

SOME RECENT DEVELOPMENTS

L'importance d'un fait se mesure donc à son rendement, c'est-à-dire, à la quantité de pensée qu'elle permet d'économiser.

Si un résultat nouveau a du prix, c'est quand, en reliant des éléments connus depuis longtemps, mais jusque-là épars et paraissant étrangers les uns aux autres, il introduit subitement l'ordre là où régnait l'apparence du désordre. Il nous permet alors de voir d'un coup d'œil chacun de ces éléments et la place qu'il occupe dans l'ensemble. Ce fait nouveau non-seulement est précieux par lui-même, mais lui seul donne leur valeur à tous les faits anciens qu'il relie.

Henri Poincaré

Atti IV congr. intern. Mat., vol. 1, *p.* 169

CHAPTER 12

THE FUNDAMENTAL THEOREM OF INFORMATION THEORY

Frequent reference has been made to Shannon's significant statement that by proper encoding it is possible to send information at a rate arbitrarily close to C through the channel, with as small a probability of error or equivocation as desired. In the context of our work, we have put in focus the accurate meaning of this statement for discrete as well as continuous channels. The major objective of this chapter is to give a detailed statement and proof of the fundamental theorem for discrete noisy memoryless channels. Unless otherwise specified, in this chapter we are concerned only with discrete noisy memoryless channels.

A glance at the above statement reveals that there are several points to be clarified before the full implication of the statement is realized. These points are as follows:

1. A quantitative definition of the words "error" and "frequency" or "probability of error" in a communication system
2. The relation between error and the equivocation entropy of the channel

The definition of error (1) requires the description of a detection or a decoding scheme. Once we have a method for decoding the received signals, then we are in a position to discuss the probability of error associated with that method. Thus, our immediate plan for the first part of this chapter consists of the following:

1. The definition of a decision (or detection) scheme (Sec. 12-1)
2. The definition of the error associated with a detection scheme (Sec. 12-2)
3. A discussion of the relation between error and equivocation (Sec. 12-3)
4. A study of the transmission of information in the extended channel (Sec. 12-4)

Subsequent to the presentation of these preliminaries, we shall turn our attention to the fundamental theorem of discrete memoryless channels.

From an organizational point of view, the chapter is divided in the following four parts: Preliminaries, Feinstein's Proof, Shannon's Proof, and Wolfowitz's Proof.

Wolfowitz's proof is the most recent of the three. Feinstein's proof is the first complete proof and uses a certain ingenious but complex procedure. Shannon's proof presents a good deal of physical insight into the problem and contains new ideas which may lead to fresh fields.

In the following presentation, we have made an effort to use the notation of the original contributors as far as possible. This conformity of notation will be found convenient for those readers who wish to study the original articles.

PRELIMINARIES

12-1. A Decision Scheme. A decision scheme is a method that a receiver employs for determining, after a word has been received, which particular message word was transmitted. Let us review briefly the mathematical model of our discrete memoryless communication system. Each transmitted and received word is an n-letter sequence like

$$u_k = x_{k1}, x_{k2}, \ . \ . \ . \ , x_{kn}$$
$$v_k = y_{k1}, y_{k2}, \ . \ . \ . \ , y_{kn}$$

where x_{ki} and y_{ki} are letters selected from a finite alphabet. At the receiver, knowing the transmission alphabet and the channel matrix, we wish to devise a detection scheme. That is, when a word v_k is received we must associate (rightfully or otherwise) one of the transmitted words with v_k. This assignment calls for a decision that may be substantiated by some statistical inference, although the possibility of a random decision is not excluded. At the transmitter, we have an alphabet of a letters at our disposal, out of which N words each composed of n letters are selected for transmission. The receiving space is partitioned into N disjoint sets $[A_1,A_2, \ . \ . \ . \ ,A_N]$ in a one-to-one correspondence with $[u_1,u_2, \ . \ . \ . \ ,u_N]$. Whenever a word $v_n \in A_j$ is received, we decide that the word u_j was transmitted. This setup is referred to as a *decision scheme*, although we have not stated the statistical criterion on which the partitioning has been based. A decision scheme generally requires (1) a given input probability distribution $P\{u\}$ and (2) a criterion for partitioning the receiving space. The maximum-likelihood principle used in Sec. 9-14 is one of the most common decision criteria; this will be described in the next section.

12-2. The Probability of Error in a Decision Scheme. In this section, we wish clearly to define what are the error and the error probability associated with a decision scheme. Error in detection occurs when a word is not received in A_k while u_k is transmitted. The probability of this error is

$$e_k = p\{u \neq u_k | v \in A_k\} = 1 - P\{u = u_k | v \in A_k\} \qquad (12\text{-}1)$$

The error E associated with a decision scheme can conveniently be considered as a random variable assuming values e_k, as stated in Eq. (12-1), with a probability $P\{v \in A_k\}$. Thus the average value of the error is

$$\bar{E} = \sum_{k=1}^{N} P\{v \in A_k\}[1 - P\{u = u_k|v \in A_k\}] \qquad (12\text{-}2)$$

The meaning of the above notation should be made clear. We find the error for a possible A_k; then the index k is changed to cover all the detection regions. It is instructive to note that the average error probability can be equivalently derived by considering an alternative method of bookkeeping. The error occurs when a word u_k is transmitted but the corresponding received word is not in A_k. This error probability can also be considered as a value taken by a random variable E.

$$P\{v \in A'_k|u = u_k\} = 1 - P\{v \in A_k|u = u_k\} \qquad (12\text{-}3)$$

The average value of E is

$$\bar{E} = \sum_{k=1}^{N} P\{u = u_k\}[1 - P\{v \in A_k|u = u_k\}] \qquad (12\text{-}4)$$

The equivalence of the expressions in Eqs. (12-2) and (12-4) can be exhibited by appropriate manipulation of the terms. Equation (12-4) can be written as

$$\begin{aligned}\bar{E} &= \sum_{k=1}^{N} P\{u_k\}\left[1 - \frac{P\{v \in A_k \cap u = u_k\}}{P\{u = u_k\}}\right] \\ &= 1 - \sum_{k=1}^{N} P\{v \in A_k \cap u = u_k\} \\ &= 1 - \sum_{k=1}^{N} P\{u = u_k|v \in A_k\}P\{v \in A_k\} \\ &= \sum_{k=1}^{N} P\{v \in A_k\}[1 - P\{u = u_k|v \in A_k\}] \qquad (12\text{-}5)\end{aligned}$$

To sum up, observe that we have arrived at a clear understanding of an average error probability for any decision scheme in general. This error probability, as anticipated earlier, is a function of the input probability distribution, the channel, and the decision criterion. In particular, the decision scheme which, upon reception of the symbol v_i, chooses that transmitted symbol u_j whose conditional probability $P\{u_j|v_i\}$ is the greatest is generally referred to as an *ideal observer*. Note that this definition is contingent upon the knowledge of the input or output message probabilities.

$$A_k = \{v_i : P\{u_k|v_i\} = \max_{\text{over } j} P\{u_j|v_i\}\}$$

12-3. A Relation between Error Probability and Equivocation. Feinstein has investigated the useful concept of uniform error bounds. A decision scheme is said to be uniform error bounding with bound λ, if there exists a number $0 \leq \lambda < 1$ such that

$$P\{v \in A_k|u_k\} \geq 1 - \lambda \qquad k = 1, 2, \ldots, N \tag{12-6}$$

Obviously, if a decision scheme has a uniform error bound, then Eqs. (12-2) and (12-4) yield

$$\bar{E} \leq \lambda \tag{12-7}$$

That is, the average detection error probability will not exceed the uniform error bound of the channel, provided that the latter exists. For a discrete memoryless channel, the following theorem holds:

Theorem. Let E be the error probability of a decision scheme with N detection regions. Then

$$H(U|V) \leq -\bar{E} \log \bar{E} - (1 - \bar{E}) \log (1 - \bar{E}) + \bar{E} \log (N - 1) \tag{12-8}$$

Proof. A proof of this theorem follows from the convexity property of the entropy function. [See Feinstein (I, pp. 35–36).] The following proof based on the law of additivity [Eq. (3-34)] is also of interest. Consider a received word v_k and the entropy $H(U|v_k)$. The original word associated with v_k will be denoted by u_{ik}. Of course, several received words may correspond to the same original word u_i. Using Eq. (3-34) and the convexity of $x \log x$,

$$\begin{aligned} H(U|v_k) = {} & -P\{u_{ik}|v_k\} \log P\{u_{ik}|v_k\} \\ & - (1 - P\{u_{ik}|v_k\}) \log (1 - P\{u_{ik}|v_k\}) \\ & + (1 - P\{u_{ik}|v_k\}) \left(\sum_{\substack{j=1 \\ j \neq ik}}^{N} \frac{P\{u_j|v_k\}}{1 - P\{u_{ik}|v_k\}} \log \frac{P\{u_j|v_k\}}{1 - P\{u_{ik}|v_k\}} \right) \\ \leq {} & -P\{u_{ik}|v_k\} \log P\{u_{ik}|v_k\} - (1 - P\{u_{ik}|v_k\}) \\ & \log (1 - P\{u_{ik}|v_k\}) + (1 - P\{u_{ik}|v_k\}) \log (N - 1) \end{aligned} \tag{12-9}$$

However,

$$H(U|V) = \sum_{k=1}^{M} P\{v_k\} H(U|v_k)$$

where M is the number of all possible words in the receiving space. Thus,

$$H(U|V) \leq -\sum_{k=1}^{M} P\{v_k\}P\{u_{ik}|v_k\} \log P\{u_{ik}|v_k\} - \sum_{k=1}^{M} P\{v_k\}(1 - P\{u_{ik}|v_k\}) \log (1 - P\{u_{ik}|v_k\}) + \sum_{k=1}^{M} P\{v_k\}(1 - P\{u_{ik}|v_k\}) \log (N-1) \quad \text{(12-10a)}$$

The summation over M possible received signals can be done conveniently by first summing over a region A_k and then proceeding over all such regions.

$$H(U|V) \leq -\sum_{k=1}^{N} P\{v \in A_k\}P\{u_k|v \in A_k\} \log P\{u_k|v \in A_k\} - \sum_{k=1}^{N} P\{v \in A_k\}(1 - P\{u_k|v \in A_k\}) \log (1 - P\{u_k|v \in A_k\}) + \sum_{k=1}^{N} P\{v \in A_k\}(1 - P\{u_k|v \in A_k\}) \log (N-1) \quad \text{(12-10b)}$$

Using the values of e_k and $\bar{E}$ as defined in Eqs. (12-1) and (12-2) and the convexity argument, one finds

$$H(U|V) \leq -\sum_{k=1}^{N} P\{v \in A_k\}(1 - e_k) \log (1 - e_k) - \sum_{k=1}^{N} P\{v \in A_k\}e_k \log e_k + \bar{E} \log (N-1) \quad \text{(12-11)}$$

or $H(U|V) \leq -(1 - \bar{E}) \log (1 - \bar{E}) - \bar{E} \log \bar{E} + \bar{E} \log (N-1)$

When the channel is not lossless, at least for some input probability distribution, the equivocation will be positive. That is, for a uniform error bound scheme the equivocation entropy will be bounded by

$$0 < H(U|V) \leq -\lambda \log \lambda - (1 - \lambda) \log (1 - \lambda) + \lambda \log (N-1) \quad \text{(12-12)}$$

$$0 < H(U|V) \leq 1 + \lambda \log (N-1)$$

To obtain a more relevant bound on equivocation, consider the case of the nth-order extension of a discrete memoryless channel with a capacity C. According to Example 3-8, for a uniform input distribution we have

$$H(U|V) = nH(X|Y) < 1 + \lambda nC \quad \text{(12-13)}$$

The equivocation per letter also remains bounded:

$$H(X|Y) < \frac{1}{n} + \lambda C \quad \text{(12-14)}$$

Furthermore, the equivocation per letter can be made arbitrarily small by choosing an appropriately small λ and an adequately large n.

12-4. The Extension of Discrete Memoryless Noisy Channels. A heuristic and a formal proof of the fundamental theorem for BSC were presented in Chap. 4 when our knowledge of probability theory was very rudimentary. From the material of Chaps. 5 to 7 we have acquired a working knowledge of the principles of probability theory. In the light of this development, we wish to present a more general proof. Our task can be divided into the following steps:

1. Define an encoding-decoding scheme and compute the error associated with a code.
2. Reconsider the transmission of information over a noisy memoryless channel driven by a discrete memoryless source. Extend the results to the nth-order extension of the source and the channel.
3. Using the laws of large numbers as described in Chap. 7, find a relation between the number of messages to be transmitted, the error of the detection scheme, and the rate of transmission.
4. Interrelate the parts discussed in 1, 2, and 3 to derive the fundamental theorem of discrete noisy memoryless channels.

Consider a discrete independent source transmitting symbols from a finite list $[a] = [1,2, \ldots ,a]$ with specified probabilities. The output of this source is fed into a discrete noisy memoryless channel with specified noise characteristics. Traditionally, we assume the input and the output to the channel as consisting of random variables X and Y, respectively. These random variables assume values from the list $[a]$ with certain probabilities. The schemes associated with these random variables will be denoted by U and V, respectively.

The different probability functions of these schemes are

$$\begin{aligned} P\{X = x\} &= f_1(x) \\ P\{Y = y|X = x\} &= f(y|x) \\ P\{X = x, Y = y\} &= f(x,y) \\ P\{Y = y\} &= f_2(y) \end{aligned} \tag{12-15}$$

where x and y both assume values from the list $[a]$.

The rate of transinformation per symbol for the channel is

$$R = \overline{I(X;Y)} = \sum_U \sum_V f(x,y) \log \frac{f(x,y)}{f_1(x)f_2(y)} \tag{12-16}$$

Next we wish to evaluate the transinformation rate for the nth-order extension of the above channel when the output of the source consists of n sequences, that is, sequences of length n with elements from $[a]$. Let X^n and Y^n stand for the input and the output, respectively, of the nth-order extension of the original channel, that is,

$$\begin{aligned} X^n &= X_1, X_2, \ldots, X_n \\ Y^n &= Y_1, Y_2, \ldots, Y_n \end{aligned} \tag{12-17}$$

To conform with our previous notation, we let U^n and V^n be the probability schemes associated with the random variables X^n and Y^n, respectively. Thus each typical sequence u of the form $x_1, x_2, \ldots, x_n$ is an n sequence with all symbols taken from $[a]$, that is, a value of X^n. Similarly, for the output sequences of the channel we use v as a specific value of the random variable Y^n. The probabilities and the information measures associated with the schemes U^n, V^n, and U^nV^n are

$$\begin{aligned} &P\{X^n = u\} = p_1(u) \\ &P\{Y^n = v|X^n = u\} = p(v|u) \\ &P\{X^n = u|Y^n = v\} = p(u|v) \\ &P\{Y^n = v, X^n = u\} = p(u,v) \\ &P\{Y^n = v\} = p_2(v) \\ &\overline{I(X^n;Y^n)} = \sum_U \sum_V p(u,v) \log \frac{p(u,v)}{p_1(u)p_2(v)} \end{aligned} \tag{12-18}$$

FEINSTEIN'S PROOF

12-5. On Certain Random Variables Associated with a Communication System.* Now that the necessary preliminaries have been covered, we proceed with Feinstein by introducing certain random variables associated with the source and the channel. The desired results are some relevant inequalities relating the probability distributions of these variables with different entropies. We assume:

1. Successive symbols transmitted by the extended source are selected independently.
2. The channel has no memory.

That is, we have

$$\begin{aligned} \text{(I)}\quad P\{X^n = u\} &= P\{X_1 = x_1, X_2 = x_2, \ldots, X_n = x_n\} \\ &= f_1(x_1)f_1(x_2) \cdots f_1(x_n) \\ \text{(II)}\quad P\{Y^n = v|X^n = u\} &= P\{Y_1 = y_1, Y_2 = y_2, \ldots, Y_n = y_n \\ &\qquad |X_1 = x_1, X_2 = x_2, \ldots, X_n = x_n\} \\ &= f(y_1|x_1)f(y_2|x_2) \cdots f(y_n|x_n) \end{aligned} \tag{12-19}$$

* The author wishes to thank Dr. A. Feinstein for helpful comments on the material of this chapter.

As a consequence of I and II we have

(III) $$P\{X^n = u|Y^n = v\} = f(x_1|y_1)f(x_2|y_2) \cdots f(x_n|y_n)$$

As before, let us denote $P\{X^n = u\}$ by $p_1(u)$ and $P\{X^n = u|Y^n = v\}$ by $p(u|v)$ and let $H(U)$ and $H(U|V)$ refer to the original scheme.*

Some important relationships can be established between the number n, the source entropy, the channel equivocation, and the rate of trans-information per symbol.

Lemma 1. Let $1 > \epsilon > 0$, $1 > \delta > 0$ be two arbitrarily chosen numbers. There exist two numbers $n_1(\epsilon,\delta)$ and $n_2(\epsilon,\delta)$ such that

$$\begin{aligned} &P\{p_1(X^n) > 2^{-n[H(U)-\epsilon]}\} < \delta \qquad n > n_1 \\ &P\{p(X^n|Y^n) < 2^{-n[H(U|V)+\epsilon]}\} < \delta \qquad n > n_2 \end{aligned} \tag{12-20}$$

Proof. Consider the random variables $p_1(X^n)$ and $p(X^n|Y^n)$. As a consequence of I and III we have

$$\log p_1(X^n) = \sum_{i=1}^{n} \log f_1(X_i) \tag{12-21}$$

and

$$\log p(X^n|Y^n) = \sum_{i=1}^{n} \log f(X_i|Y_i)$$

Because of I and II, each of these random variables consists of a sum of independent random variables with identical distributions; thus the weak law of large numbers applies. According to Sec. 7-8 for $n \to \infty$ the probability of the mean of the random variables deviating from their common mean by, at most, ϵ approaches unity. That is,

$$P\left\{-H(U) - \epsilon < \frac{1}{n}\sum_{i=1}^{n} \log f_1(X_i) < -H(U) + \epsilon\right\} \to 1 \qquad n \to \infty$$

and

$$P\left\{-H(U|V) - \epsilon < \frac{1}{n}\sum_{i=1}^{n} \log f(X_i|Y_i) < -H(U|V) + \epsilon\right\} \to 1 \tag{12-22}$$

Evidently, by properly choosing n, the probability of the mean of the above random variables deviating by more than ϵ from their common mean can be made as small as desired. More specifically, for any $\epsilon > 0$ there are numbers n_1 and n_2 such that

$$\begin{aligned} &P\left\{\left|H(U) + \frac{1}{n}\log p_1(X^n)\right| > \epsilon\right\} < \delta \qquad \text{for } n > n_1 \\ &P\left\{\left|H(U|V) + \frac{1}{n}\log p(X^n|Y^n)\right| > \epsilon\right\} < \delta \qquad \text{for } n > n_2 \end{aligned} \tag{12-23}$$

* The reader should note that $p_1(u)$ and $f_1(x_1)$ are generally different functions, as are $p(u|v)$ and $f(x_1|y_1)$, and $p(v|u)$ and $f(y_1|x_1)$.

Therefore

$$P\left\{\frac{1}{n}\log p_1(X^n) > -H(U) + \epsilon\right\} < \delta \qquad \text{for } n > n_1 \qquad (12\text{-}24)$$

or

$$P\{p_1(X^n) < 2^{-n[H(U)-\epsilon]}\} > 1 - \delta \qquad \text{for } n > n_1$$

$$P\{p(X^n|Y^n) > 2^{-n[H(U|V)+\epsilon]}\} > 1 - \delta \qquad \text{for } n > n_2 \qquad (12\text{-}25)$$

[The reader should keep in mind that in the above inequalities $p_1(X^n)$ and $p(X^n|Y^n)$ are considered random variables.]

12-6. Feinstein's Lemma

1. Let Z be a subset of the product space $U^n \otimes V^n$ satisfying the inequality

$$P\{Z\} > 1 - \delta_1 \qquad \delta_1 \text{ small positive number} \qquad (12\text{-}26)$$

2. Let $U_0 \subset U^n$ such that

$$P\{U_0\} > 1 - \delta_2 \qquad \delta_2 > 0$$

Then there exists a subset U_1 of U_0 such that for each $u_i \in U_1$ there is a set A_i of v such that

$$(A_i,u_i) \in Z$$

$$P\{A_i|u_i\} \geq 1 - a \qquad \text{for each } u_i \in U_1$$

and

$$P\{U_1\} > 1 - \delta_2 - \frac{\delta_1}{a} \qquad (12\text{-}27)$$

Proof. We begin by recalling that for any u_i

$$A_i = \{v\colon (u_i,v) \in Z\}$$

Next we define U_1 as the set of all those u_i satisfying the following two properties:

$$u_i \in U_0$$

$$P\{A_i|u_i\} \geq 1 - a \qquad (12\text{-}28)$$

So far we have defined U_1 and the A_i as described by the lemma. It remains to show that $P\{U_1\} > 1 - \delta_2 - \delta_1/a$. To this end, we may introduce an auxiliary set U_2 whose elements u_i satisfy

$$P\{A_i|u_i\} < 1 - a \qquad (12\text{-}29)$$

Thus

$$P\{v_k \in A'_k, u_k\} = P\{v_k \in A'_k|u_k\}P\{u_k\} \geq aP\{u_k\} \qquad \text{for } u_k \in U_2 \qquad (12\text{-}30)$$

We find

$$\sum_{u_k \in U_2} P\{A'_k \cap u_k\} \geq aP\{U_2\} \qquad (12\text{-}31)$$

Note that the pairs $(u_k, v_k \in A'_k)$ are not in Z. That is, their total probability is less than δ_1. Thus

$$\delta_1 \geq aP\{U_2\} \tag{12-32}$$

It remains to see how U_2, U_1, and U_0 are interrelated. For this, we note that

$$U_1 = U_0 - (U_0 \cap U_2) \tag{12-33}$$
$$P\{U_0\} = P\{U_1\} + P\{U_0 \cap U_2\}$$

In addition,

$$P\{U_0 \cap U_2\} < P\{U_2\} \leq \frac{\delta_1}{a}$$

Therefore

$$P\{U_1\} > 1 - \delta_2 - \frac{\delta_1}{a}$$

12-7. Completion of the Proof. Now we are in a position to apply Feinstein's lemma to sets Z and U_0 as described in lemma 1. Let Z be the set of all (u,v) satisfying the second inequality of Eq. (12-20); we know from lemma 1 that, for any $\delta_1 > 0$ for n sufficiently large, $P\{Z\} > 1 - \delta_1$. Furthermore, let U_0 be the set of words u satisfying the first inequality of Eq. (12-20). In view of lemma 1 for any $\delta_2 > 0$, we have

$$P\{U_0\} > 1 - \delta_2 \qquad \text{for } n \text{ sufficiently large}$$

Since Z and U_0 satisfy the hypothesis of Feinstein's basic lemma, we can assert the existence of a set $U_1 = \{u_1, \ldots, u_n\}$ and the existence of the corresponding A_i such that

$$U_1 \subset U_0 \tag{12-34a}$$
$$Z_k = \{(u_k,v) : v \in A_k\} \subset Z \qquad u_k \in U_1 \tag{12-34b}$$
$$P\{A_k|u_k\} \geq 1 - a \tag{12-34c}$$
$$P\{U_1\} > 1 - \delta_2 - \frac{\delta_1}{a} \tag{12-34d}$$

From (12-34a) and (12-34b) we see that for all $(u,v) \in Z_k$, $k = 1, 2, \ldots, N$,

$$\frac{p(u|v)}{p_1(u)} > \frac{2^{-n[H(U|V)+\epsilon_1]}}{2^{-n[H(U)-\epsilon_2]}} = 2^{n(R-\epsilon_1-\epsilon_2)} \tag{12-35}$$

This implies that for any $(u,v) \in Z_k$, $k = 1, 2, \ldots, N$,

$$\frac{p(u,v)}{p_1(u)} > 2^{n(R-\epsilon_1-\epsilon_2)}p_2(v) \tag{12-36}$$

and so for $u = u_k$

$$\frac{\sum_{v \in A_k} p(u,v)}{p_1(u)} > 2^{n(R-\epsilon_1-\epsilon_2)}P\{A_k\}$$

This of course implies

$$P\{A_k\} < 2^{-n(R-\epsilon_1-\epsilon_2)} \tag{12-37}$$

From these results and other remarks we get the following central theorem of information theory.

Theorem. With the previous notation, for any $0 < H < C$, $e > 0$, $\delta > 0$, there exists an $n > n(e,\delta)$ such that for the nth-order extension of the channel we can devise a detection scheme with bounded error; that is, we can select a set of N messages, say $M[e] = \{u_1,u_2, \ldots ,u_N\}$, such that the receiving space could be partitioned into N disjoint sets B_i with the following characteristics:

1. To each $u_i \in M[e] = M$ there corresponds a B_i.
2. $B_iB_k = \emptyset \qquad i \neq k \quad i, k = 1, 2, \ldots, N.$
3. $P\{B_i|u_i\} \geq 1 - e \qquad i = 1, 2, \ldots, N \quad 0 < e < \frac{1}{2}.$
4. $P\{B_i\} < L = 2^{-n(R-\epsilon_1-\epsilon_2)}.$
5. There does not exist a set of $N + 1$ messages satisfying conditions 1, 2, and 3.
6. $N \geq 2^{nH}.$

Proof. In several instances (see, for example, Secs. 4-16, 4-11, and 9-15) when using the concept of minimum-distance detection, we have demonstrated the possibility of devising detection schemes with disjoint B_i's and bounded error. From the foregoing material we already know that a maximal* detection scheme satisfying 1, 2, and 3 can be devised. However, no indication of the size of N has yet been given. We have no idea how many n-sequence words could be transmitted without violating the constraints 4, 3, and 6. To this end, consider the set U_1 of input words $u'_1, u'_2, \ldots, u'_{N'}$ disjoint from $M[e]$ as an enlarging set $M[a,k]$ with $a < e$, $P\{A_k\} \leq 2^{-n(R-\epsilon_1-\epsilon_2)}$, and not necessarily disjoint A_k's. We exploit the fact that an element u'_k of $M[a,k]$ cannot be used for increasing the number of words N in $M[e]$. Of course, A_k must intersect one or more of B_i's; otherwise it could have been used for enlarging M—M is used as abbreviation for $M[e]$. The following information is available concerning the set C_k of elements of A_k not in B_i's.

$$C_k = A_k - A_k \cap \left(\bigcup_{i=1}^{N} B_i\right) \tag{12-38}$$

* By a maximal set is meant the realization of an N satisfying 1, 2, 3, and 5. A maximal set which in addition satisfies $P\{B_i\} < L$, $i = 1, 2, \ldots, N$, is referred to as a maximal L-bounded set with respect to a given input probability distribution. A set satisfying 1, 3, and $P\{B_i\} < L$, $i = 1, 2, \ldots, N$, is referred to as an enlarging set with respect to a given input distribution. The concepts *maximal sets, bounded sets,* and *enlarging sets* were introduced by Feinstein. Our original plan was to apply Feinstein's lemma directly to BSC and to derive the fundamental theorem. Instead we have given here our version of Feinstein's proof, deviating very little from his notation but forgoing mathematical refinements (such as the existence and number of maximal sets, etc.). These finer points are generally evident from a physical point of view. For a more comprehensive coverage see Feinstein (I), Chap. 4, and Khinchin, Chaps. 1 and 4.

$$P\{C_k\} < P\{A_k\} < 2^{-n(R-\epsilon_1-\epsilon_2)} \tag{12-39}$$
$$P\{C_k|u'_k\} \leq 1 - e \qquad u'_k \in U_1 \tag{12-40}$$

The following inequality will provide the final step required for the proof of the theorem:

$$\begin{aligned} P\{\bigcup_{i=1}^{N} B_i\} &= \sum p_1(u)p\{\bigcup_{i=1}^{N} B_i|u\} \\ &\geq \Big(\sum_{U_1-M\cap U_1} + \sum_{M\cap U_1}\Big) P\{\bigcup_{i=1}^{N} B_i|u\}p_1(u) \\ &\geq (e-a)(P\{U_1\} - P\{M \cap U_1\}) + (1-e)P\{M \cap U_1\} \\ &\geq (e-a)\left(1 - \delta_2 - \frac{\delta_1}{a}\right) \qquad \text{since } \tfrac{1}{2} > e > a \end{aligned} \tag{12-41}$$

Thus from condition 4 and from the fact that for the elements of the first set

$$P\{\bigcup_{i=1}^{N} B_i|u\}P\{A_K \cap \bigcup_{i=1}^{N} B_i|u\} > e - a$$

we see that

$$N2^{-n(R-\epsilon_1-\epsilon_2)} > (e-a)\left(1 - \delta_2 - \frac{\delta_1}{a}\right) + (1-e)P\{M \cap U_1\} \tag{12-42}$$

Given ϵ and $H < C$, we now consider the input probability distribution $p_1(u)$ which maximizes R. Furthermore, let

$$\begin{gathered} a = \frac{e}{2} \\ C = \max \text{ of } R \\ \epsilon_1 + \epsilon_2 = \frac{C-H}{2} \\ 1 - \delta_2 - \frac{2\delta_1}{e} \geq \tfrac{1}{2} \\ n \text{ sufficiently large} \end{gathered} \tag{12-43}$$

Then

$$\begin{gathered} N > \min\,(e-a, 1-e)\left(1 - \delta_2 - \frac{2\delta_1}{e}\right) 2^{n(C+H)/2} \\ N > \min\left(\frac{e}{4}, \frac{1-e}{2}\right) \cdot 2^{nH+n(C-H)/2} \end{gathered} \tag{12-44}$$

For sufficiently large n we have

$$N \geq 2^{nH} \tag{12-45}$$

SHANNON'S PROOF

12-8. Ensemble Codes. Subsequent to Feinstein's formal proof of the fundamental theorem of encoding for discrete channels in the presence of noise, Shannon provided an alternative proof which does not employ Feinstein's inequalities. Shannon's proof is based on the concept of *ensemble codes.* He considers a class of distinct codes selected at random and evaluates the average error probability not for any specific member of this ensemble but for their totality. This average probability of error may be bounded between some limiting values which are not too difficult to derive. Thus, since there is at least one member of the ensemble of codes which has a probability of error less than the average error for the ensemble, there is a code whose error probability is less than the upper bound. In other words, if we were given an encoding machine, which selects any one of, say, k specific encoding procedures, then, on an average, the machine cannot do better or worse than certain limiting behaviors. The limiting average error values for the code ensemble, of course, depends on the over-all nature of the coding-decoding scheme. Shannon's method has the merit of quantitatively exhibiting the behavior of the probability of error as a function of word length, as will be discussed later.

In order to appreciate fully Shannon's technique, we shall devote this section to the formulation of the problem. Consider a set of M messages which are to be encoded and transmitted via a discrete memoryless channel with a finite alphabet in the presence of noise. The messages and the encoding alphabet are

$$[m] = [m_1, m_2, \ldots, m_M]$$
$$[a] = [a_1, a_2, \ldots, a_D]$$

The channel is defined as usual by its stochastic matrix:

$$[P\{Y_j|X_k\}] \qquad k, j = 1, 2, \ldots, D \tag{12-46}$$

As before, in order to combat noise, we use code words n symbols long, selected from alphabet $[a]$. We know that the per-symbol rate R of the transmission of information is closely related to M, the number of messages, and n, the word length. For the moment, assume that we have a transinformation rate in mind that is not beyond the realm of possibility. Then the word lengths are selected according to the equation

$$R = \frac{1}{n} \ln M \qquad M = e^{nR} \tag{12-47}$$

Assume that the transmitter's vocabulary has B distinct words. Our first task is to suggest a rule of correspondence between the messages and a subset of the code words. For this, Shannon chooses the following independent random encoding scheme.

Consider a wheel, the circumference of which has been partitioned into arcs numbered $u_1, u_2, \ldots, u_B$. The lengths of the arcs are selected proportional to elements of an arbitrarily specified probability row matrix:

$$[P\{u_1\}, P\{u_2\}, \ldots, P\{u_B\}]$$

The code vocabularies are composed in the following manner: We select, at random, a word u_k to correspond to the message m_1. (Messages are assumed to have equal probabilities of $1/M$.) Next, we select a word, still at random and independently of prior selections, to correspond to the message m_2, and so on. This code will be referred to as code K_1. Of course, if we do this experiment all over again it is likely to produce a code book K_2 that is not identical with K_1. Indeed there are B^M such distinct codes, some of which may be satisfactory and others rather poor. For instance, a code book in which all messages $m_1, m_2, \ldots, m_M$ have been assigned the same word u_i is a highly degenerate and useless code. Our next task is to devise a decoding scheme for every one of these B^M distinct codes and to make an estimate of the average probability of error in decoding. This is not for any particular code K_i but over the ensemble of such codes.

Consider a composite communication system having code words $u_1, u_2, \ldots, u_B$ with respective probabilities $[P\{u_1\}, P\{u_2\}, \ldots, P\{u_B\}]$, fed into the channel as described by Eq. (12-46). Knowing the noise characteristics of the channel, we can readily compute the joint probability matrix of the input and output word pairs $P\{u,v\}$. For the detection of received messages, in this composite scheme, we assume the maximum-likelihood detection criterion as before. That is, the receiving space will be partitioned into a number of regions. The regions will be assigned to the transmitted words in a one-to-one manner, based on the greatest conditional probability. A region A_i will correspond to a transmitted word u_i if, and only if, for any received word $v_k \in A_i$ we have

$$P\{u_i|v_k\} \geq P\{u_j|v_k\} \qquad j = 1, 2, \ldots, M \quad j \neq i \tag{12-48}$$

This decoding procedure will be modified to fit each of the B^M codes ($K_1, K_2, \ldots$, etc.). Suppose that a word v_0 has been received. The receiver refers to the conditional probabilities computed in the composite code

$$P\{u_\alpha|v_0\} \geq P\{u_\beta|v_0\} \geq \cdots \geq P\{u_\delta|v_0\} \geq P\{u_0|v_0\} \geq \cdots \tag{12-49}$$

written in order of magnitude, and to the particular code K that was selected at random. He then chooses the message m associated with a u in K having the highest probability in the above sequence. Now we shall consider the probability of making an error in decoding in the case where u_0 is sent and v_0 received. Suppose that a word u_0 corresponding to a message m_1 was transmitted. The receiver will make an error only if one of the u to the left of u_0 in the sequence (by order of conditional probabilities) has been assigned to a message other than m_1, or if more than one message has been assigned to u_0. The probability that a particular message, say m_2, other than m_1 was assigned a u to the left of u_0 or to u_0 in the sequence is

$$P\{u\} \qquad \text{for all } u \in S_{v_0}(u_0) \qquad (12\text{-}50)$$
$$S_{v_0}(u_0) = \{u_\alpha, u_\beta, \ldots, u_\delta, u_0\}$$

where $S_{v_0}(u_0)$ is the set of u such that $P\{u|v_0\} \geq P\{u_0|v_0\}$.

The total probability of this in the ensemble code $S_{v_0}(u_0)$ is the probability of error:

$$Q_{v_0}(u_0) = \sum_{u \in S_{v_0}(u_0)} P\{u\} \qquad (12\text{-}51)$$

The probability of having no ambiguous situations, that is, no other particular message m_2 being assigned to any words $u \in S_{v_0}(u_0)$, is

$$1 - Q_{v_0}(u_0)$$

Owing to the argument of independence of successive message encoding, the probability that no other one of the $M - 1$ messages will correspond to words in $S_{v_0}(u_0)$ is

$$[1 - Q_{v_0}(u_0)]^{M-1} \qquad (12\text{-}52)$$

Conversely, the probability of having an ambiguous or erroneous situation is

$$1 - (1 - Q_{v_0}\{u\})^{M-1}$$

This is, in a way, a measure of probability of error for any particular pair of words (u_0,v_0) in the code ensemble. The next step is to compute an average probability of error P_a for the code ensemble. Let $P\{u,v\}$ be the probability of using a particular pair of code words (u,v) in the code ensemble. The average error for the code ensemble is

$$P_a = \overline{1 - [1 - Q_v(u)]^{M-1}} = \sum_u \sum_v P\{u,v\}[1 - (1 - Q_v\{u\})^{M-1}] \qquad (12\text{-}53)$$

Note that the probability of error P_e of the best codes of the ensemble cannot be more than P_a. Thus Shannon obtains a measure of the average error probability of the code ensemble in terms of $Q_v\{u\}$. In the following section we show that P_a and $Q_v\{u\}$ are closely related to the word length n and the rate of transmission of information in the channel. A link between $Q_v\{u\}$ and the channel parameters is provided by a theorem due to Shannon.

12-9. A Relation between Transinformation and Error Probability. With the notation of the previous section, we consider the transmission of M equiprobable code words. As before, let $P\{u\}$ and $P\{u,v\}$ stand for the input and the joint probabilities of word u and pair of words (u,v), respectively. The mutual information per symbol between the words (u,v) is

$$\frac{1}{n} I(u,v) = \frac{1}{n} \log \frac{P\{u,v\}}{P\{u\}P\{v\}} \tag{12-54}$$

It is convenient to think of this quantity as a random variable Z. Each possible value of the pair (u_k,v_j) is a permissible value of the variable. Furthermore, each such value has a probability associated with it. The CDF of this variable will be designated by $F(z)$.

$$F(z) = P\left\{\frac{1}{n} I(u,v) \le z\right\} \tag{12-55}$$

The following theorem interrelates $F(z)$ with the error probability P_e of our encoding scheme.

Theorem. For a given $P\{u\}$ on a discrete memoryless channel [hence specified $F(z)$] and arbitrary $\theta > 0$, there exists a block code with prespecified M equiprobable code words such that the error probability P_e is bounded by

$$P_e \le F(R + \theta) + e^{-n\theta} \tag{12-56}$$

Proof. We divide up the (u,v) pairs in two complementary sets T and T'. If for any chosen word pair the mutual information per letter is larger than the chosen reference value of $R + \theta$, we include that pair in the set T, otherwise in T'. Analytically,

$$\begin{aligned} (u,v) \in T &\leftrightarrow \frac{1}{n} I(u;v) > R + \theta \\ (u,v) \in T' &\leftrightarrow \frac{1}{n} I(u;v) \le R + \theta \end{aligned} \tag{12-57}$$

The total probability of the random variable Z associated with the code-point pairs in T', by definition of $F(z)$, is thus equal to $F(R + \theta)$.

Now we employ the result of the previous section. That is, we consider the ensemble code and its average error probability. The average error for the code ensemble can be rewritten by dividing the summation operation described in Eq. (12-53) into two parts, first over all the elements in T and then in T'.

$$P_a = \sum_{T'} P\{u,v\}[1 - (1 - Q_v\{u\})^{M-1}] + \sum_{T} P\{u,v\}[1 - (1 - Q_v\{u\})^{M-1}] \qquad (12\text{-}58)$$

But $Q_v\{u\} \leq 1$; therefore

$$P_a \leq F(R + \theta) + \sum_{T} P\{u,v\}[1 - (1 - Q_v\{u\})^{M-1}]$$
$$P_a \leq F(R + \theta) + M \sum_{T} P\{u,v\} Q_v\{u\} \qquad (12\text{-}59)$$

[See the analogous derivation of Eq. (9-75).]

The final step required is an estimation of $Q_v\{u\}$ for $(u,v) \in T$. We note that

$$\log \frac{P\{u,v\}}{P\{u\}P\{v\}} > n(R + \theta) \qquad \text{for } (u,v) \in T$$

or

$$P\{v|u\} > e^{n(R+\theta)} P\{v\} \qquad (12\text{-}60)$$

In order to relate this inequality to $Q_v\{u\}$, observe that

$$P\{v|u'\} \geq P\{v|u\} > P\{v\}e^{n(R+\theta)} \qquad \text{for } u' \in S_v(u) \qquad (12\text{-}61)$$
$$P\{u'|v\} > P\{u'\}e^{n(R+\theta)} \qquad (12\text{-}62)$$
$$1 \geq \sum_{S_v(u)} P\{u'|v\} > e^{n(R+\theta)} Q_v\{u\}$$

Finally we find

$$e^{-n(R+\theta)} > Q_v\{u\} \qquad (12\text{-}63)$$

$$P_a \leq F(R + \theta) + M \sum_{T} P\{u,v\} e^{-n(R+\theta)}$$
$$P_a \leq R(R + \theta) + e^{nR}e^{-n(R+\theta)} \qquad (12\text{-}64)$$
$$P_a \leq F(R + \theta) + e^{-n\theta}$$

Since the average error for the ensemble code is smaller than the quantity $F(R + \theta) + e^{-n\theta}$, at least there must exist a code with an error probability less than or equal to this same quantity. Thus, the theorem is proved.

$$P_e \leq F(R + \theta) + e^{-n\theta} \qquad (12\text{-}65)$$

As n is increased, for a fixed θ (θ can be selected arbitrarily small), the term $e^{-n\theta}$ and also R become smaller and smaller. Similarly, for a fixed M, one may select n large enough to keep R and $F(R + \theta)$ arbitrarily small. Note that the probability of error decreases approximately exponentially with the increase of word length. The exponential feature of this interrelationship is an interesting result which will be further elaborated upon in the next section (see also theorems 2 and 3 of Shannon [4]).

12-10. An Exponential Bound for Error Probability. The immediate objective of this section is to integrate the derived bound on error probability in a suitable single exponential term:

$$P_e \le A_0 e^{-nB_0} \tag{12-66}$$

where A_0 and B_0 are positive constants. In this form, one can directly examine the exponential manner of achieving any desired lower error probability with increasing n. This may be done by applying an interesting inequality due to H. Chernov.* This inequality gives a bound for the probability distribution of the sum of a finite number n of identically distributed random variables in terms of n and their common moment-generating function. The inequality was first applied by Shannon to give a relation between error probability and the characteristic function of transinformation per symbol. He has also successfully applied this inequality in other instances in information-theory problems. It is hoped that the insertion of this section may prove useful for a more extensive application of Chernov's inequality in allied problems.

Simplified Chernov's Inequality. Let S_n be the sum of a finite number of independent random variables with identical moment-generating functions.

$$S = \frac{1}{n} S_n = \sum_1^n X_k$$
$$\Phi_{x_k}(t) = e^{\mu(t)} \qquad k = 1, 2, \ldots, n \tag{12-67}$$

Then $\rho(s)$, the CDF of S, will satisfy the inequality

$$\rho(\mu'(t)) \le e^{n[\mu(t) - t\mu'(t)]} \qquad \text{for } t \le 0 \tag{12-68}$$

Proof. Chernov's proof is somewhat complicated, as he actually proves stronger results than the one given here. The above simplified version of Chernov's inequality can be proved in the following simple steps.

1. Let X be a random variable with a probability density distribution $f(x)$, and $F(X)$ a real-valued function associated with X such that

$$F(x) \ge 0 \qquad \text{for } -\infty < x < \infty$$
$$F(x) \ge A > 0 \qquad \text{for } x \le a$$

Then
$$P\{X \le a\} \le \frac{1}{A} E[F(X)] \tag{12-69}$$

* H. Chernov, A Measure of Asymptotic Efficiency for Tests of a Hypothesis Based on the Sum of Observations, *Ann. Math. Statistics*, vol. 23, pp. 493–507, 1952.

The proof follows immediately from the inequalities

$$E[F(X)] = \int_{-\infty}^{\infty} F(x)f(x)\,dx \geq \int_{\{x \leq a\}} F(x)f(x)\,dx \geq A \int_{\{x \leq a\}} f(x)\,dx \tag{12-70}$$

2. As an application of the above inequality, consider the case

$$F(X) = e^{tX}$$

$$e^{tx} \geq e^{ta} > 0 \qquad \text{for } x \leq a \text{ if } t \leq 0$$

Therefore $P\{X \leq a\} \leq e^{-ta}E(e^{tX}) \qquad t \leq 0$

or $P\{X \leq a\} \leq e^{-ta}\Phi_x(t) \qquad t \leq 0 \qquad (12\text{-}71)$

3. Let $\rho(s)$ be the CDF of S; then the application of the above formula to $X = S_n$ yields

$$\rho(a) = P\{S_n \leq an\} \leq e^{-ant}(e^{\mu(t)})^n \qquad t \leq 0$$

$$\rho(a) \leq e^{n[\mu(t) - at]} \qquad t \leq 0 \tag{12-72}$$

The parameter a may be conveniently chosen:

$$\mu'(t) - a = 0$$

$$\rho(\mu'(t)) \leq e^{n[\mu(t) - t\mu'(t)]} \qquad t \leq 0 \tag{12-73}$$

As an immediate application of Chernov's inequality to the mutual information distribution, we let

$$\mu'(t) = R + \theta$$

and find

$$P_e \leq e^{n[\mu(t) - t\mu'(t)]} + e^{-n\theta} \qquad t \leq 0 \tag{12-74}$$

Furthermore, Shannon points out that the parameter θ can be suitably chosen to reduce the right-hand side of the inequality to a single exponential term:

$$\theta = t\mu'(t) - \mu(t)$$

$$P_e \leq 2e^{n[\mu(t) - t\mu'(t)]} \qquad t \leq 0 \tag{12-75}$$

This inequality relates the bound for the probability of error with the function $\mu(t)$, the logarithm of the moment-generating function of trans-information per symbol. For specified θ, $P_e \to 0$ as $n \to \infty$.

Shannon then concludes with a number of geometric results on the limiting behavior of the probability of error as n increases. This new approach seems to indicate the possibility of special applications in several different directions. These directions undoubtedly will be investigated in time. At present, Shannon's work is the principal source of reference material on this problem, although a thorough understanding of his results requires extensive preparation.

WOLFOWITZ'S PROOF

The material of the following four sections is based entirely on the contribution of J. Wolfowitz. His principal work on the subject is contained in Wolfowitz (I, II, III, IV).

12-11. The Code Book. For a given word length and error bound, Wolfowitz derives an upper bound of exponential form for the error probability. The alphabet contains a letters which may be conveniently taken as integers $[a] = [1,2,3, \ldots ,a]$. All the N encoded words $(u_1,u_2, \ldots ,u_N)$ are assumed to have equal length n and are referred to as n sequences. The received words are also n sequences. The detection scheme is based on distance criterion given in Eq. (12-80). We assume that a uniform-error-bound detection scheme exists, that is,

$$\begin{aligned} P\{v \in A'_k|u_k\} &\leq \lambda \\ P\{v \in A_k|u_k\} &\geq 1 - \lambda \end{aligned} \qquad k = 1, 2, \ldots , N \qquad (12\text{-}76)$$

Our problem is to investigate how small λ can be for a given n and N or how small n can be made for a prescribed N and λ. Consider all possible n-sequence words u to be transmitted; let $N(i|u)$ be the number of times that letter i of the alphabet occurs in a particular word u. Let $\pi = [\pi_1,\pi_2, \ldots ,\pi_a]$ be a probability distribution for the a letters of the vocabulary. Obviously, we expect on an average a number of $n\pi_i$ letters i to appear in an arbitrarily selected n sequence.

Since our final aim is to combat the effect of noise, we shall be rather selective in choosing a number of words out of the a^n possible words. We divide these words into two complementary categories, those in which the number of occurrences of any letter remains within a desired threshold value and the remainder. A good threshold level of comparison is provided by a suitable multiple of the standard deviation of the binomial distribution about $n\pi_i$ (see Sec. 7-6), that is,

$$n\pi_i \pm \sqrt{an\pi_i(1 - \pi_i)} \qquad i = 1, 2, \ldots , a \qquad (12\text{-}77)$$

On the basis of this very logical approach, Wolfowitz defines a πn sequence as a word u satisfying the inequalities

$$|N(i|u) - n\pi_i| \leq 2\sqrt{an\pi_i(1 - \pi_i)} \qquad i = 1, 2, \ldots , a \qquad (12\text{-}78)$$

This is a rather clever choice of the u words for transmission, as will be shown shortly.

Perhaps it is in order to pause a moment and ask ourselves whether it is possible to choose πn sequences, and if so, what is the total probability of the set of such words. This question is neatly answered by the following lemma:

Lemma

$$P\{\text{given } n \text{ sequence is } \pi n \text{ sequence}\} \geq \tfrac{3}{4}$$

Proof. The proof follows directly from Chebyshev's inequality (see Sec. 6-4 and Chap. 7).

$$P\{X \text{ is } \pi n \text{ sequence}\} = P\{\bigcap_{i=1}^{a} [|N(i|u) - n\pi_i| \leq 2\sqrt{an\pi_i(1-\pi_i)}]\}$$

$$= 1 - P\{\bigcup_{i=1}^{a} [|N(i|u) - n\pi_i| > 2\sqrt{an\pi_i(1-\pi_i)}]\}$$

$$\geq 1 - \sum_{1}^{a} \frac{n\pi_i(1-\pi_i)}{4an\pi_i(1-\pi_i)} = \tfrac{3}{4} \qquad (12\text{-}79)$$

Thus we have established that the supply of desired words for transmission is rather ample. Next, we need to devise a detection criterion for the transmission of these words. Let $N(i,j|u,v)$ stand for the number of times that in the kth letters of two given words u and v the letters i and j appear, respectively ($k = 1, 2, \ldots, n$). Now if u is the transmitted and v the received word, the expected number of $N(i,j|u,v)$ should be

$$N(i|u)P\{j|i\}$$

where $P\{j|i\}$ is the corresponding element of the channel probability matrix. When we look over a word pair to decide whether they are likely to correspond to each other, we again may use a threshold of comparison. Wolfowitz selects those pairs that satisfy the inequalities

$$|N(i,j|u,v) - N(i|u)P\{j|i\}| \leq \delta[N(i|u)P\{j|i\}(1 - P\{j|i\})]^{1/2}$$
$$i, j = 1, 2, \ldots, a \qquad (12\text{-}80)$$

δ is a number larger than $2a$. The reason for this selection will become more apparent later.* If a pair of words (u,v) satisfies the above requirements, then we say that v is generated (caused) by u. Of course, in the real world, v may not be caused by u, but Wolfowitz shows that this decision criterion is indeed a very intelligent one. The following lemma gives a preliminary estimate of the probability of error in our decision scheme.

12-12. A Lemma and Its Application

Lemma. Let u be any n sequence. Then

$$P\{\text{received } n \text{ sequence is generated by } u|u \text{ is transmitted}\} \geq 1 - \epsilon'$$

where $\epsilon' \leq a^2/\delta^2$.

* Using the familiar distance argument of Chap. 4, it can be seen that δ is a factor for selecting those received messages falling within a suitable radius of the transmitted message.

Proof. The desired probability is

$$\omega = P\{\bigcap_{i,j}^{a} \{|N(i,j|u,v) - N(i|u)P\{j|i\}| \leq \delta[N(i|u)P\{j|i\}(1 - P\{j|i\})]^{1/2}\}\} \quad (12\text{-}81)$$

$$\omega = 1 - P\{\bigcup_{i,j}^{a} [|N(i,j|u,v) - N(i|u)P\{j|i\}| > \delta\sqrt{N(i|u)P\{j|i\}(1 - P\{j|i\})}]\} \quad (12\text{-}82)$$

$$\omega \geq 1 - \sum_{i,j} \frac{1}{\delta^2} = 1 - \frac{a^2}{\delta^2}$$

An appropriate detection scheme can be suggested such that the probability of a correct detection is not less than ¾ and can be made higher by increasing δ.

The next natural step is to estimate the number of letters i in a word v generated by a π sequence. For this, we note that

$$N(i|v) = \sum_{j=1}^{a} N(j,i|u,v) \qquad i = 1, 2, \ldots, a \quad (12\text{-}83)$$

An upper bound for $N(i|v)$ can be obtained by recalling the upper bounds of $N(j|u)$ and $N(j,i|u,v)$ as described in Eqs. (12-78) and (12-80).

$$N(j,i|u,v) \leq [n\pi_j + 2\sqrt{an\pi_j(1 - \pi_j)}]P\{i|j\} + \delta\{[n\pi_j + 2\sqrt{an\pi_j(1 - \pi_j)}]P\{i|j\}(1 - P\{i|j\})\}^{1/2} \qquad i, j = 1, 2, \ldots, a \quad (12\text{-}84)$$

On summing and replacing $1 - \pi_j$ and $1 - P\{i|j\}$ by 1, one finds

$$N(i|v) \leq n\sum_j \pi_j P\{i|j\} + \sum_j 2\sqrt{an\pi_j}\,P\{i|j\} + \delta\sum_j [(n\pi_j + 2\sqrt{an\pi_j})P\{i|j\}]^{1/2} \quad (12\text{-}85)$$

Let

$$\sum_j \pi_j P\{i|j\} = \text{probability that element } i \text{ is in the } k\text{th place of } v = \pi'_i \qquad i = 1, \ldots, a$$

Hence, after some algebraic manipulation,

$$N(i|v) < n\pi'_i + 2\sqrt{an}\sum_j \sqrt[4]{\pi_j P\{i|j\}} + 2a\delta\sqrt{n}\sum_j \sqrt[4]{\pi_j P\{i|j\}} \quad (12\text{-}86)$$

(Note that for any number $0 < x < 1$, $\sqrt[4]{x} > \sqrt[2]{x}$.)

This upper bound of $N(i|v)$ can be simplified in several ways. For instance, one may wish to apply Hölder's inequality* to obtain

$$N(i|v) < n\pi_i' + 3a^2 \sqrt{n}\,(1 + \delta)\,\sqrt[4]{\pi_i'} \tag{12-87}$$

The corresponding lower bound can be derived in an analogous fashion:

$$n\pi_i' - 3a^2 \sqrt{n}\,(1 + \delta)\,\sqrt[4]{\pi_i'} < N(i|v) < n\pi_i' + 3a^2 \sqrt{n}\,(1 + \delta)\,\sqrt[4]{\pi_i'} \tag{12-88}$$

Thus, Wolfowitz succeeds in obtaining relatively simple and informative upper and lower bounds for $N(i|v)$. For brevity, we refer to these bounds as v_{iu} and v_{il}. Translation of this result in terms of probability is that the probability of receiving a word v, owing to the independence of its composing letters, will be bounded by the inequalities

$$\prod_i \left(\pi_i'^{\,v_{iu}}\right) < P\{v\} < \prod_i \left(\pi_i'^{\,v_{il}}\right) \tag{12-89}$$

or

$$\exp \Sigma(v_{iu} \log \pi_i') < P\{v\} < \exp \Sigma(v_{il} \log \pi_i')$$

Finally, because of Eq. (12-88), we find

$$\exp\,[-nH(\pi') - k_1(1 + \delta)\sqrt{n}] < P\{v\} < \exp\,[-nH(\pi') + k_1(1 + \delta)\sqrt{n}] \tag{12-90}$$

where

$$H(\pi') = -\Sigma\pi_i' \log \pi_i'$$

$k_1 > 0$ is Wolfowitz's constant.

12-13. Estimation of Bounds. A final phase of Wolfowitz's technique requires an estimate of the number of words in the set U of selected words and in the total set V of words generated by U. To this end, we must translate these probabilistic bounds into bounds on the number of elements of a set. The following lemma serves this purpose.

A Combinatorial Lemma. Let $A \subset U$ be some finite set A of elements u of a probability space satisfying the inequalities

$$\begin{aligned} 0 < m \le P\{u\} < M \qquad &\text{for } u \in A \\ 0 \le r \le P\{A\} \le R \le 1 \qquad & \end{aligned} \tag{12-91}$$

Then, $N(A)$, the number of elements of the set A, satisfies

$$\frac{r}{M} \le N(A) \le \frac{R}{m}$$

* See, for instance, G. H. Hardy, J. E. Littlewood, and G. Polya, "Inequalities," chap. 2, Cambridge University Press, New York, 1952.

Proof. By definition,

$$P\{A\} = \sum_{u \in A} P\{u\}$$

Then it is evident that

$$N(A) \cdot m \leq P\{A\} \leq N(A) \cdot M \qquad (12\text{-}92)$$

$$\frac{r}{m} \leq N(A) \leq \frac{R}{m}$$

Let $B_2(v|\pi)$ be the number of n sequences generated by *all* π sequences. Then in the above lemma we let

$$\begin{aligned} m &= \exp\,[-nH(\pi') - k_1(1+\delta)\sqrt{n}] \\ M &= \exp\,[-nH(\pi') + k_1(1+\delta)\sqrt{n}] \\ R &= 1 \qquad r = 9/16 \end{aligned} \qquad (12\text{-}93)$$

The following inequality thus holds:

$$9/16 \exp\,[nH(\pi') - k_1(1+\delta)\sqrt{n}] < B_2(v|\pi) < \exp\,[nH(\pi') + k_1(1+\delta)\sqrt{n}] \qquad (12\text{-}94)$$

or by letting $k_2 = k_1 - \log 9/16 > 0$:

$$\exp\,[nH(\pi') - k_2(1+\delta)\sqrt{n}] < B_2(v|\pi) < \exp\,[nH(\pi') + k_2(1+\delta)\sqrt{n}] \qquad (12\text{-}95)$$

Our next task is to find similar bounds for $B_1(\ |u)$, the number of n sequences generated by *any* π sequence. The computation can be achieved in a manner analogous to the foregoing. That is, first we find bounds on probabilities of the associated sets and then apply the last lemma. For instance, the probability of any n sequence generated by any π sequence is

$$P\{v|u\} = \prod_{i,j} P\{j|i\}^{N(i,j|u,v)}$$

In a direct manner we find

$$P\{v|u\} < \exp\Big[n \sum_{i,j} \pi_i P\{j|i\} \log P\{j|i\} - \sqrt{n}\,(2a+\delta) \sum_{i,j} \sqrt{P\{j|i\}} \log P\{j|i\}\Big]$$

$$P\{v|u\} > \exp\Big[n \sum_{i,j} \pi_i P\{j|i\} \log P\{j|i\} + \sqrt{n}\,(2a+\delta) \sum_{i,j} \sqrt{P\{j|i\}} \log P\{j|i\}\Big] \qquad (12\text{-}96)$$

Therefore,

$$\exp\Big[n \sum_i \pi_i H(\ |i) - \sqrt{n}\,(2a+\delta)k_3\Big] < B_1(\ |u) < \exp\Big[n \sum_i \pi_i H(\ |i) + \sqrt{n}\,(2a+\delta)k_3\Big] \qquad (12\text{-}97)$$

where k_3 is an appropriate positive constant and $H(\ |i)$ is the channel's conditional entropy for a specified i.

12-14. Completion of Wolfowitz's Proof. Now we are in a position to present Wolfowitz's proof of the fundamental theorem of information theory for a discrete memoryless noisy channel.

Theorem. There exists a code book containing at least N words of length n such that for any arbitrarily specified probability of error λ $(0 < \lambda \leq 1)$ there is a decoding scheme with a uniform error probability λ and

$$N \geq \exp\,(nC - k\sqrt{n}) \tag{12-98}$$

where k is a constant depending on λ but not on the channel and C is the channel capacity.

Proof. As usual, we partition the receiving space into disjoint regions $A_1, A_2, \ldots, A_N$ such that, if the sequence u_i is sent,

$$\lambda_{u_i} = 1 - P\{v_i \in A_i\} \leq \lambda \qquad i = 1, 2, \ldots, N \tag{12-99}$$

From previous discussions we know that such a code exists. The major question here is to find out if the code book can be made as large as Eq. (12-98) suggests. Attention is called to Eq. (12-82), where δ can be conveniently chosen as $a\sqrt{2/\lambda}$, or

$$\epsilon' = \frac{a^2}{\delta^2} = \frac{\lambda}{2}$$

Now let u_i $(i = 1, 2, \ldots, N)$ be π sequences achieving the maximum of transinformation C. The detection regions will be selected as

$$v \in A_i \qquad \text{if } v \text{ is an } n \text{ sequence generated by } u_i \qquad i = 1, 2, \ldots, N$$

Furthermore, we assume that the code is a maximal code in the sense that an element u_{N+1} cannot be added without raising the probability of error λ. Let u_0 be a π sequence which achieves the maximum of transinformation through the channel. If u_0 is not included in the set $u_1, u_2, \ldots, u_N$, then

$$P\{(v \text{ generated by } u_0) \in \bigcup_{k=1}^{N} A_k\} > \frac{\lambda}{2} \tag{12-100}$$

Indeed, if this was not the case, one could add u_0 to the code book and associate with it the corresponding generated words of length n which are not in $\bigcup_{k=1}^{N} A_k$. According to previous lemmas, the number of n sequences in the set $\bigcup_{k=1}^{N} A_k$ is not less than

$$\frac{3}{4}\frac{\lambda}{2}\exp\,[nH(\bar{\pi}') - k_1(1 + \delta)\sqrt{n}] \tag{12-101}$$

and not more than

$$N \exp \left[n \sum_i \bar{\pi}_i H(\ |i) + \sqrt{n}\,(2a + \delta)k_3 \right] \tag{12-102}$$

where $\bar{\pi}'$ and $\bar{\pi}$ are, respectively, the output and the input probability row vectors achieving maximum transinformation. Thus we find

$$N > \tfrac{3}{8}\lambda \exp \{nC - \sqrt{n}\,[k_1(1 + \delta) + (2a + \delta)k_3]\} \geq \exp (nC - k\sqrt{n}) \tag{12-103}$$

with $\quad k = k_1(1 + \delta) + (2a + \delta)k_3 - \log \tfrac{3}{8}\lambda$

From this proof, one can conclude that, for any specified word length n, one can find a code book containing at least $2^{n(C-\epsilon)}$ words and the probability of error is also of the exponential form $c_1 2^{-nC_2}$. For the detail and the values of constants c_1 and c_2 in terms of the previous constant, see Wolfowitz (IV).

A strong converse of the above theorem was first given by Wolfowitz. The converse of this theorem states that for a given n and λ it is impossible to devise code books with a given arbitrarily small error and containing more than $\exp (nC + k'\sqrt{n})$ words. The constant k' depends generally on λ but not on the channel. For the sake of reference, the two versions of the converse theorem are given below.

Weak Converse. The fundamental theorem does not hold for

$$\log N \leq \frac{nC + 1}{1 - \lambda} \tag{12-104}$$

A proof for the weak converse can be directly derived from Eqs. (12-12).

$$nC \geq H(U) - H(U|V) \geq \log N - \lambda \log (N - 1) - 1$$

$$\lambda \geq 1 - \frac{nC + 1}{\log N} \tag{12-105}$$

Or, if we let $N = 2^{nH}$,

$$\lambda \geq 1 - \frac{nC + 1}{nH}$$

$$\lim_{n\to\infty} \lambda \geq 1 - \frac{C}{H} \tag{12-106}$$

Obviously for sufficiently large n, $\lambda \to 0$ if $H < C$. This is not the case for $N > C$.

Strong Converse. For any specified $0 \leq \lambda < 1$ we have

$$\limsup_{n\to\infty} \frac{1}{n} \log N(n,\lambda) \leq C \tag{12-107}$$

For proof of this important theorem and its application to BSC see the cited reference and Feinstein (II).

The fundamental theorem of discrete memoryless channels and its converse have been extended to the case of channels with finite memory. Channels with memory have been studied under stationary and nonstationary regimes. These extensions require a degree of mathematical sophistication beyond the scope of this text. The mathematically inclined reader is in the fortunate position of having at his disposal a comprehensive literature on this subject. During the past few years a number of competent mathematicians in the United States, the Union of Soviet Socialist Republics, and Europe have made considerable headway in solidifying and generalizing Shannon's original ideas. Besides the mathematical references mentioned earlier in this book, a number of articles are included in the Bibliography at the end of the book. Articles by the following authors are among those of special interest: Blackwell, Breiman, and Thomasian; Dobrushin;* Feinstein; Gel'fand, Kolmogorov, and Iaglom; Gel'fand and Iaglom; A. Rényi; and Rozenblatt-Rot.

From an engineering point of view, it is of prime interest to examine the physical context of the recent mathematical developments. In light of this, the ultimate performance and the merit of our communication systems and techniques should be reevaluated. However, our journey has come to an end. It is hoped that the present work will stimulate the reader to pursue further investigations leading to possible applications of the theory.

* R. L. Dobrushin, General Formulation of Shannon's Fundamental Theorem in Information Theory, *Doklady Akad. Nauk S.S.S.R.*, vol. 126, no. 3, pp. 474–477, 1959; translation in *Automation Express*, January, 1960.

CHAPTER 13

GROUP CODES

13-1. Introduction. The fundamental theorems of information theory described in the preceding chapter are, in essence, realizability theorems. They prove the existence of a code book for the transmission of information at a rate not higher than the channel capacity with any arbitrarily small error probability. They do not show, however, methods for obtaining such valuable encoding procedures. The synthesis of encoding procedures is in itself a subject of great professional interest with a rapidly growing technical literature. In fact, a far greater number of papers on coding theory are available than on information theory per se. The contribution of these papers has created an area of investigation (generally referred to as coding theory) which appears at present to be distinct from information theory.

This chapter is not a sequel to the preceding chapter. From a pedagogical point of view, this chapter may be included at any time subsequent to Chap. 4, whenever a digression from information theory would be welcomed.

The area of information-processing machines, such as digital computers, is a most important field of application for binary codes. Fortunately, at present, there is a large variety of error-correcting binary codes available. The discovery of these codes has been greatly stimulated by the development of information theory, although coding theory seems to be somewhat directed toward the mechanical implementation of the codes.

The abundant coding literature appears to be undergoing a major fundamental organization. A shift toward general methods rather than individualized code books seems imminent. The work of a large number of scientists, such as R. W. Hamming and D. Slepian, to mention just two, has contributed to this mathematical integration.* However, relatively speaking, the mathematical theory of codes appears to be in an early stage of development.

Our objective in this chapter is to present, in brief, some developments in the field of the so-called group codes. This will provide the reader an

* For a comprehensive list of contributors see the bibliography of coding prepared by A. B. Fontaine and W. W. Peterson, Coding Newsletter 60.2.

opportunity to examine one of the possible present applications of information theory. We shall not attempt a complete discussion of the practical use of these codes. The reader interested in the theory of codes and their implementation will find adequate treatment in the literature.

We begin with a brief presentation of systematic or minimum-distance parity binary codes discovered by Hamming and supplemented by a large number of interesting papers in the past decade. The study of systematic codes may be carried on in several directions, for instance:

1. The size of the code book
2. Methods for the selection of code words
3. The error rate in the transmission

Item 1 has been extensively studied. Many combinatorial results have been derived, and more seem to be forthcoming. The study of the number of elements in different systematic codes is, in essence, of a combinatorial nature. Item 2 is to indicate how one may select a suitable code book. The group code of Slepian appears to be the first important work in this direction. Slepian's work makes use of some concepts of groups, fields, and rings. The third direction constitutes an evaluation of the rate of transmission of information for some systematic codes (for instance, the works of P. Elias and D. Slepian). As far as the objectives of this chapter are concerned, we wish briefly to acquaint the reader with these developments. To this end, an elementary acquaintance with the concepts of groups, fields, and rings is highly desirable. Sections 13-2 and 13-3 will provide such a necessary review background. Section 13-4 applies the content of the previous sections to systematic codes in general. Section 13-5 describes Hamming's codes in brief. The important concept of group codes is described in Sec. 13-6. The immediately following section is devoted to a detection scheme for group codes. Section 13-10 presents some of the results obtained on maximum size of Hamming codes. The material of the latter section is not directly related to the rest of the chapter, although of great interest to those engaged in research in coding theory.

13-2. The Concept of a Group. A set of elements denoted by G is called a group if it has a well-defined *binary operation*, which we denote by $\odot$, and an *equivalence relation* which is equal to satisfying the following conditions, known as *group axioms:*

1. Closure: For every $a, b \in G$

$$a \odot b \in G$$

2. Associative law of products: For every $a, b, c \in G$

$$(a \odot b) \odot c = a \odot (b \odot c)$$

3. Existence of an *identity element* $e \in G$: For each $a \in G$

$$a \odot e = e \odot a = a$$

4. Existence of an *inverse element:* For each $a \in G$ there exists an element $a^{-1} \in G$ such that

$$a \odot a^{-1} = a^{-1} \odot a = e$$

If, besides these axioms, the elements of the group satisfy the commutative law of products, then the group will be referred to as a commutative or *Abelian group:*

$$a \odot b = b \odot a$$

As a simple example, note that the rational numbers, with the exclusion of zero, form a group under the ordinary multiplication and definition of the equality of numbers. The number 1 is the identity element of the group, and the exclusion of 0 guarantees the existence of an inverse for every member of the group. Furthermore, these numbers form an Abelian group.

As a second example, one can verify that the set of the following six matrices forms a group under ordinary matrix product and equality.

$$\begin{bmatrix} 1 & 0 \\ 0 & 1 \end{bmatrix} \quad \begin{bmatrix} 0 & 1 \\ 1 & 0 \end{bmatrix} \quad \begin{bmatrix} 1 & 0 \\ -1 & -1 \end{bmatrix} \quad \begin{bmatrix} 0 & 1 \\ -1 & -1 \end{bmatrix} \quad \begin{bmatrix} -1 & -1 \\ 1 & 0 \end{bmatrix} \quad \begin{bmatrix} -1 & -1 \\ 0 & 1 \end{bmatrix}$$

In fact, axioms 1 and 2 can be directly verified. The first matrix may be selected as the identity element; since all matrices are nonsingular, the verification of the fourth axiom is immediate.

Two groups, G and G', are said to be *simply isomorphic* or *equivalent* if there is a one-to-one correspondence between their elements such that

$$\begin{array}{ll} \text{if} & a \in G \to a' \in G' \\ \text{and} & b \in G \to b' \in G' \\ \text{then} & a \odot b \in G \to a' \odot b' \in G' \end{array}$$

If a nonempty subset of the elements of a group under the same product law themselves form a group, they are referred to as a *subgroup* of the original group.

The associative law of products implies that we can form different powers of elements of a group:

$$a \odot a = a^2$$
$$a \odot a^2 = a^2 \odot a = a^3$$
$$\cdots\cdots\cdots\cdots$$

A group which consists of a single element and its powers is called a *cyclic group.* Evidently, cyclic groups are Abelian. *The order of a group* is the number of elements of the group. A group of finite order is referred to as a *finite group.*

Equivalence Relation. Given a set S, we say that a relation of equivalence has been established on elements of S if it is possible to assert whether any two arbitrary elements a and b are equivalent or not. Two arbitrary elements a and b of S are said to have an equivalence relation ($a \sim b$ in symbols) if the relation satisfies the following conditions:

1. Reflexive: that is, true of a and a
2. Symmetric: that is, true for b and a if it is true for a and b
3. Transitive: that is, true for a and c if it is true for a and b and for b and c

An equivalence relation divides the elements of a set into disjoint subsets of equivalent elements.

Now we shall apply the concept of equivalence to groups. Let H be a subgroup of a group G. Two elements a and b of G are said to be *right congruent* if, and only if, there exists an element c in H such that

$$a = b \odot c$$

Similarly, by left congruence we mean $a = c \odot b$.

A congruence relationship with respect to a subgroup H of a group G satisfies all three requirements of equivalence. Thus, one can see that the elements of G are divided into equivalent classes. These equivalent classes are referred to as *cosets.*

For the proof, let

$$a \in G \qquad b \in G$$

Then $a \sim b$ is an equivalence relation if and only if $ab^{-1} \in H$. In fact,

1. $a \in G$ implies $a \sim a$ and $aa^{-1} = e \in H$.
2. $ab^{-1} \in H$ implies $ba^{-1} \in H$ as $ba^{-1} = (ab^{-1})^{-1}$.
3. $\left.\begin{array}{l} ab^{-1} \in H \\ bc^{-1} \in H \end{array}\right\}$ imply $ac^{-1} \in H$ since $(ab^{-1})(bc^{-1}) = ac^{-1}$.

It can be shown that the order of a subgroup H of a finite group G is a divisor of the order of G. In fact, let the order of G and H be g and h, respectively. Let the elements of H be

$$\{H\} = \{a_1,a_2, \ . \ . \ . \ ,a_h\}$$

If b_1 is an element of G but not of H, then the set H_1 of order h has all distinct elements which are also distinct from the elements of H.

$$\{H_1\} = \{b_1 \odot a_1, b_1 \odot a_2, \ . \ . \ . \ , b_1 \odot a_h\}$$

If $\{H\}$ and $\{H_1\}$ do not exhaust G, select an element $b_2 \in G$ but $b_2 \notin H \cup H_1$ and consider the set H_2:

$$\{H_2\} = \{b_2 \odot a_1, b_2 \odot a_2, \ . \ . \ . \ , b_2 \odot a_h\}$$

Of course, elements of $\{H_2\}$ are distinct from elements of $\{H_1\}$ and $\{H\}$. Continuing this process, we find that the elements of G will be divided into k cosets each having h elements. Thus

$$g = hk$$

13-3. Fields and Rings. In the preceding section, we have outlined the definition of a group and a subgroup. This section presents the basic definition of a *field* and a *ring*. While the content of Sec. 13-2 is essential for the study of the elements of group codes, the material of this section is rather optional. Its inclusion may prove to be a convenient reference item for those who wish to consult some of the most recent articles on group codes.

A system of elements is said to form a field if (the equality of the elements) two operations (say addition and multiplication) are defined and if for any $a, b, c, x \in F$ the following laws are valid. (The laws given here are not presented in the form of a minimal set of postulates. Slight redundancy has been injected for pedagogical reasons.)

1. Closure: $a + b \in F$
2. Associativity:

$$(a + b) + c = a + (b + c)$$

3. Additive identity: There exists $z \in F$ such that

$$x + z = z + x = x$$

4. Additive inverse: For each $x \in F$ there exists $x^* \in F$

$$x + x^* = x^* + x = z$$

5. Commutativity:

$$a + b = b + a$$

1. Closure: $a \cdot b \in F$
2. Associativity:

$$(a \cdot b) \cdot c = a \cdot (b \cdot c)$$

3. Multiplicative identity: There exists $u \in F$ such that

$$x \cdot u = u \cdot x = x$$

4. Multiplicative inverse: For each $x \neq z$, there exists an inverse $x^{-1} \in F$ such that

$$x \cdot x^{-1} = x^{-1} \cdot x = u$$

5. Commutativity:

$$a \cdot b = b \cdot a$$

Distributive Laws:

$$a \cdot (b + c) = a \cdot b + a \cdot c$$
$$(b + c) \cdot a = b \cdot a + c \cdot a$$

A simple example of a field can be given by the set of all complex numbers $a + \sqrt{-1}\, b$ under ordinary addition and multiplication. The identity element z in this case is $(0 + \sqrt{-1}\ 0)$ and $u = 1 + \sqrt{-1}\ 0$.

As another example, let us investigate whether under ordinary matrix operations the set of all matrices of the type

$$A = \begin{bmatrix} x & y \\ -y & x \end{bmatrix} \qquad x, y \text{ real numbers}$$

forms a field. To this end, let $B = \begin{bmatrix} a & b \\ -b & a \end{bmatrix}$ and note that the closure and associative laws are satisfied:

$$A + B = B + A = \begin{bmatrix} x + a & y + b \\ -(y + b) & x + a \end{bmatrix}$$

$$AB = BA = \begin{bmatrix} xa - yb & xb + ya \\ -(xb + ya) & xa - yb \end{bmatrix}$$

As to the zero and unit element, one may write

$$z = \begin{bmatrix} 0 & 0 \\ 0 & 0 \end{bmatrix} \qquad u = \begin{bmatrix} 1 & 0 \\ 0 & 1 \end{bmatrix}$$

The multiplicative inverse law has not yet been examined. For the moment, let B be the inverse element, $AB = u$; then, since the multiplication and the unit matrix are already defined, we find

$$\begin{cases} xa - yb = 1 \\ xb + ya = 0 \end{cases} \qquad \begin{cases} a = \dfrac{x}{x^2 + y^2} \\ b = \dfrac{-y}{x^2 + y^2} \end{cases}$$

Thus if B is selected as the inverse matrix in the familiar sense, the inverse matrix is also of the desired form. Since this inverse exists for every A except $A = z$, the above set of matrices forms a field.

The idea of a mathematical system called a *ring* simply follows from the concept of a field. All that is required is to relax the laws of multiplicative identity, inverse, and commutativity. Removal of laws 3 to 5 in the right-hand column in the list of field postulates leads to a ring. Obviously, a ring is a more general relationship than that of a field. For instance, the set of all $n \times n$ matrices with real elements forms a ring.

13-4. Algebra for Binary n-digit Words. In the light of the above development, it becomes apparent that elements of the set S_n consisting of all n-digit sequences form a group under $\oplus$. In fact, for any n-sequence word pairs a and b that are members of S_n we have

$$\begin{gathered} a \odot b = a \oplus b \in S_n \\ (a \odot b) \odot c = a \oplus (b \oplus c) \in S_n \\ e \odot a = e \oplus a = a \qquad \text{where } e = (000 \cdots 0) \\ a \odot a^{-1} = a \oplus a^{-1} = e \\ a^{-1} = a \in S_n \end{gathered} \tag{13-1}$$

Note that the above n sequences form an Abelian group of order 2^n.

In order to verify whether the ring algebra holds, take the above $\oplus$ definition for the addition operation and define the product of two words $a \in S_n$ and $b \in S_n$ as a word $a \cdot b$ whose digits are the product of the matching digits of a and b. With this rule in mind, we note that

$$\begin{aligned} a \cdot b &= b \cdot a \in S_n \\ (a \cdot b) \cdot c &= a \cdot (b \cdot c) \\ a \cdot (b \oplus c) &= a \cdot b \oplus a \cdot c \end{aligned} \tag{13-2}$$

The inversion law is not satisfied. The algebra is a commutative ring algebra.

If the word ϕ containing n zero digits is taken as the point of origin of an n-dimensional euclidean space $\phi = (0,0, \ldots ,0)$, the weight of an element $a \in S_n$ is defined as the number of 1's in a:

$$D(a,\phi) = D(a \oplus \phi) = ||a||^* \tag{13-3}$$

Then the following relation is self-evident:

$$||(a \oplus c) \oplus (b \oplus c)|| = ||(a \oplus b) \oplus (c \oplus c)|| = ||a \oplus b|| \tag{13-4}$$

$D(a,b)$ is also referred to as the Hamming distance between points a and b. Let $D(u,v)$ be the Hamming distance between u and v; then

$$\begin{aligned} D(u,v) &= D(v,u) = ||u \oplus v|| \\ D(u,u) &= ||u \oplus u|| = ||\phi|| = 0 \end{aligned} \tag{13-5}$$

The following relations are easy to prove. The reader may wish to offer a proof as an exercise.

$$\begin{aligned} ||a \cdot b|| &\leq \min [||a||,||b||] \\ ||a \oplus b|| &= ||a|| + ||b|| - 2||a \cdot b|| \end{aligned} \tag{13-6}$$

A set of n-digit words with k parity checks may form a subgroup of the Abelian group S_n, as will be discussed shortly. For example, in the case of $n = 5$ and $k = 3$, we have 2^3 words in the code book which form a subgroup of all possible 32 five-digit words.

$$\begin{aligned} e &= 0\,0\,0\,0\,0 \qquad & u_4 &= 1\,1\,0\,0\,1 \\ u_1 &= 0\,0\,1\,1\,1 & u_5 &= 0\,0\,1\,0\,0 \\ u_2 &= 1\,1\,1\,1\,0 & u_6 &= 1\,1\,1\,0\,1 \\ u_3 &= 0\,0\,0\,1\,1 & u_7 &= 1\,1\,0\,1\,0 \end{aligned}$$

The above eight words could be considered as points in the five-dimensional euclidean space. Thus, according to Sec. 13-2, the group can be

* For brevity we make use of $||a||$ in lieu of "weight of a." This notation should not be confused with the ordinary euclidean norm.

expressed in terms of its cosets as

e	u_1	u_2	u_3	u_4	u_5	u_6	u_7
00000	00111	11110	00011	11001	00100	11101	11010
10000	10111	01110	10011	01001	10100	01101	01010
01000	01111	10110	01011	10001	01100	10101	10010
00010	00101	11100	00001	11011	00110	11111	11000

Example 13-1. Verify the different operations defined in Sec. 13-4 for the binary numbers

$$a = 11101 \qquad b = 01000 \qquad c = 01001$$

Solution. It is easy to show that a, b, and c are elements of the binary group S_5 and that the rules of ring algebra hold.

$$\begin{aligned}
a \oplus b &= 10101 \qquad a \oplus c = 10100 \qquad b \oplus c = 00001 \\
(a \oplus b) \oplus c &= 10101 \oplus 01001 = 11100 \\
a \oplus (b \oplus c) &= 11101 \oplus 00001 = 11100 \\
e \oplus a &= 00000 \oplus 11101 = 11101 \\
a \oplus a^{-1} &= 11101 \oplus 11101 = 00000 \\
a \cdot b &= 01000 \qquad b \cdot c = 01000 \qquad a \cdot c = 01001 \\
(a \cdot b) \cdot c &= (01000) \cdot (01001) = 01000 \\
a \cdot (b \cdot c) &= (01000) \cdot (01000) = 01000 \\
a \cdot (b \oplus c) &= (11101) \cdot (00001) = 00001 \\
a \cdot b \oplus a \cdot c &= 01000 \oplus 01001 = 00001
\end{aligned}$$

To verify the observations about the weight of these elements, one writes

$$\|a\| = D(a,\phi) = 4 \qquad \|b\| = 1 \qquad \|c\| = 2$$
$$\|a \oplus b\| = 3 \qquad \|a \oplus c\| = 2 \qquad \|b \oplus c\| = 1$$
$$\|(a \oplus c) \oplus (b \oplus c)\| = \|10100 \oplus 00001\| = \|10101\| = 3$$

Also note that, in accordance with Eq. (13-6), we have

$$\|a \cdot b\| = 1 \qquad 1 = \min\,[4,1] \qquad 3 = 4 + 1 - 2 \times 1$$

13-5. Hamming's Codes. Before proceeding to a discussion of group codes, it now may be worthwhile to pause and apply our newly acquired concepts to Hamming's codes. This slight digression should shed more light on the material discussed in the latter part of Chap. 4.

We consider n-digit binary words as points in the n-dimensional euclidean space. By a set $A(n,d)$ we imply the optimal set of such points with a minimum mutual distance d, that is, a set containing A elements such that no additional n-digit word can be incorporated in the set without destroying the property of minimal mutual distance d among its elements. For example, the elements below constitute a set $A_1(6,3)$.

000000	101010
010101	111111

Of course, generally we may have several optimal sets $A_1(n,d)$, $A_2(n,d)$, etc. For instance, the set $A_2(6,3)$ below is also an optimal set:

000000	111000
010101	001011
100110	111111

Among all optimal sets $A(n,d)$, the one that contains the largest number of elements will be denoted by $B(n,d)$. For example, it can be shown without difficulty that there are eight elements in $B(6,3)$:

000000	010101	111000	101101
100110	110011	011110	001011

In the light of the developments of Secs. 13-1 to 13-3, we now reexamine these optimal sets of elements and the technique associated with the Hamming codes.

A set S of binary n sequences is said to be a Hamming d-minimum-distance set if for every two distinct words u and v in S we have

$$||u \oplus v|| \geq d \qquad d \text{ is a positive integer}$$

Let S_n denote the set of all 2^n binary words of length n. Each word has a point representation in n space and may be considered as a vertex of a unit cube in euclidean n space. A set of words $S \subset S_n$ may be called a δ-error-correcting or δ-error-detecting code if for every pair of distinct elements u_i and u_j in S we have, respectively,

$$D(u_i,u_j) \geq d = 2\delta + 1 \qquad (13\text{-}7)$$
$$D(u_i,u_j) \geq d = 2\delta$$

The set of all n sequences S_n forms an Abelian group under the modulo 2 addition. If the set S is a subgroup of the group S_n, then S is called a group code. As will be shown later, a d-minimum-distance code need not be a group code in general.

Hamming's parity distance code (called systematic code) can be effectively used for transmitting information in the presence of independent noise. The general concept of the transmission method is to consider the correction of δ or less independent errors for a specified word length. Then, by letting $d = 2\delta + 1$, choosing *the* maximal set $B(n,d)$, and accepting the minimum-distance technique in decoding, one is able to correct up to δ single errors. A point a in the n space which is closer than δ to a transmitted signal u_k will be decoded as u_k. A discussion of Hamming's systematic codes and a tabulation of $B(n,d)$ for $n \leq 17$ and $\delta \leq 6$ are given by A. E. Laemmel* (see also Secs. 4-13 and 4-14).

* A. E. Laemmel, Efficiency of Noise-reducing Codes, "London Symposium on Communication Theory 1952," pp. 111–118, Academic Press, Inc., New York, 1953. See also N. Wax, *IRE Trans. on Inform. Theory*, vol. IT-5, pp. 168–174, 1959.

In the above treatment we have discussed the class of error-correcting codes. The same general idea is applicable to the class of error-detecting codes. That is, given a δ-error-detecting scheme, the minimum distance among the words of this code is not less than $e = 2\delta$. The elimination of a character from every code word can at most reduce their distances by one unit. Thus we obtain a code $(n - 1, d = 2\delta - 1)$. In fact, the relation between the two classes is

$$A(n,e) = A(n - 1, e - 1)$$

Some of Hamming's results are as follows:

Hamming's theorem:

$$\begin{aligned} &\text{(I)} && B(n,1) = 2^n \\ &\text{(II)} && B(n,2) = 2^{n-1} \\ &\text{(III)} && B(n,2k) = B(n-1, 2k-1) \\ &\text{(IV)} && B(n, 2k+1) \leq \frac{2^n}{1 + \binom{n}{1} + \cdots + \binom{n}{k}} \end{aligned} \tag{13-8}$$

The proof of the validity of these relations follows directly, for instance, from error-correcting considerations as discussed in Chap. 4.

A proof of (IV) may be suggested by contradiction. Let S be the set of all the representative points of the code book and $S_x \subset S$ and $S_y \subset S$ two subsets containing all points in S_n not farther than δ from two distinct code points x and y, respectively. Evidently $S_x \cap S_y = \phi$; otherwise their intersection would contain at least one point $z \in S$ such that

$$||z - x|| \leq \delta \qquad ||z - y|| \leq \delta$$

Such a point z cannot belong to S since by hypothesis $||x - y|| > 2\delta$. Note that each sphere contains $1 + \binom{n}{1} + \binom{n}{2} + \cdots + \binom{n}{\delta}$ points; hence the number of points in S is precisely the number of such spheres. Since the total number of points in S_n is 2^n, then the number of spheres cannot exceed the right-hand side of (IV).

Hamming's code can best be defined in terms of its parity-check matrix $[A] = [a]$. This is a matrix with element values restricted to 0 and 1, having $n - k$ rows and k columns. Each word to be transmitted over BSC consists of $n - k$ parity checks and k information digits. Without loss of generality, we may assign the last $n - k$ digits to parity checks.

$$u = \alpha_1, \alpha_2, \ldots, \alpha_k; \beta_1, \beta_2, \ldots, \beta_{n-k}$$

For a given parity-check matrix and a k sequence, a check digit β_i of u is obtained through the following equation:

$$\beta_i = \sum_{j=1}^{k} a_{ij}\alpha_j \qquad i = 1, 2, \ldots, n-k \quad (\text{mod } 2) \tag{13-9}$$

Symbolically we may write

$$[\beta] = [A][\alpha] \tag{13-10}$$

where the matrices $[\beta]$, $[A]$, and $[\alpha]$ are $(n-k) \times 1$, $(n-k) \times k$, and $k \times 1$, respectively. When $[\alpha]$ runs over all possible k sequences, Eq. (13-10) provides the parity-check parts of the words to be transmitted.

As an example, consider the case of a single-error correcting block code

$$n = 7 \qquad k = 4 \qquad n - k = 3$$

The matrix A may be selected to be

$$[A] = \begin{bmatrix} 1 & 0 & 1 & 1 \\ 1 & 1 & 1 & 0 \\ 1 & 1 & 0 & 1 \end{bmatrix}$$

For instance, if the information digits of a word are specified as $\alpha_1\alpha_2\alpha_3\alpha_4$, then the parity checks will be given by

$$\begin{aligned} \beta_1 &= \alpha_1 + \alpha_3 + \alpha_4 \\ \beta_2 &= \alpha_1 + \alpha_2 + \alpha_3 \\ \beta_3 &= \alpha_1 + \alpha_2 + \alpha_4 \end{aligned}$$

The problem of devising systematic codes with a specified mutual distance is thus equivalent to that of finding a suitable parity-check matrix A such that it generates a set of words with the required property. This problem has been discussed at length in the coding literature (see, for instance, M. Golay, G. E. Sacks, R. Chien, E. J. McCluskey, Jr.,* and A. G. Koinheim).

As a second example, consider the case of $n = 5$, $k = 2$, $d = 3$. The following 3×2 matrix is a suitable parity-check matrix.

$$[A] = \begin{bmatrix} 1 & 1 \\ 1 & 0 \\ 0 & 1 \end{bmatrix}$$

* McCluskey states that a parity-check matrix is suitable for a minimum-distance code d if, and only if,

1. The weight of each column is greater than or equal to $d - 1$.
2. The weight of the sum of j columns is greater than or equal to $d - j$.

The test procedure for this is by no means brief. McCluskey notes that the synthesis of the parity-check matrix is equivalent to a linear programming problem.

The check and the information digits are related by equations

$$\beta_1 = \alpha_1 + \alpha_2$$
$$\beta_2 = \alpha_1$$
$$\beta_3 = \alpha_2$$

When all possible four binary words are considered, the following four words with a minimum mutual distance of 3 are found.

$$\begin{array}{c} \\ u_1 \\ u_2 \\ u_3 \\ u_4 \end{array} \begin{array}{c} \begin{array}{ccccc} \alpha_1 & \alpha_2 & \beta_1 & \beta_2 & \beta_3 \end{array} \\ \begin{bmatrix} 0 & 0 & 0 & 0 & 0 \\ 0 & 1 & 1 & 0 & 1 \\ 1 & 0 & 1 & 1 & 0 \\ 1 & 1 & 0 & 1 & 1 \end{bmatrix} \end{array} \tag{13-11}$$

While any $A(n,d)$ may be selected as a distance code, it may not necessarily form a group code. The selection of the parity matrix for a group code and the associated detection scheme will be further described in the next section.

13-6. Group Codes. While studying Hamming's parity-check codes, Slepian considered the interesting problem of finding a subset $S \subset S_n$ such that the elements of S form a subgroup of S_n. Such codes are referred to as *systematic group codes*. Errors in these codes can be rectified by the parity-check method. The existence of such a subgroup and its determination will become apparent shortly.

Let $[A]$ be an appropriate matrix for a parity-check code with the general code word $u = (\alpha_1,\alpha_2, \ldots ,\alpha_k; \beta_1,\beta_2, \ldots ,\beta_{n-k})$. We show that if $[\alpha]$ runs over all possible 2^k information sequences the code thus obtained is a subgroup of S_n. Indeed, if u^1 and u^2 are two code words of S, their sum modulo 2 is also a code word of S. Symbolically we may write

$$u^1 = \begin{bmatrix} \alpha^1 \\ \beta^1 \end{bmatrix} = \begin{bmatrix} \alpha^1 \\ A\alpha^1 \end{bmatrix} \qquad u^2 = \begin{bmatrix} \alpha^2 \\ \beta^2 \end{bmatrix} = \begin{bmatrix} \alpha^2 \\ A\alpha^2 \end{bmatrix} \tag{13-12}$$

where α^1 and α^2 are two column matrices describing any two information sequences and β^1 and β^2 are their corresponding parity columns; hence

$$u^1 \oplus u^2 = \begin{bmatrix} \alpha^1 \oplus \alpha^2 \\ \beta^1 \oplus \beta^2 \end{bmatrix} = \begin{bmatrix} \alpha^1 \oplus \alpha^2 \\ A(\alpha^1 \oplus \alpha^2) \end{bmatrix} \tag{13-13}$$

Furthermore, when the k sequence $(00 \cdots 0)$ is taken as the information part of a code word, the check digits of that word will also be a zero $(n - k)$ sequence. If the zero n sequence is taken as the identity element, then every code word in S is its own inverse. Elements of S thus fulfill all the requirements of a group.

Next we show that, if $S \subset S_n$ is an optimal code with the minimum distance d, and a any binary n sequence not in S, then $S_1 = a \oplus S$ also constitutes an optimal code with minimum distance d. The validity of this statement can be examined by letting

$$\begin{aligned} \{S\} &= \{b_1, b_2, \ldots, b_m\} \\ \{S_1\} &= \{a \oplus b_1, a \oplus b_2, \ldots, a \oplus b_m\} \end{aligned} \tag{13-14}$$

and reasoning along the following lines. If S_1 is not optimal, one may find at least one element ($x \in S_n$, $x \notin S_1$) such that its distance from every element of S_1 is more than d. That is,

$$D(x \oplus a \oplus b_k) \geq d \qquad k = 1, 2, \ldots, m \tag{13-15}$$

This inequality implies that

$$D(x \oplus b_k) \geq d \qquad k = 1, 2, \ldots, m$$

a fact that is in contradiction to the optimal character of S.

The 2^n elements of S_n can be expressed in terms of the elements of S and its cosets as shown below:

$$\begin{array}{lllll} e = b_0 = u_1 & u_2 & u_3 & \cdots & u_N \\ b_1 & b_1 \oplus u_2 & b_1 \oplus u_3 & \cdots & b_1 \oplus u_N \\ b_2 & b_2 \oplus u_2 & b_2 \oplus u_3 & \cdots & b_2 \oplus u_N \\ \cdots & \cdots & \cdots & \cdots & \cdots \\ b_m & b_m \oplus u_2 & b_m \oplus u_3 & & b_m \oplus u_N \end{array} \tag{13-16}$$

where $N = 2^k - 1$.

In this development every element of S_n appears once and only once. Each row is a coset, and the elements directly under e are the coset leaders. Coset leaders may be chosen one by one, subject only to the constraints of Eq. (13-17).

$$\begin{aligned} & b_1 \notin S \\ & b_2 \notin (S \cup S_1) && \text{where } S_1 = b_1 \oplus S \\ & b_3 \notin (S \cup S_1 \cup S_2) && \text{where } S_2 = b_2 \oplus S_1 \end{aligned} \tag{13-17}$$

Observe that if a coset leader is replaced by any element in the coset the same coset will result. For example, the following two cosets are identical.

$$\begin{array}{lllll} b_2 & b_2 \oplus u_2 & b_2 \oplus u_3 & \cdots & b_2 \oplus u_N \\ b_2 \oplus u_3 & b_2 \oplus u_3 \oplus u_2 & b_2 \oplus u_3 \oplus u_3 & \cdots & b_2 \oplus u_3 \oplus u_N \end{array} \tag{13-18}$$

Thus, without loss of generality, we may conveniently arrange the elements of each coset so that the weight of none of its elements is less than that of the coset leader.

$$||b_k|| \leq ||b_k + u_j|| \qquad j = 2, \ldots, N \tag{13-19}$$

When the words of a group code book are arranged in the way described above, we say that the words are in *standard order*. The expansion given at the end of Sec. 13-4 is that of a standard array. Of course, we have not yet described what decoding technique is going to be employed for transmission over the binary channel. This is described below.

13-7. A Detection Scheme for Group Codes. Assume that the vocabulary of a transmitter consists of $N = 2^\mu$ binary n sequences:

$$S = \{u_1 = e, u_2, \ldots, u_N\}$$

Furthermore, let these words form an optimal set of elements with minimum mutual distance d. For convenience of calculation, we assume that all words $u \in S$ are transmitted with equal probability over a BSC. At the receiving end, we may receive any one of 2^n distinct words $v \in S_n$. We know that S is a subgroup of S_n. Thus, the words of S_n may be developed in a standard array according to elements of S. Our problem now is to devise a suitable detection. At the receiving end, if a word in the ith column of the standard array is received, let the decoder conjecture that the word u_i was transmitted. In the following discussion it will be shown that this conjecture is in agreement with what we have called a maximum-likelihood (minimum-distance) decoder. Indeed, if the word $b_i \oplus u_j$ is received, the detector will search for a u_k such that

$$D(b_i \oplus u_j, u_k) \le D(b_i \oplus u_j, u_r) \qquad r = 1, 2, \ldots, N \qquad (13\text{-}20)$$

(Of course the equality may hold for several words; in this case some ambiguity will remain in the method.) But

$$D(b_i \oplus u_j, u_k) = D(b_i \oplus u_j \oplus u_k, e) = ||b_i \oplus u_j \oplus u_k|| \ge ||b_i|| \qquad (13\text{-}21)$$

where the last step follows because $u_j \oplus u_k$ is some other element of the subgroup and the array (13-16) is in standard order. Thus if the word u_j is selected to correspond to $b_i \oplus u_j$, then Eq. (13-19) will be satisfied.

$$D(b_i \oplus u_j, u_j) = ||b_i|| \le D(b_i \oplus u_j, u_k) \qquad (13\text{-}22)$$

Any element of a standard array is at least as close to the element on the top of its column as to any other transmitted word. The probability of error for the transmission over a BSC can be readily computed. If a word u_i is transmitted and the word $b_k \oplus u_i$ received, the detector does not commit any error. This is, for instance, the case when the noise is additive and has the structure of one of the coset leaders. The probability of such a situation is

$$\begin{aligned} Q_i &= P\{u = u_i | v = u_i \cup (u_i \oplus b_1) \cup (u_i \oplus b_2) \cup \cdots \\ &\qquad\qquad\qquad\qquad\qquad \cup (u_i \oplus b_m)\} \\ &= \sum_{k=0}^{m} p^{||b_k||} q^{n-||b_k||} \end{aligned} \qquad (13\text{-}23)$$

Since the coset leaders are of minimal weight, in this detection scheme the probability of correct detection Q_i is made as large as possible (for $p \leq \frac{1}{2}$). Furthermore, we assume that all words are transmitted with equal probability. Thus the average error probability becomes $1 - Q_i$.

Following the above considerations, as first pointed out by Slepian, one arrives at the interesting result that for a group alphabet the suggested detection scheme is a maximum-likelihood detector, providing the least average probability of error for the transmission of a specified $N(n,k)$ vocabulary over BSC. For a minute examination of this statement along with the computation of a bound on the corresponding average error probability the reader is referred to Feinstein (I).

13-8. Slepian's Technique for Single-error Correcting Group Codes. We begin with a simple illustration describing Slepian's (n,k) code. Consider Hamming's single-error correcting procedure for $n = 5$ and $k = 2$. There are four possible information 2 sequences:

$$(00, 01, 10, 11)$$

A suitable 3×2 parity-check matrix is selected as

$$[A] = \begin{bmatrix} 1 & 0 \\ 0 & 1 \\ 1 & 1 \end{bmatrix}$$

The parity checks are found to be

$$\begin{bmatrix} 1 & 0 \\ 0 & 1 \\ 1 & 1 \end{bmatrix} \begin{bmatrix} 0 & 0 & 1 & 1 \\ 0 & 1 & 0 & 1 \end{bmatrix} = \begin{bmatrix} 0 & 0 & 1 & 1 \\ 0 & 1 & 0 & 1 \\ 0 & 1 & 1 & 0 \end{bmatrix}$$

Thus, the four 5 sequences to be transmitted over the BSC are

$$\{S\} = \begin{bmatrix} u_1 \\ u_2 \\ u_3 \\ u_4 \end{bmatrix} = \begin{bmatrix} 0 & 0 & 0 & 0 & 0 \\ 0 & 1 & 0 & 1 & 1 \\ 1 & 0 & 1 & 0 & 1 \\ 1 & 1 & 1 & 1 & 0 \end{bmatrix}$$

Note that

$$\|u_k\| \geq 3 \qquad k = 2, 3, 4$$

If we have the definite knowledge that not more than a single error has occurred, we can certainly detect and correct that error. When more than a single error may occur, this detector still may be used. For this, we select the $\binom{5}{1} = 5$ words with the lowest possible weight as coset leaders and develop the S_5 group according to the elements of S.

u_1	u_2	u_3	u_4
0 0 0 0 0	0 1 0 1 1	1 0 1 0 1	1 1 1 1 0
0 0 0 0 1	0 1 0 1 0	1 0 1 0 0	1 1 1 1 1
0 0 0 1 0	0 1 0 0 1	1 0 1 1 1	1 1 1 0 0
0 0 1 0 0	0 1 1 1 1	1 0 0 0 1	1 1 0 1 0
0 1 0 0 0	0 0 0 1 1	1 1 1 0 1	1 0 1 1 0
1 0 0 0 0	1 1 0 1 1	0 0 1 0 1	0 1 1 1 0
1 0 0 1 0	1 1 0 0 1	0 0 1 1 1	0 1 1 0 0
1 1 0 0 0	1 0 0 1 1	0 1 1 0 1	0 0 1 1 0

For the seventh coset we select as a leader an element of low weight which has not yet appeared in the array. The same rule, of course, applies to the leader of the last coset. If a word in a column u_k ($k = 1, 2, 3, 4$) is received, the decoder conjectures that u_k was transmitted. In doing so, the decoder corrects all possible $\binom{5}{1} = 5$ single-error and 2 out of $\binom{5}{2} = 10$ possible double-error patterns. The probability of error associated with this detection scheme for every transmitted word is

$$\begin{aligned} 1 - Q_1 &= 1 - (q^5 + 5q^4p + 2q^3p^2) \\ &= p^5 + 5p^4(1-p) + 10p^3(1-p)^2 + 8p^2(1-p)^3 \end{aligned}$$

The average probability of error for this decoding scheme is $1 - Q_1$. The selection of elements of lowest weight in the above situation has brought forth the largest possible Q_1, that is, the code with the lowest error probability among all possible (5,2) codes. Also note that

$$2^{5-2} - \sum_{i=0}^{3} \binom{3}{i} < 2$$

Now we are in a position to focus attention on Slepian's important statement, that if positive integers n and k satisfy the inequality

$$k \geq 2^{n-k} - \sum_{i=0}^{3} \binom{n-k}{i}$$

then a simple parity-check code with a maximum-likelihood detection rule can be described which cannot be excelled by any other code of 2^k n-sequence binary words. No other code of the same size has a smaller average probability of error.

A proof of this statement may be given in the light of the following lemma, which was first proved by Hamming in a simpler form.

Hamming's Lemma. The necessary and sufficient condition for an (n,k) group code to be δ-error correcting is that, except for the null sequence, every sequence to be transmitted has a weight not less than $2\delta + 1$. To prove the necessity, note that in a group code the null sequence is a word, and so must be at a distance of $2\delta + 1$ or more from all other words. The sufficiency follows from the fact that in a group code, $u_i \oplus u_j = u_k$ is an element of the coset S (recall that S is a subgroup of S_n). Then,

$$\|u_i \oplus u_j\| = \|u_k\| \geq 2\delta + 1$$

That is, all elements of S are at least $2\delta + 1$ units apart.

Denote the $N = 2^k$ words to be transmitted by

$$(u_1, u_2, \ldots, u_N)$$

where u_1 is the null n sequence. If the condition of the lemma is satisfied, we select words with weights not exceeding δ as coset leaders. Thus, if this array is used as the detection scheme, we shall have

$$\|u_j + b_i\| \geq \|u_j\| - \|b_i\| \geq \delta + 1 > \|b_i\| \tag{13-24}$$

That is, the development of the group is that of a standard array. Conversely, if b_i is a coset leader with $\|b_i\| \leq \delta$, then the elements of that coset should satisfy the condition

$$\|b_i + u_j\| > \|b_i\| \qquad j = 1, 2, \ldots, N \tag{13-25}$$

Now suppose that one of the u elements, say u_j, has a weight of 2δ or less. It is not difficult to show by contradiction that, if the digits of b_i and u_j are conveniently chosen, their sum may lead to a sequence with a higher weight than the coset leader.

For an (n,k) single-error correcting group code, all transmitted words are of weight 3 or more, except the zero sequence. The coset leaders are selected from elements with lowest weight, and thus a standard array is formed. If all words are equiprobable, the source entropy per binary symbol becomes $(1/n) \log 2^k = k/n$. The probability of error for any one of the transmitted messages is

$$P\{v \in v'_j | u_j\} = 1 - Q_j \tag{13-26}$$

The over-all error probability and equivocation are, respectively,

$$\bar{E} = \sum_j \frac{1}{N} (1 - Q_j) = 1 - \frac{1}{N} \sum_j Q_j \tag{13-27}$$

$$H(Y|X) = -\sum_j Q_j \log Q_j$$

The value of Q_j depends on the weight of the coset leaders as described in

Eq. (13-23). If coset leaders contain γ_i elements of weight i, then

$$Q_1 = \sum_{i=0}^{n} \gamma_i p^i q^{n-i} \tag{13-28}$$

Of course γ_i and the total number of coset leaders must satisfy the conditions

$$\gamma_i \leq \binom{n}{i}$$

$$\sum_{i=0}^{n} \gamma_i = 2^{n-k} \tag{13-29}$$

For given values of (n,k) up to $n = 12$ and $k = 10$, Slepian evaluates the γ coefficients which correspond to the highest Q_i in each case. Table T-4 gives the values of coefficients γ_i for different values of n and k. Table T-5 suggests suitable parity-check matrices.

For a more detailed proof of this technique see Slepian (I, particularly Sec. 2.7).

The following example of group encoding and decoding may prove useful in illustrating the described technique.

Example 13-2. Determine a group code (6,3) by Slepian's technique.

Solution. First select a suitable parity-check matrix. According to Table T-5, we may choose

$$\begin{bmatrix} 4 & 1 & 2 \\ 5 & 1 & 3 \\ 6 & 2 & 3 \end{bmatrix} \qquad \begin{aligned} \beta_4 &= \alpha_1 + \alpha_2 \\ \beta_5 &= \alpha_1 + \alpha_3 \\ \beta_6 &= \alpha_2 + \alpha_3 \end{aligned} \qquad \text{or} \qquad [A] = \begin{bmatrix} 1 & 1 & 0 \\ 1 & 0 & 1 \\ 0 & 1 & 1 \end{bmatrix}$$

When the information vector covers the eight possible columns of the following message matrix, we obtain the corresponding eight check vectors in a 3 × 8 matrix.

$$\begin{bmatrix} 1 & 1 & 0 \\ 1 & 0 & 1 \\ 0 & 1 & 1 \end{bmatrix} \begin{bmatrix} 0 & 0 & 0 & 1 & 0 & 1 & 1 & 1 \\ 0 & 0 & 1 & 0 & 1 & 0 & 1 & 1 \\ 0 & 1 & 0 & 0 & 1 & 1 & 0 & 1 \end{bmatrix} = \begin{bmatrix} 0 & 0 & 1 & 1 & 1 & 1 & 0 & 0 \\ 0 & 1 & 0 & 1 & 1 & 0 & 1 & 0 \\ 0 & 1 & 1 & 0 & 0 & 1 & 1 & 0 \end{bmatrix}$$

Thus, the encoded messages are

$$\begin{bmatrix} u_1 \\ u_2 \\ u_3 \\ u_4 \\ u_5 \\ u_6 \\ u_7 \\ u_8 \end{bmatrix} = \left[\begin{array}{ccc|ccc} 0 & 0 & 0 & 0 & 0 & 0 \\ 0 & 0 & 1 & 0 & 1 & 1 \\ 0 & 1 & 0 & 1 & 0 & 1 \\ 1 & 0 & 0 & 1 & 1 & 0 \\ 0 & 1 & 1 & 1 & 1 & 0 \\ 1 & 0 & 1 & 1 & 0 & 1 \\ 1 & 1 & 0 & 0 & 1 & 1 \\ 1 & 1 & 1 & 0 & 0 & 0 \end{array}\right]$$

The rows of this matrix are now used as a coset for developing the group of 2^6 elements. This can be done by selecting the columns of the following matrix as coset leaders and computing each coset, as before.

Coset leaders

1	2	3	4	5	6	7
0	0	0	0	0	1	1
0	0	0	0	1	0	0
0	0	0	1	0	0	0
0	0	1	0	0	0	0
0	1	0	0	0	0	0
1	0	0	0	0	0	1

The elements of cosets are obtained by modulo 2 addition, as shown below.

000000	001011	010101	100110	011110	101101	110011	111000
000001	001010	010100	100111	011111	101100	110010	111001
000010	001001	010110	100100	011100	101111	110001	111010
000100	001111	010001	100010	011010	101001	110111	111100
001000	000011	011101	101110	010110	100101	111011	110000
010000	011011	000101	110110	001110	111101	100011	101000
100000	101011	110101	000110	111110	001101	010011	011000
100001	101010	010100	000111	111111	001100	010010	011001

The detection scheme for this standard array is rather simple. If a message 100010 is received, it will be decoded as the message 100110. The corresponding noise vector is 000100.

The probability of a correct detection over a BSC with $P\{1|1\} = p$ is equal to the probability of the noise being of the form of a coset leader:

$$Q_1 = \sum_{i=0}^{n} \gamma_i p^i q^{n-i}$$

where $\gamma_i \leq \binom{n}{i}$ is the number of coset leaders of weight i and is given in Table T-4:

$$Q_1 = q^6 + 6pq^5 + p^2q^4$$

This decoding procedure corrects all single errors and one double error out of a possible $\binom{6}{2} = 15$ double errors.

13-9. Further Notes on Group Codes. In this section, as a supplement to the material of the previous section, we discuss rules for selecting suitable $[A]$ matrices. The topic has been discussed frequently in numerous papers on coding. The first discussion appears as early as Hamming's paper of 1950. The problem has been explored by D. Slepian, G. Sacks, W. Peterson, E. J. McCluskey, R. C. Bose and R. R. Kuebler, Jr.,* and many others.

R. C. Bose and D. K. R. Chaudhuri (see footnote, page 460) have shown that the necessary and sufficient condition for the existence of a δ-error correcting (n,k) binary group code is the existence of an n by $n - k$

* *Ann. Math. Statistics*, vol. 31, no. 1, pp. 113–134, March, 1960.

binary matrix of rank $n - k$ such that any sets of its 2δ rows are mutually independent. This ensures the feasibility of selecting all n sequences of weight less than or equal to δ as coset leaders. (For a full discussion see "Error Correcting Codes" by W. W. Peterson.)

The following method for the selection of a suitable k by $n - k$ parity-check matrix $A' = A^t$ and a corresponding maximum-likelihood detection scheme appears in Slepian (II). Integers n and k are chosen so that

$$k \geq 2^{n-k} - \sum_{i=0}^{3} \binom{n-k}{i} \tag{13-30}$$

1. The first row of $[A']$ consists of $n - k$ 1's. The second row will be a binary $n - k$ sequence containing one 0 only. The succeeding rows are arbitrarily selected among all possible remaining distinct $n - k$ sequences with only one 0. Then we start using $n - k$ sequences with only two 0's and so on. This procedure is continued until k rows are obtained. The $k \times (n - k)$ matrix thus obtained may be used as a permissible parity-check matrix as described earlier (see also Sec. 4-14).

2. Suppose that a word v is received:

$$v = \alpha_1, \alpha_2, \ldots, \alpha_k; \beta_1, \beta_2, \ldots, \beta_{n-k}$$

Calculate

$$f_j = \beta_j \oplus \sum_{i=1}^{k} a_{ij} \oplus \alpha_i \qquad j = 1, 2, \ldots, n - k \tag{13-31}$$

The sequence $f = f_1 f_2 \cdots f_{n-k}$ may or may not coincide with a row (say the rth row) of $[A']$. Then the following rules apply.

1. If the two are identical, then α_r is an erroneously received digit.

2. f is not a row of $[A']$ and does not contain exactly three 1's. The erroneously received digits are in the position where f has 1's.

3. f is not a row of $[A']$ but has exactly three 1's. Then, search for a row (say the ith row) of $[A']$ having exactly four 1's, three of them in the same positions as the 1's of f. If the 1 which is not in common is in the jth column, then α_i and β_j were erroneously received.

The proof of the above, although straightforward, would require considerable time and effort. The method in essence is similar to the one described in Sec. 4-14.

The extension of the above systematic approach to the so-called Reed-Muller code* was also given by Slepian. This coding technique, which was discovered independently by Reed and Muller, in brief can be considered as a particular class of Hamming codes for which the length

* See the bibliography at the end of the book.

of the binary sequence is selected as

$$n = 2^m$$

$$\text{Number of information digits} = k = \sum_{i=0}^{r} \binom{n}{i} \qquad \text{(13-32)}$$

Number of check digits $= n - k$ $\quad m$ and r positive integers $\quad r \leq m$

The minimum distance between every two code words is 2^{m-r}. Then up to $2^{m-r-1} - 1$ erroneous digits can be corrected. A systematic procedure for the determination of an $m \times 2^m$ matrix for parity checking and devising a maximum-likelihood detection scheme is given in Slepian (II).

Algebraic Operations of Group Codes. In a recent article, Slepian* examines the concepts of equivalence and algebraic operations of group codes. Two (n,k) codes are said to be equivalent if one can be obtained from the other by a fixed permutation of the places of every word. For example, the following three (3,2) group code alphabets are equivalent:

0	0	0		0	0	0		0	0	0
1	0	0		0	1	0		0	0	1
0	1	1		1	0	1		1	1	0
1	1	1		1	1	1		1	1	1

Two equivalent group codes have the same probability of error Q_1 when all the words are transmitted with equal probability.

The 2^k words of an (n,k) group code can be obtained from any set of its k linearly independent words. For this reason, a set of k linearly independent words may be generally referred to as a generating matrix of the code. The relation between a $k \times n$ generating matrix Ω and an $(n - k) \times n$ binary matrix Λ of rank $n - k$ is given by

$$\Lambda\Omega^t = 0$$

Thus, every parity-check matrix $[\Lambda]$ of a group code (n,k) can be considered as a generating matrix for the dual code $(n, n - k)$.

Two generating matrices Ω_1 and Ω_2 are said to be equivalent if they lead to equivalent group codes. The following results are due to D. Slepian.

PROPOSITION 1. Every $k \times n$ binary generating matrix Ω is equivalent to the partitioned matrix

$$[I_k|M]$$

where I_k is the $k \times k$ unit matrix and M is a $k \times (n - k)$ binary matrix.

PROPOSITION 2. The necessary and sufficient condition for two binary $k \times n$ matrices $[\Omega_1]$ and $[\Omega_2]$ to generate equivalent (n,k) codes is that their columns can be placed into a one-to-one correspondence that preserves mod 2 addition of the columns.

* D. Slepian, Some Further Theory of Group Codes, *Bell System Tech. J.*, October, 1960.

An example of proposition 1 for (5,3) follows:

$$\Omega_1 = \left[\begin{array}{ccc|cc} 1 & 0 & 0 & 1 & 0 \\ 0 & 1 & 0 & 1 & 1 \\ 0 & 0 & 1 & 0 & 1 \end{array}\right]$$

In order to obtain a generating matrix equivalent to Ω_1, one can make any mod 2 linear combination of the columns of Ω_1 which does not obviate the condition of one-to-one correspondence. For instance, we may consider

$$\Omega_2 = \begin{bmatrix} 1 & 0 & 1 & 1 & 1 \\ 1 & 1 & 0 & 1 & 0 \\ 0 & 0 & 1 & 1 & 0 \end{bmatrix}$$

A useful application of the concept of equivalent generating matrices can be given by defining algebraic operations on group codes. Let Ω_a and Ω_b be generating matrices for two group codes (n,k) and (n',k'), respectively. The generating matrix Ω_c,

$$\Omega_c = \left[\begin{array}{c|c} \Omega_a & 0 \\ \hline 0 & \Omega_b \end{array}\right]$$

designates an $(n + n', k + k')$ group code which will be denoted by $\mathcal{C}$ and referred to as the sum of the two group codes $\mathcal{A}$ and $\mathcal{B}$.

$$\mathcal{C} = \mathcal{A} + \mathcal{B} \tag{13-33a}$$

Each word from $\mathcal{C}$ is a sequence of a word from $\mathcal{A}$ and $\mathcal{B}$. Owing to the independence of the noise, it is clear that the probability of no error described in Eq. (13-28) for $\mathcal{C}$ is the product of the corresponding terms for $\mathcal{A}$ and $\mathcal{B}$.

When an (n,k) code $\mathcal{A}$ can be made equivalent to the sum of two (n,k) codes it is called a *decomposable* code. Otherwise the code is said to be *indecomposable*. A further interesting result of Slepian's is that "every (n,k) code $\mathcal{A}$ is equivalent to a sum of indecomposable codes":

$$\mathcal{A} : \mathcal{A}_1 + \mathcal{A}_2 + \cdots + \mathcal{A}_n$$

The Kronecker product of two generating matrices Ω_a and Ω_b gives a matrix Ω_c which can be used as a generating matrix for a new group code (nn',kk').

$$\Omega_c = \text{Kronecker product of } \Omega_a \text{ and } \Omega_b = \left[\begin{array}{c|c|c|c} a_{11}\Omega_b & a_{12}\Omega_b & \cdots & a_{1n}\Omega_b \\ \hline a_{21}\Omega_b & a_{22}\Omega_b & \cdots & a_{2n}\Omega_b \\ \hline \cdots & \cdots & \cdots & \cdots \\ \hline a_{k1}\Omega_b & a_{k2}\Omega_b & \cdots & a_{kn}\Omega_b \end{array}\right]$$

(where $[a_{ij}] = \Omega_a$).

The resulting code is referred to as the product of the codes $\mathcal{A}$ and $\mathcal{B}$.

$$\mathcal{C} = \mathcal{A} \cdot \mathcal{B} \tag{13-33b}$$

The basic concepts of the algebraic operations of sum, product, and decomposition of codes are of considerable significance in devising efficient codes. They appear to be in harmony with the modern "system" point of view. The extent of the practical usefulness of the theory remains to be seen.

13-10. Some Bounds on the Number of Words in a Systematic Code. The prime objective of this section is to present some further bounds on the function $B(n,d)$. There is a considerable amount of literature available giving interesting bounds and interrelations among the number of elements of different systematic codes with related n and d. Hamming's theorem for $e = 1$ gives

$$B(n,3) \leq \frac{2^n}{n+1}$$

The equality is valid when $2^n/(n+1)$ is an integer. Such codes are referred to as close-packed codes. Examples of close-packed codes are obtained by selecting $n = 2^k - 1$, where k is a suitable positive integer.* As the name indicates, in a close-packed code all words lie within or on one of the disjoint detection spheres centered about a transmitted word with a radius δ (see Sec. 13-5).

An interesting example of a close-packed group code for (23,7) was first derived by Golay, who also observed that $B(23,7) = 2^{11}$. Shapiro and Slotnick† have proved that the only close-packed double-error correcting code is the trivial code (5,5) with two points, and aside from the trivial code (7,7) with two points, the Golay code is the only close-packed triple-error correcting code. In other words, the equation

$$1 + \binom{n}{1} + \binom{n}{2} + \binom{n}{3} = 2^k$$

has only two sets of admissible solutions (7,7) and (23,7). Golay's code considered as a (23,12) group code corrects all patterns of three or fewer independent errors.

Golay's (23,7) code is also referred to as a *lossless* code. A systematic δ-error correcting code (n,d) using an alphabet of 2^m symbols is said to be lossless if it satisfies the Diophantine equation:

$$\sum_{j=0}^{\delta} \binom{n}{j} (2^m - 1)^j = 2^{mr} \qquad r \leq n \tag{13-34}$$

(n, m, r, δ are appropriate integers.)

* For close-packed codes see S. P. Lloyd, Binary Block Coding, *Bell System Tech. J.*, vol. 36, pp. 517–535, March, 1957.

† See particularly their Theorems 5 and 6 (see also Lloyd, *ibid.*, p. 532).

Thus, the success of Golay's code is based on the relation

$$\sum_{j=0}^{3} \binom{23}{j} (2^1 - 1)^j = 2^{11} \tag{13-35}$$

According to R. T. Chien,* an IBM 704 was given the task of solving the above Diophantine equation for $n \leq 100$ and arbitrary p^m-nary basis. The only other solution for $\delta > 1$ was the lossless ternary (11,5) code with $r = 5$:

$$\binom{11}{0} + \binom{11}{1} 2 + \binom{11}{2} 4 = 3^5$$

The following interesting results are due to M. Plotkin.†

Plotkin's Theorems. Let $B(n,d)$ be, as before, the largest possible number of words in a minimum-distance (n,d) code; then,

$$\begin{aligned} B(n,d) &\leq 2B(n-1, d) \\ B(2n,2d) &\geq B(n,2d) \cdot B(n,d) \\ B(4m,2m) &\leq 8m \end{aligned} \tag{13-36}$$

If $4m - 1$ is a prime number, then

$$B(4m,2m) = 8m \tag{13-37}$$

While the proof of some of the suggested bounds on $A(n,d)$ and $B(n,d)$ is quite interesting, we have no intention of tabulating more bounds along with their proofs. However, a significant remark made by R. C. Bose and S. S. Shrikhande‡ should be brought forward. These authors first observed that some of the coding results cited in this section can be obtained by established methods of combinatorial analysis. For instance, Plotkin's sharp results are similar to those obtained by Bose in 1939 while studying the so-called "balanced incomplete block design." The latter problem is defined by Bose as follows:

Given an arrangement of v objects into b sets satisfying the following conditions: Each set contains exactly k objects, each object occurs in exactly r different sets, and any arbitrary pair of objects occurs in exactly λ different sets.

$$\begin{aligned} bk &= vr \\ \lambda(v-1) &= r(k-1) \end{aligned} \tag{13-38}$$

It is shown in the cited paper that, for instance, Plotkin's problem is equivalent to the problem of balanced incomplete block design. For

* Paper presented to URSI-IRE, Oct. 12, 1959, San Diego, Calif.

† M. Plotkin, "Binary Codes with Specified Minimum Distance," *Univ. Penn., Moore School Elec. Eng., Rept.* 51-20, 1951.

‡ R. C. Bose and S. S. Shrikhande, A Note on a Result in the Theory of Code Construction, *Inform. and Control,* vol. 2, pp. 183–194, 1959.

this, one has to let

$$v = b = 4m - 1 \qquad r = k = 2m - 1 \qquad \lambda = m - 1 \tag{13-39}$$

As a further point of classical interest, the cited paper observes that the statistical problem of balanced incomplete block design was in turn shown by J. A. Todd to be analogous to certain well-known problems on the existence of Hadamard matrices of certain order. Thus certain types of problems of bounds on $A(n,d)$ can be equivalently translated into a problem on Hadamard matrices. This is indeed an interesting basic observation which suggests some strong links between certain coding problems and some established parts of classical mathematics.

A Hadamard matrix is a square matrix of order α whose elements are $+1$ or -1 and any two rows (subsequently any two columns) are orthogonal. It was shown by Paley* in 1933 that, besides the trivial case of $\alpha = 2$, the only Hadamard matrices are those for which α is a multiple of 4. The result that the problem of balanced incomplete block design is equivalent to the existence of a Hadamard matrix is due to Todd.† He has shown that the two problems are equivalent when

$$\begin{aligned} v = b = 4m - 1 \qquad r = k = 2m - 1 \\ \lambda = m - 1 \qquad \alpha = 4m \end{aligned} \tag{13-40}$$

For methods of constructing Hadamard matrices the reader is referred to the references cited in Bose and Shrikhande.

An extension of Hamming's results to p-nary codes has also appeared in the literature (see, for instance, Shapiro and Slotnick, *loc. cit.*). Let p denote any prime number and d the distance between two points (the number of coordinates in which they differ); then, the maximum number of points (n-tuples of symbols) with minimum mutual distance d will be denoted by $A^{(p)}(n,d)$. The Hamming relations can be generalized in a straightforward manner. For instance, the following inequality is valid:

$$A^{(p)}(n, 2e + 1) \leq \frac{p^n}{1 + (p-1)\binom{n}{1} + \cdots + (p-1)^e\binom{n}{e}} \tag{13-41}$$

We close this introductory chapter on group coding by emphasizing that an adequate coverage of coding theory requires time and space beyond the scope of this book. The reader is referred to the broad literature on coding theory, particularly forthcoming books on the subject.

Two of the most important areas which remain to be fully investigated are as follows:

* R. E. A. C. Paley, On Orthogonal Matrices, *J. Math. Phys.*, vol. 12, pp. 311–320, 1933.

† J. A. Todd, A Combinatorial Problem, *J. Math. Phys.*, vol. 12, pp. 321–333, 1933.

1. The evaluation of sharp bounds on the probability of error for different (n,k) codes and their relations to the channel capacity

2. Extension of the existing coding techniques to any finite, semicontinuous, or continuous channels with or without memory

While remarkable progress has been made (particularly during the past 5 years), much more integration is needed. "Our understanding of group alphabets is still fragmentary," as was pointed out by D. Slepian.

PROBLEMS*

13-1. Consider the following six matrices:

$$\begin{bmatrix} 1 & 0 \\ 0 & 1 \end{bmatrix} \quad \begin{bmatrix} 1 & 0 \\ 0 & -1 \end{bmatrix} \quad \begin{bmatrix} -\frac{1}{2} & \frac{1}{2}\sqrt{3} \\ \frac{1}{2}\sqrt{3} & \frac{1}{2} \end{bmatrix} \quad \begin{bmatrix} -\frac{1}{2} & -\frac{1}{2}\sqrt{3} \\ -\frac{1}{2}\sqrt{3} & \frac{1}{2} \end{bmatrix}$$

$$\begin{bmatrix} -\frac{1}{2} & \frac{1}{2}\sqrt{3} \\ -\frac{1}{2}\sqrt{3} & -\frac{1}{2} \end{bmatrix} \quad \begin{bmatrix} -\frac{1}{2} & -\frac{1}{2}\sqrt{3} \\ \frac{1}{2}\sqrt{3} & -\frac{1}{2} \end{bmatrix}$$

Show that they form a group under ordinary matrix multiplication, the first matrix being the identity element.

13-2. Same question as in Prob. 13-1.

$$\begin{bmatrix} 1 & 0 \\ 0 & 1 \end{bmatrix} \quad \begin{bmatrix} 0 & 1 \\ 1 & 0 \end{bmatrix} \quad \begin{bmatrix} 1 & 0 \\ -1 & -1 \end{bmatrix} \quad \begin{bmatrix} 0 & 1 \\ -1 & -1 \end{bmatrix} \quad \begin{bmatrix} -1 & -1 \\ 1 & 0 \end{bmatrix} \quad \begin{bmatrix} -1 & -1 \\ 0 & 1 \end{bmatrix}$$

13-3. Show the validity of the following identities for the words a, b, and c which are binary n sequences.

(*a*) If $\|a \cdot b\| = \|a \cdot c\| = \|a\|$, then $\|b \cdot c\| \geq \|a\|$.

(*b*) If $a \cdot b = a$ and $\|a\| = \|b\|$, then $a = b$.

(*c*) The necessary and sufficient condition for $\|a \cdot b\| = \|a\|$ is that $a \cdot b = a$.

13-4. Devise a single-error correcting group code and the associated coding scheme for

(*a*) $n = 7 \quad k = 2$

(*b*) $n = 7 \quad k = 3$

(*c*) $n = 8 \quad k = 2$

(*d*) $n = 8 \quad k = 4$

In each case compute the error probability Q_1 from Table T-4.

13-5. Let the words of a single-error correcting (n,k) group code be designated by $U_1 = (0,0,0, \ldots ,0)$, U_2, U_3, $\ldots$, U_N, where $N = 2^k$. Show that any word U_j can be obtained as a linear combination of k independent words (U_1 excluded), that is,

$$U_j = \lambda_2 U_2 + \lambda_3 U_3 + \cdots + \lambda_{k+1} U_{k+1} \qquad j = 1, 2, \ldots , N$$

where $(\lambda_2, \lambda_3, \ldots , \lambda_{k+1})$ are binary elements.

* For a supply of problems on coding, see a forthcoming monograph entitled "Error Correcting Codes" by W. W. Peterson (John Wiley & Sons, Inc., New York).

APPENDIX

ADDITIONAL NOTES AND TABLES

During the past decade the field of information theory has grown by leaps and bounds. There have been new mathematical contributions, as well as a host of applications to physical and social problems. Coding theory has already become a professional field. Applications to detection problems, such as radar detection theory, embrace another professional area. The areas of application to some aspects of linguistics, electronics, computers, optics, psychology, and others seem to be following the same pattern of specialization. The mathematical foundation of information theory constitutes another specialized territory.

The aim of this book has been an introductory presentation of the essentials of information theory. We have attempted to emphasize the fundamentals that are indispensable for an understanding of the subject prior to its application. The areas of specialized applications are outside the scope of the present work. Fortunately, an increasing supply of literature dealing with the application of information theory to many specialized fields is available, and more books on the general subject are forthcoming. It is hoped that the reader will be stimulated to a more inclusive study of the subject.

The following diversified notes may be of additional interest. The notes are not necessarily complete or self-sustaining. They are inserted as a few examples of the multitude of topics available for further reading. For each note, adequate reference for further pursuit is provided. The bibliography at the end of the book also may prove helpful for this purpose.

N-1. The Gambler with a Private Wire. J. L. Kelly, Jr.,* has suggested an interesting model which presents the problem of the rate of transmission of information in a different way.

Consider the case of a gambler with a private wire who places bets on the outcomes of a game of chance. We assume that the side information which he receives has a probability p of being true and of $1 - p$ of being false. Let the original capital of the gambler be V_0, and V_K his capital after the Kth betting. Since the gambler is not certain that the side

* J. L. Kelly, Jr., A New Interpretation of Information Rate, *Bell System Tech. J.*, vol. 35, pp. 917–926, 1956.

information is entirely reliable, he places only a fraction e of his capital on each bet. Thus, subsequent to n bettings, assuming the independence of the successive tips, his capital is

$$V_n = (1 + e)^w (1 - e)^l V_0 \tag{N1-1}$$

where w is the number of times he won and $l = n - w$ the number of times he lost. These numbers are, in general, values taken by two random variables denoted by W and L. According to the law of large numbers (Sec. 7-8),

$$\lim_{n\to\infty} \frac{1}{n} W = p$$
$$\lim_{n\to\infty} \frac{1}{n} L = q = 1 - p \tag{N1-2}$$

The problem with which the gambler is faced is the determination of e leading to the maximum of the average exponential rate of growth of his capital. That is, he wishes to maximize the value of

$$G = \lim_{n\to\infty} \frac{1}{n} \log \frac{V_n}{V_0} \tag{N1-3}$$

with respect to e, assuming a fixed original capital and specified p.

$$G = \lim_{n\to\infty} \left[\frac{W}{n} \log (1 + e) + \frac{L}{n} \log (1 - e)\right]$$
$$G = p \log (1 + e) + q \log (1 - e) \tag{N1-4}$$

The maximum of G occurs when

$$\frac{p}{1+e} - \frac{q}{1-e} = 0$$
$$e = p - q \tag{N1-5}$$

Therefore,

$$\max G = p \log 2p + q \log 2q = 1 + p \log p + q \log q \tag{N1-6}$$

Thus, under these rather natural hypotheses, the maximum possible average exponential gain of the gambler coincides with the numerical value of the channel capacity. If the channel were noiseless, the gambler would obviously risk all his capital at each betting. This is, of course, in agreement with Eq. (N1-1). Also, if he knew the value of p beforehand, he would be able to use this knowledge to his advantage and bet all his capital (or none). But the reliability of the tip is not known to him.

According to Kelly, here we have an example of a real-life situation where considerations similar to the concept of source, channel, rate of transinformation, and channel capacity are valid. In the above reference, Kelly extends these results to more general cases of a gambler placing bets on outcomes of several games of chance. The gambler receives independent tips on each game conditional on the result of another game.

The situation is analogous to a discrete independent source driving a discrete memoryless noisy channel.

In conclusion, our acquired knowledge of information theory, which was based primarily on Shannon's communication model, can well be applied to other mathematical models arising from real-life problems.

Bellman and Kalaba have successfully applied the theory of dynamic programming to Kelly's model. They have also extended the problem to a type of problem which could be referred to as *stochastic learning processes*. In all these generalizations, the problem envisaged is the determination of some optimal policy for the gambler such that the expected value of the logarithm of his capital after n bettings is maximized. Since the constraints of the problem are of the type of linear inequalities, the maximization procedure can manifestly be done by Bellman's dynamical programming techniques. The reader interested in applications of the theory of dynamic programming to the study of communication processes may be interested in the articles by the following authors, as well as a number of references given there: R. Bellman and R. Kalaba, H. Robbins, and M. B. Marcus.

N-2. Some Remarks on Sampling Theorem. The literature available on sampling theorems is quite extensive. Because of lack of space, only a few of the existing references are cited here.

1. Reconstruction of a band-limited function $(-\omega,+\omega)$ from its sampled derivatives has been done by L. J. Fogel,* D. L. Jagerman,* F. E. Bond and C. R. Cahn,* and D. A. Linden and N. M. Abramson.* The latter authors show that for a continuous band-limited function one has

$$\xi(t) = \sum_{k=-\infty}^{\infty} \left[x(k\tau) + (t - k\tau)x'(k\tau) + \frac{(t-k\tau)^2}{2} x''(k\tau) + \cdots + \frac{(t-k\tau)^R}{R!} x^{(R)}(k\tau) \right] \left[\frac{\sin \pi(t/\tau - k)}{\pi(t/\tau - k)} \right]^{R+1}$$

(See *Inform. and Control*, vol. 4, pp. 95–96, 1961.) k is an integer (N2-1)

2. As has been pointed out by several authors, the sampling intervals need not be uniformly distributed (see, for instance, J. L. Yen,† who discusses some special nonuniform sampling).

* L. J. Fogel, A Note on the Sampling Theorem, *IRE Trans. on Inform. Theory*, vol. IT-1, pp. 47–48, 1955; D. L. Jagerman and L. J. Fogel, Some General Aspects of the Sampling Theorem, *IRE Trans. on Inform. Theory*, vol. IT-2, pp. 139–146, 1956; D. A. Linden and N. M. Abramson, A Generalization of the Sampling Theorem, *Inform. and Control*, vol. 3, no. 1, pp. 26–31, 1960; F. E. Bond and C. R. Cahn, On Sampling the Zeros of Bandwidth Limited Signals, *IRE Trans. on Inform. Theory*, vol. IT-4, pp. 110–113, September, 1958.

† J. L. Yen, On Nonuniform Sampling of Bandwidth-limited Signals, *IRE Trans. on Circuit Theory*, vol. CT-3, pp. 251–257, December, 1956.

3. A. V. Balakrishnan* has generalized the sampling theorem to the case of a continuous-parameter stochastic process. The main result of his paper is as follows:

Theorem. Let $\{x(t)\}$ be a real- or complex-valued stochastic process, stationary of second order, possessing a spectral density which vanishes outside the interval of $[-2\pi W, 2\pi W]$. Then $\{x(t)\}$ has the following representation:

$$\{x(t)\} = \underset{N\to\infty}{\text{l.i.m.}} \sum_{-N}^{N} \left\{x\left(\frac{n}{2W}\right)\right\} \frac{\sin \pi(2Wt - n)}{\pi(2Wt - n)}$$

A method of proof consists in applying the sampling function for nonrandom functions to the covariance function of the process.

4. Sampling theorem in n-dimensional space: The suggested proof for the sampling theorem can be directly generalized to a band-limited function defined in an n-dimensional space (n positive integer). To this end, one usually employs the n-dimensional Fourier integral pairs. Let $f(x_1,x_2, \ldots ,x_n)$ be a function of n real variables; the Fourier transform of $f(x_1,x_2, \ldots ,x_n)$ is defined by

$$F(y_1,y_2, \ldots ,y_n) = \underbrace{\int \cdots \int}_{\leftarrow n \rightarrow} f(x_1,x_2, \ldots ,x_n) \exp\,[-j(x_1y_1 + x_2y_2 + \cdots + x_ny_n)]\, dx_1 \cdots dx_n \tag{N2-2}$$

when this integral exists. The inverse Fourier transform is defined by

$$f(x_1,x_2, \ldots ,x_n) = (2\pi)^{-n} \underbrace{\int \cdots \int}_{\leftarrow n \rightarrow} F(y_1,y_2, \ldots ,y_n) \exp\,[+j(x_1y_1 + x_2y_2 + \cdots + x_ny_n)]\, dy_1 \cdots dy_n \tag{N2-3}$$

In the light of this definition, one may state a generalized form of the sampling theorem in n-dimensional space.

Theorem. Let $f(t_1,t_2, \ldots ,t_n)$ be a function of n real variables, whose n-dimensional Fourier integral exists and is identically zero outside an n-dimensional rectangle symmetrical about the origins, that is,

$$F(y_1,y_2, \ldots ,y_n) = 0 \qquad \text{for } |y_k| > |\omega_k| \quad k = 1, 2, \ldots , n$$

Then

$$f(t_1,t_2, \ldots ,t_n) = \sum_{m_1=-\infty}^{m_1=+\infty} \cdots \sum_{m_n=-\infty}^{m_n=+\infty} f\left(\pi\frac{m_1}{\omega_1}, \ldots , \pi\frac{m_n}{\omega_n}\right) \frac{\sin \omega_1(t_1 - m_1/\omega_1)}{\omega_1(t_1 - m_1/\omega_1)} \cdots \frac{\sin \omega_n(t_n - m_n/\omega_n)}{\omega_n(t_n - m_n/\omega_n)} \tag{N2-4}$$

* A. V. Balakrishnan, A Note on the Sampling Principle for Continuous Signals, *IRE Trans. on Inform. Theory*, vol. IT-3, pp. 143–146, 1957.

The proof of this theorem follows directly from the proof of the sampling theorem in the time domain. For details, see the report by E. Parzen.* Other generalizations of sampling theorems have been obtained by A. Kohlenberg (*J. Appl. Phys.*, vol. 24, 1953) and D. Gabor (London Symposium on Information Theory, 1960).

5. According to Kolmogorov (I), the sampling theorem was also used in communication problems by the Russian scientist Kotel'nikov.†

6. An interesting informal derivation of the sampling theorem, along with several other results, has been given by the late Balth Van Der Pol (*Ann. Computation Lab., Harvard Univ.*, vol. 29, pt. I, pp. 3–25, 1959).

Let $p(x)$ be an entire function with simple roots at $\{. . . ,a_1,a_2, . . .\}$ and $f(x)$ a band-limited time function with a cutoff angular frequency of π; then, assuming that $f(x)$ and $p(x)$ have no common root, in an informal manner, Van der Pol writes

$$\frac{f(x)}{p(x)} = \sum_{k=-\infty}^{\infty} \frac{f(a_k)}{p'(a_k)(x - a_k)} \qquad k \text{ integer} \tag{N2-5}$$

If $p(x)$ is selected to be equal to $\sin \pi x$, that is,

$$\{\cdot \cdot \cdot ,-2,-1,0,1,2, \cdot \cdot \cdot\}$$

one finds

$$\frac{f(x)}{\sin \pi x} = \sum_{k=-\infty}^{\infty} \frac{f(k)}{\pi \cos \pi k} \frac{1}{x - k}$$

$$f(x) = \sum_{k=-\infty}^{\infty} f(k) \frac{\sin \pi(x - k)}{\pi(x - k)}$$

Several interesting classical results could also be derived from the sampling theorem. For instance, if $f(x)$ is taken to be unity, we find

$$1 = \sum_{k=-\infty}^{\infty} \frac{\sin \pi(t - k)}{\pi(t - k)}$$

or

$$\frac{1}{\sin \pi t} = \frac{1}{\pi} \sum_{k=-\infty}^{\infty} \frac{(-1)^k}{t - k} \tag{N2-6}$$

N-3. Analytic Signals and the Uncertainty Relation. While this topic is of basic interest in the communication sciences, adequate space for its

* E. Parzen, "A Simple Proof and Some Extensions of the Sampling Theorem," *Stanford Univ. Tech. Rept. 7, Dept. of Statistics*, December, 1956.

† T. A. Kotel'nikov, Material for the First All-Union Conference on Questions of Communications, 1933.

presentation is not available here. Nonetheless, we wish to introduce the reader to the existence of the topic. D. Gabor* focused attention in this direction in 1946, followed by E. Wolf in 1947.

Let $f(t)$ be a normalized function of the real variable t, in L_2, and $F(\omega)$ its Fourier integral transform [see Eqs. (9-18)]. Then $|f(t)|^2$ and $|F(\omega)|^2$ may be considered as probability density distributions for two random variables, say T and Ω. The standard deviation of each of these random variables in terms of their moments is

$$\begin{aligned}(\sigma_T)^2 &= \overline{T^2} - \bar{T}^2 = \int_{-\infty}^{\infty} (t - \bar{T})^2 |f(t)|^2 \, dt \\ (\sigma_\Omega)^2 &= \overline{\Omega^2} - \bar{\Omega}^2 = \int_{-\infty}^{\infty} (\omega - \bar{\Omega})^2 |F(\omega)|^2 \, d\omega\end{aligned} \tag{N3-1}$$

What is referred to as the uncertainty relation in quantum mechanics implies that

$$\sigma_T \sigma_\Omega \geq \tfrac{1}{2} \tag{N3-2}$$

For the proof of this relation, which is mathematically straightforward, the reader is referred to a text on quantum mechanics.

In the communication sciences we deal primarily with a real time function $f(t)$, which in turn implies

$$\begin{aligned}|F(\omega)| &= |F^*(-\omega)| \\ \arg F(\omega) &= -\arg F(-\omega) \\ \bar{\Omega} &= 0\end{aligned}$$

When $f(t)$ is real, by applying the Fourier integral transform formula it can be shown that the function $f(t)$ satisfying the equality sign in the uncertainty relation is

$$\begin{aligned}f(t) &= \left(\frac{1}{\sqrt{2\pi}\,\sigma_T}\right)^{1/2} \exp\left[-\frac{(t - \bar{T})^2}{4\sigma_T^2}\right] \\ F(\omega) &= \left(\frac{2\sigma_T}{\sqrt{2\pi}}\right)^{1/2} \exp\left(-\sigma_T^2\omega^2\right)\end{aligned} \tag{N3-3}$$

From a communication-theory point of view, D. Gabor finds it convenient to introduce the concept of the *analytic signal* $f_+(t)$, which is a complex function associated with a real time function $f(t)$:

$$f(t) = \sqrt{2}\,\mathrm{Re}\, f_+(t) \tag{N3-4}$$

* The concept of analytic signals was introduced in communication theory by Gabor and Ville. For the definition and properties of analytic signals, see Gabor, Ville, and Cherry: D. Gabor, Theory of Communications, *J. Inst. elec. Engrs. (London)*, vol. 93, pt. III, pp. 429–457, November, 1946; J. Ville, Théorie et application de la notion de signal analytique, *Câbles et transm.*, January, 1948, pp. 61–74; C. Cherry, Quelques remarques sur le temps considéré comme variable complexe, *Onde eléctrique*, vol. 34, pp. 7–13, January, 1954.

The analytic signal has the interesting property that its Fourier transform $F_+(\omega)$ is identically zero for $\omega < 0$, that is,*

$$f_+(t) = \frac{1}{\sqrt{\pi}} \int_0^\infty F(\omega) e^{j\omega t}\, d\omega$$
$$F_+(\omega) = \sqrt{2}\, F(\omega) \qquad \omega \geq 0$$
$$F_+(\omega) \equiv 0 \qquad \omega < 0$$

Now the previously discussed probability distributions lend themselves to further simplification. In fact, for the new random variables T_+ and Ω_+,

$$\sigma_{T_+}\sigma_{\Omega_+} \geq \tfrac{1}{2} \tag{N3-5}$$

E. Wolf has pointed out that, if σ_{T_+} exists, then $\sigma_{T_+} = \sigma_T$. Conversely, the existence of σ_T implies the existence of σ_{T_+} if $F(0) = 0$. Subject to this requirement, the uncertainty relation becomes

$$\sigma_T\sigma_{\Omega_+} > \tfrac{1}{2} \tag{N3-6}$$

There are a large number of articles available on this subject in the engineering as well as the mathematical literature. To mention one, the reader is referred to Silverman and Kay. In this article, they derive the following interesting inequality:

$$\sigma_T\sigma_{\Omega_+} \geq \tfrac{1}{2}|1 - 2|F(0)|^2\bar{\Omega}_+| \equiv \tfrac{1}{2}|1 - |F_+(0)|^2\bar{\Omega}_+| \tag{N3-7}$$

which reduces to the familiar uncertainty relation when $F(0) = 0$.

Similar results but from a slightly different point of view have also appeared in the literature. See, for instance, D. G. Lampard.†

A function $f(t)$ which is not identically zero cannot be band-limited in the time and frequency domains simultaneously. A proof of this statement can be obtained directly from a classic theorem of Paley and Wiener.‡ The theorem states:

Let $\phi(\omega)$ be a real nonnegative square-integrable function not equivalent to zero in $[-\infty < \omega < +\infty]$. A necessary and sufficient condition that there should exist a function $f(t)$ defined in $-\infty < t < \infty$ and

* The connection between analytic signals, network theory, Hilbert transforms, and classical function theory is discussed in detail by Oswald and Zemanian: J. R. V. Oswald, The Theory of Analytic Band-limited Signals Applied to Carrier Systems, *IRE Trans. on Circuit Theory*, vol. CT-3, pp. 245–251, December, 1956; A. H. Zemanian, Network Realizability in the Time Domain, *IRE Trans. on Circuit Theory*, vol. CT-6, pp. 288–291, September, 1959.

† D. G. Lampard, Definitions of "Bandwidth" and "Time Duration" of Signals Which Are Connected by an Identity, *IRE Trans. on Circuit Theory*, vol. CT-3, pp. 286–288, December, 1956.

‡ R. E. A. C. Paley and N. Wiener, Fourier Transforms in the Complex Domain, *Am. Math. Soc. Colloq.*, vol. 19, p. 16, 1934.

identically zero for some range $t \geq t_0$, and such that

$$\phi(\omega) = |\text{Fourier integral of } f(t)| = |F(\omega)|$$

is that

$$\int_{-\infty}^{\infty} \frac{|\log \phi(\omega)|}{1 + \omega^2} \, d\omega < \infty \tag{N3-8}$$

It can be seen that, if $F(\omega)$ is identically zero over a finite range, over that range the above integral is not finite. (An interesting alternative proof based on the sampling theorem is given by R. E. Wernikoff in the reference cited below.*)

N-4. Elias's Proof of the Fundamental Theorem for BSC. Elias has suggested a proof for the fundamental theorem in the case of the two most common types of discrete noisy channels without memory, BSC and BEC. His method of proof is based on what is referred to as random block coding. The messages are encoded in blocks each n digits long. When a word is received, the receiver lists the most probable word, that is, the word in the code that differs in the least number of places from the received sequence. If there is more than one such word, the decision will necessarily be ambiguous. The plausibility of this method is, of course, due to the fact that in a BSC with $p < \frac{1}{2}$ the probability of a number of k errors occurring in a word $p^k q^{n-k}$ is a monotonically decreasing function of k (see Secs. 4-11 and 4-16).

Let each n-digit message contain m information digits and $n - m$ parity digits. Thus, the rate of transmission of information R is m/n. We assume without loss of generality that the number of messages to be transmitted is

$$M = 2^m = 2^{nR}$$

and the parity check used allows correction of all sets of k_1 or fewer errors in each n-digit block. If all the input block words are transmitted with equal probability, the transinformation per symbol will be

$$R = \frac{1}{n} \log M \leq \frac{1}{n} \log \left[2^n \Big/ \sum_{j=0}^{k_1} \binom{n}{j} \right] = 1 - \frac{1}{n} \log \sum_{j=0}^{k_1} \binom{n}{j} \tag{N4-1}$$

(See, for instance, the sphere-packing argument of Secs. 13-5 and 4-13.) The equality can hold only under the most favorable circumstances of lossless coding. The minimum ambiguity probability Q_{opt} corresponding to those received words that are not within the disjoint spheres of radii k_1

* R. E. Wernikoff, Time-limited and Band-limited Functions, *Mass. Inst. Technol., Research Lab. Electronics, Quart. Progr. Rept.*, January, 1957, pp. 72–74.

centered at the transmitted words can be directly computed from the tail of the binomial distribution:

$$Q_{\text{opt}} = \sum_{j=k_1+1}^{n} \binom{n}{j}$$

Next, consider all possible such parity codes selected at random. That is, we select at random 2^{nR} n-symbol words from a total of 2^n possible such words. Of course, since a word may be selected twice or more as a code word for a given message, the ensemble of these codes may contain some very poor codes. However, the average ambiguity of all these codes cannot be less than the ambiguity Q_b of the best of them. Elias has proved that for a fixed k_1 and

$$p < p_1 = \frac{k_1}{n} < p_{\text{crit}} = \frac{\sqrt{p}}{\sqrt{p} + \sqrt{q}} \tag{N4-2}$$

we may write

$$Q_b \leq Q_{\text{av}} \leq p^{np_1}q^{nq_1} \binom{n}{np_1} \left[\frac{pq_1}{p_1 - p} + \frac{1}{1 - (q/p)(p_1/q_1)^2} \right]$$
$$\geq Q_{\text{opt}} \geq p^{np_1}q^{nq_1} \binom{n}{np_1} \frac{q_1}{p_1 + 1/n} \tag{N4-3}$$

Using Stirling's approximation, these inequalities can be rewritten as

$$Q_b \leq Q_{\text{av}} \leq \left[\frac{pq_1}{p_1 - p} + \frac{1}{1 - (q/p)(p_1/q_1)^2} \right] \exp_2 \left\{ -n \left[C + R + (p_1 - p) \log \frac{q}{p} \right] \right\}$$
$$\geq Q_{\text{opt}} \approx \frac{pq_1}{p_1 - p} \exp_2 \left\{ -n \left[-C + R + (p_1 - p) \log \frac{q}{p} \right] \right\} \tag{N4-4}$$

Elias concludes that Q_b, the ambiguity probability of the best code, exponentially depends on n. This result may be finally expressed in the form

$$k_2 2^{-nB_2} \leq Q_{\text{av}} \leq k_1 2^{-nB_1} \tag{N4-5}$$

The probability of error, on an average, is bounded above and below by exponential functions. The terms B_1 and B_2 are independent of n and depend solely on the channel parameters and the specified R. k_1 and k_2 are nearly constant terms. Thus, for a given channel and transmission rate, the probability of error can be made arbitrarily small by increasing the block length. Similar results hold for the case $p_1 > p_{\text{crit}}$. The mathematical derivation of these formulas is somewhat complex. The reader is referred to Elias's article in the Proceedings of the Third London Symposium on Information Theory, 1955. A mathematically more refined treatment of Elias's original exponential bounds is presented in Feinstein (I).

Similar results for a BSC were also independently derived by several other authors, for instance, G. A. Barnard (Third London Symposium on Information Theory, pp. 96–102, 1955) and E. N. Gilbert (*Bell Tel. Labs. Tech. Mem.*, June, 1956). An alternative proof was also given by J. M. Wozencraft (*Mass. Inst. Technol., Research Lab. Electronics, Quart. Progr. Rept.*, Jan. 15, 1958, pp. 90–95). Wozencraft's proof is

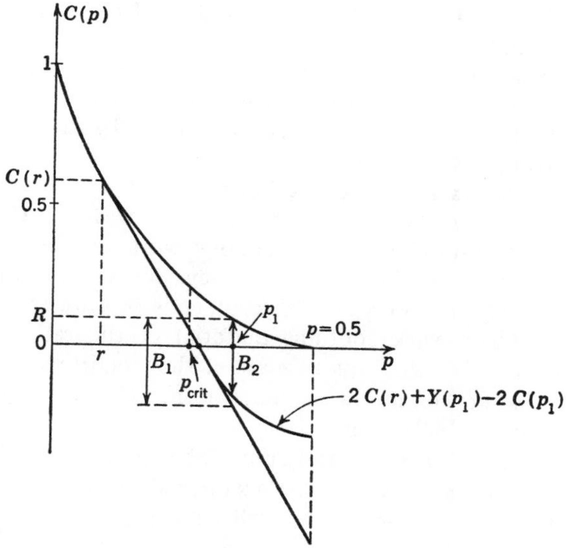

FIG. N4-1. A graphical method for determination of the error-probability bounds.

motivated by Shannon's method, as described in Chap. 12. He assumes maximum-likelihood decoding procedure and computes accordingly the probability of error. A bound for the probability of error is obtained by using Shannon's method, which, in lieu of Elias's use of approximation by Stirling's formula, applies Chernov's inequality to mutual information.

A geometrical interpretation of the dependence of B_1 and B_2 on channel probability p is illustrated in Fig. N4-1. The curve represents the capacity of a BSC for various values of p.

$$C = 1 + p \log p + q \log q = 1 - H(p) \tag{N4-6}$$

The equation of the tangent to the curve at a point with abscissa r is

$$Y = C(r) + (p - r)[\log r - \log (1 - r)] \tag{N4-7}$$

The exponent B_1 is given by the difference of the ordinates of this curve and the tangent at $r = p$ for any desired transmission rate when $p_1 < p_{crit}$. This is illustrated in Fig. N4-1 for a specified transmission rate R. For $p_1 < p_{crit}$, the exponent B_2 is equal to B_1, but for $p_1 > p_{crit}$ a slight modification of the above procedure is in order. We note that in this range

$$B_2 = C(r) - [2C(r) + Y(p_1) - 2C(p_1)] < B_1$$

Thus, for a geometrical interpretation one has to trace a new curve and measure the difference of its ordinate with that of the capacity curve, as illustrated in Fig. N4-1.*

Also, note that in region $p_1 < p_{crit}$ the two exponents are identical; that is, the upper and the lower bounds are proportional. In that region the ratio of the two bounds is proportional to n. In the range $p_1 > p_{crit}$, the bounds diverge exponentially.

N-5. Further Remarks on Coding Theory

The Bose-Chaudhuri t-error Correcting Group Codes. An interesting generalization of Hamming's and Slepian's work has been derived by R. C. Bose and D. K. R. Chaudhuri.† They have derived the necessary and sufficient conditions for a binary group code (n,k) to be a δ-error correcting code. Their work contains material of theoretical as well as practical interest. We shall quote some of their main results with little (if any) indication of their method of derivation. For the proof the reader is referred to the original article.

Theorem 1. The necessary and sufficient condition for the existence of a δ-error correcting (n,k) binary group code is the existence of an $n \times r$ matrix $[A]$ with $r = n - k$ such that any 2δ row vectors of A are mutually independent.

Theorem 2. If $n = 2^m - 1$, there exists a δ-error correcting binary group code (n,k) with

$$k \geq n - \delta m$$

The proof of Theorem 1 is based in part on Hamming's lemma, which states: The necessary and sufficient condition for a binary δ-error correcting group code (n,k) is that each word (except the null word) have a norm of $2\delta + 1$. Theorem 2 is a sharper statement than the result obtained earlier by Varsamov.‡ The latter author has shown that

* See Elias, cited above, and R. M. Fano, The Statistical Theory of Information, *Nuovo cimento*, ser. X, vol. 13, suppl. 2, pp. 353–372.

† R. C. Bose and D. K. R. Chaudhuri, On a Class of Error Correcting Binary Group Codes, *Inform. and Control*, vol. 3, no. 1, pp. 68–79, 1960.

‡ R. R. Varsamov, The Evaluation of Signals in Codes with Correction of Errors, *Doklady Akad. Nauk S.S.S.R.* new series, vol. 117, pp. 739–741, 1957.

if k satisfies the inequality

$$S_r^{2\delta-1} + \binom{k-1}{1} S_r^{2\delta-2} + \cdots + \binom{k-1}{2\delta-2} S_r^{1} + \binom{k-1}{2\delta-1} < 2^r$$

where $$S_r^q = 1 + \binom{r}{1} + \binom{r}{2} + \cdots + \binom{r}{q} \tag{N5-1}$$

then a δ-error correcting binary group code (n,k) exists.

The merit of the proof suggested by Bose and Chaudhuri lies particularly in the fact that it provides a constructive method for the codes. Also, the implementation of these codes does not seem to be too complex (see, for instance, Peterson*). Peterson points out that these codes have a *cyclic* property and hence can be implemented with what is called a *shift-register generator*, as demonstrated earlier by Prange.† (The cyclic property implies that, if a word $u = a_1, a_2, \ldots, a_n \in S$, the words obtained by cyclically shifting digits of u in some fashion are also members of S, for example, $u_1 = a_n, a_1, a_2, \ldots, a_{n-1} \in S$. An early study of shift-register generators can be found in a report by N. Zierler.‡)

Dependent Error Correction. In all error correcting schemes thus far presented, the occurrence of error was assumed to be a statistically independent phenomenon. In many data-processing systems the occurrence of an error in a particular binary digit is conditional on the occurrence of error in the preceding digits. Several interesting procedures for the detection and correction of interdependent errors have appeared in the literature for special error patterns, although a general solution has not yet been devised. An interesting class of these codes has been investigated by N. M. Abramson, P. Fire, and D. W. Hagelbarger. The latter author devises codes for correction and detection of a "burst" of errors (for example, when lightning may knock out several adjacent telegraph pulses). A brief discussion of Abramson's approach is presented below without reference to the practically important problem of instrumentation.

Abramson has suggested a code which corrects single or double adjacent error (SEC-DAEC). Let m be the number of information digits and k the number of parity checks; the number of distinct single and double adjacent errors in a word with $n = m + k$ digits is $n + n = 2n$. The parity-check number k then must satisfy

$$\begin{aligned} 2^k &\geq 2(m+k) + 1 \\ m &\leq 2^{k-1} - k - \tfrac{1}{2} \end{aligned} \tag{N5-2}$$

* W. W. Peterson, "Error Correcting Codes," John Wiley & Sons, Inc., New York.

† E. Prange, "Some Cyclic Error-correcting Codes with Simple Decoding Algorithms," Air Force Cambridge Research Center AFCRC-TN-58-156, April, 1958.

‡ N. Zierler, Several Binary-sequence Generators, *Mass. Inst. Technol., Lincoln Lab. Tech. Rept.* 95, September, 1955.

When the number of parity digits satisfies the equality

$$m = m_0 = 2^{k-1} - k - 1$$

the code is referred to as a complete code. The following values are given in Abramson.

k	4	5	6	7	8	9	10
m_0	3	10	25	56	119	246	501

To devise a complete SEC-DAEC, one has to set up a set of $2n + 1$ binary equations whose solution determines the single or the double adjacent-error position. A systematic method for setting up these equations and an instrumentation technique based on the use of *shift register* are given in the above cited reference.

Convolution Codes. Most of the work available on coding theory is related, in one way or another, to systematic block codes. A completely different type of coding was suggested by J. M. Wozencraft in 1957. In these codes, the information message is a single sequence of binary digits. The parity checks assume some specific pattern; for example, they may be interlaced by information digits. Each check digit is determined by the preceding digits through a checking equation. Of course, theoretically, for a long message, it seems that one has to take into consideration the effect of all transmitted digits. But, for all practical purposes, as far as the error probability is concerned, one may confine oneself to a suitable number of immediately preceding digits. A most significant property of convolution codes is the fact that in a certain sense the "average amount of computation per digit" for encoding-decoding, hence the equipment needed for implementation of codes, is quite realistic. The encoding-decoding procedure for convolution codes and the computation of error probability have been accomplished in the past 3 years. A monograph giving a full description of these codes is in preparation by J. M. Wozencraft of Massachusetts Institute of Technology. (See also M. A. Epstein.)

N-6. Partial Ordering of Channels. A recent contribution of C. E. Shannon* has provided a basis of comparison for some communication channels through their stochastic matrices. While it is early to speculate on possible applications of Shannon's original idea, it appears that it will encompass some important areas of investigation. Algebraic operations on stochastic matrices and their physical interpretation seem to indicate an area where information theory and systems theory considerations could join forces.

Consider two discrete memoryless channels with m input and m output

* C. E. Shannon, "A Note on a Partial Ordering for Communication Channels, Information and Control," vol. 1, pp. 390–397, Academic Press, Inc., New York, 1958.

symbols. The cascading of these channels is equivalent to a channel whose stochastic matrix K is the product of the two corresponding stochastic matrices K_1 and K_2. In the following, for simplicity, we consider first the cascading of such channels. The results can be generalized in a direct fashion.

A channel K_1 is said to include another channel K_2 if there exists a channel K such that K_2 can be obtained by cascading K_1 and K. The stochastic matrix of K_2 is the product of the stochastic matrices of K_1 and K. The channel-inclusion relation is denoted by $K_1 \supseteq K_2$. If for two channels neither $K_1 \supseteq K_2$ nor $K_2 \subseteq K_1$, then the two channels are mutually exclusive or not comparable. We also say that such channels have no partial ordering. The following properties may be derived:

1. Transitive property. If $K_1 \supseteq K_2$ and $K_2 \supseteq K_3$, then

$$K_1 \supseteq K_3$$

2. Multiplication. If

$$K_1 \supseteq K_3$$
$$K_2 \supseteq K_4$$

then

$$K_1K_2 \supseteq K_3K_4$$

3. Convexity. Let K_1, K_2, and K_3 be the stochastic matrices of three channels such that

$$K_1 \supseteq K_2$$
$$K_1 \supseteq K_3$$

and K the stochastic matrix of a new channel where

$$K = \lambda K_2 + (1 - \lambda)K_3$$
$$0 \leq \lambda \leq 1$$

Then

$$K_1 \supseteq K$$

The above definition and properties can next be generalized for defining a partial ordering of two finite channels K_1 and K_2 (not necessarily $m \times m$). We say that $K_1 \supseteq K_2$ if K_1 could be derived from K_2 by a *pre* and a *post* cascading channel, that is,

$$K_1 = AK_2B \tag{N6-1}$$

From a physical point of view, as Shannon has described, partial ordering means roughly that some sort of operation is applied to a channel K_2 in order that it look like K_1 (for instance, cascading of A, K_2, and B).

As an exercise, the reader may wish to derive the necessary and sufficient conditions for two given binary channels to have a partial ordering. Shannon has derived an interesting theorem on two channels with a partial-ordering relation, namely,

Theorem. Let

1. $(m_1, m_2, \ldots, m_N)$ be a set of n-sequence words transmitted with specified probabilities $(p_1, p_2, \ldots, p_N)$ over a discrete memoryless channel K_1.

2. p_{e1} be the probability of error for this channel under some specified decoding scheme.

3.

$$K_1 \supseteq K_2$$

Then there exists a set of N n-sequence messages and a decoding scheme for K_2 such that, if messages are transmitted with the same input probabilities as before, the error probability (p_{e2}) will not increase, that is,

$$p_{e2} \leq p_{e1}$$

For a mathematical proof of this intuitively plausible statement, see Shannon (*loc. cit.*).

N-7. Information Theory and Radar Problems. The central problem in this area is the detection of radar signals of known characteristics in the presence of noise. For instance, if a signal $x(t)$ is transmitted and $y(t)$ is received, assuming some additive noise, we have

$$y(t) = x(t - \tau) + \text{noise}$$

A first problem is the evaluation of delay time τ and the radar range by comparison of x and y. Generally the noise is taken to be gaussian and x and y as random variables with specified characteristics. Thus, subsequent to certain plausible assumptions, one is able to find some type of probability distribution for the delay and the range.

The role of information theory, although perhaps conceptually enlightening from a procedural point of view, is a secondary one. The most complex part of the problem is in formulating and solving an input-output type of probability problem for some linear or nonlinear system. When the problem is solved, that is, a probability distribution function for the unknown is derived, the entropy associated with that distribution reveals a certain measure of uncertainty about the searched quantity. This problem falls in the general field of statistical extraction of signal from noisy background and design of optimum filters. Thus it appears that, while the problem occupies an important place in the statistical theory of communication, it is not immediately related to our subject. Furthermore, several sources with adequate coverage of this subject are available. For the benefit of the reader the following list of references is included. A treatment of the now classic work of Woodward and Davis on radar problems appears in P. M. Woodward (Chaps. 5–7).

For a concise presentation of radar detection theory based on the maximum-likelihood technique, see W. B. Davenport, Jr., and W. L. Root (Chap. 14).

For a comprehensive theoretical study of the subject, including the work of Middleton and Van Meter, and a list of the most recent contributions, see D. Middleton.

Table T-1. Normal Probability Integral

$$\phi_{(z)} = \frac{1}{\sqrt{2\pi}} \int_0^z e^{-\frac{1}{2}t^2}\, dt$$

z	$\phi_{(z)}$	z	$\phi_{(z)}$	z	$\phi_{(z)}$	z	$\phi_{(z)}$
0.00	0.0000	0.65	0.2422	1.30	0.4032	1.95	0.4744
0.01	0.0040	0.66	0.2454	1.31	0.4049	1.96	0.4750
0.02	0.0080	0.67	0.2486	1.32	0.4066	1.97	0.4756
0.03	0.0120	0.68	0.2517	1.33	0.4082	1.98	0.4761
0.04	0.0160	0.69	0.2549	1.34	0.4099	1.99	0.4767
0.05	0.0199	0.70	0.2580	1.35	0.4115	2.00	0.4772
0.06	0.0239	0.71	0.2611	1.36	0.4131	2.02	0.4783
0.07	0.0279	0.72	0.2642	1.37	0.4147	2.04	0.4793
0.08	0.0319	0.73	0.2673	1.38	0.4162	2.06	0.4803
0.09	0.0359	0.74	0.2703	1.39	0.4177	2.08	0.4812
0.10	0.0398	0.75	0.2734	1.40	0.4192	2.10	0.4821
0.11	0.0438	0.76	0.2764	1.41	0.4207	2.12	0.4830
0.12	0.0478	0.77	0.2794	1.42	0.4222	2.14	0.4838
0.13	0.0517	0.78	0.2823	1.43	0.4236	2.16	0.4846
0.14	0.0557	0.79	0.2852	1.44	0.4251	2.18	0.4854
0.15	0.0596	0.80	0.2881	1.45	0.4265	2.20	0.4861
0.16	0.0636	0.81	0.2910	1.46	0.4279	2.22	0.4868
0.17	0.0675	0.82	0.2939	1.47	0.4292	2.24	0.4875
0.18	0.0714	0.83	0.2967	1.48	0.4306	2.26	0.4881
0.19	0.0753	0.84	0.2995	1.49	0.4319	2.28	0.4887
0.20	0.0793	0.85	0.3023	1.50	0.4332	2.30	0.4893
0.21	0.0832	0.86	0.3051	1.51	0.4345	2.32	0.4898
0.22	0.0871	0.87	0.3078	1.52	0.4357	2.34	0.4904
0.23	0.0910	0.88	0.3106	1.53	0.4370	2.36	0.4909
0.24	0.0948	0.89	0.3133	1.54	0.4382	2.38	0.4913
0.25	0.0987	0.90	0.3159	1.55	0.4394	2.40	0.4918
0.26	0.1026	0.91	0.3186	1.56	0.4406	2.42	0.4922
0.27	0.1064	0.92	0.3212	1.57	0.4418	2.44	0.4927
0.28	0.1103	0.93	0.3238	1.58	0.4429	2.46	0.4931
0.29	0.1141	0.94	0.3264	1.59	0.4441	2.48	0.4934
0.30	0.1179	0.95	0.3289	1.60	0.4452	2.50	0.4938
0.31	0.1217	0.96	0.3315	1.61	0.4463	2.52	0.4941
0.32	0.1255	0.97	0.3340	1.62	0.4474	2.54	0.4945
0.33	0.1293	0.98	0.3365	1.63	0.4484	2.56	0.4948
0.34	0.1331	0.99	0.3389	1.64	0.4495	2.58	0.4951
0.35	0.1368	1.00	0.3413	1.65	0.4505	2.60	0.4953
0.36	0.1406	1.01	0.3438	1.66	0.4515	2.62	0.4956
0.37	0.1443	1.02	0.3461	1.67	0.4525	2.64	0.4959
0.38	0.1480	1.03	0.3485	1.68	0.4535	2.66	0.4961
0.39	0.1517	1.04	0.3508	1.69	0.4545	2.68	0.4963
0.40	0.1554	1.05	0.3531	1.70	0.4554	2.70	0.4965
0.41	0.1591	1.06	0.3554	1.71	0.4564	2.72	0.4967
0.42	0.1628	1.07	0.3577	1.72	0.4573	2.74	0.4969
0.43	0.1664	1.08	0.3599	1.73	0.4582	2.76	0.4971
0.44	0.1700	1.09	0.3621	1.74	0.4591	2.78	0.4973
0.45	0.1736	1.10	0.3643	1.75	0.4599	2.80	0.4974
0.46	0.1772	1.11	0.3665	1.76	0.4608	2.82	0.4976
0.47	0.1808	1.12	0.3686	1.77	0.4616	2.84	0.4977
0.48	0.1844	1.13	0.3708	1.78	0.4625	2.86	0.4979
0.49	0.1879	1.14	0.3729	1.79	0.4633	2.88	0.4980
0.50	0.1915	1.15	0.3749	1.80	0.4641	2.90	0.4981
0.51	0.1950	1.16	0.3770	1.81	0.4649	2.92	0.4982
0.52	0.1985	1.17	0.3790	1.82	0.4656	2.94	0.4984
0.53	0.2019	1.18	0.3810	1.83	0.4664	2.96	0.4985
0.54	0.2054	1.19	0.3830	1.84	0.4671	2.98	0.4986
0.55	0.2088	1.20	0.3849	1.85	0.4678	3.00	0.49865
0.56	0.2123	1.21	0.3869	1.86	0.4686	3.20	0.49931
0.57	0.2157	1.22	0.3888	1.87	0.4693	3.40	0.49966
0.58	0.2190	1.23	0.3907	1.88	0.4699	3.60	0.499841
0.59	0.2224	1.24	0.3925	1.89	0.4706	3.80	0.499928
0.60	0.2257	1.25	0.3944	1.90	0.4713	4.00	0.499968
0.61	0.2291	1.26	0.3962	1.91	0.4719	4.50	0.499997
0.62	0.2324	1.27	0.3980	1.92	0.4726	5.00	0.499997
0.63	0.2357	1.28	0.3997	1.93	0.4732		
0.64	0.2389	1.29	0.4015	1.94	0.4738		

TABLE T-2. NORMAL DISTRIBUTIONS

z	$P\{0 < X \leq z\}$ 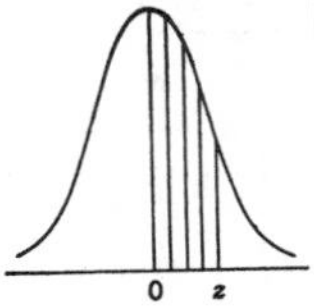	$P\{-z < X \leq z\}$ 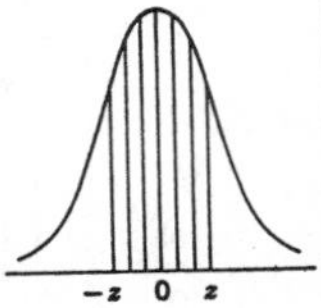	$P\{\lvert X\rvert > z\}$ 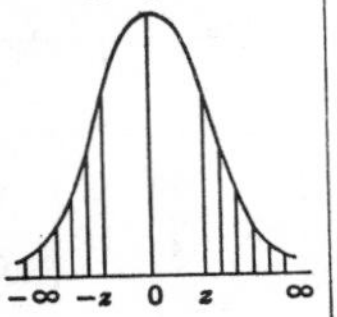	$P\{X \leq z\}$ 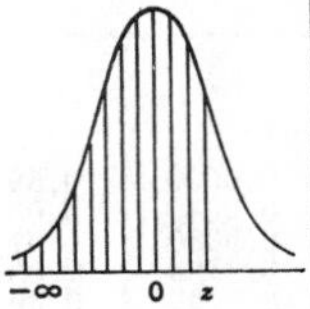	$P\{X > z\}$ 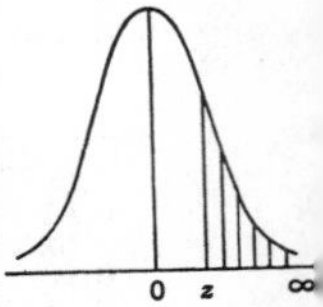
1.000	0.34134	0.68268	0.31732	0.84134	0.15866
1.960	0.47500	0.95000	0.05000	0.97500	0.02500
2.000	0.47725	0.95450	0.04550	0.97725	0.02275
2.576	0.49500	0.99000	0.01000	0.99500	0.00500
3.000	0.49865	0.99730	0.00270	0.99865	0.00135
3.291	0.49950	0.99900	0.00100	0.99950	0.00050

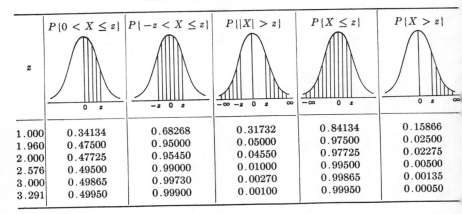

TABLE T-3. A SUMMARY OF SOME COMMON PROBABILITY FUNCTIONS

Probability function*	Mean $E(X)$	Variance $\sigma^2 = E(X^2) - [E(X)]^2$	Characteristic function $E(e^{jtX})$
$\binom{n}{k} p^k q^{n-k}$ $k = 0, 1, 2, \ldots, n$	np	npq	$(pe^{jt} + q)^n$
$\frac{e^{-\lambda}\lambda^k}{k!}$ $k = 0, 1, 2, \ldots, n$	λ	λ	$e^{\lambda}(e^{jt} - 1)$
$\frac{1}{\sigma\sqrt{2\pi}} e^{-\frac{1}{2}[x-m/\sigma]^2}$ $\sigma > 0 \quad -\infty < x < \infty$	m	σ^2	$e^{jtm - \frac{1}{2}t^2\sigma^2}$
$f(x) = \frac{1}{\pi}\frac{1}{1+x^2}$ $-\infty < x < \infty$	0	See Sec. 6-8	$e^{-\lvert t \rvert}$
$f(x) = \frac{1}{b-a}$ $a < x < b$	$\frac{a+b}{2}$	$\frac{(b-a)^2}{12}$	$\frac{e^{jtb} - e^{jta}}{jt(b-a)}$
$f(x) = \frac{1}{2a}$ $-a < x < a$	0	$\frac{a^2}{3}$	$\frac{\sin at}{at}$
$f(x) = 1 - \lvert x \rvert$ $\lvert x \rvert < 1$	0	$\frac{1}{6}$	$\frac{2}{t^2}(1 - \cos t)$
$\lambda e^{-\lambda x} \quad \lambda > 0$ $0 < x < \infty$	$\frac{1}{\lambda}$	$\frac{1}{\lambda^2}$	$\left(1 - \frac{jt}{\lambda}\right)^{-1}$
$f(x) = \frac{a}{2e^{-a\lvert x \rvert}}$ $-\infty < x < \infty \quad a > 0$	0	2	$\frac{a^2}{a^2 + t^2}$

* Assumed to be zero outside the domain of definition.

TABLE T-4. PROBABILITY OF NO ERROR FOR BEST GROUP CODE*

$$Q_1 = \Sigma \gamma_i p^i q^{n-i}$$

	i	$\binom{n}{i}$	$k = 2$ γ_i	$k = 3$ γ_i	$k = 4$ γ_i	$k = 5$ γ_i	$k = 6$ γ_i	$k = 7$ γ_i	$k = 8$ γ_i	$k = 9$ γ_i	$k = 10$ γ_i
$n = 4$	0	1	1								
	1	4	3								
$n = 5$	0	1	1	1							
	1	5	5	3							
	2	10	2								
$n = 6$	0	1	1	1	1						
	1	6	6	6	3						
	2	15	9	1							
$n = 7$	0	1	1	1	1	1					
	1	7	7	7	7	3					
	2	21	18	8							
	3	25	6								
$n = 8$	0	1	1	1	1	1	1				
	1	8	8	8	8	7	3				
	2	28	28	20	7						
	3	56	27	3							
$n = 9$	0	1	1	1	1	1	1	1			
	1	9	9	9	9	9	7	3			
	2	36	36	33	22	6					
	3	84	64	21							
	4	126	18								
$n = 10$	0	1	1	1	1	1	1	1	1		
	1	10	10	10	10	10	10	7	3		
	2	45	45	45	39	21	5				
	3	120	110	64	14						
	4	210	90	8							
$n = 11$	0	1	1	1	1		1	1	1	1	
	1	11	11	11	11		11	11	7	3	
	2	55	55	55	55		20	4			
	3	165	165	126	61						
	4	330	226	63							
	5	462	54								
$n = 12$	0	1	1	1				1	1	1	1
	1	12	12	12				12	12	7	3
	2	66	66	66				19	3		
	3	220	220	200							
	4	495	425	233							
	5	792	300								

* Reproduced from D. Slepian, *Bell System Tech. J.*, vol. 35, p. 213, January, 1956, with kind permission of *Bell System Technical Journal*.

TABLE T-5. PARITY-CHECK RULES FOR BEST GROUP CODE*

	$k = 2$	$k = 3$	$k = 4$	$k = 5$	$k = 6$	$k = 7$	$k = 8$	$k = 9$	$k = 10$
$n = 4$	3 2 4 1 2								
$n = 5$	3 1 2 4 2 5 1	4 1 2 5 1 3							
$n = 6$	3 2 4 1 2 5 1 6 1	4 1 2 5 1 3 6 2 3	5 1 2 3 6 1 2 4						
$n = 7$	3 1 4 1 5 1 6 1 2 7 2	4 1 3 5 1 2 6 1 2 3 7 1 2 3	5 1 3 4 6 1 2 4 7 1 2 3	6 1 7 1					
$n = 8$	3 1 4 1 5 2 6 2 7 1 2 8 1 2	4 1 5 1 2 6 1 3 7 2 3 8 1 2 3	5 1 3 4 6 1 2 4 7 1 2 3 8 1 2 3 4	6 1 3 4 7 1 2 4 8 1 2 3	7 1 8 1				
$n = 9$	3 1 4 1 5 1 6 2 7 2 8 1 2 9 1 2	4 1 5 2 6 1 2 7 1 3 8 2 3 9 1 2 3	5 1 3 4 6 1 2 4 7 1 2 3 8 1 2 3 9 1 2 3	6 1 3 4 5 7 1 2 4 5 8 1 2 3 5 9 1 2 3 4	7 1 3 4 8 1 2 4 9 1 2 3	8 1 9 1			

Table T-5. Parity-check Rules for Best Group Code (*Continued*)

	$k = 2$	$k = 3$	$k = 4$	$k = 5$	$k = 6$	$k = 7$	$k = 8$	$k = 9$	$k = 10$
$n = 10$	3 1 4 1 5 1 6 2 7 2 8 1 2 9 1 2 10 1 2	4 1 5 2 6 3 7 1 2 8 1 3 9 2 3 10 1 2 3	5 3 4 6 1 2 3 7 1 2 4 8 1 3 4 9 2 3 4 10 1 2 3 4	6 1 3 4 5 7 1 2 4 5 8 1 2 3 5 9 1 2 3 4 10 1 2 3 4 5	7 1 3 4 5 8 1 2 4 5 9 1 2 3 5 6 10 1 2 3 4 6	8 1 3 4 9 1 2 4 10 1 2 3	9 1 10 1		
$n = 11$	3 1 4 1 5 1 6 2 7 2 8 2 9 1 2 10 1 2 11 1 2	4 3 5 3 6 2 7 1 3 8 1 3 9 1 2 10 1 2 3 11 1 2 3	5 1 3 6 2 4 7 1 4 8 2 3 9 1 3 4 10 2 3 4 11 1 2 3 4		7 1 3 4 5 6 8 1 2 4 5 6 9 1 2 3 5 6 10 1 2 3 4 6 11 1 2 3 4 5	8 1 3 4 5 9 1 2 4 5 7 10 1 2 3 5 6 11 1 2 3 4 6 7	9 1 3 4 10 1 2 4 11 1 2 3	10 1 11 1	
$n = 12$	3 1 4 1 5 1 6 1 7 2 8 2 9 2 10 2 11 1 2 12 1 2	4 1 5 2 6 3 7 1 2 8 1 2 9 1 3 10 2 3 11 1 2 3 12 1 2 3				8 1 3 4 5 6 9 1 2 4 5 6 10 1 2 3 5 6 7 11 1 2 3 4 6 7 12 1 2 3 4 5 7	9 1 2 3 5 6 7 8 10 1 2 3 4 6 11 1 2 4 5 7 12 1 3 4 5 8	10 1 2 3 11 1 2 4 12 1 3 4	11 1 12 1

* Reproduced from D. Slepian, *Bell System Tech. J.*, vol. 35, pp. 216–217, January, 1956, with kind permission of *Bell System Technical Journal*.

TABLE T-6. LOGARITHM TO THE BASE 2

N	Log N	N	Log N	N	Log N	N	Log N
1	0.000000	51	5.672425	101	6.658211	151	7.238404
2	1.000000	52	5.700439	102	6.672425	152	7.247927
3	1.584962	53	5.727920	103	6.686500	153	7.257388
4	2.000000	54	5.754887	104	6.700439	154	7.266786
5	2.321928	55	5.781359	105	6.714245	155	7.276124
6	2.584962	56	5.807355	106	6.727920	156	7.285402
7	2.807355	57	5.832890	107	6.741467	157	7.294620
8	3.000000	58	5.857981	108	6.754887	158	7.303780
9	3.169925	59	5.882643	109	6.768184	159	7.312883
10	3.321928	60	5.906890	110	6.781359	160	7.321928
11	3.459431	61	5.930737	111	6.794415	161	7.330916
12	3.584962	62	5.954196	112	6.807355	162	7.339850
13	3.700440	63	5.977280	113	6.820179	163	7.348728
14	3.807355	64	6.000000	114	6.832890	164	7.357552
15	3.906890	65	6.022367	115	6.845490	165	7.366322
16	4.000000	66	6.044394	116	6.857981	166	7.375039
17	4.087463	67	6.066089	117	6.870364	167	7.383704
18	4.169925	68	6.087462	118	6.882643	168	7.392317
19	4.247927	69	6.108524	119	6.894817	169	7.400879
20	4.321928	70	6.129283	120	6.906890	170	7.409391
21	4.392317	71	6.149747	121	6.918863	171	7.417852
22	4.459431	72	6.169925	122	6.930737	172	7.426264
23	4.523562	73	6.189824	123	6.942514	173	7.434628
24	4.584962	74	6.209453	124	6.954196	174	7.442943
25	4.643856	75	6.228818	125	6.965784	175	7.451211
26	4.700439	76	6.247927	126	6.977280	176	7.459431
27	4.754887	77	6.266786	127	6.988684	177	7.467605
28	4.807355	78	6.285402	128	7.000000	178	7.475733
29	4.857981	79	6.303780	129	7.011227	179	7.483815
30	4.906890	80	6.321928	130	7.022367	180	7.491853
31	4.954196	81	6.339850	131	7.033423	181	7.499846
32	5.000000	82	6.357552	132	7.044394	182	7.507794
33	5.044394	83	6.375039	133	7.055282	183	7.515699
34	5.087463	84	6.392317	134	7.066089	184	7.523562
35	5.129283	85	6.409391	135	7.076815	185	7.531381
36	5.169925	86	6.426264	136	7.087462	186	7.539158
37	5.209453	87	6.442943	137	7.098032	187	7.546894
38	5.247927	88	6.459431	138	7.108524	188	7.554588
39	5.285402	89	6.475733	139	7.118941	189	7.562242
40	5.321928	90	6.491853	140	7.129283	190	7.569855
41	5.357552	91	6.507794	141	7.139551	191	7.577428
42	5.392317	92	6.523562	142	7.149747	192	7.584962
43	5.426264	93	6.539158	143	7.159871	193	7.592457
44	5.459431	94	6.554588	144	7.169925	194	7.599912
45	5.491853	95	6.569855	145	7.179909	195	7.607330
46	5.523562	96	6.584962	146	7.189824	196	7.614709
47	5.554589	97	6.599912	147	7.199672	197	7.622051
48	5.584962	98	6.614709	148	7.209453	198	7.629356
49	5.614710	99	6.629356	149	7.219168	199	7.636624
50	5.643856	100	6.643856	150	7.228818	200	7.643856

TABLE T-6. LOGARITHM TO THE BASE 2 (*Continued*)

N	Log N	N	Log N	N	Log N	N	Log N
201	7.651051	251	7.971543	301	8.233619	351	8.455327
202	7.658211	252	7.977280	302	8.238404	352	8.459431
203	7.665336	253	7.982993	303	8.243174	353	8.463524
204	7.672425	254	7.988684	304	8.247927	354	8.467605
205	7.679480	255	7.994353	305	8.252665	355	8.471675
206	7.686500	256	8.000000	306	8.257387	356	8.475733
207	7.693487	257	8.005624	307	8.262094	357	8.479780
208	7.700439	258	8.011227	308	8.266786	358	8.483815
209	7.707359	259	8.016808	309	8.271463	359	8.487840
210	7.714245	260	8.022367	310	8.276124	360	8.491853
211	7.721099	261	8.027906	311	8.280770	361	8.495855
212	7.727920	262	8.033423	312	8.285402	362	8.499846
213	7.734709	263	8.038918	313	8.290018	363	8.503825
214	7.741467	264	8.044394	314	8.294620	364	8.507794
215	7.748192	265	8.049848	315	8.299208	365	8.511752
216	7.754887	266	8.055282	316	8.303780	366	8.515699
217	7.761551	267	8.060696	317	8.308339	367	8.519636
218	7.768184	268	8.066089	318	8.312883	368	8.523561
219	7.774787	269	8.071462	319	8.317412	369	8.527476
220	7.781359	270	8.076815	320	8.321928	370	8.531381
221	7.787902	271	8.082149	321	8.326429	371	8.535275
222	7.794415	272	8.087463	322	8.330916	372	8.539158
223	7.800899	273	8.092757	323	8.335390	373	8.543031
224	7.807354	274	8.098032	324	8.339850	374	8.546894
225	7.813781	275	8.103287	325	8.344296	375	8.550746
226	7.820179	276	8.108524	326	8.348728	376	8.554588
227	7.826548	277	8.113742	327	8.353146	377	8.558420
228	7.832890	278	8.118941	328	8.357552	378	8.562242
229	7.839203	279	8.124121	329	8.361943	379	8.566054
230	7.845490	280	8.129283	330	8.366322	380	8.569855
231	7.851749	281	8.134426	331	8.370687	381	8.573647
232	7.857981	282	8.139551	332	8.375039	382	8.577428
233	7.864186	283	8.144658	333	8.379378	383	8.581200
234	7.870364	284	8.149747	334	8.383704	384	8.584962
235	7.876516	285	8.154818	335	8.388017	385	8.588714
236	7.882643	286	8.159871	336	8.392317	386	8.592457
237	7.888743	287	8.164907	337	8.396604	387	8.596189
238	7.894817	288	8.169925	338	8.400879	388	8.599912
239	7.900866	289	8.174925	339	8.405141	389	8.603626
240	7.906890	290	8.179909	340	8.409390	390	8.607330
241	7.912889	291	8.184875	341	8.413628	391	8.611024
242	7.918863	292	8.189824	342	8.417852	392	8.614709
243	7.924812	293	8.194757	343	8.422064	393	8.618385
244	7.930737	294	8.199672	344	8.426264	394	8.622051
245	7.936638	295	8.204571	345	8.430452	395	8.625708
246	7.942514	296	8.209453	346	8.434628	396	8.629356
247	7.948367	297	8.214319	347	8.438791	397	8.632995
248	7.954196	298	8.219168	348	8.442943	398	8.636624
249	7.960002	299	8.224001	349	8.447083	399	8.640244
250	7.965784	300	8.228818	350	8.451211	400	8.643856

TABLE T-6. LOGARITHM TO THE BASE 2 (*Continued*)

N	Log *N*	*N*	Log *N*	*N*	Log *N*	*N*	Log *N*
401	8.647458	451	8.816983	501	8.968666	551	9.105908
402	8.651051	452	8.820178	502	8.971543	552	9.108524
403	8.654636	453	8.823367	503	8.974414	553	9.111135
404	8.658211	454	8.826548	504	8.977279	554	9.113742
405	8.661778	455	8.829722	505	8.980139	555	9.116343
406	8.665335	456	8.832889	506	8.982993	556	9.118941
407	8.668885	457	8.836050	507	8.985841	557	9.121533
408	8.672425	458	8.839203	508	8.988684	558	9.124121
409	8.675956	459	8.842350	509	8.991521	559	9.126704
410	8.679480	460	8.845490	510	8.994353	560	9.129283
411	8.682994	461	8.848622	511	8.997179	561	9.131857
412	8.686500	462	8.851748	512	9.000000	562	9.134426
413	8.689997	463	8.854868	513	9.002815	563	9.136991
414	8.693486	464	8.857980	514	9.005624	564	9.139551
415	8.696967	465	8.861086	515	9.008428	565	9.142107
416	8.700439	466	8.864186	516	9.011227	566	9.144658
417	8.703903	467	8.867278	517	9.014020	567	9.147205
418	8.707359	468	8.870364	518	9.016808	568	9.149747
419	8.710806	469	8.873444	519	9.019590	569	9.152285
420	8.714245	470	8.876516	520	9.022367	570	9.154818
421	8.717676	471	8.879583	521	9.025139	571	9.157346
422	8.721099	472	8.882643	522	9.027906	572	9.159871
423	8.724513	473	8.885696	523	9.030667	573	9.162391
424	8.727920	474	8.888743	524	9.033423	574	9.164907
425	8.731318	475	8.891783	525	9.036173	575	9.167418
426	8.734709	476	8.894817	526	9.038918	576	9.169925
427	8.738092	477	8.897845	527	9.041659	577	9.172427
428	8.741466	478	8.900866	528	9.044394	578	9.174925
429	8.744833	479	8.903881	529	9.047123	579	9.177419
430	8.748192	480	8.906890	530	9.049848	580	9.179909
431	8.751544	481	8.909893	531	9.052568	581	9.182394
432	8.754887	482	8.912889	532	9.055282	582	9.184875
433	8.758223	483	8.915879	533	9.057991	583	9.187352
434	8.761551	484	8.918863	534	9.060696	584	9.189824
435	8.764871	485	8.921840	535	9.063395	585	9.192292
436	8.768184	486	8.924812	536	9.066089	586	9.194757
437	8.771489	487	8.927777	537	9.068778	587	9.197216
438	8.774786	488	8.930737	538	9.071462	588	9.199672
439	8.778077	489	8.933690	539	9.074141	589	9.202123
440	8.781359	490	8.936637	540	9.076815	590	9.204571
441	8.784634	491	8.939579	541	9.079484	591	9.207014
442	8.787902	492	8.942514	542	9.082149	592	9.209453
443	8.791162	493	8.945443	543	9.084808	593	9.211888
444	8.794415	494	8.948367	544	9.087462	594	9.214319
445	8.797661	495	8.951284	545	9.090112	595	9.216745
446	8.800899	496	8.954196	546	9.092757	596	9.219168
447	8.804130	497	8.957102	547	9.095397	597	9.221587
448	8.807354	498	8.960001	548	9.098032	598	9.224001
449	8.810571	499	8.962896	549	9.100662	599	9.226412
450	8.813781	500	8.965784	550	9.103287	600	9.228818

TABLE T-6. LOGARITHM TO THE BASE 2 (*Continued*)

N	Log N	N	Log N	N	Log N	N	Log N
601	9.231221	651	9.346513	701	9.453270	751	9.552669
602	9.233619	652	9.348728	702	9.455327	752	9.554588
603	9.236014	653	9.350939	703	9.457380	753	9.556506
604	9.238404	654	9.353146	704	9.459431	754	9.558420
605	9.240791	655	9.355351	705	9.461479	755	9.560332
606	9.243174	656	9.357552	706	9.463524	756	9.562242
607	9.245552	657	9.359749	707	9.465566	757	9.564149
608	9.247927	658	9.361943	708	9.467605	758	9.566053
609	9.250298	659	9.364134	709	9.469641	759	9.567956
610	9.252665	660	9.366322	710	9.471675	760	9.569855
611	9.255028	661	9.368506	711	9.473705	761	9.571752
612	9.257387	662	9.370687	712	9.475733	762	9.573647
613	9.259743	663	9.372865	713	9.477758	763	9.575539
614	9.262094	664	9.375039	714	9.479780	764	9.577428
615	9.264442	665	9.377210	715	9.481799	765	9.579315
616	9.266786	666	9.379378	716	9.483815	766	9.581200
617	9.269126	667	9.381542	717	9.485829	767	9.583082
618	9.271463	668	9.383704	718	9.487840	768	9.584962
619	9.273795	669	9.385862	719	9.489848	769	9.586839
620	9.276124	670	9.388017	720	9.491853	770	9.588714
621	9.278449	671	9.390169	721	9.493855	771	9.590587
622	9.280770	672	9.392317	722	9.495855	772	9.592457
623	9.283088	673	9.394462	723	9.497851	773	9.594324
624	9.285402	674	9.396604	724	9.499846	774	9.596189
625	9.287712	675	9.398743	725	9.501837	775	9.598052
626	9.290018	676	9.400879	726	9.503825	776	9.599912
627	9.292321	677	9.403012	727	9.505811	777	9.601770
628	9.294620	678	9.405141	728	9.507794	778	9.603626
629	9.296916	679	9.407267	729	9.509774	779	9.605479
630	9.299208	680	9.409390	730	9.511752	780	9.607330
631	9.301496	681	9.411511	731	9.513727	781	9.609178
632	9.303780	682	9.413628	732	9.515699	782	9.611024
633	9.306061	683	9.415741	733	9.517669	783	9.612868
634	9.308339	684	9.417852	734	9.519636	784	9.614709
635	9.310612	685	9.419960	735	9.521600	785	9.616548
636	9.312883	686	9.422064	736	9.523561	786	9.618385
637	9.315149	687	9.424166	737	9.525520	787	9.620219
638	9.317412	688	9.426264	738	9.527476	788	9.622051
639	9.319672	689	9.428360	739	9.529430	789	9.623881
640	9.321928	690	9.430452	740	9.531381	790	9.625708
641	9.324180	691	9.432541	741	9.533329	791	9.627533
642	9.326429	692	9.434628	742	9.535275	792	9.629356
643	9.328674	693	9.436711	743	9.537218	793	9.631177
644	9.330916	694	9.438791	744	9.539158	794	9.632995
645	9.333155	695	9.440869	745	9.541096	795	9.634811
646	9.335390	696	9.442943	746	9.543031	796	9.636624
647	9.337621	697	9.445014	747	9.544964	797	9.638435
648	9.339850	698	9.447083	748	9.546894	798	9.640244
649	9.342074	699	9.449148	749	9.548821	799	9.642051
650	9.344296	700	9.451211	750	9.550746	800	9.643856

TABLE T-6. LOGARITHM TO THE BASE 2 (*Continued*)

N	Log N	N	Log N	N	Log N	N	Log N
801	9.645658	851	9.733015	901	9.815383	951	9.893301
802	9.647458	852	9.734709	902	9.816983	952	9.894817
803	9.649256	853	9.736401	903	9.818582	953	9.896332
804	9.651051	854	9.738092	904	9.820178	954	9.897845
805	9.562844	855	9.739780	905	9.821773	955	9.899356
806	9.654636	856	9.741466	906	9.823367	956	9.900866
807	9.656424	857	9.743151	907	9.824958	957	9.902375
808	9.658211	858	9.744833	908	9.826548	958	9.903881
809	9.659995	859	9.746514	909	9.828136	959	9.905386
810	9.661778	860	9.748192	910	9.829722	960	9.906890
811	9.663557	861	9.749869	911	9.831307	961	9.908392
812	9.665335	862	9.751544	912	9.832889	962	9.909893
813	9.667111	863	9.753216	913	9.834471	963	9.911391
814	9.668884	864	9.754887	914	9.836050	964	9.912889
815	9.670656	865	9.756556	915	9.837627	965	9.914385
816	9.672425	866	9.758223	916	9.839203	966	9.915879
817	9.674192	867	9.759888	917	9.840777	967	9.917372
818	9.675956	868	9.761551	918	9.842350	968	9.918863
819	9.677719	869	9.763212	919	9.843920	969	9.920352
820	9.679479	870	9.764871	920	9.845490	970	9.921840
821	9.681238	871	9.766528	921	9.847057	971	9.923327
822	9.682994	872	9.768184	922	9.848622	972	9.924812
823	9.684748	873	9.769837	923	9.850186	973	9.926295
824	9.686500	874	9.771489	924	9.851748	974	9.927777
825	9.688250	875	9.773139	925	9.853309	975	9.929258
826	9.699997	876	9.774786	926	9.854868	976	9.930737
827	9.691743	877	9.776433	927	9.856425	977	9.932214
828	9.693486	878	9.778077	928	9.857980	978	9.933690
829	9.695228	879	9.779719	929	9.859534	979	9.935164
830	9.696967	880	9.781359	930	9.861086	980	9.936637
831	9.698704	881	9.782998	931	9.862637	981	9.938109
832	9.700439	882	9.784634	932	9.864186	982	9.939579
833	9.702172	883	9.786269	933	9.865733	983	9.941047
834	9.703903	884	9.787902	934	9.867278	984	9.942514
835	9.705632	885	9.789533	935	9.868822	985	9.943979
836	9.707359	886	9.791162	936	9.870364	986	9.945443
837	9.709083	887	9.792790	937	9.871905	987	9.946906
838	9.710806	888	9.794415	938	9.873443	988	9.948367
839	9.712526	889	9.796039	939	9.874981	989	9.949826
840	9.714245	890	9.797661	940	9.876516	990	9.951284
841	9.715961	891	9.799281	941	9.878050	991	9.952741
842	9.717676	892	9.800899	942	9.879583	992	9.954196
843	9.719388	893	9.802516	943	9.881113	993	9.955649
844	9.721099	894	9.804130	944	9.882643	994	9.957102
845	9.722807	895	9.805743	945	9.884170	995	9.958552
846	9.724513	896	9.807354	946	9.885696	996	9.960001
847	9.726218	897	9.808964	947	9.887220	997	9.961449
848	9.727920	898	9.810571	948	9.888743	998	9.962896
849	9.729620	899	9.812177	949	9.890264	999	9.964340
850	9.731318	900	9.813781	950	9.891783	1000	9.965784

TABLE T-7. ENTROPY OF A DISCRETE BINARY SOURCE

P	$-\operatorname{Log} P$	$-P \operatorname{Log} P$	H	$-Q \operatorname{Log} Q$	$-\operatorname{Log} Q$
0.0001	13.287712	0.001329	0.001473	0.000144	0.000144
0.0005	10.965784	0.005483	0.006204	0.000721	0.000722
0.0010	9.965784	0.009966	0.011408	0.001442	0.001443
0.0015	9.380822	0.014071	0.016234	0.002162	0.002166
0.0020	8.965784	0.017932	0.020814	0.002882	0.002888
0.0025	8.643856	0.021610	0.025212	0.003602	0.003611
0.0030	8.380822	0.025142	0.029464	0.004322	0.004335
0.0035	8.158429	0.028555	0.033595	0.005041	0.005058
0.0040	7.965784	0.031863	0.037622	0.005759	0.005782
0.0045	7.795859	0.035081	0.041559	0.006477	0.006507
0.0050	7.643856	0.038219	0.045415	0.007195	0.007232
0.0055	7.506353	0.041285	0.049198	0.007913	0.007957
0.0060	7.380822	0.044285	0.052915	0.008630	0.008682
0.0065	7.265345	0.047225	0.056572	0.009347	0.009408
0.0070	7.158429	0.050109	0.060172	0.010063	0.010134
0.0075	7.058894	0.052942	0.063721	0.010780	0.010861
0.0080	6.965784	0.055726	0.067222	0.011495	0.011588
0.0085	6.878321	0.058466	0.070676	0.012211	0.012315
0.0090	6.795859	0.061163	0.074088	0.012926	0.013043
0.0095	6.717857	0.063820	0.077460	0.013640	0.013771
0.0100	6.643856	0.066439	0.080793	0.014355	0.014500
0.0110	6.506353	0.071570	0.087352	0.015782	0.015958
0.0120	6.380822	0.076570	0.093778	0.017208	0.017417
0.0130	6.265345	0.081449	0.100082	0.018633	0.018878
0.0140	6.158429	0.086218	0.106274	0.020056	0.020340
0.0150	6.058894	0.090883	0.112361	0.021477	0.021804
0.0160	5.965784	0.095453	0.118350	0.022897	0.023270
0.0170	5.878321	0.099931	0.124248	0.024316	0.024737
0.0180	5.795859	0.104325	0.130059	0.025733	0.026205
0.0190	5.717857	0.108639	0.135788	0.027149	0.027675
0.0200	5.643856	0.112877	0.141441	0.028563	0.029146
0.0210	5.573467	0.117043	0.147019	0.029976	0.030619
0.0220	5.506353	0.121140	0.152527	0.031388	0.032094
0.0230	5.442222	0.125171	0.157969	0.032797	0.033570
0.0240	5.380822	0.129140	0.163346	0.034206	0.035047
0.0250	5.321928	0.133048	0.168661	0.035613	0.036526
0.0260	5.265345	0.136899	0.173917	0.037018	0.038006
0.0270	5.210897	0.140694	0.179116	0.038422	0.039488
0.0280	5.158429	0.144436	0.184261	0.039825	0.040972
0.0290	5.107803	0.148126	0.189352	0.041226	0.042457

TABLE T-7. ENTROPY OF A DISCRETE BINARY SOURCE (*Continued*)

P	$-\operatorname{Log} P$	$-P \operatorname{Log} P$	H	$-Q \operatorname{Log} Q$	$-\operatorname{Log} Q$
0.0300	5.058894	0.151767	0.194392	0.042625	0.043943
0.0310	5.011588	0.155359	0.199382	0.044023	0.045431
0.0320	4.965784	0.158905	0.204325	0.045420	0.046921
0.0330	4.921390	0.162406	0.209220	0.046815	0.048412
0.0340	4.878321	0.165863	0.214071	0.048208	0.049905
0.0350	4.836501	0.169278	0.218878	0.049600	0.051399
0.0360	4.795859	0.172651	0.223642	0.050991	0.052895
0.0370	4.756331	0.175984	0.228364	0.052380	0.054392
0.0380	4.717857	0.179279	0.233046	0.053767	0.055891
0.0390	4.680382	0.182535	0.237688	0.055153	0.057392
0.0400	4.643856	0.185754	0.242292	0.056538	0.058894
0.0410	4.608232	0.188938	0.246858	0.057921	0.060397
0.0420	4.573467	0.192086	0.251388	0.059303	0.061902
0.0430	4.539519	0.195199	0.255882	0.060683	0.063409
0.0440	4.506353	0.198280	0.260341	0.062061	0.064917
0.0450	4.473931	0.201327	0.264765	0.063438	0.066427
0.0460	4.442222	0.204342	0.269156	0.064814	0.067939
0.0470	4.411195	0.207326	0.273514	0.066188	0.069452
0.0480	4.380822	0.210279	0.277840	0.067560	0.070967
0.0490	4.351074	0.213203	0.282134	0.068931	0.072483
0.0500	4.321928	0.216096	0.286397	0.070301	0.074001
0.0510	4.293359	0.218961	0.290630	0.071668	0.075520
0.0520	4.265345	0.221798	0.294833	0.073035	0.077041
0.0530	4.237864	0.224607	0.299007	0.074400	0.078564
0.0540	4.210897	0.227388	0.303152	0.075763	0.080088
0.0550	4.184425	0.230143	0.307268	0.077125	0.081614
0.0560	4.158429	0.232872	0.311357	0.078485	0.083141
0.0570	4.132894	0.235575	0.315419	0.079844	0.084670
0.0580	4.107803	0.238253	0.319454	0.081201	0.086201
0.0590	4.083141	0.240905	0.323462	0.082557	0.087733
0.0600	4.058894	0.243534	0.327445	0.083911	0.089267
0.0625	4.000000	0.250000	0.337290	0.087290	0.093109
0.0650	3.943416	0.256322	0.346981	0.090659	0.096962
0.0675	3.888969	0.262505	0.356524	0.094019	0.100824
0.0700	3.836501	0.268555	0.365924	0.097369	0.104697
0.0725	3.785875	0.274476	0.375185	0.100709	0.108581
0.0750	3.736966	0.280272	0.384312	0.104039	0.112475
0.0775	3.689660	0.285949	0.393308	0.107360	0.116379
0.0800	3.643856	0.291508	0.402179	0.110671	0.120294
0.0825	3.599462	0.296956	0.410927	0.113972	0.124220

TABLE T-7. ENTROPY OF A DISCRETE BINARY SOURCE (*Continued*)

P	$-\text{Log } P$	$-P \text{ Log } P$	H	$-Q \text{ Log } Q$	$-\text{Log } Q$
0.0850	3.556393	0.302293	0.419556	0.117263	0.128156
0.0875	3.514573	0.307525	0.428070	0.120544	0.132104
0.0900	3.473931	0.312654	0.436470	0.123816	0.136062
0.0925	3.434403	0.317682	0.444760	0.127078	0.140030
0.0950	3.395929	0.322613	0.452943	0.130329	0.144010
0.0975	3.358454	0.327449	0.461020	0.133571	0.148001
0.1000	3.321928	0.332193	0.468996	0.136803	0.152003
0.1025	3.286304	0.336846	0.476871	0.140024	0.156016
0.1050	3.251539	0.341412	0.484648	0.143236	0.160040
0.1075	3.217591	0.345891	0.492329	0.146438	0.164076
0.1100	3.184425	0.350287	0.499916	0.149629	0.168123
0.1125	3.152003	0.354600	0.507411	0.152811	0.172181
0.1150	3.120294	0.358834	0.514816	0.155982	0.176251
0.1175	3.089267	0.362989	0.522132	0.159143	0.180332
0.1200	3.058894	0.367067	0.529361	0.162294	0.184425
0.1225	3.029146	0.371070	0.536505	0.165434	0.188529
0.1250	3.000000	0.375000	0.543564	0.168564	0.192645
0.1275	2.971431	0.378857	0.550542	0.171684	0.196773
0.1300	2.943416	0.382644	0.557438	0.174794	0.200913
0.1325	2.915936	0.386361	0.564255	0.177893	0.205064
0.1350	2.888969	0.390011	0.570993	0.180982	0.209228
0.1375	2.862496	0.393593	0.577654	0.184061	0.213404
0.1400	2.836501	0.397110	0.584239	0.187129	0.217591
0.1425	2.810966	0.400563	0.590749	0.190186	0.221791
0.1450	2.785875	0.403952	0.597185	0.193233	0.226004
0.1475	2.761213	0.407279	0.603549	0.196270	0.230228
0.1500	2.736966	0.410545	0.609840	0.199295	0.234465
0.1525	2.713119	0.413751	0.616061	0.202311	0.238715
0.1550	2.689660	0.416897	0.622213	0.205315	0.242977
0.1575	2.666576	0.419986	0.628295	0.208309	0.247251
0.1600	2.643856	0.423017	0.634310	0.211293	0.251539
0.1625	2.621488	0.425992	0.640257	0.214265	0.255839
0.1650	2.599462	0.428911	0.646138	0.217227	0.260152
0.1675	2.577767	0.431776	0.651954	0.220178	0.264478
0.1700	2.556393	0.434587	0.657705	0.223118	0.268817
0.1725	2.535332	0.437345	0.663392	0.226047	0.273169
0.1750	2.514573	0.440050	0.669016	0.228966	0.277534
0.1775	2.494109	0.442704	0.674577	0.231873	0.281912
0.1800	2.473931	0.445308	0.680077	0.234769	0.286304
0.1825	2.454032	0.447861	0.685516	0.237655	0.290709

Table T-7. Entropy of a Discrete Binary Source (*Continued*)

P	$-\text{Log } P$	$-P \text{ Log } P$	H	$-Q \text{ Log } Q$	$-\text{Log } Q$
0.1850	2.434403	0.450365	0.690894	0.240529	0.295128
0.1875	2.415037	0.452820	0.696212	0.243393	0.299560
0.1900	2.395929	0.455226	0.701471	0.246245	0.304006
0.1925	2.377070	0.457586	0.706672	0.249086	0.308466
0.1950	2.358454	0.459899	0.711815	0.251916	0.312939
0.1975	2.340075	0.462165	0.716900	0.254735	0.317427
0.2000	2.321928	0.464386	0.721928	0.257542	0.321928
0.2050	2.286304	0.468692	0.731816	0.263124	0.330973
0.2100	2.251539	0.472823	0.741483	0.268660	0.340075
0.2150	2.217591	0.476782	0.750932	0.274150	0.349235
0.2200	2.184425	0.480573	0.760167	0.279594	0.358454
0.2250	2.152003	0.484201	0.769193	0.284992	0.367732
0.2300	2.120294	0.487668	0.778011	0.290344	0.377070
0.2350	2.089267	0.490978	0.786626	0.295648	0.386468
0.2400	2.058894	0.494134	0.795040	0.300906	0.395929
0.2450	2.029146	0.497141	0.803257	0.306116	0.405451
0.2500	2.000000	0.500000	0.811278	0.311278	0.415037
0.2550	1.971431	0.502715	0.819107	0.316392	0.424688
0.2600	1.943416	0.505288	0.826746	0.321458	0.434403
0.2650	1.915936	0.507723	0.834198	0.326475	0.444184
0.2700	1.888969	0.510022	0.841465	0.331443	0.454032
0.2750	1.862496	0.512187	0.848548	0.336362	0.463947
0.2800	1.836501	0.514220	0.855451	0.341230	0.473931
0.2850	1.810966	0.516125	0.862175	0.346049	0.483985
0.2900	1.785875	0.517904	0.868721	0.350817	0.494109
0.2950	1.761213	0.519558	0.875093	0.355535	0.504305
0.3000	1.736966	0.521090	0.881291	0.360201	0.514573
0.3050	1.713119	0.522501	0.887317	0.364816	0.524915
0.3100	1.689660	0.523795	0.893173	0.369379	0.535332
0.3150	1.666576	0.524972	0.898861	0.373890	0.545824
0.3200	1.643856	0.526034	0.904381	0.378347	0.556393
0.3250	1.621488	0.526984	0.909736	0.382752	0.567041
0.3300	1.599462	0.527822	0.914926	0.387104	0.577767
0.3350	1.577767	0.528552	0.919953	0.391402	0.588574
0.3400	1.556393	0.529174	0.924819	0.395645	0.599462
0.3450	1.535332	0.529689	0.929523	0.399834	0.610433
0.3500	1.514573	0.530101	0.934068	0.403967	0.621488
0.3550	1.494109	0.530409	0.938454	0.408046	0.632629
0.3600	1.473931	0.530615	0.942683	0.412068	0.643856
0.3650	1.454032	0.530722	0.946755	0.416034	0.655171

TABLE T-7. ENTROPY OF A DISCRETE BINARY SOURCE (*Continued*)

P	$-\text{Log } P$	$-P \text{ Log } P$	H	$-Q \text{ Log } Q$	$-\text{Log } Q$
0.3700	1.434403	0.530729	0.950672	0.419943	0.666576
0.3750	1.415037	0.530639	0.954434	0.423795	0.678072
0.3800	1.395929	0.530453	0.958042	0.427589	0.689660
0.3850	1.377070	0.530172	0.961497	0.431325	0.701342
0.3900	1.358454	0.529797	0.964800	0.435002	0.713119
0.3950	1.340075	0.529330	0.967951	0.438621	0.724993
0.4000	1.321928	0.528771	0.970951	0.442179	0.736966
0.4050	1.304006	0.528122	0.973800	0.445678	0.749038
0.4100	1.286304	0.527385	0.976500	0.449116	0.761213
0.4150	1.268817	0.526559	0.979051	0.452493	0.773491
0.4200	1.251539	0.525646	0.981454	0.455808	0.785875
0.4250	1.234465	0.524648	0.983708	0.459061	0.798366
0.4300	1.217591	0.523564	0.985815	0.462251	0.810966
0.4350	1.200913	0.522397	0.987775	0.465378	0.823677
0.4400	1.184425	0.521147	0.989588	0.468441	0.836501
0.4450	1.168123	0.519815	0.991254	0.471439	0.849440
0.4500	1.152003	0.518401	0.992774	0.474373	0.862496
0.4550	1.136062	0.516908	0.994149	0.477241	0.875672
0.4600	1.120294	0.515335	0.995378	0.480043	0.888969
0.4650	1.104697	0.513684	0.996462	0.482778	0.902389
0.4700	1.089267	0.511956	0.997402	0.485446	0.915936
0.4750	1.074001	0.510150	0.998196	0.488046	0.929611
0.4800	1.058894	0.508269	0.998846	0.490577	0.943416
0.4850	1.043943	0.506313	0.999351	0.493038	0.957356
0.4900	1.029146	0.504282	0.999711	0.495430	0.971431
0.4950	1.014500	0.502177	0.999928	0.497751	0.985645
0.5000	1.000000	0.500000	1.000000	0.500000	1.000000

BIBLIOGRAPHY

REFERENCE BOOKS ON PROBABILITY THEORY

Bartlett, M. S.: "An Introduction to Stochastic Processes," Cambridge University Press, New York, 1955.

Carnap, Rudolf: "Logical Foundations of Probability," University of Chicago Press, Chicago, 1950.

Cramèr, Harald: "Mathematical Methods of Statistics," Princeton University Press, Princeton, N.J., 1946.

Darmois, G.: "Calcul des probabilités," Centre de documentation universitaire, Sorbonne, Paris, 1954.

Derman, C., and M. Klein: "Probability and Statistical Inference," Oxford University Press, New York, 1958.

Doob, John L.: "Stochastic Processes," John Wiley & Sons, Inc., New York, 1953.

Dugué, D.: "Traité de statistique théorique et appliquée," Masson et Cie, Paris, 1958.

Feller, William: "Probability Theory and Its Applications," John Wiley & Sons, Inc., New York, 1950.

Fortet, R.: "Calcul des probabilités, I," Centre national de la recherche scientifique, Paris, 1950.

Frazer, D. A. S.: "Statistics: An Introduction," John Wiley & Sons, Inc., New York, 1958.

Jeffreys, Harold: "Theory of Probability," 2d ed., Oxford University Press, New York, 1948.

Kemeny, J. G., and J. L. Snell: "Finite Markov Chains," D. Van Nostrand Company, Inc., Princeton, N.J., 1960.

Kolmogorov, Andrei N. (I): "Foundations of the Theory of Probability," Chelsea Publishing Company, New York, 1950.

Lehman, E. L.: "Theory of Testing Hypotheses," John Wiley & Sons, Inc., New York, 1960.

Loève, Michel: "Probability Theory," D. Van Nostrand Company, Inc., Princeton, N.J., 1955.

Mood, Alexander M.: "Introduction to the Theory of Statistics," McGraw-Hill Book Company, Inc., New York, 1950.

Parzen, E.: "Modern Probability Theory and Its Applications," John Wiley & Sons, Inc., New York, 1960.

Uspensky, J. V.: "Introduction to Mathematical Probability," McGraw-Hill Book Company, Inc., New York, 1937.

Wald, A.: "Statistical Decision Functions," John Wiley & Sons, Inc., New York, 1950.

Wilks, S. S.: "Mathematical Statistics," Princeton University Press, Princeton, N.J., 1943.

REFERENCE BOOKS ON STATISTICAL THEORY OF COMMUNICATIONS

Baghdady, R. J. (ed.): "Lectures on Communication System Theory," McGraw-Hill Book Company, Inc., New York, 1960.

Bell, D. A.: "Information Theory," Sir Isaac Pitman & Sons, Ltd., London, 1953.

Bendat, J. S.: "Principles and Applications of Random Noise Theory," John Wiley & Sons, Inc., New York, 1958.

Blanc-Lapierre, André, and Robert Fortet: "Théorie des fonctions aléatoires," Masson et Cie, Paris, 1953.

Brillouin, L. (I): "Science and Information Theory," Academic Press, Inc., New York, 1956.

Cherry, E. C.: "On Human Communication," John Wiley & Sons, Inc., New York, 1957.

Cherry, E. C. (ed.): "Information Theory," Third London Symposium, September, 1955, Academic Press, Inc., New York, 1956.

Davenport, W. B., Jr., and W. L. Root: "Introduction to Random Signals and Noise," McGraw-Hill Book Company, Inc., New York, 1958.

Fano, R. M. (I): Statistical Theory of Communication, unpublished notes, Massachusetts Institute of Technology, Cambridge, Mass., 1954.

Feinstein, A. (I): "Foundations of Information Theory," McGraw-Hill Book Company, Inc., New York, 1958.

Freeman, J. J.: "Principles of Noise," John Wiley & Sons, Inc., New York, 1958.

Goldman, S.: "Information Theory," Prentice-Hall, Inc., Englewood Cliffs, N.J., 1953.

Goode, H. H., and R. E. Machol: "System Engineering," McGraw-Hill Book Company, Inc., New York, 1957.

Grabbe, E. M., S. Ramo, and D. E. Wooldridge: "Handbook of Automation Computation and Control," vols. I and II, John Wiley & Sons, Inc., New York, 1958, 1959.

Harman, W. W.: Statistical Communication Theory, unpublished notes, Royal Technical University of Denmark, 1959.

Haus, H. A., and R. B. Adler: "Circuit Theory of Linear Noisy Networks," The Technology Press, M.I.T., Cambridge, Mass., and John Wiley & Sons, Inc., New York, 1959.

Iaglom, A. M., and I. M. Iaglom: "Probabilité et information," Dunod, Paris, 1959.

Jackson, W. (ed.): "Symposium of Information Theory—Report of Proceedings," First London Symposium, September, 1950, Ministry of Supply, London.

Jackson, W. (ed.): "Communication Theory," Second London Symposium, September, 1952, Academic Press, Inc., New York, 1953.

James, H. M., N. B. Nichols, and R. S. Phillips: "Theory of Servomechanisms," M.I.T. Radiation Laboratory Series, vol. 25, McGraw-Hill Book Company, Inc., New York, 1947.

Kharkevich, A. A.: "Outline of General Communication Theory," Gosudar. Izdat. Tekh. Teor. Lit., Moskva, 1955.

Khinchin, A. I.: "Mathematical Foundations of Information Theory," Dover Publications, New York, 1957.

Khinchin, A. I., D. A. Fadiev, A. N. Kolmogorov, A. Rényi, and J. Balatoni: "Arbeiten zur Informationstheorie I. Mathematische Forschungsberichte," V. E. B. Deutscher Verlag der Wissenschaften, Berlin, 1957.

Kotel'nikov, T. A.: "The Theory of Optimum Noise Immunity," Power Engineering Press, Moscow, 1956 (Russian). English translation by R. A. Silverman, McGraw-Hill Book Company, Inc., New York, 1959.

Kullback, S. (I): "Information Theory and Statistics," John Wiley & Sons, Inc., New York, 1959.

Laning, J. Halcombe, Jr., and Richard H. Battin: "Random Processes in Automatic Control," McGraw-Hill Book Company, Inc., New York, 1956.

Lawson, J. L., and G. E. Uhlenbeck: "Threshold Signals," M.I.T. Radiation Laboratory Series, vol. 24, McGraw-Hill Book Company, Inc., New York, 1950.

Lee, Y. W.: "Statistical Theory of Communications," John Wiley & Sons, Inc., New York, 1960.

Levin, B. R.: "Theory of Random Processes and Its Application in Radio Engineering," Sovietskoe Radio, Moscow, 1957.

McCarthy, J., and C. E. Shannon (eds.): "Automata Studies," Princeton University Press, Princeton, N.J., 1956.

Machol, R. E. (ed.): "Information and Decision Processes," McGraw-Hill Book Company, Inc., New York, 1960.

Meyer-Eppler, W.: "Grundlagen und Anwendungen der Informationstheorie," Springer-Verlag, Berlin, 1959.

Middleton, D.: "An Introduction to Statistical Communication Theory," McGraw-Hill Book Company, Inc., New York, 1960.

Pugachev, V. S.: "Random Functions and Their Applications in Automatic Control Theory," Goztekhizdat, Moscow, 1957.

Quastler, H.: "Information Theory in Psychology: Problems and Methods," Free Press, Glencoe, Ill., 1956.

Riordan, J.: "An Introduction to Combinatorial Analysis," John Wiley & Sons, Inc., New York, 1958.

Schwartz, M.: "Information Transmission, Modulation, and Noise," McGraw-Hill Book Company, Inc., New York, 1959.

Shannon, C. E., and W. Weaver: "The Mathematical Theory of Communication," University of Illinois Press, Urbana, Ill., 1949.

Solodovnikov, V. V.: "Introduction to Statistical Dynamics of Automatic Control Systems," Goztekhiteorizdat, Moscow-Leningrad, 1952.

Wax, Nelson: "Selected Papers on Noise and Stochastic Processes," Dover Publications, New York, 1954.

Wiener, N.: "Cybernetics," Technology Press, M.I.T., Cambridge, Mass., and John Wiley & Sons, Inc., New York, 1948.

Wiener, N.: "Extrapolation, Interpolation, and Smoothing of Stationary Time Series," John Wiley & Sons, Inc., New York, 1949.

Woodward, P. M.: "Probability and Information Theory, with Application to Radar," Pergamon Press Ltd., London, 1955.

PAPERS ON INFORMATION AND CODING THEORY

Abramson, N. M.: A Class of Systematic Codes for Non-independent Errors, *IRE Trans. on Inform. Theory,* vol. IT-5, pp. 150–157, December, 1959.

Aslund, N.: Informationsteoriens fundamentalsatser, Seminar notes, Stockholm, 1958.

Bakhmet'ev, M. M.: Calculation of Entropy for Special Probability Distributions, *Radiotekh. i Elektron.*, vol. 1, pp. 613–622, May, 1956.

Bakhmet'ev, M. M., and R. R. Vasil'ev: Information Criteria for Estimation of Telemetering Systems, *Avtomat. i Telemekh.*, vol. 18, no. 4, pp. 371–375, 1957.

Bar-Hillet, Y.: An Examination of Information Theory, *Phil.-Scr.*, vol. 22, pp. 86–105, 1955.

Bar-Hillet, Y., and R. Carnap: Semantic Information, London Information Theory Symposium, 1953, pp. 503–512.

Barnard, G. A., III: Statistical Calculation of Word Entropies for Four Western Languages, *IRE Trans. on Inform. Theory*, vol. IT-1, pp. 49–54, March, 1955.

Bellman, R., and R. E. Kalaba: On the Role of Dynamic Programming in Statistical Communication Theory, *IRE Trans. on Inform. Theory*, vol. IT-3, p. 197, September, 1957.

Bennett, W. R.: Methods of Solving Noise Problems, *Proc. IRE*, vol. 44, pp. 609–638, May, 1956.

Bennett, W. R., and D. Slepian: unpublished notes on communication theory, Bell Telephone Laboratories, Inc., 1952.

Berkowitz, R. S.: Methods of Sampling Band-limited Functions, *Proc. IRE*, vol. 44, pp. 231–235, February, 1956.

Birdsall, T. G., and M. P. Ristenbatt: Introduction to Linear Shift-register Generated Sequences, *Univ. Mich., Research Inst. Tech. Rept.* 90, 1958.

Blachman, N. M.: Minimum Cost Encoding of Information, *Trans. IRE on Inform. Theory*, vol. IT-3, pp. 139–149, 1954.

Blackwell, D.: The Entropy of Functions of Finite-state Markov Chains, *Trans. First Prague Conf. on Inform. Theory*, November, 1956, pp. 13–20.

Blackwell, D., L. Breiman, and A. J. Thomasian: Proof of Shannon's Transmission Theorem for Finite-state Indecomposable Channels, *Ann. Math. Statistics*, vol. 29, no. 4, pp. 1209–1220, December, 1958.

Blokh, E. L., and A. A. Kharkevich: Geometrical Proof of Shannon's Theorem, *Radiotekh.*, vol. 11, pp. 5–16, November, 1956.

Bose, R. C., and R. R. Kuebler, Jr.: A Geometry of Binary Sequences Associated with Group Alphabets in Information Theory, *Ann. Math. Statistics*, vol. 31, pp. 113–139, March, 1960.

Bose, R. C., and S. S. Shrikhande: A Note on a Result in the Theory of Code Construction, *Inform. and Control*, vol. 2, pp. 183–194, 1959.

Brillouin, L. (II): Information and Entropy, I, *J. Appl. Phys.*, vol. 22, no. 3, p. 334, 1951; Physical Entropy and Information, II, vol. 22, no. 3, p. 338, 1951.

Broglie, L. De: La Cybernétique, Report of Paris Conference, 1951, published by *Revue d'optique*, Paris.

Brookes, B. C.: An Introduction to the Mathematical Theory of Information, *Math. Gaz.*, vol. 40, pp. 170–180, 1956.

Calabi, L., and H. G. Haefeli: A Class of Binary Systematic Codes Correcting Errors at Random and in Bursts, *IRE Trans. on Circuit Theory*, vol. CT-6, pp. 79–94, May, 1959.

Campbell, L. L.: Error Rates in Pulse Position Coding, *IRE Trans. on Inform. Theory*, vol. IT-3, pp. 18–24, March, 1957.

Chang, S. L.: Theory of Information Feedback Systems, *IRE Trans. on Inform. Theory*, vol. IT-2, pp. 29–40, September, 1956.

Cheatham, T. P., Jr., and W. G. Tuller: Communication Theory and Network Synthesis, *Proc. Symposium on Inform. Theory*, Polytechnic Institute of Brooklyn, 1954, pp. 11–17.

Chien, R.: On the Characteristics of Error-correcting Codes, Correspondence, *IRE Trans. on Inform. Theory*, vol. IT-5, p. 91, June, 1959.

Costas, J. P.: Poisson, Shannon, and the Radio Amateur, *Proc. IRE*, vol. 47, no. 12, pp. 2058–2068, December, 1959.

Darmois, G.: Le Role des méthodes statistiques dans l'exploitation des mesures, Actes des journées d'étude "Mesures et connaîssance," Paris, December, 1957.

Davis, H.: Radar Problems and Information Theory, *IRE Conv. Record*, pt. 8, pp. 39–47, 1953.

Dolukhanov, M. P.: Some Possible Methods of Entropy Measurement on Discrete and Continuous Information Sources, *Radiotekh. i Elektron.*, vol. 4, no. 4, pp. 559–565, 1959.

Elias, P. (I): Coding for Two Noisy Channels, Information Theory, London Symposium, pp. 61–77, 1955.

Elias, P. (II): Information Theory, chap. 16 in "Handbook of Automation, Computation and Control," vol. 1, pp. 16-01 to 16-48, John Wiley & Sons, Inc., New York, 1958.

Elspas, B.: The Theory of Autonomous Linear Sequential Networks, *IRE Trans. on Circuit Theory*, vol. CT-6, no. 1, pp. 45–60, March, 1959.

Epstein, M. A.: Algebraic Decoding for a Binary Erasure Channel, *Mass. Inst. Technol., Research Lab. Electronics, Tech. Rept.* 340, Mar. 14, 1958.

Erohin, V.: *E*-entropy of a Discrete Random Variable, *Theor. Veroiat. Primenen* (Theory of Probability and Its Application), vol. 3, no. 1, pp. 106–113, 1958.

Fadiev, D. A.: On the Notion of Entropy of a Finite Probability Space (in Russian), *Uspekhi Mat. Nauk*, vol. 11, no. 1 (67), pp. 227–231, 1956.

Fano, R. M. (II): The Transmission of Information, *Mass. Inst. Technol., Research Lab. Electronics, Tech. Rept.* 65, 1949; *Rept.* 149, 1950.

Feinstein, A. (II): On the Coding Theorem and Its Converse for Finite-memory Channels, *Inform. and Control*, vol. 2, no. 1, pp. 25–44, 1959.

Féron, R.: Information, régression, corrélation, *Univ. Paris, Publ. Inst. Statist.*, vol. 5, pp. 111–215, 1956.

Fleischer, I.: The Central Concepts of Communication Theory for Infinite Alphabets, *J. Math. Phys.*, vol. 37, pp. 223–228, 1958.

Fleishman, B. S., and G. B. Linkovskii: Maximum Entropy of an Unknown Discrete Distribution with a Given First Moment, *Radiotekh. i Elektron.*, vol. 3, no. 4, pp. 554–556, 1958.

Fontaine, A. B., and W. W. Peterson (I): On Coding for the Binary Symmetric Channel, *Poughkeepsie IBM Rept.* RC-43, Feb. 21, 1958 (also *Trans. AIEE*, November, 1958, pp. 648–656).

Fontaine, A. B., and W. W. Peterson (II): Group Code Equivalence and Optimum Codes, *IRE Trans. on Inform. Theory*, vol. IT-5, pp. 60–70, 1959.

Gabor, D.: Theory of Communication, *J. Inst. Elec. Engrs.* (*London*), vol. 93, pp. 111, 429, 1946.

Gabor, D., and A. Gabor: An Essay on the Mathematical Theory of Freedom, *Proc. Roy. Statist. Soc.*, ser. A, vol. 117, pp. 31–72, 1954.

Gavrilov, M. A., and G. A. Shastova: The Theory of Signal Formation and Noise Reduction, *Izdatel'stvo Akademii Nauk S.S.S.R.*, Moscow and Leningrad, 1957.

Gel'fand, I. M.: On Some Problems of Functional Analysis, *Uspekhi Mat. Nauk S.S.S.R.*, new series, vol. 11, no. 6, pp. 3–12, 1956 (Russian).

Gel'fand, I. M., A. N. Kolmogorov, and A. M. Iaglom: On the General Definition of the Amount of Information, *Doklady Akad. Nauk S.S.S.R.*, new series, vol. 11, pp. 745–748, 1956.

Gilbert, E. N. (I): A Comparison of Signalling Alphabets, *Bell System Tech. J.*, vol. 31, pp. 504–522, 1952.

Gilbert, E. N. (II): An Outline of Information Theory, *Am. Statistician*, vol. 12, pp. 13–19, February, 1958.

Gilbert, E. N., and E. F. Moore: Variable Length Binary Encodings, *Bell System Tech. J.*, vol. 38, pp. 933–968, July, 1959.

Gnedenko, B. V.: Some Russian Articles about Information Theory, *Trans. First Prague Conf. on Inform. Theory*, November, 1956, pp. 21–28.

Golay, M. J. E. (I): Notes on Digital Coding, *Proc. IRE*, vol. 37, p. 657, 1949.

Golay, M. J. E. (II): Notes on the Penny-weighing Problem, Lossless Symbol Coding with Non-primes, etc., *IRE Trans. on Inform. Theory,* vol. IT-4, pp. 103–109, September, 1958.

Golomb, S. W., B. Gordon, and L. R. Welch: Comma-free Codes, *Can. J. Math.,* vol. 10, no. 2, pp. 202–209, 1958.

Good, I. J.: Some Terminology and Notation in Information Theory, *Proc. Inst. Elec. Engrs. (London),* vol. 103C, pp. 200–204, 1955.

Good, I. J.: Rational Decisions, *J. Roy. Statist. Soc.,* ser. B, vol. 14, no. 1, pp. 107–114, 1952.

Green, J. H., Jr., and R. L. San Soucie: An Error-correcting Encoder and Decoder of High Efficiency, *Proc. IRE,* vol. 46, pp. 1741–1744, October, 1958.

Green, P. E., Jr.: Information Theory in the U.S.S.R., *1957 IRE Wescon Conv. Record,* pt. 2, pp. 67–83. Contains a few references on coding.

Green, P. E., Jr.: A Bibliography of Soviet Literature on Noise, Correlation, and Information Theory, *IRE Trans. on Inform. Theory,* vol. IT-2, pp. 91–94, June, 1956.

Hagelbarger, D. W.: Recurrent Codes: Easily Mechanized, Burst Correcting, and Binary Codes, *Bell System Tech. J.,* vol. 38, pp. 969–984, July, 1959.

Hamming, R. W.: Error Detecting and Error Correcting Codes, *Bell System Tech. J.,* vol. 29, pp. 147–150, 1950; *Bell Labs. Record,* vol. 28, pp. 193–198, 1950.

Harris, B., A. Hauptschein, K. C. Morgan, and L. S. Schwartz: Binary Communication Feedback Systems, *Trans. AIEE, Communs. and Electronics,* no. 40, pp. 960–969, January, 1959.

Hirschman, I. I.: A Note on Entropy, *Am. J. Math.,* vol. 79, pp. 152–156, 1957.

Huffman, D. A. (I): A Method for the Construction of Minimum Redundancy Codes, *Proc. IRE,* September, 1952, pp. 1098–1101.

Huffman, D. A. (II): The Synthesis of Linear Sequential Coding Networks, *Proc. Symposium on Inform. Theory,* London, 1955.

Huggins, W. H.: Signal Theory, *IRE Trans. on Circuit Theory,* vol. CT-3, pp. 210–216, December, 1956.

IRE Standards, Information Theory: Definitions of Terms, *Proc. IRE,* vol. 46, pp. 1646–1648, 1958.

Joshi, D. D. (I): A Note on Upper Bounds for Minimum Distance Codes, *Inform. and Control,* vol. 1, no. 3, pp. 289–295, September, 1958.

Joshi, D. D. (II): L'Information en statistique, mathématique et dans la théorie des communications, *Univ. Paris, Inst. de statist.,* vol. 7, fas. 2, pp. 83–159, 1959.

Kalaba, R. E.: see R. Bellman.

Kautz, W. H.: Unit-distance Error-checking Codes, *IRE Trans. on Electronic Computers,* vol. EC-7, no. 2, pp. 179–180, June, 1958.

Kelly, J. L., Jr. (I): A New Interpretation of Information Rate, *Bell System Tech. J.,* vol. 35, p. 917, 1956.

Kelly, J. L., Jr. (II): A Class of Codes for Signaling on a Noisy Continuous Channel, *IRE Trans. on Inform. Theory,* vol. IT-6, pp. 22–24, March, 1960.

Kharkevich, A. A.: On Optimum Codes, *Elektrosvyaz,* vol. 10, no. 2, pp. 65–70, 1956.

Kim, W. H., and C. V. Freiman: Multi-error Correcting Codes for a Binary Asymmetric Channel, *IRE Trans. on Inform. Theory,* vol. IT-5, May, 1959, special supplement.

Kolmogorov, A. N. (II): On the Shannon Theory of Information Transmission in the Case of Continuous Signals, *IRE Trans. on Inform. Theory,* vol. IT-2, p. 102, December, 1956.

Kryala, A.: Eindeutigkeitsbedingungen bei den Probensätzen der Informationstheorie, *NTZ,* vol. 11, pp. 533–534, November, 1960.

Kullback, S. (II): An Application of Information Theory to Multivariate Analysis, *Ann. Math. Statistics,* vol. 27, pp. 122–146, 1956.

Laemmel, A. E.: Efficiency of Noise-reducing Codes, in Willis Jackson (ed.), "Communication Theory," pp. 111–118, Academic Press, Inc., New York, 1953.
Lee, C. Y.: Some Properties of Nonbinary Error-correcting Codes, *IRE Trans. on Inform. Theory*, vol. IT-4, no. 2, pp. 77–82, June, 1958.
Leifer, M., and W. F. Schreiber: Communication Theory, in "Advances in Electronics," pp. 306–343, Academic Press, Inc., New York, 1951.
Linkovskii, G. B.: see Fleishman.
Lloyd, S. P.: Binary Block Coding, *Bell System Tech. J.*, vol. 36, p. 517, 1957.
Loeb, J.: Canaux binaries en cascade, *Ann. télécommun.*, vol. 13, no. 1–2, pp. 42–44, 1958 (E102454-1958).
Löfgren, L.: Automata of High Complexity and Methods of Increasing Their Reliability by Redundancy, *Inform. and Control*, vol. 1, pp. 127–147, 1958.
Luce, R. D.: A Survey of the Theory of Selective Information and Some of Its Behavioral Applications, Behavioral Models Project, *Columbia Univ. Rept.* 8, 1956.
McCluskey, E. J.: Error Correcting Codes—A Linear Programming Approach, *Bell System Tech. J.*, vol. 38, pp. 1485–1512, November, 1959.
MacDonald, J. E., Jr.: Constructive Coding Methods for the Binary Asymmetric Independent Data Transmission Channel, M.E.E. thesis, Syracuse University, January, 1958.
McGaughan, H. S.: Statistical Theory of Communication, unpublished notes, Cornell University, Ithaca, N.Y.
McMillan, B. (I): Two Inequalities Implied by Unique Decipherability, *IRE Trans. on Inform. Theory*, vol. IT-2, pp. 115–116, December, 1956.
McMillan, B. (II): The Basic Theorems of Information Theory, *Ann. Math. Statistics*, vol. 24, p. 196, 1952.
Mandelbrot, B. (I): On Recurrent Noise Limiting Coding, P.I.B. Symposium of Information Network, 1954, pp. 205–221.
Mandelbrot, B. (II): "Jeux de communication," vol. 2, pts. 1, 2, L'Université de Paris, Institut de Statistique, Paris, 1953.
Mandelbrot, B. (III): "Diagnostic et transduction en l'absence de bruit," p. 73, L'Université de Paris, Institut de Statistique, 1955.
Marcum, J. L.: A Statistical Theory of Target Detection by Pulsed Radar, *Rand Corp. Repts.* RM-753, R. 113, 1948; RM-754, 1947.
Marcus, M. B.: The Utility of a Communication Channel and Applications to Suboptimal Information Handling Procedures, *IRE Trans. on Inform. Theory*, vol. IT-4, pp. 147–151, December, 1958.
Meter, D. Van, and D. Middleton: Modern Statistical Approaches to Reception in Communication Theory, *IRE Trans. on Inform. Theory*, sec. 3.5, p. 119, September, 1954.
Middleton, D., and D. Van Meter: Detection and Extraction of Signals in Noise from the Point of View of Statistical Decision Theory, *J. Soc. Ind. Appl. Math.*, pt. 1, vol. 3, pp. 192–253, 1955; pt. 2, vol. 4, pp. 86–119, 1956.
Moore, E. F.: see E. N. Gilbert.
Muller, D. E.: Metric Properties of Boolean Algebra and Their Application to Switching Circuits, *Univ. Illinois, Digital Computer Lab. Rept.* 46, April, 1953.
Muroga, S.: On the Capacity of a Discrete Channel, *J. Phys. Soc. Japan*, 8L484, 1953.
Paley, R. E. A. C.: On Orthogonal Matrices, *J. Math. and Phys.*, vol. 12, pp. 311–320, 1933.
Perez, A.: Notions généralisées d'incertitude, d'entropie et d'information du point de vue de la théorie de martingales, *Trans. First Prague Conf. on Inform. Theory*, pp. 183–208, November, 1956.

Peterson, W. W., and M. O. Rabin: On Codes for Checking Logical Operations, *IBM Journal*, vol. 3, pp. 163–168, April, 1959.

Pierce, J. R.: What Good Is Information Theory to Engineers? *1957 IRE Natl. Conv. Record*, pt. 2, pp. 51–55.

Plotkin, M.: Binary Codes with Specified Minimum Distance, *Univ. Penna., Moore School Research Div. Rept.* 51–20, 1951.

Polytechnic Institute of Brooklyn: *Proc. Symposium on Information Theory*, 1954, p. 328.

Powers, K. H.: A Unified Theory of Information, *Mass. Inst. Technol., Research Lab. Electronics, Tech. Rept.* 311, February, 1956.

Prange, B.: Cyclic Error-correcting Codes in Two Symbols, Air Force Cambridge Research Center, *Tech. Note* AFCRC-TN-57-103, September, 1957.

Quastler, H. (ed.): "Information Theory and Psychology," Free Press, Glencoe, Ill., 1955.

Quastler, H. (ed.): "Essays on the Use of Information Theory and Biology," University of Illinois Press, Urbana, Ill., 1953.

Reed, I. S.: A Class of Multiple-error-correcting Coding and Decoding Schemes, *IRE Trans. on Inform. Theory*, vol. IT-4, pp. 38–49, 1954.

Reed, I. S., and G. Solomon: Polynomial Codes over Finite Fields, *J. Soc. Ind. Appl. Math.*, vol. 8, no. 2, pp. 300–304, 1960.

Reza, F. M.: see the footnote.*

Rice, S. O.: Communication in the Presence of Noise—Probability of Error for Two Encoding Schemes, *Bell System Tech. J.*, vol. 29, pp. 60–93, 1950.

Rozenblatt-Rot, M. (I): Entropy of Stochastic Processes, *Doklady Akad. Nauk S.S.S.R.*, vol. 112, pp. 16–19, January, 1957 (Russian).

Rozenblatt-Rot, M. (II): Theory of Transmission of Information through Stochastic Communication Channels, *Doklady Akad. Nauk S.S.S.R.*, vol. 112, pp. 202–205, Jan. 11, 1957.

Sacks, G. E.: Multiple Error Correction by Means of Parity Checks, *IRE Trans. on Inform. Theory*, vol. IT-4, pp. 145–147, December, 1958.

Sardinas, A. A., and G. W. Patterson: A Necessary and Sufficient Condition for Unique Decomposition of Coded Messages, *IRE Conv. Record*, pt. 8, pp. 104–109, March, 1953.

Schouten, J. F.: Ignorance, Knowledge and Information, London Symposium on Information Theory, 1955, pp. 37–43.

Schutzenberger, M. P. (I): "On Some Measures of Information Used in Statistics," Third London Symposium on Information Theory, pp. 18–25, Butterworth & Co. (Publishers) Ltd., London, 1955.

* The material in Chaps. 2, 5, 3, 8, and 10 is based on work which originally appeared in the reports listed below.

(I): "An Introduction to Probability Theory: Stochastic Processes," U.S. Air Force Rome Air Development Center, RADC TN-59–56, April, 1959.

(II): "An Introduction to Probability Theory: Discrete Schemes," U.S. Air Force Rome Air Development Center, RADC TN-59-129, August, 1959.

(III): "An Introduction to Probability Theory: Continuous Schemes," U.S. Air Force Rome Air Development Center, RADC-TN-60-67, 1960.

(IV): "Basic Concepts of Information Theory: Finite Scheme," U.S. Air Force Cambridge Research Center, Syracuse University Institute Rept. AF CRC-TN-59-588, September, 1959.

(V): "Basic Concepts of Information Theory: Entropy in a Continuous Channel," U.S. Air Force Cambridge Research Center, Syracuse University Institute Rept. AF CRC-TR-60.F, 1960.

Schutzenberger, M. P. (II): Sur un probléme du codage binaire, *Univ. Paris, Publ. Inst. Statist.*, vol. 2, pp. 125–127, 1953.

Schutzenberger, M. P. (III): Une théorie algébrique du codage, *Compt. rend. acad. sci. Paris*, vol. 242, pp. 862–864, 1956.

Schwartz, L. S., S. S. L. Cheng, B. Harris, A. Hauptschein, J. Metzner, and K. C. Morgan: Theory of Digital Communication Systems, unpublished notes, New York University, 1960.

Shannon, C. E. (I): A Mathematical Theory of Communication, *Bell System Tech. J.*, vol. 27, pp. 379–423, 623–656, 1948.

Shannon, C. E. (II): Communication in the Presence of Noise, *Proc. IRE*, vol. 37, pp. 10–21, 1949.

Shannon, C. E. (III): Probability of Error for Optimal Codes in a Gaussian Channel, *Bell System Tech. J.*, vol. 38, no. 3, pp. 611–656, 1959.

Shannon, C. E. (IV): Certain Results in Coding Theory for Noisy Channels, *Inform. and Control*, vol. 1, no. 1, pp. 6–25, September, 1957.

Shapiro, H. S., and D. L. Slotnick: On the Mathematical Theory of Error-correcting Codes, *IBM J. Research and Develop.*, vol. 3, no. 1, pp. 25–34, January, 1959.

Shastova, G. A.: The Noise-proof Feature of the Hamming Code, *Radiotekh i Elektron.*, vol. 3, no. 1, pp. 19–26, 1958 (Russian); translation in vol. 3, no. 1, pp. 24–37, 1958.

Sherman, S.: Non-near-square Error Criteria, *IRE Trans. on Inform. Theory*, vol. IT-4, pp. 125–127, September, 1958.

Siebert, W. McC.: A Radar Detection Philosophy, *IRE Trans. on Inform. Theory*, vol. IT-2, no. 3, pp. 204–221, September, 1956.

Siforov, V. I.: Error Reduction of Systems with Correcting Codes, *Radiotekh. i Elektron.*, vol. 1, pp. 131–142, 1956.

Siforov, V. I., and Y. B. Sindler: Conditions for the Equivalence of the Statistical Properties of Radio Communication Systems with a Large Number of Random Parameters, *Doklady Akad. Nauk S.S.S.R.*, vol. 116, pp. 956–958, Oct. 21, 1957.

Silverman, R. A.: On Binary Channels and Their Cascades, *IRE Trans. on Inform. Theory*, vol. IT-1, pp. 19–27, December, 1955.

Silverman, R. A., and S. H. Chang: Topics in the Theory of Discrete Information Channels, *N.Y. Univ. Research Rept.* EM-152, April, 1960.

Silverman, R. A., and I. Kay: On the Uncertainty Relation for Real Signals, *Inform. and Control*, vol. 1, no. 1, pp. 64–75, 1957.

Sindler, Y. B.: see V. I. Siforov.

Slepian, D. (I): A Class of Binary Signalling Alphabets, *Bell System Tech. J.*, vol. 35, pp. 203–234, January, 1956.

Slepian, D. (II): A Note on Two Binary Signalling Alphabets, *IRE Trans. on Inform. Theory*, vol. IT-2, pp. 84–86, June, 1956.

Soest, J. L. Van: Some Consequences of the Finiteness of Information, London Symposium on Information Theory, 1955, pp. 3–7.

Stern, T. E., and B. Friedland: Application of Modular Sequential Circuits to Single Error-correcting *p*-nary Codes, *IRE Trans. on Inform. Theory*, vol. IT-5, pp. 114–123, September, 1959.

Storer, J. E., and R. Turyn: Optimum Finite Code Groups, *Proc. IRE*, vol. 46, p. 1649, September, 1958.

Stumpers, F. L.: A Bibliography of Information Theory, Communication Theory—Cybernetics, *IRE Trans. on Inform. Theory*, vol. IT-2, November, 1953; vol. IT-1, pp. 31–47, September, 1955; vol. IT-3, pp. 150–166, June, 1957.

Stutt, C. A.: Information Rate in a Continuous Channel for Regular-Simplex Codes, *IRE Trans. on Inform. Theory*, vol. IT-6, pp. 516–522, December, 1960.

Swerling, P.: Detection of Fluctuating Pulsed Signals in the Presence of Noise, *IRE Trans. on Inform. Theory*, vol. IT-3, pp. 175–177, September, 1957 (E100718-1958).

Szilard, L.: Uber die Entropieverminderung in einem Thermodynamischen System bei Eingriffen Intelligenter Wesen, *Z. Physik*, vol. 53, p. 840, 1929.

Takano, K.: On the Basic Theorems of Information Theory, *Ann. Inst. Statist. Math. (Tokyo)*, vol. 9, pp. 53–77, 1958.

Thomasian, A. J.: Elementary Proof of the AEP of Information Theory, *Ann. Math. Statist.*, vol. 31, no. 2, pp. 452–456, 1960.

Tikhomiro, V. M.: On the ϵ-entropy of Some Classes of Analytical Functions, *Doklady Akad. Nauk S.S.S.R.*, vol. 117, no. 2, pp. 191–194, 1957.

Tuller, W. G.: *IRE Trans. on Inform. Theory*, vol. IT-6, pp. 25–51, March, 1950.

Ville, J. A. (I): Application of Information Theory to Amplitude Compression, *Câbles et transm.*, vol. 12, no. 2, pp. 144–147, April, 1958.

Ville, J. A. (II): Leçons sur quelques aspects nouveaux de la théorie des probabilités, *Ann. inst. Henri Poincaré*, vol. 14, pp. 61–143, 1954.

Watanabe, S. (I): Toward a Mathematical Theory of Knowledge, unpublished notes, IBM.

Watanabe, S. (II): Binary Coding and Isometric Transformations in a Finite Metric Boolean Algebra, *Rep. Unio. Electrio. Commun.*, vol. 7, pp. 17–40, 1955; *Math. Rev.*, vol. 17, p. 1179, 1956.

Wiesner, J. B.: Communication Theory and Transmission of Information, in L. N. Ridenour (ed.), "Modern Physics for the Engineer," pp. 427–454, McGraw-Hill Book Company, Inc., New York, 1955.

Wolfowitz, J. (I): The Coding of Messages Subject to Chance Errors, *Illinois J. Math.*, pp. 591–606, 1957.

Wolfowitz, J. (II): An Upper Bound on the Rate of Transmission of Messages, *Illinois J. Math.*, pp 137–141, 1958.

Wolfowitz, J. (III): The Maximum Achievable Length of an Error Correcting Code, *Illinois J. Math.*, pp. 454–458, 1958.

Wolfowitz, J. (IV): unpublished notes of a forthcoming monograph, Cornell University.

Wozencraft, J. M.: Sequential Decoding for Reliable Communication, *IRE Conv. Record*, vol. 5, pt. 2, pp. 11–25, March, 1957; also *Mass. Inst. Technol., Research Lab. Electronics, Rept.* 325, Aug. 9, 1957.

Yockey, H. P., R. L. Platzman, and H. Quastler (eds.): "Symposium on Information Theory in Biology," Pergamon Press Ltd., London, 1958.

Youla, D. C.: The Use of the Method of Maximum Likelihood in Estimating Continuous-modulated Intelligence Which Has Been Corrupted by Noise, *IRE Trans. on Inform. Theory*, vol. IT-3, pp. 90–106, March, 1954.

Yovits, M. C., and S. Cameron (eds.): "Self-organizing Systems," Proceedings of an Interdisciplinary Conference, Pergamon Press, Ltd., London, 1960.

Zakai, M.: A Class of Definitions of "Duration" or "Uncertainty" and the Associated Uncertainty Relations, *J. Inform. Control*, vol. 3, no. 2, pp. 101–115, June, 1960.

Zaremba, S. K. (I): A Covering Theorem for Abelian Groups, *J. London Math. Soc.*, vol. 26, 1951.

Zaremba, S. K. (II): Covering Problems Concerning Abelian Groups, *J. London Math. Soc.*, vol. 27, 1952.

Zierler, N. (I): Several Binary Sequence Generators, *Mass. Inst. Technol., Lincoln Lab., Tech. Rept.* 95, September, 1955.

Zierler, N. (II): Linear Recurring Sequences, *J. Soc. Ind. Appl. Math.*, vol. 7, pp. 31–48, March, 1959.

NAME INDEX

Abramson, N., 13, 452, 461, 462, 483

Balakrishnan, A. V., 453
Barnard, G. A., 161, 165, 183, 459, 484
Bartlett, M. S., 364, 481, 484
Battin, R. H., 358, 359, 373, 483
Bayes, 46–49
Bell, D. A., 13, 183–185, 482
Bellman, R., 452, 484
Bennett, W. R., 358, 484
Blachman, N. M., 484
Blackwell, D., 331, 393, 423, 484
Blanc-Lapierre, 13, 359, 482
Bochner, S., 370
Bond, F. E., 452
Bose, R. C., 442, 447, 460, 484
Breiman, L., 331, 423, 484
Brillouin, L., 13, 130, 484
Broglie, L. de, 484
Brookes, B. C., 484
Burton, N. G., 184

Cahn, C. R., 332, 452
Campbell, L. L., 484
Chang, S. H., 14, 119–121, 489
Chaudhuri, D. K. R., 442, 460
Chebyshev, 227–229, 335
Chernov, H., 414–415, 459
Cherry, E. C., 11, 13, 455, 482
Chien, R., 434, 447, 484

Darmois, G., 481, 484
Davenport, W. B., Jr., 338, 359, 361, 363, 373, 464, 482
Dobrushin, R. L., 13, 423
Dolukhanov, M. P., 484
Doob, J. L., 20, 338, 368–369, 481
Driml, M., 393

Elias, P., 13, 161, 165, 176–180, 183, 457–460, 485
Epstein, M. A., 462, 482, 485

Fadiev, D. A., 13, 81, 126, 482, 485
Fano, R. M., 13, 138–140, 143, 156, 161, 400, 482, 485
Feinstein, A., 7, 13, 15, 81, 126, 143, 161, 166, 375, 393, 397, 403–408, 423, 438, 485
Feller, W., 481
Feron, R., 485
Fire, P., 461
Fisher, R. A., 20, 82
Fleishman, B. S., 130, 485
Fogel, L. J., 452
Fontaine, A. B., 424, 485
Freeman, J. J., 363, 482
Friedman, E. A., 185

Gabor, D., 11, 13, 305, 454, 455, 485
Gavrilov, M. A., 13, 485
Gel'fand, I. M., 13, 14, 289, 295–296, 423, 485
Gilbert, E. N., 13, 158–160, 165, 183, 185, 188, 330, 459, 485
Golay, M. J. E., 152, 434, 446, 447, 485
Goldman, S., 13, 320, 332, 482
Good, I. J., 13, 486
Goode, H. H., 482
Green, P. E., Jr., 13, 486

Hagelbarger, D. W., 461, 486
Hamming, R. W., 13, 166–176, 424, 425, 431–435, 439, 486
Hartley, R. V. H., 7, 11, 290
Hirschman, I. I., 486
Huffman, D. A., 13, 155–158, 167, 185, 188, 486

Iaglom, A. M., 13, 14, 16, 289, 295–298, 423, 482, 485
Iaglom, I. M., 16, 482
Ignatyev, N. K., 13

Jackson, W., 482
Jagerman, D. L., 452
Joshi, D. D., 180–183, 486

Kalaba, R. E., 452, 484
Kay, I., 456, 489
Kelley, J. L., Jr., 13, 450–452, 486
Kemeny, J. G., 378, 380, 394, 481
Kerrich, J. E., 17
Keubler, R. R., Jr., 442
Kharkevich, A. A., 13, 482, 484, 486
Khinchin, A. I., 13, 15, 81, 82, 120, 126, 161, 166, 370, 375, 386–387, 391–393, 482
Kohlenberg, A., 454
Koinheim, A. G., 434
Kolmogorov, A. N., 13, 20, 268, 289–290, 337, 358, 393, 423, 454, 481, 485, 486
Kotelnikov, V. A., 13, 454, 482
Kraft, L., 143, 147
Kryala, A., 486

Kullback, S., 14, 295, 483, 486
Küpfmüller, K., 11, 305

Laemmel, A. E., 432, 487
Lampard, D. G., 456
Laning, J. H., Jr., 358, 359, 373, 483
Lawson, J. L., 363, 483
Lee, C. Y., 487
Leifer, M., 487
Licklider, J. C. R., 184
Linden, D. A., 452
Linkovskii, G. B., 130, 485
Lloyd, S. P., 446, 487
Loeb, J., 13, 120, 487
Loève, M., 91, 334, 481

McCluskey, E. J., Jr., 434, 442, 487
Machol, R. E., 482
Mackay, D. M., 11
McMillan, B., 13, 15, 115, 143, 147, 161, 375, 384–388, 393, 487
Mandelbrot, B., 13, 143, 147, 487
Marcum, J. L., 487
Marcus, M. B., 452, 487
Meshkovski, K. A., 13
Meyer-Eppler, W., 13, 483
Middleton, D., 338, 359, 373, 464, 487
Miller, G. A., 185
Moore, E. F., 158–160, 185, 188, 485
Muller, D. E., 443, 487
Muroga, S., 13, 117, 487

Nyquist, H., 11, 305

Oswald, J. R. V., 456

Paley, R. E. A. C., 448, 456, 487
Parzen, E., 454
Patterson, G. W., 143, 488
Perez, A., 13, 393, 487
Peterson, W. W., 13, 424, 442, 449, 461, 485, 488
Pierce, R., 166, 488
Plotkin, M., 447, 488
Poincaré, H., 189, 395
Prange, B., 461, 488
Pugachev, V. S., 247, 483

Quastler, H., 488

Reed, I. S., 443, 483, 488
Rice, S. O., 330, 359, 363, 488
Root, W. L., 338, 359, 361, 363, 373, 464, 482
Rozenblat-Rot, M., 13, 393, 423, 488

Sacks, G. E., 434, 442, 488
Sardinas, A. A., 143, 488
Schouten, J. F., 488
Schreiber, W. F., 487
Schutzenberger, M. P., 13, 81–82, 488, 489
Shannon, C. E., 1, 4, 12–15, 76, 77, 81, 82, 86, 106, 109, 110, 117, 119, 126, 138–140, 151–153, 154, 161, 165, 166, 184, 268, 282, 290, 291, 322–330, 359, 397, 409–415, 462–464, 489
Shapiro, H. S., 446, 448, 489
Shastova, G. A., 176, 485
Shirkhande, S. S., 447, 484
Siforov, V. I., 13, 14, 489
Silverman, R. A., 13, 14, 119, 120–121, 456, 489
Sindler, Y. B., 489
Slepian, D., 13, 166, 183, 424, 425, 438–447, 468–470, 489
Slotnick, D. L., 446, 448, 489
Snell, J. L., 378, 380, 394, 481
Soest, J. L. van, 489
Stumpers, F. L., 12, 13, 489
Stutt, C. A., 489
Swerling, P., 440
Szilard, L., 143, 490

Takano, K., 490
Thomasian, A., 13, 330–331, 334, 423, 484, 490
Tikhomiro, V. M., 490

Uhlenbeck, G. E., 363, 483

Van Der Pol, B., 454
Van Meter, D., 464, 487
Varsamov, R. R., 13, 460, 461
Ville, J. A., 455, 490

Wax, N., 432, 483
Wernikoff, R. E., 457
Wiener, N., 4, 12, 77, 82, 86, 337, 358, 375, 456, 490
Wolf, E., 455, 456
Wolfowitz, J., 13, 15, 161, 166, 397, 416–423, 490
Woodward, P. M., 13, 464, 483
Wozencraft, J. M., 13, 459, 462, 490

Yen, J. L., 452

Zemanian, A. H., 456
Zierler, N., 461, 490

SUBJECT INDEX

Abstract space (*see* Vector space)
Alphabet, 90, 132, 142, 322
Analytic signals, 454–457
Asymptotic equipartition property, 387–388
Autocorrelation, 350–351
Autocovariance, 350
Average amount of information, 8, 14, 79
(*See also* Entropy; Transinformation)
Axioms of probability, 37, 40–41

Balanced incomplete block design, 447–448
Band-limited signals, 300–307
Bayes' law, 46–49
BEC, 114, 176–179
Binary channels, 114–121
Binary pulse width communications, 122–124
Binary source, 8, 90
entropy table, 476–480
Binary unit of information, 7
Binomial distribution, 63–64, 467
moments, 234
normal approximation, 259–262
two-dimensional, 236
Bivariate discrete distribution, 61–63
Bivariate normal distribution, 206–208
approximated by normal distribution, 259–262
moments, 236
Block code, 176–179, 323
Bochner's theorem, 370
Bounds on systematic codes, 171, 433, 446–449

Capacity (channel), 14–15, 108–110
BEC, 114
binary channels, 115–118, 120
BSC, 114
Gaussian, with additive noise, 285–287, 297–299
$m \times m$ channels, 117, 119–120
symmetric channel, 111–113

Cauchy distribution, 198–199, 467
moment, 236
Central-limit theorem, 257–258
Central moments, 225
Channel, 10, 14
additive, 283–285
continuous, 267–268
discrete, with independent input-output, 100–101
noise-free, 99–100
extension, 114, 154–155, 402–403
Gaussian, 282–283
with memory, 380–393
stationary, 391–393
Characteristic function, 238–241
table, 467
Chebyshev inequality, 227–229
Chernov's inequality, 414–415
Codes, Bose-Chaudhuri, 460–461
close-packed, 446
convolution, 462
ensemble, 409–412
error-correcting and error-detecting, 166–176
Gilbert-Moore, 158–160
group (*see* Group codes)
Hamming, 171–176, 431–435
Huffman, 155–158
irreducible, 136
minimum-redundancy, 155–158
Reed-Muller, 443
separable, 136–137
Shannon's binary, 151–153
Shannon-Fano, 138–142
systematic (*see* Systematic codes)
uniquely decipherable, 136
Combination, 50–51
Combinatorial analysis, 49–52
Combinatorial lemma, 419–420
Common probability functions, 467
Complement, of an event, 35
of a set, 23
Complete finite scheme, 78
Conditional entropy, 94–96

Conditional probability, 42–44, 204–206
Continuous channels, 267–268
Convergence in probability, 363–365
Converse of fundamental theorem, strong, 422
  weak, 422
Convexity, 87–89
Convolution of distribution functions, 242–243
Convolution codes, 462
Correlation coefficient, 230–232
Correlation function, 349–352, 356–357
Coset, 427
Covariance, 350
Cross-correlation function, 351
Cumulative distribution function (CDF), 59–60, 193–195
Cylinder set, 385

Deciphering, 132
Decision scheme, 398
Decoder, 4, 132
Decomposable codes, 445
Dependent error correction, 461–462
Detection scheme for group codes, 437–438, 443
Discrete random variable, 58–60
Distance defined for code words, 168, 430
Domain, 30
Dualization (De Morgan's law), 26

Efficiency, 108–110
  of encoder, 133–135
Elias' iteration method, 176–180
Enciphering (*see* Encoding)
Encoder, 4, 132–135
Encoding, English alphabet, 183–185
  fundamental theorems in presence of noise, 160–165
  McMillan's theorem, 147–148
  noiseless theorem, 142–147
    fundamental, 154–155
English alphabet encoded, 183–185
Ensemble averages, 343
Ensemble codes, 409–412
Entropy (communication), 76–77, 87
  additive property, 84
  band-limited, 320–321
  basic inequalities, 101–102
  binary source table, 476–480
  conditional, 94–96
  continuity, 83
  continuous channels, 267–268
  continuous multivariate, 293
  convexity, 83
  equivocation of, 98
  extremal property, 83, 86
Entropy (communication), marginal, 94
  Markov chains, 380–385
  maximization, 278–282
  related random variables, 287–289
  set-theory interpretation, 106–108
  stationary source, 384–388
  uniqueness, 124–126
Equivalent sets, 32
Equivocation, 98
  relation to error probability, 400–401
Ergodic hypothesis, 351
Ergodic process, 347–349
Ergodic response of linear systems, 359–363
Error probability in decision scheme, 398–400
  for best group codes, 437, 440
  relation, to equivocation, 400–401
    to transinformation, 412–414
  Shannon's exponential bound, 414–415
  Shannon's lower bound, 325–327
  Shannon's upper bound, 327–329
Events, 35
  certain, 35, 39
  disjoint, 37
  frequency, 38–40
  impossible, 35, 39
  incompatible, 35
  independent, 44
  simultaneous, 35
Expectation, of a product, 222–223
  of a sum, 222–223
Expected value, 67, 220–222
Extension of a source and a channel, discrete memoryless, 114, 154–155, 402–403

Feinstein's lemma, 405–406
Field, 428
Fisher's quantity of information, 82
Fundamental theorem, 15
  channels with additive noise, 329–331
  continuous memoryless, 328–331
  converse, 422
  discrete noiseless, 154–155
  proof, for BSC, Elias', 457–460
      heuristic, 161–166
      Joshi's, 180–183
    Feinstein's, 403–408
    Shannon's, 409–415
    Wolfowitz's, 416–423

Gaussian process, 367–368
Gilbert-Moore encoding, 158–160, 188
Group, abelian, 426
  cyclic, 426
  decomposable, 445

Group, indecomposable, 445
order of, 426–427
Group axioms, 425–426
Group codes, 435–446
algebraic operations, 444–446
detection scheme, 437–438, 443
Group expansion, 427–428

Hadamard matrices, 448
Hamming's lemma, 439
Hamming's single-error correcting codes, 171–176, 431–435
Hilbert space, 311–312
BL^2 space, 315–317
Fourier series, 313–314
L^2 space, 314–315

Indecomposable codes, 445
Independent events, 44
Independent random variables, 201–204
Information, average amount of, 8, 14, 79
binary unit of, 7
(*See also* Entropy; Transinformation)
Inner product, 310
Inner-product space, 310–311
Intersection, of events, 35
of sets, 23
Irreducibility, 136

Laws of large numbers, 263–264
strong, 264
weak, 263
Limit in probability, 363–365
Logarithm to the base, 2, 471–475
Lossless code, 446

Marginal distribution, 202–204
Marginal entropy, 94
Markov process, 54–58, 376–377
absorbing chain, 376
ergodic chain, 376
regular chain, 376
standard, 369
Martingale, 369
Maximum-likelihood decoding, 323–325
Minimum-distance decoder, 323–325
Minimum redundancy code, 155–156
Moment generating function, 239–241
Moments, 68, 224–226
Multinormal distribution, 250–252

Norm, 310
Normal distribution, 196–198, 467
approximation, to binomial distribution, 259–262
to Poisson distribution, 262–263
moments, 235
Normal distribution, multidimensional, 250–252
process, 352–353
Normal probability table, 465–466
Normed space, 310
Null set (empty set), 22

Optimal decoding, 323
Optimal Hamming code, 431–433
Orthogonality, 313, 321
Outcome, 35

Parity check, even and odd, 167
matrix, 433–435, 443–445
Partial ordering of channels, 462–463
Pascal triangle, 51–52
Permutation, 49
Plotkin's theorems, 447
Poisson distribution, 65–66, 467
approximated by normal distribution, 262–263
moments, 233
process, 353–354
Positive definite functions, 250–251, 370–371
Power (average), 315–319
Power spectra, 357–359
Prefix property, 136
Probability, addition rule, 40–42
axioms of, 37, 40–41
conditional, 42–44, 204–206
convergence in, 363–365
cumulative distribution, 191–194
density, 194–198
of a sum, 242–244
limit in, 363–365
marginal distribution, 61–62, 202–204
measure, 36–38
Product or intersection, of events, 35
of sets, 23
Product rule, 44

Quadratic forms, 250–251

Radar problems, 464
Random variables, 58–59
discrete, 58–60
functions of, 208–214
independent, 201–204
multidimensional, 200–202
transformation, 208–215
uncorrelated, 232, 350
Random-walk problem, 258–259
Range, 30
Rayleigh distribution, 215
Redundancy, 108–110
of encoding, 133–135

Reed-Muller code, 443
Regression line, 232
Ring, 428–429

Sample space, 34, 35
  continuous, 191–192
  discrete, 35
Sampling theorem, 300
  derivative, 452
  frequency domain, 301–303
  n-dimensional, 453
  nonuniform, 452
  physical interpretation, 305
  stochastic process, 453
  time domain, 303–304
Self-information, 8, 105
  (*See also* Entropy)
Separability, 136
Sets, algebra, 24–26
  complement or negative, 23
  cylinder, 385
  definition, 21–22
  denumerable, 32
  equivalent, 32
  intersection or product, 23
  mutually exclusive, 24
  operations, 23–24
  union or sum, 23
Shannon-Fano encoding, 138–140
Shannon-Hartley law, 282–320
Sheffer-stroke diagram, 28–29, 70
Shift-register generator, 461
Signal space, band-limited, 315–320
  Fourier series, 313–314
Signals, band-limited, 300–307
Single-error detection and correction, 168–176
Slepian's best group code table, 468
  (*See also* Group codes)
Slepian's codes, 438–442
Slepian's parity-check table, 469–470
Source, 9, 10, 14–15
  binary, 89–91
  stationary, 391–393
Sphere-packing bound, 181, 319, 320, 433, 446
Standard array, 436–437
Standard deviation, 225, 226
Standard process, 368
Standard random variable, 255
  example, 6–7
State diagram, 52–58
Stationary process, 342–346
  Gaussian, 367–368
  in wide sense, 345
Stationary source, 391–393
Statistical average (*see* Expectation)
Statistical independence, 44, 201, 204
Stirling's approximation, 161, 181, 182
Stochastic convergence, 363–365
Stochastic differentiation, 365–367
Stochastic integration, 365–467
Stochastic limit, 363–365
Stochastic process, standard, 368
    Markov, 369
    martingale, 369
Subgroup, 426
Subset, 21
Sum or union, of events, 35
  of sets, 23
Systematic codes, 170–176
  bounds on, 171, 433, 446–449

Time average, 344
Transinformation (or mutual information), 9, 14, 92, 104–105
  continuous channels, 275–278
  general definition, 289–290
  of two Gaussian vectors, 295–296
Transition probability matrix, 54–56
Tree diagram, 52–58

Uncertainty (*see* Entropy)
Uncertainty relation, 454–457
Uncorrelated random variables, 232, 350
Uniform probability distribution, 467
Unique decipherability, 136
Universal set, 22

Variance, 225, 226
Vector space, axioms, 309
  basis, 312
  inner-product, 310–311
  linear, 310
  normed, 310
Venn diagram, 22

Weight of a binary word, 430

A CATALOG OF SELECTED

DOVER BOOKS

IN SCIENCE AND MATHEMATICS

A CATALOG OF SELECTED
DOVER BOOKS
IN SCIENCE AND MATHEMATICS

QUALITATIVE THEORY OF DIFFERENTIAL EQUATIONS, V.V. Nemytskii and V.V. Stepanov. Classic graduate-level text by two prominent Soviet mathematicians covers classical differential equations as well as topological dynamics and ergodic theory. Bibliographies. 523pp. 5⅜ x 8½. 65954-2 Pa. $14.95

MATRICES AND LINEAR ALGEBRA, Hans Schneider and George Phillip Barker. Basic textbook covers theory of matrices and its applications to systems of linear equations and related topics such as determinants, eigenvalues and differential equations. Numerous exercises. 432pp. 5⅜ x 8½. 66014-1 Pa. $10.95

QUANTUM THEORY, David Bohm. This advanced undergraduate-level text presents the quantum theory in terms of qualitative and imaginative concepts, followed by specific applications worked out in mathematical detail. Preface. Index. 655pp. 5⅜ x 8½. 65969-0 Pa. $14.95

ATOMIC PHYSICS (8th edition), Max Born. Nobel laureate's lucid treatment of kinetic theory of gases, elementary particles, nuclear atom, wave-corpuscles, atomic structure and spectral lines, much more. Over 40 appendices, bibliography. 495pp. 5⅜ x 8½. 65984-4 Pa. $13.95

ELECTRONIC STRUCTURE AND THE PROPERTIES OF SOLIDS: The Physics of the Chemical Bond, Walter A. Harrison. Innovative text offers basic understanding of the electronic structure of covalent and ionic solids, simple metals, transition metals and their compounds. Problems. 1980 edition. 582pp. 6⅛ x 9¼. 66021-4 Pa. $16.95

BOUNDARY VALUE PROBLEMS OF HEAT CONDUCTION, M. Necati Özisik. Systematic, comprehensive treatment of modern mathematical methods of solving problems in heat conduction and diffusion. Numerous examples and problems. Selected references. Appendices. 505pp. 5⅜ x 8½. 65990-9 Pa. $12.95

A SHORT HISTORY OF CHEMISTRY (3rd edition), J.R. Partington. Classic exposition explores origins of chemistry, alchemy, early medical chemistry, nature of atmosphere, theory of valency, laws and structure of atomic theory, much more. 428pp. 5⅜ x 8½. (Available in U.S. only) 65977-1 Pa. $11.95

A HISTORY OF ASTRONOMY, A. Pannekoek. Well-balanced, carefully reasoned study covers such topics as Ptolemaic theory, work of Copernicus, Kepler, Newton, Eddington's work on stars, much more. Illustrated. References. 521pp. 5⅜ x 8½. 65994-1 Pa. $12.95

PRINCIPLES OF METEOROLOGICAL ANALYSIS, Walter J. Saucier. Highly respected, abundantly illustrated classic reviews atmospheric variables, hydrostatics, static stability, various analyses (scalar, cross-section, isobaric, isentropic, more). For intermediate meteorology students. 454pp. 6⅛ x 9¼. 65979-8 Pa. $14.95

ASYMPTOTIC METHODS IN ANALYSIS, N.G. de Bruijn. An inexpensive, comprehensive guide to asymptotic methods–the pioneering work that teaches by explaining worked examples in detail. Index. 224pp. 5⅜ x 8½. 64221-6 Pa. $7.95

OPTICAL RESONANCE AND TWO-LEVEL ATOMS, L. Allen and J. H. Eberly. Clear, comprehensive introduction to basic principles behind all quantum optical resonance phenomena. 53 illustrations. Preface. Index. 256pp. 5⅜ x 8½. 65533-4 Pa. $8.95

COMPLEX VARIABLES, Francis J. Flanigan. Unusual approach, delaying complex algebra till harmonic functions have been analyzed from real variable viewpoint. Includes problems with answers. 364pp. 5⅜ x 8½. 61388-7 Pa. $9.95

ATOMIC SPECTRA AND ATOMIC STRUCTURE, Gerhard Herzberg. One of best introductions; especially for specialist in other fields. Treatment is physical rather than mathematical. 80 illustrations. 257pp. 5⅜ x 8½. 60115-3 Pa. $7.95

APPLIED COMPLEX VARIABLES, John W. Dettman. Step-by-step coverage of fundamentals of analytic function theory–plus lucid exposition of five important applications: Potential Theory; Ordinary Differential Equations; Fourier Transforms; Laplace Transforms; Asymptotic Expansions. 66 figures. Exercises at chapter ends. 512pp. 5⅜ x 8½. 64670-X Pa. $12.95

ULTRASONIC ABSORPTION: An Introduction to the Theory of Sound Absorption and Dispersion in Gases, Liquids and Solids, A.B. Bhatia. Standard reference in the field provides a clear, systematically organized introductory review of fundamental concepts for advanced graduate students, research workers. Numerous diagrams. Bibliography. 440pp. 5⅜ x 8½. 64917-2 Pa. $11.95

UNBOUNDED LINEAR OPERATORS: Theory and Applications, Seymour Goldberg. Classic presents systematic treatment of the theory of unbounded linear operators in normed linear spaces with applications to differential equations. Bibliography. 199pp. 5⅜ x 8½. 64830-3 Pa. $7.95

LIGHT SCATTERING BY SMALL PARTICLES, H.C. van de Hulst. Comprehensive treatment including full range of useful approximation methods for researchers in chemistry, meteorology and astronomy. 44 illustrations. 470pp. 5⅜ x 8½. 64228-3 Pa. $12.95

CONFORMAL MAPPING ON RIEMANN SURFACES, Harvey Cohn. Lucid, insightful book presents ideal coverage of subject. 334 exercises make book perfect for self-study. 55 figures. 352pp. 5⅝ x 8¼. 64025-6 Pa. $11.95

OPTICKS, Sir Isaac Newton. Newton's own experiments with spectroscopy, colors, lenses, reflection, refraction, etc., in language the layman can follow. Foreword by Albert Einstein. 532pp. 5⅜ x 8½. 60205-2 Pa. $12.95

GENERALIZED INTEGRAL TRANSFORMATIONS, A.H. Zemanian. Graduate-level study of recent generalizations of the Laplace, Mellin, Hankel, K. Weierstrass, convolution and other simple transformations. Bibliography. 320pp. 5⅜ x 8½. 65375-7 Pa. $8.95

THE ELECTROMAGNETIC FIELD, Albert Shadowitz. Comprehensive undergraduate text covers basics of electric and magnetic fields, builds up to electromagnetic theory. Also related topics, including relativity. Over 900 problems. 768pp. 5⅝ x 8¼. 65660-8 Pa. $18.95

FOURIER SERIES, Georgi P. Tolstov. Translated by Richard A. Silverman. A valuable addition to the literature on the subject, moving clearly from subject to subject and theorem to theorem. 107 problems, answers. 336pp. 5⅜ x 8½. 63317-9 Pa. $9.95

THEORY OF ELECTROMAGNETIC WAVE PROPAGATION, Charles Herach Papas. Graduate-level study discusses the Maxwell field equations, radiation from wire antennas, the Doppler effect and more. xiii + 244pp. 5⅜ x 8½. 65678-0 Pa. $6.95

DISTRIBUTION THEORY AND TRANSFORM ANALYSIS: An Introduction to Generalized Functions, with Applications, A.H. Zemanian. Provides basics of distribution theory, describes generalized Fourier and Laplace transformations. Numerous problems. 384pp. 5⅜ x 8½. 65479-6 Pa. $11.95

THE PHYSICS OF WAVES, William C. Elmore and Mark A. Heald. Unique overview of classical wave theory. Acoustics, optics, electromagnetic radiation, more. Ideal as classroom text or for self-study. Problems. 477pp. 5⅜ x 8½. 64926-1 Pa. $13.95

CALCULUS OF VARIATIONS WITH APPLICATIONS, George M. Ewing. Applications-oriented introduction to variational theory develops insight and promotes understanding of specialized books, research papers. Suitable for advanced undergraduate/graduate students as primary, supplementary text. 352pp. 5⅜ x 8½. 64856-7 Pa. $9.95

A TREATISE ON ELECTRICITY AND MAGNETISM, James Clerk Maxwell. Important foundation work of modern physics. Brings to final form Maxwell's theory of electromagnetism and rigorously derives his general equations of field theory. 1,084pp. 5⅜ x 8½. 60636-8, 60637-6 Pa., Two-vol. set $25.90

AN INTRODUCTION TO THE CALCULUS OF VARIATIONS, Charles Fox. Graduate-level text covers variations of an integral, isoperimetrical problems, least action, special relativity, approximations, more. References. 279pp. 5⅜ x 8½. 65499-0 Pa. $8.95

HYDRODYNAMIC AND HYDROMAGNETIC STABILITY, S. Chandrasekhar. Lucid examination of the Rayleigh-Benard problem; clear coverage of the theory of instabilities causing convection. 704pp. 5⅜ x 8¼. 64071-X Pa. $14.95

CALCULUS OF VARIATIONS, Robert Weinstock. Basic introduction covering isoperimetric problems, theory of elasticity, quantum mechanics, electrostatics, etc. Exercises throughout. 326pp. 5⅜ x 8½. 63069-2 Pa. $9.95

DYNAMICS OF FLUIDS IN POROUS MEDIA, Jacob Bear. For advanced students of ground water hydrology, soil mechanics and physics, drainage and irrigation engineering and more. 335 illustrations. Exercises, with answers. 784pp. 6⅛ x 9¼. 65675-6 Pa. $19.95

NUMERICAL METHODS FOR SCIENTISTS AND ENGINEERS, Richard Hamming. Classic text stresses frequency approach in coverage of algorithms, polynomial approximation, Fourier approximation, exponential approximation, other topics. Revised and enlarged 2nd edition. 721pp. 5⅜ x 8½. 65241-6 Pa. $15.95

THEORETICAL SOLID STATE PHYSICS, Vol. 1: Perfect Lattices in Equilibrium; Vol. II: Non-Equilibrium and Disorder, William Jones and Norman H. March. Monumental reference work covers fundamental theory of equilibrium properties of perfect crystalline solids, non-equilibrium properties, defects and disordered systems. Appendices. Problems. Preface. Diagrams. Index. Bibliography. Total of 1,301pp. 5⅜ x 8½. Two volumes. Vol. I: 65015-4 Pa. $16.95
Vol. II: 65016-2 Pa. $16.95

OPTIMIZATION THEORY WITH APPLICATIONS, Donald A. Pierre. Broad spectrum approach to important topic. Classical theory of minima and maxima, calculus of variations, simplex technique and linear programming, more. Many problems, examples. 640pp. 5⅜ x 8½. 65205-X Pa. $16.95

THE CONTINUUM: A Critical Examination of the Foundation of Analysis, Hermann Weyl. Classic of 20th-century foundational research deals with the conceptual problem posed by the continuum. 156pp. 5⅜ x 8½. 67982-9 Pa. $6.95

ESSAYS ON THE THEORY OF NUMBERS, Richard Dedekind. Two classic essays by great German mathematician: on the theory of irrational numbers; and on transfinite numbers and properties of natural numbers. 115pp. 5⅜ x 8½. 21010-3 Pa. $5.95

THE FUNCTIONS OF MATHEMATICAL PHYSICS, Harry Hochstadt. Comprehensive treatment of orthogonal polynomials, hypergeometric functions, Hill's equation, much more. Bibliography. Index. 322pp. 5⅜ x 8½. 65214-9 Pa. $9.95

NUMBER THEORY AND ITS HISTORY, Oystein Ore. Unusually clear, accessible introduction covers counting, properties of numbers, prime numbers, much more. Bibliography. 380pp. 5⅜ x 8½. 65620-9 Pa. $10.95

THE VARIATIONAL PRINCIPLES OF MECHANICS, Cornelius Lanczos. Graduate level coverage of calculus of variations, equations of motion, relativistic mechanics, more. First inexpensive paperbound edition of classic treatise. Index. Bibliography. 418pp. 5⅜ x 8½. 65067-7 Pa. $12.95

MATHEMATICAL TABLES AND FORMULAS, Robert D. Carmichael and Edwin R. Smith. Logarithms, sines, tangents, trig functions, powers, roots, reciprocals, exponential and hyperbolic functions, formulas and theorems. 269pp. 5⅜ x 8½. 60111-0 Pa. $6.95

THEORETICAL PHYSICS, Georg Joos, with Ira M. Freeman. Classic overview covers essential math, mechanics, electromagnetic theory, thermodynamics, quantum mechanics, nuclear physics, other topics. First paperback edition. xxiii + 885pp. 5⅜ x 8½. 65227-0 Pa. $21.95

ORDINARY DIFFERENTIAL EQUATIONS, Morris Tenenbaum and Harry Pollard. Exhaustive survey of ordinary differential equations for undergraduates in mathematics, engineering, science. Thorough analysis of theorems. Diagrams. Bibliography. Index. 818pp. 5⅜ x 8½. 64940-7 Pa. $18.95

STATISTICAL MECHANICS: Principles and Applications, Terrell L. Hill. Standard text covers fundamentals of statistical mechanics, applications to fluctuation theory, imperfect gases, distribution functions, more. 448pp. 5⅜ x 8½.
65390-0 Pa. $11.95

ORDINARY DIFFERENTIAL EQUATIONS AND STABILITY THEORY: An Introduction, David A. Sánchez. Brief, modern treatment. Linear equation, stability theory for autonomous and nonautonomous systems, etc. 164pp. 5⅜ x 8¼.
63828-6 Pa. $6.95

THIRTY YEARS THAT SHOOK PHYSICS: The Story of Quantum Theory, George Gamow. Lucid, accessible introduction to influential theory of energy and matter. Careful explanations of Dirac's anti-particles, Bohr's model of the atom, much more. 12 plates. Numerous drawings. 240pp. 5⅜ x 8½. 24895-X Pa. $7.95

THEORY OF MATRICES, Sam Perlis. Outstanding text covering rank, nonsingularity and inverses in connection with the development of canonical matrices under the relation of equivalence, and without the intervention of determinants. Includes exercises. 237pp. 5⅜ x 8½. 66810-X Pa. $8.95

GREAT EXPERIMENTS IN PHYSICS: Firsthand Accounts from Galileo to Einstein, edited by Morris H. Shamos. 25 crucial discoveries: Newton's laws of motion, Chadwick's study of the neutron, Hertz on electromagnetic waves, more. Original accounts clearly annotated. 370pp. 5⅜ x 8½. 25346-5 Pa. $10.95

INTRODUCTION TO PARTIAL DIFFERENTIAL EQUATIONS WITH APPLICATIONS, E.C. Zachmanoglou and Dale W. Thoe. Essentials of partial differential equations applied to common problems in engineering and the physical sciences. Problems and answers. 416pp. 5⅜ x 8½. 65251-3 Pa. $11.95

BURNHAM'S CELESTIAL HANDBOOK, Robert Burnham, Jr. Thorough guide to the stars beyond our solar system. Exhaustive treatment. Alphabetical by constellation: Andromeda to Cetus in Vol. 1; Chamaeleon to Orion in Vol. 2; and Pavo to Vulpecula in Vol. 3. Hundreds of illustrations. Index in Vol. 3. 2,000pp. 6⅛ x 9¼.
23567-X, 23568-8, 23673-0 Pa., Three-vol. set $44.85

CHEMICAL MAGIC, Leonard A. Ford. Second Edition, Revised by E. Winston Grundmeier. Over 100 unusual stunts demonstrating cold fire, dust explosions, much more. Text explains scientific principles and stresses safety precautions. 128pp. 5⅜ x 8½. 67628-5 Pa. $5.95

AMATEUR ASTRONOMER'S HANDBOOK, J.B. Sidgwick. Timeless, comprehensive coverage of telescopes, mirrors, lenses, mountings, telescope drives, micrometers, spectroscopes, more. 189 illustrations. 576pp. 5⅝ x 8¼. (Available in U.S. only) 24034-7 Pa. $11.95

GEOMETRY OF COMPLEX NUMBERS, Hans Schwerdtfeger. Illuminating, widely praised book on analytic geometry of circles, the Moebius transformation, and two-dimensional non-Euclidean geometries. 200pp. 5⅝ x 8¼. 63830-8 Pa. $8.95

MECHANICS, J.P. Den Hartog. A classic introductory text or refresher. Hundreds of applications and design problems illuminate fundamentals of trusses, loaded beams and cables, etc. 334 answered problems. 462pp. 5⅜ x 8½. 60754-2 Pa. $11.95

TOPOLOGY, John G. Hocking and Gail S. Young. Superb one-year course in classical topology. Topological spaces and functions, point-set topology, much more. Examples and problems. Bibliography. Index. 384pp. 5⅜ x 8¼. 65676-4 Pa. $10.95

STRENGTH OF MATERIALS, J.P. Den Hartog. Full, clear treatment of basic material (tension, torsion, bending, etc.) plus advanced material on engineering methods, applications. 350 answered problems. 323pp. 5⅜ x 8½. 60755-0 Pa. $9.95

ELEMENTARY CONCEPTS OF TOPOLOGY, Paul Alexandroff. Elegant, intuitive approach to topology from set-theoretic topology to Betti groups; how concepts of topology are useful in math and physics. 25 figures. 57pp. 5⅜ x 8½. 60747-X Pa. $3.95

ADVANCED STRENGTH OF MATERIALS, J.P. Den Hartog. Superbly written advanced text covers torsion, rotating disks, membrane stresses in shells, much more. Many problems and answers. 388pp. 5⅜ x 8½. 65407-9 Pa. $10.95

COMPUTABILITY AND UNSOLVABILITY, Martin Davis. Classic graduate-level introduction to theory of computability, usually referred to as theory of recurrent functions. New preface and appendix. 288pp. 5⅜ x 8½. 61471-9 Pa. $8.95

GENERAL CHEMISTRY, Linus Pauling. Revised 3rd edition of classic first-year text by Nobel laureate. Atomic and molecular structure, quantum mechanics, statistical mechanics, thermodynamics correlated with descriptive chemistry. Problems. 992pp. 5⅜ x 8½. 65622-5 Pa. $19.95

AN INTRODUCTION TO MATRICES, SETS AND GROUPS FOR SCIENCE STUDENTS, G. Stephenson. Concise, readable text introduces sets, groups, and most importantly, matrices to undergraduate students of physics, chemistry, and engineering. Problems. 164pp. 5⅜ x 8½. 65077-4 Pa. $7.95

THE HISTORICAL BACKGROUND OF CHEMISTRY, Henry M. Leicester. Evolution of ideas, not individual biography. Concentrates on formulation of a coherent set of chemical laws. 260pp. 5⅜ x 8½. 61053-5 Pa. $8.95

THE PHILOSOPHY OF MATHEMATICS: An Introductory Essay, Stephan Körner. Surveys the views of Plato, Aristotle, Leibniz & Kant concerning propositions and theories of applied and pure mathematics. Introduction. Two appendices. Index. 198pp. 5⅜ x 8½. 25048-2 Pa. $8.95

THE DEVELOPMENT OF MODERN CHEMISTRY, Aaron J. Ihde. Authoritative history of chemistry from ancient Greek theory to 20th-century innovation. Covers major chemists and their discoveries. 209 illustrations. 14 tables. Bibliographies. Indices. Appendices. 851pp. 5⅜ x 8½. 64235-6 Pa. $18.95